IEE Electromagnetic Waves Series 25

Series Editors: Professors P. J. B. Clarricoats,
E. D. R. Shearman and J. R. Wait

PRINCIPLES of microwave CIRCUITS

Other volumes in this series

Volume 1	**Geometrical theory of diffraction for electromagnetic waves**	
	Graeme L. James	
Volume 2	**Electromagnetic waves and curved structures**	
	Leonard Lewin, David C. Chang and Edward F. Kuester	
Volume 3	**Microwave homodyne systems**	
	Ray J. King	
Volume 4	**Radio direction-finding**	
	P. J. D. Gething	
Volume 5	**ELF communications antennas**	
	Michael L. Burrows	
Volume 6	**Waveguide tapers, transitions and couplers**	
	F. Sporleder and H. G. Unger	
Volume 7	**Reflector antenna analysis and design**	
	P. J. Wood	
Volume 8	**Effects of the troposphere on radio communications**	
	Martin P. M. Hall	
Volume 9	**Schumann resonances in the earth-ionosphere cavity**	
	P. V. Bliokh, A. P. Nikolaenko and Y. F. Filippov	
Volume 10	**Aperture antennas and diffraction theory**	
	E. V. Jull	
Volume 11	**Adaptive array principles**	
	J. E. Hudson	
Volume 12	**Microstrip antenna theory and design**	
	J. R. James, P. S. Hall and C. Wood	
Volume 13	**Energy in electromagnetism**	
	H. G. Booker	
Volume 14	**Leaky feeders and subsurface radio communications**	
	P. Delogne	
Volume 15	**The handbook of antenna design, volume 1**	
	Editors: A. W. Rudge, K. Milne, A. D. Olver, P. Knight	
Volume 16	**The handbook of antenna design, volume 2**	
	Editors: A. W. Rudge, K. Milne, A. D. Olver, P. Knight	
Volume 17	**Surveillance radar performance prediction**	
	P. Rohan	
Volume 18	**Corrugated horns for microwave antennas**	
	P. J. B. Clarricoats and A. D. Olver	
Volume 19	**Microwave antenna theory and design**	
	Editor: S. Silver	
Volume 20	**Advances in radar techniques**	
	Editor: J. Clarke	
Volume 21	**Waveguide handbook**	
	N. Marcuvitz	
Volume 22	**Target adaptive matched illumination radar**	
	D. T. Gjessing	
Volume 23	**Ferrites at microwave frequencies**	
	A. J. Baden Fuller	
Volume 24	**Propagation of short radio waves**	
	Editor: D. E. Kerr	

PRINCIPLES of microwave CIRCUITS

Edited by
C.G. Montgomery
R.H. Dicke
E.M. Purcell

Peter Peregrinus Ltd on behalf of The Institution of Electrical Engineers

Published by Peter Peregrinus Ltd., London, United Kingdom

© 1987 Errata and preface to this reprint, Peter Peregrinus Ltd.

This book was first published in 1948 by the McGraw-Hill Book Company Inc.

British Library Cataloguing in Publication Data

Principles of microwave circuits.—[Rev.ed.]
—(IEE electromagnetic waves series; 25).
1. Microwave circuits
I. Montgomery, C. G. II. Dicke, R. H.
III. Purcell, Edward M. IV. Series
621.381′32   TK7876

ISBN 0 86341 100 2

Printed in England by Short Run Press Ltd., Exeter

PRINCIPLES OF MICROWAVE CIRCUITS

EDITORIAL STAFF

C. G. Montgomery
D. D. Montgomery

CONTRIBUTING AUTHORS

E. R. Beringer
R. H. Dicke
N. Marcuvitz
C. G. Montgomery
E. M. Purcell

Foreword

THE tremendous research and development effort that went into the development of radar and related techniques during World War II resulted not only in hundreds of radar sets for military (and some for possible peacetime) use but also in a great body of information and new techniques in the electronics and high-frequency fields. Because this basic material may be of great value to science and engineering, it seemed most important to publish it as soon as security permitted.

The Radiation Laboratory of MIT, which operated under the supervision of the National Defense Research Committee, undertook the great task of preparing these volumes. The work described herein, however, is the collective result of work done at many laboratories, Army, Navy, university, and industrial, both in this country and in England, Canada, and other Dominions.

The Radiation Laboratory, once its proposals were approved and finances provided by the Office of Scientific Research and Development, chose Louis N. Ridenour as Editor-in-Chief to lead and direct the entire project. An editorial staff was then selected of those best qualified for this type of task. Finally the authors for the various volumes or chapters or sections were chosen from among those experts who were intimately familiar with the various fields, and who were able and willing to write the summaries of them. This entire staff agreed to remain at work at MIT for six months or more after the work of the Radiation Laboratory was complete. These volumes stand as a monument to this group.

These volumes serve as a memorial to the unnamed hundreds and thousands of other scientists, engineers, and others who actually carried on the research, development, and engineering work the results of which are herein described. There were so many involved in this work and they worked so closely together even though often in widely separated laboratories that it is impossible to name or even to know those who contributed to a particular idea or development. Only certain ones who wrote reports or articles have even been mentioned. But to all those who contributed in any way to this great cooperative development enterprise, both in this country and in England, these volumes are dedicated.

L. A. DuBridge.

Preface

IN THE engineering application of low-frequency currents, an important step forward was the development of the impedance concept and its utilization through the theory of linear networks. It was almost inevitable that this concept would be generalized and become useful in the application of microwaves. This volume is devoted to an exposition of the impedance concept and to the equivalent circuits of microwave devices. It is the intention to emphasize the underlying principles of these equivalent circuits and the results that may be obtained by their use. Specific devices are not discussed except as illustrations of the general methods under consideration. These devices and the details of the design procedure are treated in other volumes of this series. The solutions of the boundary-value problems which give the susceptances of microwave-circuit elements are likewise omitted. The results of such calculations that have been performed up to the present time are compiled in Vol. 10, the *Waveguide Handbook*, and these results are used freely. Although the work of the Radiation Laboratory at MIT was the development of military radar equipment, the principles discussed in this volume can be applied to microwave equipment of all kinds.

The publishers have agreed that ten years after the date on which each volume of this series is issued, the copyright thereon shall be relinquished, and the work shall become part of the public domain.

THE AUTHORS.

NEW HAVEN, CONN.,
February, 1947.

Preface to the 1987 edition

This volume is an unabridged reprint of Principles of Microwave Circuits which was published by McGraw Hill as Volume 8 of the MIT Radiation Laboratory Series in 1948. The principal contributors to the Volume were the distinguished physicists C. G. Montgomery, R. H. Dicke and E. M. Purcell. Additional contributions were made by R. Beringer and N. Marcuvitz. Since 1948, a number of textbooks dealing with microwave circuits have been published but they have all built on material contained in Volume 8. The Dover reprint of this volume has long been out-of-print and in view of the great interest shown in the IEE reprints of Volume 10 (The Waveguide Handbook) and Volume 12 (Microwave Antenna Theory and Design) the decision was made to add Volume 8 to the Electromagnetic Waves Series. A number of corrections have been incorporated, should others be extant the IEE would be glad to receive notice of these so that later generations of microwave engineers may benefit.

Queen Mary College
University of London
1987

Peter Clarricoats
Series Editor

Contents

FOREWORD BY L. A. DuBRIDGE. vii

PREFACE. ix

CHAP. 1. INTRODUCTION BY E. M. PURCELL. 1

    1·1. Microwaves. 1
    1·2. Microwave Circuits 3
    1·3. Microwave Measurements. 5
    1·4. The Aims of Microwave Circuit Analysis 8
    1·5. Linearity 9
    1·6. Dissipation 9
    1·7. Symmetry. 9

CHAP. 2. ELECTROMAGNETIC WAVES BY C. G. MONTGOMERY 10

  THE FIELD RELATIONS 10

    2·1. Maxwell's Equation 10
    2·2. Poynting's Vector and Energy Theorems 14
    2·3. Solutions of Maxwell's Equations 16

  PURELY TRANSVERSE ELECTROMAGNETIC WAVES. 17

    2·4. Uniform Plane Waves 17
    2·5. Nonuniform Transverse-electromagnetic Plane Waves . . . 19
    2·6. TEM-waves between Parallel Plates 22
    2·7. TEM-waves between Coaxial Cylinders 23
    2·8. Spherical TEM-waves 25
    2·9. Uniform Cylindrical Waves. 26
    2·10. Babinet's Principle. 28

  ELECTROMAGNETIC WAVES WITH LONGITUDINAL COMPONENTS . . . 30

    2·11. General Procedure. 30
    2·12. The Normal Modes of Rectangular Pipes 33
    2·13. The Normal Modes in Round Pipes 38
    2·14. Higher Modes in Coaxial Cylinders. 41
    2·15. Normal Modes for Other Cross Sections 42
    2·16. Transmission Losses 45
    2·17. Cylindrical Cavities 48
    2·18. Energy Density and Power Flow in Waveguides. 50
    2·19. Summary of Results 54

CONTENTS

CHAP. 3. WAVEGUIDES AS TRANSMISSION LINES BY C. G. MONT-
GOMERY 60

    3·1. Some General Properties of Guided Waves 60
    3·2. Low-frequency Transmission Lines. 64
    3·3. The Transformation of Impedances 67
    3·4. Power Flow. 69
    3·5. The Combination of Admittances 70
    3·6. Transmission-line Charts 71
    3·7. Impedance Concept in Waveguide Problems. 75
    3·8. Equivalent T-network of a Length of Waveguide. 77
    3·9. Transmission-line Equations for the H_{10}-mode. 79

CHAP. 4. ELEMENTS OF NETWORK THEORY BY C. G. MONTGOMERY . 83

    4·1. Elementary Considerations 83
    4·2. The Use of Matrices in Network Theory 87
    4·3. Fundamental Network Theorems 90
    4·4. The Synthesis Problem and Networks with One Terminal Pair . 95
    4·5. The Circuit Parameters of Two-terminal-pair Networks 99
    4·6. Equivalent Circuits of Two-terminal-pair Networks 104
    4·7. Symmetrical Two-terminal-pair Networks. 110
    4·8. Chains of Four-terminal Networks. 112
    4·9. Filters 115
    4·10. Series and Parallel Connection of Networks. 119
    4·11. Three-terminal-pair Networks. 121
    4·12. Circuits with N Terminal Pairs 124
    4·13. Resonant Circuits 127

CHAP. 5. GENERAL MICROWAVE CIRCUIT THEOREMS BY R. H.
DICKE. 130

    5·1. Some General Properties of a Waveguide Junction 130

  THE TERMINATION OF A SINGLE TRANSMISSION LINE 132

    5·2. Poynting's Energy Theorem for a Periodic Feld 132
    5·3. Uniqueness of Terminal Voltages and Currents 134
    5·4. Connections between Impedance and Stored and Dissipated
        Energy. 135
    5·5. Field Quantities in a Lossless Termination. 136
    5·6. Wave Formalism. 137
    5·7. Connection between the Reflection Coefficient and Stored Energy. 138

  THE JUNCTION OF SEVERAL TRANSMISSION LINES 139

    5·8. Extension of the Uniqueness Theorem to N-terminal-pair Junc-
        tions. 139
    5·9. Impedance and Admittance Matrix. 140
    5·10. Symmetry of Impedance and Admittance Matrices. 141
    5·11. Physical Realizability. 142
    5·12. The Polyterminal-pair Lossless Junction 143
    5·13. Definition of Terminal Voltages and Currents for Waveguides
        with More than One Propagating Mode. 144

- 5·14. Scattering Matrix 146
- 5·15. Symmetry. 148
- 5·16. Energy Condition 148
- 5·17. Transformation of the Scattering Matrix under a Shift in Position of the Terminal Reference Planes 149
- 5·18. The T-matrix of a Series of Junctions Connected in Cascade . . 150
- 5·19. The Scattering Matrix of a Junction with a Load Connected to One of the Transmission Lines. 151

FREQUENCY DEPENDENCE OF A LOSSLESS JUNCTION 151

- 5·20. Variational Energy Integral. 151
- 5·21. Application to Impedance and Admittance Matrix. 152
- 5·22. Application to Scattering Matrix. 153
- 5·23. Transmission-line Termination. 154
- 5·24. Foster's Reactance Theorem. 156
- 5·25. Frequency Variation of a Lossless Junction with Two Transmission Lines. 158

CHAP. 6. WAVEGUIDE CIRCUIT ELEMENTS BY C. G. MONTGOMERY . . 162

- 6·1. Obstacles in a Waveguide. 162

THIN DIAPHRAGMS AS SHUNT REACTANCES 163

- 6·2. Shunt Reactances 163
- 6·3. The Inductive Slit. 164
- 6·4. The Capacitive Diaphragm 166
- 6·5. The Thin Inductive Wire. 167
- 6·6. Capacitive Tuning Screw 168
- 6·7. Resonant Irises 169
- 6·8. Diaphragms in Waveguides of Other Cross Sections 171
- 6·9. The Interaction between Two Diaphragms 173
- 6·10. Babinet's Principle. 174
- 6·11. The Susceptance of Small Apertures 176

IMPEDANCE MATCHING WITH SHUNT SUSCEPTANCES 179

- 6·12. Calculation of the Necessary Susceptance. 179
- 6·13. Screw Tuners 181
- 6·14. Cavity Formed by Shunt Reactances. 182

CHANGES IN THE CHARACTERISTIC IMPEDANCE OF A TRANSMISSION LINE . 187

- 6·15. Diameter Changes in Coaxial Lines. 187
- 6·16. Change in the Dimensions of a Rectangular Waveguide. . . . 188
- 6·17. Quarter-wavelength Transformers 189
- 6·18. Tapered Sections of Line 191
- 6·19. The Cutoff Wavelength of Capacitively Loaded Guides. . . . 192

BRANCHED TRANSMISSION LINES 193

- 6·20. Shunt Branches in Coaxial Lines. 193
- 6·21. Series Branches in Coaxial Lines. 195
- 6·22. Series Branches and Choke Joints in Waveguide. 197

DISCONTINUITIES WITH SHUNT AND SERIES ELEMENTS 198
 6·23. Obstacles of Finite Thickness 198
 6·24. Radiation from Thick Holes. 201
 6·25. Bends and Corners in Rectangular Waveguide. 201
 6·26. Broadbanding. 203

Chap. 7. RESONANT CAVITIES AS MICROWAVE CIRCUIT ELEMENTS by Robert Beringer. 207

EQUIVALENT CIRCUIT OF A SINGLE-LINE LOSSLESS CAVITY-COUPLING SYSTEM 208
 7·1. Impedance Functions of Lossless Lumped Circuits 209
 7·2. Impedance Functions of Lossless Distributed Circuits 211
 7·3. Impedance-function Synthesis of a Short-circuited Lossless Transmission Line 213

EQUIVALENT CIRCUIT OF A SINGLE-LINE CAVITY-COUPLING SYSTEM WITH LOSS. 214
 7·4. Foster's Theorem for Slightly Lossy Networks. 215
 7·5. The Impedance Functions of Simple Series- and Parallel-resonant Circuits. 217
 7·6. The Equivalent Circuit of a Loop-coupled Cavity 218
 7·7. Impedance Functions Near Resonance 225
 7·8. Coupling Coefficients and External Loading. 228
 7·9. General Formulas for Q-Factors 230
 7·10. Iris-coupled, Short-circuited Waveguide. 231

CAVITY-COUPLING SYSTEMS WITH TWO EMERGENT TRANSMISSION LINES . 234
 7·11. General Representation of Lossless Two-terminal-pair Networks. 234
 7·12. Introduction of Loss 237
 7·13. Representation of a Cavity with Two Loop-coupled Lines. . . 237
 7·14. Transmission through a Two-line Cavity-coupling System. . . . 237

Chap. 8. RADIAL TRANSMISSION LINES by N. Marcuvitz. 240

 8·1. The Equivalent-circuit Point of View. 240
 8·2. Differences between Uniform and Nonuniform Regions. 241
 8·3. Impedance Description of Uniform Lines 248
 8·4. Field Representation by Characteristic Modes. 252
 8·5. Impedance Description of a Radial Line 256
 8·6. Reflection Coefficients in Radial Lines 265
 8·7. Equivalent Circuits in Radial Lines 267
 8·8. Applications. 271
 8·9. A Coaxial Cavity 273
 8·10. Capacitively Loaded Cavity. 274
 8·11. Capacitively Loaded Cavity with Change in Height 275
 8·12. Oscillator Cavity Coupled to Rectangular Waveguide. 277

Chap. 9. WAVEGUIDE JUNCTIONS WITH SEVERAL ARMS by C. G. Montgomery and R. H. Dicke. 283

T-JUNCTIONS 283
 9·1. General Theorems about T-junctions. 283

CONTENTS

9·2. The Choice of an Equivalent Circuit. Transformation of Reference Planes 286
9·3. The E-plane T-junction at Long Wavelengths 288
9·4. E-plane T-junction at High Frequencies 291
9·5. H-plane T-junctions 294
9·6. A Coaxial-line T-junction 295
9·7. The T-junction with a Small Hole 296

WAVEGUIDE JUNCTIONS WITH FOUR ARMS 298

9·8. The Equivalent Circuit of a Four-junction 298
9·9. Directional Couplers 299
9·10. The Scattering Matrix of a Directional Coupler 301
9·11. The Arbitrary Junction of Four Transmission Lines 303
9·12. The Magic T 306
9·13. Ring Circuits 308
9·14. Four-junctions with Small Holes 311
9·15. Degenerate Four-junctions 313
9·16. A Generalization of the Theory of Four-terminal Networks to Four-terminal-pair Networks 315

RADIATION AND SCATTERING BY ANTENNAS 317

9·17. Representation in Terms of Plane Waves 317
9·18. Representation in Terms of Spherical Waves 319
9·19. Solutions of the Vector Wave Equations 322
9·20. Scattering Matrix of Free Space 324
9·21. Scattering Matrix of a Simple Electric Dipole 325
9·22. The General Antenna 326
9·23. The General Scattering Problem 327
9·24. Minimum-scattering Antenna 329

CHAP. 10. MODE TRANSFORMATIONS BY E. M. PURCELL AND R. H. DICKE 334

10·1. Mode Transducers 335
10·2. General Properties of Mode Transducers 340
10·3. The Problem of Measurement 343
10·4. Mode Filters and Mode Absorbers 347
10·5. The TE-mode in Round Guide 349
10·6. Permissible Transformations of a Scattering Matrix 351
10·7. Quarter-wave Pipe 354
10·8. Rotary Phase Shifter 355
10·9. A Rectangular-to-round Transducer 358
10·10. Discontinuity in Round Guide 359
10·11. Principal Axes in Round Guide 360
10·12. Resonance in a Closed Circular Guide 361

CHAP. 11. DIELECTRICS IN WAVEGUIDES BY C. G. MONTGOMERY . . 365

11·1. Waveguides Filled with Dielectric Materials 365
11·2. Reflection from a Change in Dielectric Constant 369
11·3. Dielectric Plates in Waveguides 374
11·4. The Nature of Dielectric Phenomena 376

11·5.	Ferromagnetism at Microwave Frequencies	382
11·6.	Guides Partially Filled with Dielectric	385
11·7.	Dielectric Post in Waveguide	389
11·8.	Cavities Containing Dielectrics	390
11·9.	Propagation in Ionized Gases	393
11·10.	Absorbing Materials for Microwave Radiation	396

CHAP. 12. THE SYMMETRY OF WAVEGUIDE JUNCTIONS BY R. H. DICKE . . . 401

12·1.	Classes of Symmetry	401
12·2.	Symmetry of the Thin Iris	403

MATRIX ALGEBRA . . . 405

12·3.	The Eigenvalue Problem	405
12·4.	Symmetrical Matrices	407
12·5.	Rational Matrix Functions, Definitions	409
12·6.	Commuting Matrices	410
12·7.	Cayley-Hamilton's Theorem	410

SYMMETRIES OF MAXWELL'S EQUATIONS . . . 411

12·8.	The symmetry of a Reflection in a Plane	412
12·9.	Symmetry Operators	414
12·10.	Field Distributions Invariant under Axial and Point Reflections	416

WAVEGUIDE JUNCTIONS WITH TWO OR THREE ARMS . . . 417

12·11.	The Thick Iris	417
12·12.	The Symmetrical Y-junction	420
12·13.	Experimental Determination of S_1 and S_2	427
12·14.	Symmetrical T-junctions	430
12·15.	The Shunt T-junction	432
12·16.	The Use of the T-junction as an Element of a Tuner	435
12·17.	Directional Couplers	437
12·18.	The Single-hole Directional Coupler	437
12·19.	The Biplanar Directional Coupler	445
12·20.	The Magic T	447
12·21.	The Synthesis Problem	448
12·22.	Coupling-hole Magic T's	451
12·23.	Magic T with a Single Symmetry Plane	452
12·24.	Synthesis of Magic T with a Single Symmetry Plane in Coaxial Lines	454
12·25.	The Star	455
12·26.	The Turnstile Junction	459
12·27.	Purcell's Junction	466

FREQUENCY DEPENDENCE OF SYMMETRICAL JUNCTIONS . . . 476

12·28.	The Eigenvalue Formulation	476
12·29.	Wideband Symmetrical Junctions	479

INDEX . . . 481

CHAPTER 1

INTRODUCTION

By E. M. Purcell

1·1. Microwaves.—The microwave region of the electromagnetic spectrum is commonly taken to include frequencies of 10^9 cycles per second and higher. The upper frequency limit exists only as an active frontier lying, at the time of writing, not far below 10^{11} cycles per second. It is not necessary here to attempt to fix the boundaries more precisely. Instead, it may be asked why it is profitable and proper to single out for special treatment this section of the r-f spectrum. The answer to this question will provide a broad definition of "microwaves" and may serve to indicate the scope and purpose of this book.

The distinguishing features of this region of the spectrum are most striking from the point of view of the electrical engineer. Indeed, from the point of view of the physicist concerned with the properties of matter and radiation, boundaries such as those suggested above would appear wholly arbitrary. If he were obliged to choose a name for the region, his choice might suggest very long, rather than very short, wavelengths. In the centimeter wavelength range, the interaction of radiation with matter appears to afford, with certain notable exceptions, less abundant evidence of the structure of molecules and atoms than the spectroscopist finds at much shorter wavelengths. On the other hand, in the study of the dielectric constants and magnetic permeabilities of matter in the bulk no special distinction is made between microwave frequencies and much lower radio frequencies. This is not to suggest that the microwave region is uninteresting to the physicist but rather that its aspect and extent are various, depending on the nature of the problem at hand. The engineer, however, is concerned with the techniques of producing, controlling, transmitting, and detecting electromagnetic energy, and in the microwave region these techniques take on a novel and characteristic form.

The low-frequency end of the microwave spectrum marks roughly the point at which many of the familiar techniques of the radio-frequency art become difficult or ineffective. Perhaps fortuitously, perhaps inevitably, approximately at this point those methods and devices which exploit the shortness of the wavelength become practical and effective. A simple resonant circuit, for example, for a frequency of 30 Mc/sec might consist of a coil and condenser, as in Fig. 1·1a.

It is not hard to show that if all linear dimensions of this circuit were reduced by a factor of 100, the resonant frequency would be increased by the same factor, to 3000 Mc/sec. This method of scaling is not practical for reasons other than the ridiculously small size of the resulting object. In the circuit shown in Fig. 1·1b, the resistance of the wire forming the coil is ten times as effective in damping the oscillations of the circuit; that is, the Q of circuit b is one-tenth that of circuit a. Moreover, the amount of energy that can be stored in the circuit without dielectric breakdown—often an important consideration—is smaller for circuit b

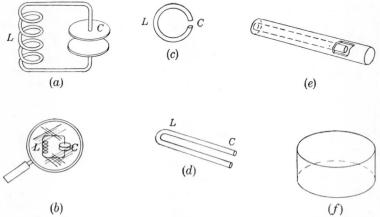

Fig. 1·1.—Resonant circuits at low frequencies and at microwave frequencies.

by a factor of about 1000. It would be better, of course, to reduce the number of turns in the coil, as shown in Fig. 1·1c where the inductance of a single turn is combined with the capacitance of a gap of reasonable size. One further step leads to the reduction of the area of the single turn and to the widening of the condenser gap, as shown in Fig. 1·1d.

Thus the disadvantages of circuit b have been overcome to a considerable degree. The circuit d differs from b, however, in one important respect: its physical dimensions are *not small* compared with its resonant wavelength. Consequently, the circuit, if excited, will lose energy by radiation; the condenser now acts as an antenna, and the circuit behaves as if a series resistance had been inserted to absorb energy and damp the oscillations. This loss of energy can be avoided by enclosing the entire circuit within conducting walls, which might be done in the manner shown in Fig. 1·1e where the two conductors of circuit d have become coaxial cylinders which, provided certain requirements on the thickness and conductivity of the walls are met, confine all electric and magnetic fields to the region between the two cylinders. Another possible solution is shown in Fig. 1·1f. This resonant circuit is simply a hollow metal cylinder. It will merely be asserted here that if the dimensions of the cylinder are correctly chosen, the cylindrical cavity will display an electrical resonance at the desired frequency.

The forms of the resonant circuits e and f are characteristic of microwave devices, which are usually distinguished by two features: (1) the physical size of the circuit elements is comparable with the wavelength involved, and (2) the electromagnetic fields are totally confined within conducting walls, except where it is desired that energy be radiated into space. As has already been suggested, the second of these features is an unavoidable consequence of the first. It is worth remarking that the features of the circuit of Fig. 1·1b which made it impractical in the microwave region and which led to devices comparable to a wavelength in size will eventually be encountered again as higher and higher frequencies are reached. It will no longer be possible, even by the use of microwave techniques, to design circuits of convenient physical size that permit transmission of high power or storage of much energy or that have reasonably low loss. These difficulties are not equally fundamental, nor will they necessarily become acute at the same point in the spectrum. Nevertheless, considerations of this sort will ultimately determine the practical upper frequency limit of what is here called the microwave region and will stimulate the search for other methods of handling electromagnetic energy.

In this book, then, the word "microwave" will be used to imply, not necessarily a particular range of frequencies, but a characteristic technique and a point of view. A resonator of the form of Fig. 1·1f comes within the scope of this book, even if it is 20 ft in diameter, with a corresponding fundamental resonant frequency of 40 Mc/sec. It may be regarded as more or less accidental that the methods of analysis and measurement to be discussed find their widest application at centimeter and millimeter wavelengths.

1·2. Microwave Circuits.—The example of the resonant circuit, discussed above in rather oversimplified language, will already have suggested to the reader that the application of the ordinary low-frequency terminology of *inductance* and *capacitance* (and even the word *circuit* itself) to the objects of Fig. 1·1d, e, and f is of uncertain validity. In the progression from circuits a and b through to f, the concepts of capacitance and inductance lose their identity. In the hairpin-shaped resonator d, for instance, although it is both useful and meaningful to regard one end of the structure as a condenser and the other as an inductance, these concepts cannot be made quantitative except by an arbitrary and artificial convention. Inductance and capacitance are thoroughly disguised in the cylindrical resonator f. To a varying extent, the notions of current, voltage, and resistance have likewise lost their uniqueness.

Evidently, if the electrical properties of such structures are to be investigated, a direct analysis of the electromagnetic field must be made.

In the study of low-frequency circuits, it is possible to avoid an analysis of the field only because all relevant properties of a circuit element, such as a coil or condenser, can be described by one or two numbers, and different elements can be combined in various ways without impairing the validity of the description. A simple coil is characterized by two quantities: inductance and resistance. These two quantities suffice to determine, at any frequency, the relation between the current flowing at the terminals of the coil and the voltage between these terminals. Since current and voltage have here a perfectly definite meaning, and since the relation of one to the other at any frequency is all that need be known, it is not necessary to inquire into the structure of the complicated magnetic field surrounding the coil unless, on occasion, it is desired to calculate, rather than to measure, the inductance of the coil. To be able in this manner to avoid solving Maxwell's equations for each new structure saves an enormous amount of work and makes feasible the analysis of quite complicated circuits. Examples in which this abbreviated description in terms of inductance, capacitance, and resistance is inadequate are, however, not hard to find, even at ordinary radio frequencies. The apparent inductance of the coil of Fig. 1·1a would, in fact, be found to vary noticeably with the frequency. To attribute this variation, as is the custom, to the effect of the "distributed capacitance" of the windings is to acknowledge that the behavior of the coil and the associated electromagnetic field cannot be exactly described by a single number, except at a single frequency. In this special case, for purposes of circuit analysis, the statement that the coil is a linear device with two terminals presenting a certain impedance Z is a complete and accurate description. At microwave frequencies the problem is to analyze an electrical structure that cannot be broken down into such simple elements and for which it is not apparent that voltage and current have a unique meaning.

There is at least one way out of this difficulty. Renouncing all attempts to describe circuits in terms of voltage, current, and impedance and restricting the emphasis to those quantities which appear explicitly in the equations of the electromagnetic field, the generality of Maxwell's equations can be utilized. Each new problem could then be stated as a boundary-value problem; that is, a solution of Maxwell's equations satisfying certain prescribed conditions appropriate to the particular circuit at hand would be sought. Although it would be easy to state the problem in this way, in all but the simplest cases the actual solution would be hopelessly difficult. Furthermore, any modification of the original circuit would usually lead to an entirely new problem. It would not be possible to combine several elements in a new way and to predict and understand the behavior of the combination without re-examining in detail the behavior of each element. If no course other than this were open, it is likely that knowledge of this art would consist of a meager collection of completely solved special cases, together with an extensive assortment of practical devices, designed by trial and error and imperfectly understood.

Fortunately, the situation is not nearly so hopeless. In the first place, the foregoing discussion has overemphasized the complexity of microwave circuits. If electronic devices and antennas are excluded, the components of microwave circuits consist chiefly of cavity resonators and transmission lines of various types. The analysis of the electrical properties of cavities of various shapes has received much attention in recent years, especially in connection with the development of velocity-modulation tubes. A cavity resonator by itself is a rather simple device, in the sense that the properties that are usually of interest can be described by a very few parameters, much as the properties of a low-frequency resonant circuit are summarized in the statement of the resonant frequency, the value of Q, and the impedance at resonance. The uniform transmission line is perhaps even more familiar to most engineers. It will be part of the purpose of this book to show how the standard methods for the analysis of uniform transmission lines can be applied to the propagation of energy through hollow pipes of various sorts—in other words, to generalize the notion of a transmission line. This is not difficult. It can be said that certain isolated elements of the microwave circuit problem are rather easily handled. The main problem is encountered when things are connected together or when nonuniformities are introduced into a previously uniform line. In the following chapters methods for attacking this problem will be worked out.

The direction that the development of microwave circuit analysis and design techniques takes is strongly influenced by (1) the kind of measurements that can be made at microwave frequencies, (2) the nature of the question to be answered, and (3) the existing well-tried and powerful methods for the solution of low-frequency circuit problems, from which it is, naturally, expedient to borrow as often as possible.

1·3. Microwave Measurements.—At microwave frequencies neither the circuit elements to be examined nor the measuring apparatus itself can conveniently be made small compared with a wavelength. This circumstance calls for new experimental techniques and, more important for this discussion, shifts attention from the conventional circuit quantities of voltage, current, and resistance to other quantities more directly accessible to measurement.

At very low frequencies, the voltage between the two conductors of a two-wire transmission line might be measured at some point along the line by connecting an a-c voltmeter, of either the dynamometer or the iron-vane type, between the two conductors. This becomes impractical even in the audio-frequency range because of the inductance of the coil through which the current must flow. It is then necessary to resort to instruments in which the high-frequency circuit more nearly approaches a simple resistance element, the current through which is measured by indirect means—perhaps by a thermal effect or by the use of a rectifier. For example, a fine wire of suitably high resistance might be stretched between the conductors, its rise in temperature as a result of the Joule heat developed determined by measuring its d-c resistance, and the result compared with the temperature rise produced in the same environment by a known direct current passing through the wire. With suitable corrections for the difference between the d-c resistance of the wire and its resistance at the frequency in question, this method is sound and reliable up to frequencies for which the length of the wire is an appreciable fraction of a wavelength. At frequencies higher than this the inductance of the wire cannot be neglected, which is another way of saying that the wire itself is becoming a transmission line. The current that flows in the wire is not the same at different points along the wire, and it is not surprising to find that the average temperature rise of the wire is not related in a simple way to the voltage between the two terminals. Since the wire must necessarily span the distance between the two conductors of the main transmission line, it is clear that a fundamental difficulty stands in the way of any attempt to devise an instrument which may be legitimately called a "voltmeter" when applied to circuits whose dimensions are comparable with a wavelength.

The root of the difficulty goes deeper than the foregoing remarks suggest. The potential difference between two points ordinarily means the line integral of the electric field strength, $\int \mathbf{E} \cdot d\mathbf{s}$, taken at one instant of time, along some path joining the two points. This concept is unique and useful only if the value of the line integral is independent of the path. When the path necessarily extends over a distance not small compared with a wavelength, the line integral is not, in general, independent of the path, and the significance of the term "voltage" is lost.

This suggests that attention be directed to the electric field. Although it is not common engineering practice to measure the electric field strength at a point in absolute terms, it is not hard to think of ways in which it might be done. For instance, a tiny dielectric rod might be suspended in the field in such a way that the torque tending to line it up with the field could be directly measured. From a knowledge of the shape of the rod and the dielectric constant of the material of which it is made, the value of the field strength in volts per meter could be computed. Usually it is much more convenient and just as useful to measure the ratio of the electric field strengths at two points in a system. A familiar example of this procedure is the use of a "probe" to examine the variation of the field strength along a coaxial transmission line. This probe usually consists of a short length of auxiliary line, the central conductor of which projects through a small hole into a slot in the outer conductor of the transmission line under examination. The other end of the auxiliary line terminates in some detecting device of a rectifying or of a thermal type. If precautions are taken to ensure that the probe itself causes only a negligible disturbance in the line under test, the excitation of the auxiliary line will be proportional to the field in the main line, which allows comparison of the field at one location in the main line with the field at some other point.

In much the same way, relative measurements of the magnetic field strength can be made, and these, in turn, may be considered to replace the measurement of current. In microwave circuits, current usually appears as a volume or surface current density, and whereas these quantities have a perfectly unique meaning, it is more natural to associate with the field the related quantities that can be measured.

The usual measurements of field quantities just described are relative ones. The quantities that are customarily measured in absolute terms are frequency or wavelength, and power. The r-f power delivered to a load can be measured directly by calorimetric means; the rate of evolution of heat in some object is inferred from its temperature rise and may be indicated in various ways. To be sure, absolute measurements could be made, if necessary, of other electrical quantities. A method of measuring the electric field strength has already been suggested; it would also be possible in principle and, in certain instances, in practice to measure the radiation pressure associated with the electromagnetic field at a boundary. However, it is approximately correct to say that the *watt* is the only electrical unit that is of direct importance here.

In the microwave region both frequency and wavelength can be measured with great precision. The determination of wavelength involves the accurate measurement of one or more lengths associated with a simple resonant circuit such as a cylindrical cavity or a transmission line. Microwave frequencies, on the other hand, can be compared directly with lower frequencies by standard methods of frequency multiplication. This rather cumbersome and inflexible method is preferred where extreme absolute accuracy is required, as in the establishment of frequency standards. For most work, the measurement of wavelength is much more convenient. Moreover, the physical quantity wavelength is directly and intimately associated with the dimensions of the microwave circuit, and it becomes natural to think in terms of wavelength rather than frequency.

Although this book is not directly concerned with the extensive subject of microwave measurement techniques—to which Vol. 11 of the Series is devoted—the reader will more readily understand the emphasis and approach adopted in the following chapters if the situation just described is kept in mind. For example, the term *impedance* will be used again and again to denote a dimensionless ratio (the *normalized impedance*) and only rarely to denote the ratio of a potential difference in volts to a current in amperes.

1·4. The Aims of Microwave Circuit Analysis.—A microwave circuit is a region enclosed by metallic walls of any shape and communicating with the exterior only by way of a number of transmission lines or waveguides, which may be called the terminals of the circuit. This definition is not so narrow as it might at first appear, for it is easy to extend the notion of transmission lines as terminals to include waveguides carrying many modes, and even antennas.

The final object of microwave circuit analysis, as of low-frequency circuit analysis, is to provide a complete description of what goes on at the *terminals* of the circuit. Suppose that a given circuit has certain arbitrarily selected impedances connected to all pairs of terminals but one. If we are able to predict correctly the impedance that will be measured at the remaining terminal pair—no matter which terminal this happens to be and no matter what the frequency—our description of the circuit can be said to be complete. The methods by which such a prediction can be made, once the properties of the component parts of the circuit are given, and the various ways in which the results of the analysis can be expressed are the main topics of this book.

The properties of the circuit *elements* will not be derived *ab initio*, for the most part. Such problems involve extensive theoretical calculations. The results, however, of such calculations, as well as numerous experimental results, are summarized in Vol. 10, the *Waveguide Handbook*. The limitation in scope of the present volume is thus characteristic of a book on network analysis in which one would not expect to find a derivation of the inductance of a coil of a certain shape. This illustration fails to suggest, however, the variety and novelty of the problems that can be solved and the degree to which the usefulness of the methods to be described here is enhanced by the availability of solutions to those problems.

One often has to deal with a circuit whose complete description cannot easily be deduced. Nevertheless it may be possible to make certain restricted statements, based on very general considerations, about the behavior of the circuit. Such observations prove very useful, not only in reducing the number of parameters that have to be determined experimentally to complete the description of the circuit, but in disclosing basic similarities between circuits superficially different and in avoiding vain efforts to contrive circuits having properties prohibited by one of these general principles. For these reasons much attention will be devoted in this book to those general theorems which can be shown to apply to certain classes of circuits. One topic, for example, will be the properties of a lossless three-terminal-pair junction, by which is meant an enclosure of arbitrary shape, with perfectly conducting walls, provided with three outlets.

1·5. Linearity.—It is probably obvious that the program outlined does not include *nonlinear* circuit elements in its scope. The situation in this respect is similar to that which prevails at lower frequencies, except that, at least until now, the need for general techniques of nonlinear circuit analysis has been less acute in the microwave region. The microwave oscillator, the gas switch, and the crystal detector are the only nonlinear microwave devices of present importance, and each of these must be treated in a special way and is so treated elsewhere in this series. It is, of course, just this restriction to linear circuit elements which makes possible the sort of analysis with which we shall be concerned.

1·6. Dissipation.—The reader will find that much more attention is devoted to *dissipationless* networks than would be justified in a study of low-frequency circuits. It is one of the attractive features of the microwave region that most microwave circuit elements show such a small loss that the error made in assuming that all conductors are perfect conductors is usually negligible—except, of course, in the case of highly resonant circuits. The reason for this, broadly speaking, is that low-frequency circuit elements and microwave circuit elements do not differ much in physical size, and the Q of elements of the same size varies directly as the square root of the frequency, if skin resistance only is concerned. Ferromagnetic materials are not employed as a rule, and the dielectric materials are used sparingly.

1·7. Symmetry.—One other conspicuous feature of microwave circuits is the symmetry properties possessed by many widely used circuit elements. By simply joining a number of similar waveguides together in a geometrically symmetrical way, it is relatively easy to construct a circuit whose low-frequency equivalent would consist of a complicated network of accurately matched elements. In other words, geometrical and electrical symmetry are here naturally and closely allied. The result is to stimulate the investigation of symmetrical networks of many types, including those with as many as four or even six terminal pairs, and to simplify the analysis of many useful devices.

CHAPTER 2

ELECTROMAGNETIC WAVES

By C. G. Montgomery

THE FIELD RELATIONS

2·1. Maxwell's Equations.—In electromagnetic theory, the interaction between charges and their mutual energy is expressed in terms of four field vectors **E** and **B**, **D** and **H**. The first pair define the force on a charge density ρ moving with a velocity **v** by the equation

$$\mathbf{F} = \rho[\mathbf{E} + (\mathbf{v} \times \mathbf{B})].$$

The vectors **E** and **B** satisfy the field equations

$$\operatorname{curl} \mathbf{E} = -\frac{\partial \mathbf{B}}{\partial t}, \qquad \operatorname{div} \mathbf{B} = 0. \tag{1}$$

The second pair of field vectors are determined by the charges and currents present and satisfy the equations

$$\operatorname{curl} \mathbf{H} = \mathbf{J} + \frac{\partial \mathbf{D}}{\partial t}, \qquad \operatorname{div} \mathbf{D} = \rho, \tag{2}$$

where **J** is the current density and ρ the charge density. The set of Eqs. (1) and (2) is known as Maxwell's equations of the electromagnetic field. Sometimes the force equation is included as a member of the set.

The connection between **D** and **E**, and **B** and **H**, depends upon the properties of the medium in which the fields exist. For free space, the connection is given by the simple relations

$$\mathbf{D} = \epsilon_0 \mathbf{E}, \qquad \mathbf{B} = \mu_0 \mathbf{H}.$$

For material mediums of the simplest type the relations are of the same form but with other characteristic parameters

$$\mathbf{D} = \epsilon \mathbf{E}, \qquad \mathbf{B} = \mu \mathbf{H}.$$

The symbol ϵ denotes the permittivity of the medium, and μ the permeability. For crystalline mediums which are anisotropic, the scalars ϵ and μ must be generalized to dyadic quantities. Then **D** and **E** no longer have the same direction in space. This more complex relationship will not be dealt with here. If the medium is a conductor, then Ohm's law is

generally valid and
$$\mathbf{J} = \sigma \mathbf{E},$$

where σ is the conductivity. The properties of the medium are thus expressed in terms of the three parameters σ, ϵ, and μ. Each of these parameters may be a function of frequency, but the variation is usually small and will nearly always be neglected. These parameters are, moreover, not functions of the fields. The conductivity varies from 10^{-14} mho per meter in the best insulators to 5.8×10^7 mhos per meter for pure copper. The quantity ϵ/ϵ_0 is called the specific inductive capacity. It is denoted by k_e and usually lies between 1 and 10. For most substances, μ/μ_0 is very close to unity, but may be as great as 1000 for soft iron. In this case, μ/μ_0 is also a function of $\mathbf{H}$. For a medium in which ϵ and μ are constant and Ohm's law applies, the field equations become

$$\left. \begin{array}{ll} \operatorname{curl} \mathbf{E} = -\mu \dfrac{\partial \mathbf{H}}{\partial t}, & \operatorname{curl} \mathbf{H} = \sigma \mathbf{E} + \epsilon \dfrac{\partial \mathbf{E}}{\partial t}, \\ \operatorname{div} (\mu \mathbf{H}) = 0, & \operatorname{div} (\epsilon \mathbf{E}) = \rho. \end{array} \right\} \quad (3)$$

The units in which these equations are expressed are rationalized units; that is, the unit electric field has been so chosen as to eliminate the factors of 4π that occur when Maxwell's equations are written in electrostatic units, for example. The practical rationalized system, the mks system of units, will be used here. Thus $\mathbf{E}$ and $\mathbf{B}$ are measured in volts per meter and webers per square meter respectively, $\mathbf{H}$ in amperes per meter, $\mathbf{D}$ in coulombs per square meter, ρ in coulombs per cubic meter, and $\mathbf{J}$ in amperes per square meter. In the mks system μ and ϵ have dimensions, and have numerical values that are not unity. For free space μ and ϵ will be written as μ_0 and ϵ_0, and these quantities have the values

$$\mu_0 = 1.257 \times 10^{-6} \quad \text{henry per meter,}$$
$$\epsilon_0 = 8.854 \times 10^{-12} \quad \text{farad per meter.}$$

The velocity of light in free space is $1/\sqrt{\mu_0 \epsilon_0}$ and is equal to 2.998×10^8 meters per second. A quantity that frequently appears in the theory is $\sqrt{\mu/\epsilon}$. For free space, this is equal to 376.7 ohms. The mks system of units is particularly suitable for radiation problems. As will be shown later in this section, a plane wave in free space whose electric field amplitude is 1 volt per meter has a magnetic field amplitude of 1 amp per meter, and the power flow is $\frac{1}{2}$ watt per square meter. Of more importance, perhaps, is the fact that in practical units, the values of impedance encountered in radiation theory are neither very large nor very small numbers but have the same range of values as the impedances encountered in the study of low-frequency circuits. The dimensions of quantities in the mks system may be chosen in a convenient manner. To the basic dimensions of mass, length, and time is added the electric charge as the basic electrical dimension. Table 2·1 shows the dimensions and practical units for the quantities that are of most importance. It is of interest to note that no fractional exponents occur in this table. It is this circumstance, in fact, which urges the adoption of the charge Q as the fourth basic dimension for physical quantities.

TABLE 2·1.—DIMENSIONS AND UNITS OF ELECTRICAL QUANTITIES

Quantity	Dimensions	Practical unit
Length	L	Meter
Mass	M	Kilogram
Time, t	T	Second
Power	ML^2T^{-3}	Watt
Charge	Q	Coulomb
Current, I	QT^{-1}	Ampere
Resistance, R	$ML^2T^{-1}Q^{-2}$	Ohm
Electric potential, V	$ML^2T^{-2}Q^{-1}$	Volt
Electric field, E	$MLT^{-2}Q^{-1}$	Volt/meter
Displacement, D	$L^{-2}Q$	Coulomb/square meter
Conductivity, σ	$M^{-1}L^{-3}TQ^2$	Mho/meter
Dielectric constant, ϵ	$M^{-1}L^{-3}T^2Q^2$	Farad/meter
Capacitance, C	$M^{-1}L^{-2}T^2Q^2$	Farad
Magnetic intensity, H	$L^{-1}T^{-1}Q$	Ampere/meter
Magnetic induction, B	$MT^{-1}Q^{-1}$	Weber/square meter
Permeability, μ	MLQ^{-2}	Henry/meter
Inductance, L	ML^2Q^{-2}	Henry

Maxwell's equations are to be applied with the following boundary conditions at the junction of two mediums, provided that $\mathbf{J}$ and ρ show no discontinuities at the boundary

$$E_t = E'_t, \qquad H_t = H'_t, \\ \epsilon E_n = \epsilon' E'_n, \qquad \mu H_n = \mu' H'_n, \qquad (4)$$

where the primed symbols refer to one medium and the unprimed ones refer to the other. The subscripts t and n refer to the tangential and normal components of the fields at the surface of discontinuity. When a current sheet is present, the boundary condition must be modified to be

$$H_t - H'_t = K \qquad (5)$$

for the discontinuity in H across the boundary, where K is the surface current density. If a surface charge of density ξ is present, then

$$\epsilon E_n - \epsilon' E'_n = \xi.$$

For perfect conductors, σ becomes infinite, and a finite current can be supported with a zero electric field. The magnetic field also vanishes within the conductor, and the boundary conditions are

$$E_t = 0, \quad H_n = 0, \quad H_t = K. \tag{6}$$

Such a surface may be called a sheet of zero impedance or an electric wall. For certain calculations it is convenient to employ the complementary set of boundary conditions

$$E_n = 0, \quad H_t = 0. \tag{7}$$

A surface that imposes these boundary conditions is called a sheet of infinite impedance or a "magnetic wall." In practice, a good conductor is a close approximation to an electric wall which imposes Conditions (6). No materials exist for which Conditions (7) are satisfied, and those conditions are employed as a means of expressing certain symmetry conditions on the field.

There remains one additional fundamental equation which expresses the conservation of charge

$$\text{div } \mathbf{J} + \frac{\partial \rho}{\partial t} = 0. \tag{8}$$

In fact, it is logical to consider that the fundamental equations consist of the curl equations of Eqs. (3) together with Eq. (8). From these three equations, the two divergence equations can be easily derived.

• The set of field relations is now complete and self-consistent. If the charge density ρ and the current density $\mathbf{J}$ are given, the equations can be solved and the field vectors obtained at all points. Conversely, for a given configuration of conductors and dielectrics, a possible field and the currents and charges necessary to maintain that field can be found. This latter procedure is the one which will be more often adopted, principally because it is by far the simpler. By a combination of such elementary solutions, a situation of any degree of complexity can be represented.

For fields that vary harmonically with time, Maxwell's equations take a simpler form,

$$\mathbf{E} = \mathbf{E}e^{j\omega t}, \quad \mathbf{H} = \mathbf{H}e^{j\omega t},$$

where the amplitudes of $\mathbf{E}$ and $\mathbf{H}$ are to be regarded as complex numbers and thus include the phases of the field quantities. Equations (3) become

$$\text{curl } \mathbf{E} = -j\mu\omega\mathbf{H}, \quad \text{curl } \mathbf{H} = \mathbf{J} + j\omega\epsilon\mathbf{E} = (\sigma + j\omega\epsilon)\mathbf{E}, \tag{9}$$

the other pair of equations being unchanged. If such simple harmonic solutions are obtained, solutions with an arbitrary time dependence can be constructed by means of the usual Fourier transformation theory. The monochromatic solutions, however, are the solutions of greatest interest.

In many cases, it is convenient to modify this notation and obtain a form of Eqs. (9) that is apparently even simpler. If $\epsilon_1 = \epsilon - j\sigma/\omega$, the second of Eqs. (9) becomes simply

$$\operatorname{curl} \mathbf{H} = j\omega\epsilon_1 \mathbf{E}. \tag{10}$$

In fact, Eq. (10) may be regarded as more general than Eqs. (9), since the form of Eq. (10) is a suitable representation of any loss of energy that is proportional to the square of the electric field, such as, for example, that part of the dielectric loss which is to be explained by permanent dipoles whose motion is restricted by the viscous nature of the dielectric.

2·2. Poynting's Vector and Energy Theorems.—In the study of very-high-frequency phenomena, it is convenient to think in terms of the electromagnetic fields rather than in terms of the flow of currents and charges. The field should be regarded as containing energy which flows along a transmission line in the field rather than in the currents. This concept is helpful in two ways. First, at very high frequencies the flow of current on a surface is not divergenceless. The displacement currents are large in magnitude and difficult to visualize effectively. Second, a close relationship exists between a waveguide transmission line and a radiating antenna. In order to obtain a good physical picture of energy transfer, it is necessary to invoke the intermediate action of the electromagnetic fields rather than to try to conceive of interaction at a distance between currents on the surfaces of conductors. Philosophical difficulties may be created in the minds of some people by this emphasis on the picture of a physical quantity, such as energy, existing in empty space. It is better to disregard these difficulties, at least for the moment, in order to gain, from the field picture, greater insight into the problems at hand.

The amount of energy that exists in an electromagnetic field may be found by considering the manner by which energy can be transferred from the field into mechanical work, by means of the forces and motions of charges and charged bodies. Thus from the force equation given in Sec. 2·1, expressions for stored electric and magnetic energy can be deduced. This will not be done here; rather, the discussion will be confined to the concept of the flow of electromagnetic energy through a closed surface.

If from Eqs. (3) the following expression is formed,

$$\mathbf{H} \cdot \operatorname{curl} \mathbf{E} - \mathbf{E} \cdot \operatorname{curl} \mathbf{H} = \operatorname{div} (\mathbf{E} \times \mathbf{H}),$$

then

$$\operatorname{div} (\mathbf{E} \times \mathbf{H}) = -\mathbf{E} \cdot \mathbf{J} - \mu \mathbf{H} \cdot \frac{\partial \mathbf{H}}{\partial t} - \epsilon \mathbf{E} \cdot \frac{\partial \mathbf{E}}{\partial t}.$$

The integral of this expression over a volume V, bounded by the surface S, gives

$$\int_S (\mathbf{E} \times \mathbf{H})_n \, dS = -\int_V \mathbf{E} \cdot \mathbf{J} \, dV - \frac{\partial}{\partial t} \int_V \left(\frac{1}{2} \epsilon E^2 + \frac{1}{2} \mu H^2 \right) dV. \tag{11}$$

This equation is taken to represent the conservation of electromagnetic energy in the volume V. The first term on the right is the work done per second by the impressed forces (currents); $\frac{1}{2}\epsilon E^2$ is the electric energy per unit volume; and $\frac{1}{2}\mu H^2$ is the magnetic energy. The left-hand term is then the flow of energy through the surface S enclosing the volume V. It is in the form of the flux of a vector

$$\mathbf{S} = \mathbf{E} \times \mathbf{H}, \tag{12}$$

which is taken to be the density of energy flow and is called Poynting's vector. It is recognized that this identification is not unique, since, for example, the curl of any field vector may be added to $\mathbf{S}$ without affecting

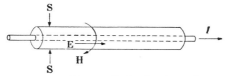

FIG. 2·1.—Energy dissipation in a wire.

the validity of Eq. (11). The arbitrariness in the interpretation of $\mathbf{S}$ will, however, not lead to difficulty, since only the surface integral of $\mathbf{S}$ will be used.

As an example of the use of Poynting's vector, let us consider a current flowing through a long wire that has a finite conductivity. For the closed surface, let us choose a small cylinder enclosing the wire as shown in Fig. 2·1. The electric field on this surface will be constant and parallel to the wire. The magnetic field lines will be circles about the wire, and the field strength will be $H = I/2\pi r$, where r is the radius of the cylinder. Poynting's vector will be perpendicular to the cylindrical surface and directed toward the wire. Its value will be

$$S = \frac{EI}{2\pi r}.$$

If l is the length of the surface, then the normal flux of S will be simply EIl. This represents, then, the rate of energy flow into the wire, and it is just what would be calculated from more elementary considerations of the Joule heating produced by the electric current.

It is known that this energy is transferred to the wire by collisions of the electrons with the atoms of the wire. It is useful, however, to disregard considerations of this mechanism and to think of the energy as being stored in the electromagnetic field and flowing into the wire at the definite rate given by Eq. (11).

In a similar manner, for periodic fields represented by complex quantities, the following expression can be formed

$$\mathbf{H}^* \operatorname{curl} \mathbf{E} - \mathbf{E} \cdot \operatorname{curl} \mathbf{H}^* = \operatorname{div} (\mathbf{E} \times \mathbf{H}^*).$$

From Eqs. (9), after transformation to the integral form and division by 2, there results

$$\tfrac{1}{2} \int_S (\mathbf{E} \times \mathbf{H}^*)_n \, dS = -\tfrac{1}{2} \int_V \mathbf{E} \cdot \mathbf{J}^* \, dV$$
$$+ \tfrac{1}{2} j\omega \int_V (\epsilon |E|^2 - \mu |H|^2) \, dV. \quad (13)$$

The real part of the first term on the right is the average work done per second by the impressed currents, and this must be equal to the power flowing across S. The average electric and magnetic energies stored per unit volume are respectively $\tfrac{1}{2}\epsilon |E|^2$ and $\tfrac{1}{2}\mu |H|^2$. Thus a complex Poynting vector may be defined as

$$\mathbf{S} = \tfrac{1}{2} \mathbf{E} \times \mathbf{H}^*. \quad (14)$$

The real part of the normal flux of the vector **S** through a surface is the average power flow through the surface. It is to be noted that Eq. (13) is not derivable from Eq. (11) but represents an entirely distinct energy theorem.

2·3. Solutions of Maxwell's Equations.—General solutions of Maxwell's equations can be expressed in several forms. Perhaps the one that gives the simplest physical picture is the solution in which **E** and **H** are expressed in terms of the retarded potentials which, in turn, are related to the magnitudes of the currents and charges present. Much more information may be obtained by considering particular solutions and deducing the properties of more complex situations from the behavior of the simpler ones. Since the equations are linear, a linear combination of particular solutions is also a solution. A series of cases of increasing complexity, leading up to the case of the propagation of energy through metal pipes, will now be considered.

Electromagnetic fields and energy propagation will be spoken of in terms of waves. This means simply that **E** or **H** will be expressed in terms of the coordinates and will have a time dependence of $e^{j\omega t}$. Since **E** and **H** are complex numbers, their absolute values or moduli are called the "amplitudes" of the waves, and their arguments the "phases" of the waves.

If a metal surface is present in a field, E_t is zero or very small over the surface of the metal. The Poynting vector **S** can have only a very small component into the metal, and only a small amount of energy flows through the surface. Thus it is possible to enclose an electromagnetic field within a metal tube and transmit power from place to place without too much loss. Some loss will be present, to be sure, since currents must flow in the tube walls to maintain the fields, and some power will therefore be converted into heat. Let us investigate, then, the conditions necessary for propagation of this type to take place.

Either **E** or **H** can be eliminated from Eqs. (9) by taking the curl of the equations and employing the vector identity

$$\text{curl curl } \mathbf{E} = \text{grad div } \mathbf{E} - \nabla^2 \mathbf{E}.$$

Since the divergences of the field quantities are zero,

$$\nabla^2 \mathbf{E} + \omega^2 \mu \epsilon \mathbf{E} = 0,$$

and an identical equation for **H** is obtained. These equations are known as the wave equations and must always be satisfied, whatever boundary conditions are imposed. The wave equations are then necessary conditions on **E** and **H**. They are not, however, sufficient; the fundamental Maxwell's equations must be satisfied, and they are not derivable from the wave equations. Maxwell's equations give the necessary connection between **E** and **H**.

It is helpful to classify the particular solutions of Maxwell's equations according to whether or not components of **E** or **H** exist in the direction of propagation. Let us take the z-axis in the direction of propagation. First, it can be shown that no purely longitudinal waves may exist. Let us take $E_x = E_y = H_x = H_y = 0$. Then

$$\text{curl}_z \mathbf{E} = -j\omega\mu H_z = \frac{\partial E_y}{\partial x} - \frac{\partial E_x}{\partial y} = 0,$$

and

$$\text{curl}_z \mathbf{H} = j\omega\epsilon E_z = \frac{\partial H_y}{\partial x} - \frac{\partial H_x}{\partial y} = 0,$$

and all components vanish. Second, purely transverse waves exist in which neither **E** nor **H** has longitudinal components. Such waves are called transverse-electromagnetic (TEM) waves or principal waves. Third, transverse waves may exist in which only **E** has longitudinal components. Such waves are called transverse-magnetic (TM) waves or E-waves. In a fourth class are waves in which only **H** has longitudinal components. These are called transverse-electric (TE) waves or H-waves. This classification is inclusive, since all possible solutions of Maxwell's equations may be built up of linear combinations of elementary solutions of the latter three types. The TEM-waves will first be discussed; then the TE- and TM-cases will be considered.

PURELY TRANSVERSE ELECTROMAGNETIC WAVES

2.4. Uniform Plane Waves.—A wave is said to be plane if the equiphase surfaces, at a given instant of time, are planes. A wave is uniform if there is no change in amplitude along the equiphase surface. Let the xy-plane be taken as the equiphase surface. Then $\partial/\partial x = \partial/\partial y = 0$, and E_z and H_z must be zero. A uniform plane wave is therefore said to be a transverse-electromagnetic (*TEM*) wave. Equations (9) reduce to

$$\frac{\partial E_x}{\partial z} = -j\omega\mu H_y, \qquad \frac{\partial H_y}{\partial z} = -(\sigma + j\omega\epsilon)E_x.$$

If H_y is eliminated,

$$\frac{\partial^2 E_x}{\partial z^2} - j\omega\mu(\sigma + j\omega\epsilon)E_x = 0. \tag{15}$$

An identical equation exists for H_y. These equations are known as the "telegrapher's" equations.

The solution of Eq. (15) is $E_x = Ee^{-\gamma z}$, where E may be taken as real, and

$$\gamma = \sqrt{j\omega\mu\sigma - \omega^2\epsilon\mu} = \alpha + j\beta. \tag{16}$$

The quantity γ is called the "propagation constant" of the wave; its real part α is called the "attenuation constant"; and the imaginary part β is called the "phase constant." The sign of the square root in the expression for γ is to be taken so that α is positive for a wave traveling in the positive z-direction and negative for a wave in the negative z-direction. The phase velocity v of the wave is then ω/β, and the wavelength $\lambda = 2\pi/\beta$. The quantity β is called the "wave number" although, strictly speaking, it is the number of radians per unit length and not the number of wavelengths. If the conductivity of the medium, σ, is small, $\beta \approx \omega \sqrt{\epsilon\mu}$ and $v \approx 1/\sqrt{\epsilon\mu}$. If σ is not negligible, the exact expressions are

$$\alpha = [\tfrac{1}{2}\omega\mu(\sqrt{\sigma^2 + \omega^2\epsilon^2} - \omega\epsilon)]^{1/2},$$
$$\beta = [\tfrac{1}{2}\omega\mu(\sqrt{\sigma^2 + \omega^2\epsilon^2} + \omega\epsilon)]^{1/2}.$$

The solution of the second telegrapher's equation is

$$H_y = -\frac{\gamma}{j\omega\mu} E_x.$$

The ratio E_x/H_y is called the "wave impedance" and may be denoted by Z_w.

$$Z_w = \sqrt{\frac{j\omega\mu}{\sigma + j\omega\epsilon}}. \tag{17}$$

When $\sigma = 0$, $Z_w = \sqrt{\mu/\epsilon}$; and for free space, $Z_w = 377$ ohms. For σ not equal to zero,

$$Z_w = \frac{1}{\sqrt{\sigma^2 + \omega^2\epsilon^2}} (\beta + j\alpha).$$

It is to be noted that, for *TEM*-waves, Z_w depends only on the properties of the medium. The expression for Z_w given by Eq. (17) is a combination of the constants of the medium which often occurs. It is called the "intrinsic impedance" and is denoted by ζ. Thus

$$\zeta = \sqrt{\frac{j\omega\mu}{\sigma + j\omega\epsilon}}.$$

For the special case of *TEM*-waves,

$$Z_w = \zeta.$$

The inverse of ζ is the intrinsic admittance and is denoted by η.

Since γ may be positive or negative, depending on the sign of the square root in Eq. (16), E_x can be written in the general form

$$E_x = E_1 e^{-\gamma z} + E_2 e^{\gamma z},$$

and a similar equation can be written for H_y. These equations represent two waves traveling in opposite directions.

A general solution of Maxwell's equations may be built up out of a set of elementary uniform plane waves, the directions and amplitudes being chosen in such a way as to satisfy the boundary conditions. This is an excellent artifice for some purposes, but because the description of the field in these terms becomes quite complicated except in the simplest cases, this method will not be discussed further here. In a similar manner, the six components of **E** and **H** might be replaced by wave impedances $Z_{ij} = E_i/H_j$, i, j referring to some coordinate of the space that is being considered.

In general, the problem is not simplified in this way. However, in certain cases the wave impedance is a useful concept, and a complete description of a situation may often be obtained in terms of wave impedances and propagation constants. A very important example is the treatment of the problem of reflection and refraction of plane waves at the boundary between two mediums. For a complete discussion of this problem the reader is referred to other texts.[1]

2·5. Nonuniform Transverse-electromagnetic Plane Waves.—Let us remove the restriction that the fields be uniform in a transverse plane, but let $H_z = E_z = 0$. The electromagnetic equations may be broken up into two sets, the first of which governs the propagation in the z-direction

$$\left. \begin{array}{ll} \dfrac{\partial E_x}{\partial z} = -j\omega\mu H_y, & \dfrac{\partial E_y}{\partial z} = j\omega\mu H_x, \\ \dfrac{\partial H_x}{\partial z} = (\sigma + j\omega\epsilon)E_y, & \dfrac{\partial H_y}{\partial z} = -(\sigma + j\omega\epsilon)E_x. \end{array} \right\} \quad (18)$$

The second set governs the distribution of **E** and **H** in a transverse plane

[1] See, for example, S. A. Schelkunoff, *Electromagnetic Waves*, Van Nostrand, New York, 1943, Chap. 8, p. 251.

$$\frac{\partial E_x}{\partial x} + \frac{\partial E_y}{\partial y} = 0, \qquad \frac{\partial H_x}{\partial x} + \frac{\partial H_y}{\partial y} = 0, \\ \frac{\partial E_x}{\partial y} = \frac{\partial E_y}{\partial x}, \qquad \frac{\partial H_x}{\partial y} = \frac{\partial H_y}{\partial x}. \bigg\} \qquad (19)$$

From the first of these sets, identical equations in any of the four variables E_x, E_y, H_x, or H_y can be obtained by elimination. For example

$$\frac{\partial^2 E_x}{\partial z^2} = j\omega\mu(\sigma + j\omega\epsilon)E_x.$$

The solution may be $E_x = E(x,y)e^{-\gamma z}$ where γ has the same value as before. It has been shown, therefore, that all plane TEM-waves have the same propagation constant and hence the same velocity and attenuation as uniform plane waves. Moreover $E_x = \dfrac{\gamma}{j\omega\mu} H_y$ and

$$Z_{xy} = \frac{E_x}{H_y} = \sqrt{\frac{j\omega\mu}{\sigma + j\omega\epsilon}} = \zeta.$$

Hence the wave impedances of all plane TEM-waves are identical, and equal to the intrinsic impedance of the medium.

If E_y had been found from Maxwell's equations, and if it had been assumed that $E_y = E(x,y)e^{-\gamma z}$, then the wave impedance would have been found to be

$$Z_{yx} = \frac{E_y}{H_x} = -\frac{j\omega\mu}{\gamma} = -Z_{xy} = -\zeta.$$

This may appear to be a contradiction, but the situation is readily understood if the z-component of the Poynting vector is formed

$$S_z = \tfrac{1}{2}(\mathbf{E} \times \mathbf{H}^*)_z = \tfrac{1}{2}(E_x H_y^* - E_y H_x^*) \\ = \tfrac{1}{2}(Z_{xy} H_y H_y^* - Z_{yx} H_x H_x^*), \\ S_z = \tfrac{1}{2} Z_{xy}(|H_x|^2 + |H_y|^2) = \tfrac{1}{2} Z_{xy} |\mathbf{H}|^2.$$

The negative sign for the wave impedance is thus merely the result of the choice of the positive directions of the coordinate axes and represents no significant new fact.

The field distribution in the transverse plane is governed by Eqs. (19). They show that either $\mathbf{E}$ or $\mathbf{H}$ or both can be derived from a scalar potential or from a stream function. Let, for example,

$$\mathbf{E} = -\operatorname{grad} \phi$$

so that

$$E_x = -\frac{\partial \phi}{\partial x}, \qquad E_y = -\frac{\partial \phi}{\partial y},$$

where ϕ is a scalar potential. The curl equation is therefore satisfied,

$$\frac{\partial E_x}{\partial y} = \frac{\partial E_y}{\partial x} = -\frac{\partial^2 \phi}{\partial x\, \partial y}.$$

The other equation for **E** gives

$$\frac{\partial^2 \phi}{\partial x^2} + \frac{\partial^2 \phi}{\partial y^2} = 0.$$

Thus **E** is determined as though the field were a static one, and all of the techniques for solving static problems become available. Since $Z_{xy} = -Z_{yx}$,

$$\frac{E_x}{E_y} = -\frac{H_y}{H_x},$$

or the electric lines of force must be perpendicular to the magnetic lines of force. However, the electric lines are perpendicular to the lines $\phi = $ constant; hence the equipotential lines must represent magnetic lines.

One very important consequence of these conditions may be shown easily. Suppose that there is a cylindrical metal tube whose wall is represented by the curve $\phi = a$. Now if ϕ has no singularities within the region bounded by $\phi = a$, then ϕ must be a constant throughout the region and E must be zero. Therefore no purely transverse electromagnetic wave can propagate down a hollow pipe. If there is another conductor, however, within the region $\phi = a$, then a finite value of E is possible, since ϕ may have singularities within the inner conductor which allow the boundary conditions to be satisfied. This mode of propagation will be studied in more detail in Sec. 2·7.

Suppose that there is a field distribution representing a *TEM*-wave. Then without disturbing the field, a conductor can be inserted along any curve $\phi = $ constant so that it is everywhere perpendicular to the electric lines. This is a useful device and will often be employed.

A stream function may also be employed to specify the fields. If

$$H_x = \frac{\partial A}{\partial y}, \qquad H_y = -\frac{\partial A}{\partial x},$$

then the equation

$$\frac{\partial H_x}{\partial x} + \frac{\partial H_y}{\partial y} = \frac{\partial^2 A}{\partial x\, \partial y} - \frac{\partial^2 A}{\partial x\, \partial y} = 0$$

is satisfied. Hence A must be determined by the equation

$$\frac{\partial^2 A}{\partial x^2} + \frac{\partial^2 A}{\partial y^2} = 0.$$

The scalar A may be regarded as the z-component of a vector **A** whose other components are zero. Then

$$\mathbf{H} = \operatorname{curl} \mathbf{A}.$$

For this special case **A** might be considered as the vector potential or the Hertz vector.

2·6. *TEM*-waves between Parallel Plates.

If there are two infinite conducting planes at $y = 0$ and $y = b$ (Fig. 2·2), the electric field must be normal to these planes and the magnetic field must be entirely tangential. The solution of Maxwell's equations given previously for uniform plane waves is the solution in the region between the planes. It was found that

$$E_y = E e^{-\gamma z} e^{j\omega t}, \qquad H_x = -\frac{E_y}{\zeta} = -\frac{E}{\zeta} e^{-\gamma z} e^{j\omega t},$$

which hold for $0 \leq y \leq b$; for y outside this region, $E_y = H_x = 0$.

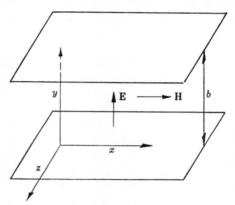

Fig. 2·2.—Infinite conducting planes.

Conductors perpendicular to **E** have been inserted along lines $\phi = $ constant, and this has not altered the values of the fields. As before

$$\gamma = j\sqrt{\omega^2 \epsilon \mu - j\omega\mu\sigma}.$$

Currents must flow in the plates to maintain the fields. They are determined by the boundary condition expressed in Eq. (5). The linear current density that flows in the direction of propagation is therefore

$$K = -H_x = \frac{E}{\zeta} e^{-\gamma z} e^{j\omega t}.$$

The power flow between the plates in the z-direction is found by calculation of the Poynting vector

$$S_z = \frac{1}{2}(\mathbf{E} \times \mathbf{H}^*)_z = \frac{E^2}{2}\frac{1}{\zeta^*} e^{-(\gamma+\gamma^*)z}.$$

If $\gamma = \alpha + j\beta$,

$$S_z = \frac{E^2}{2}\frac{1}{\zeta^*} e^{-2\alpha z}.$$

If S_z is integrated from $y = 0$ to $y = b$, then the power flow across a unit

length in the x-direction is
$$P = b \text{ Re } (S_z) = \frac{E^2}{2} \frac{b}{\text{Re } (\zeta^*)} e^{-2\alpha z},$$

where Re (ζ^*) represents the real part of the intrinsic impedance. If the medium is lossless,
$$\zeta = \text{Re } (\zeta) = \sqrt{\frac{\mu}{\epsilon}}.$$

The power is thus attenuated at the rate of 2α per unit length.

This power flow may be expressed in terms of the currents in the conducting plates and quantities that may be called impedances. Thus
$$P = \tfrac{1}{2}KK^*b \text{ Re } (\zeta).$$

To make this analogous to the power flow in a low-frequency circuit, an impedance may be defined that is different from the intrinsic impedance ζ and might be called the "current impedance" Z_I,
$$Z_I = b \text{ Re } (\zeta).$$

It should be noted that the dimensions of this quantity are not ohms but ohm-meters. In a similar and wholly arbitrary fashion an impedance may be defined in terms of voltage. Let us take V for the voltage according to
$$V = \int_0^b E_y \, dy = bE_y.$$

Since there are no longitudinal components of the fields, this integral is independent of the choice of path of integration between the plates. Thus V is uniquely defined. The "voltage impedance" may now be defined as
$$Z_V = \text{Re}\left(\frac{VV^*}{2P}\right) = b \text{ Re } (\zeta^*),$$

and is thus equal to Z_I. It should also be noticed that a third kind of impedance may be defined by
$$\frac{V}{K} = \frac{bE_y}{H_x} = b\zeta.$$

Thus, for this simple case of a parallel-plate transmission line, there are several definitions of impedance that lead to the same result, in a fashion identical with that which prevails for low-frequency transmission along a wire. For the more complicated modes of propagation it will be found that these simple relations are no longer true.

2·7. TEM-waves between Coaxial Cylinders.—Another simple case of considerable interest is the propagation of waves between conducting coaxial cylinders. The fields are no longer uniform, but a *TEM*-wave is permitted because the electrostatic potential may have a singularity within the inner conductor. Let the inner and outer radii be a and b respectively (Fig. 2·3). Then

$$E_z = H_z = H_r = E_\phi = 0,$$

and the equations

$$\frac{\partial E_r}{\partial z} = -j\omega\mu H_\phi, \qquad \frac{\partial H_\phi}{\partial z} = -(\sigma + j\omega\epsilon)E_r$$

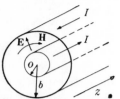

FIG. 2·3.—*TEM*-waves between coaxial cylinders.

determine the propagation in the z-direction with the same propagation constant γ and wave impedance as those obtained in general for *TEM*-waves. The transverse variation of the fields is determined from

$$\frac{1}{r}\frac{\partial}{\partial r}(rH_\phi) - \frac{1}{r}\frac{\partial H_r}{\partial \phi} = (\sigma + j\omega\epsilon)E_z,$$

or

$$\frac{\partial}{\partial r}(rH_\phi) = 0,$$

and rH_ϕ is therefore a constant. It is convenient to express this constant in terms of the total current I flowing on the conductors. There is flowing on the outer conductor a current density K,

$$K = H_\phi(b).$$

The current I is then

$$\int_0^{2\pi} Kb \, d\phi = 2\pi b H_\phi(b),$$

and hence

$$H_\phi = \frac{I}{2\pi r}.$$

The total current on the inner conductor is also I. The radial electric field is

$$E_r = \frac{I}{2\pi r}\zeta = \frac{I}{2\pi r}\sqrt{\frac{j\omega\mu}{\sigma + j\omega\epsilon}}.$$

The directions of E_r, H_ϕ, and I for a wave traveling in the z-direction are shown in Fig. 2·3.

The Poynting vector is

$$S_z = \frac{1}{2}E_r H_\phi^* = \frac{1}{2}\zeta \frac{II^*}{(2\pi r)^2}$$

and the total power is

$$P = \int_a^b \operatorname{Re}|S_z|2\pi r \, dr = \frac{\operatorname{Re}(\zeta)|I|^2}{4\pi}\ln\frac{b}{a}.$$

As in the preceding section, several impedances can be defined

$$Z_I = \frac{\text{Re }(\zeta)}{2\pi} \ln \frac{b}{a} = \int_a^b \frac{E_r \, dr}{I} = \frac{V}{I} = Z_V,$$

which are again equal only because of the simplicity of the field configuration. The impedances defined here are all measured in ohms. Here also the integral in the definition of V is unique, since the fields are entirely in the transverse plane.

The fields can also be expressed in terms of the fields at the surface of the center conductor. Let these fields be E_a and H_a. Then

$$E_r = \frac{a}{r} E_a,$$

$$H_r = \frac{a}{r} H_a.$$

The total power flow is then

$$P = \text{Re }(\zeta)\pi a^2 H_a^2 \ln \frac{b}{a} = \frac{\pi a^2 E_a^2 \ln \frac{b}{a}}{\text{Re }(\zeta)}.$$

In terms of the coaxial impedance $Z_I = Z_V = Z_0$,

$$P = 2\pi^2 a^2 Z_0 H_a^2 = 2\pi^2 a^2 \frac{Z_0}{[\text{Re}(\zeta)]^2} E_a^2.$$

A coaxial line, operated in this mode of transmission, is thus very similar to a low-frequency circuit. The voltage V, the current I, and a unique characteristic impedance Z_0 can be defined.

It should be pointed out that this is not the mode of transmission for a single wire with a finite conductivtiy in free space, in the ordinary low-frequency approximation. In the case of a single wire, the electric field is in the direction of the wire and the mode of propagation is a transverse-magnetic mode with respect to the direction of the wire.

2·8. Spherical *TEM*-waves.— Transverse-electromagnetic spherical waves are analogous to plane waves. In the spherical co-ordinates r, θ, ϕ, shown in Fig. 2·4, the vector operations are defined by

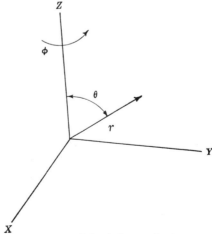

Fig. 2·4.—Spherical co-ordinates.

$$(\operatorname{curl} \mathbf{A})_r = \frac{1}{r \sin \theta} \left[\frac{\partial}{\partial \theta} (\sin \theta A_\phi) - \frac{\partial A_\theta}{\partial \phi} \right],$$

$$(\operatorname{curl} \mathbf{A})_\theta = \frac{1}{r \sin \theta} \left[\frac{\partial A_r}{\partial \phi} - \sin \theta \frac{\partial}{\partial r} (r A_\phi) \right],$$

$$(\operatorname{curl} \mathbf{A})_\phi = \frac{1}{r} \left[\frac{\partial}{\partial r} (r A_\theta) - \frac{\partial A_r}{\partial \theta} \right];$$

$$\operatorname{div} \mathbf{A} = \frac{1}{r^2 \sin \theta} \left[\sin \theta \frac{\partial}{\partial r} (r^2 A_r) + r \frac{\partial}{\partial \theta} (\sin \theta A_\theta) + r \frac{\partial A_\phi}{\partial \phi} \right].$$

If $H_r = E_r = 0$, the curl equations of the electromagnetic field reduce to

$$j\omega\mu r H_\theta = \frac{\partial}{\partial r} (r E_\phi), \qquad j\omega\mu r H_\phi = - \frac{\partial}{\partial r} (r E_\theta),$$

$$(\sigma + j\omega\epsilon) r E_\theta = - \frac{\partial}{\partial r} (r H_\phi), \qquad (\sigma + j\omega\epsilon) r E_\phi = \frac{\partial}{\partial r} (r H_\theta).$$

If these are compared with the set of Eqs. (18), it is seen that they are identical provided that

rE_ϕ and rH_θ are replaced by E_y and H_x,
rE_θ and rH_ϕ are replaced by E_x and H_y.

The divergence equations are

$$\frac{\partial}{\partial \theta} (\sin \theta E_\theta) + \frac{\partial}{\partial \phi} E_\phi = 0, \qquad \frac{\partial}{\partial \theta} (\sin \theta H_\theta) + \frac{\partial}{\partial \phi} H_\phi = 0.$$

$$\frac{\partial}{\partial \theta} (\sin \theta E_\phi) = \frac{\partial E_\theta}{\partial \phi}, \qquad \frac{\partial}{\partial \theta} (\sin \theta H_\phi) = \frac{\partial H_\theta}{\partial \phi}.$$

These are again exactly analogous to Eqs. (19) and show that either $\mathbf{E}$ or $\mathbf{H}$ or both are derivable from either a potential function or a stream function. It can therefore be concluded that spherical *TEM*-waves have the same wave impedance and propagation constant as plane *TEM*-waves.

2·9. Uniform Cylindrical Waves.—The solutions of the electromagnetic equations for the special cases of purely transverse waves will be completed by considering uniform cylindrical waves. Thus

$$\frac{\partial}{\partial \phi} = \frac{\partial}{\partial z} = 0,$$

$$E_r = H_r = 0.$$

The electromagnetic equations break up into two independent sets, as before,

$$\frac{\partial E_z}{\partial r} = j\omega\mu H_\phi, \qquad \frac{\partial}{\partial r} (r H_\phi) = (\sigma + j\omega\epsilon) r E_z, \qquad (20)$$

and
$$\frac{\partial}{\partial r}(rE_\phi) = -j\omega\mu r H_z, \qquad \frac{\partial H_z}{\partial r} = -(\sigma + j\omega\epsilon)E_\phi. \tag{21}$$

Only the first set, Eqs. (20), will be considered. Elimination of H_ϕ results in
$$r\frac{d^2 E_z}{dr^2} + \frac{dE_z}{dr} - j\omega\mu(\sigma + j\omega\epsilon)rE_z = 0.$$

If this equation is divided through by a new variable $x = \gamma r$, it becomes
$$\frac{d^2 E_z}{dx^2} + \frac{1}{x}\frac{dE_z}{dx} - E_z = 0.$$

The solution of this equation is a Bessel function of order zero with the argument jx.[1] Any pair of independent solutions may be chosen, in terms of which the fields may be expressed. Let us take
$$E_z = A H_0^{(1)}(jx) + B J_0(jx),$$
where A and B are arbitrary constants and $H_0^{(1)}$ is a Hankel function of the first kind
$$H_0^{(1)}(x) = J_0(x) + j N_0(x).$$

To find the physical meaning of these solutions, let us consider them separately. If
$$E_z = A H_0^{(1)}(jx) = A H_0^{(1)}(j\gamma r),$$
then
$$H_\phi = \frac{j\gamma A}{j\omega\mu}\frac{dH_0^{(1)}(j\gamma r)}{d(j\gamma r)} = -\frac{jA}{\zeta}H_1^{(1)}(j\gamma r),$$
from Eqs. (20), and the wave impedance Z_w is
$$\frac{E_z}{H_\phi} = Z_w = \zeta j \frac{H_0^{(1)}(j\gamma r)}{H_1^{(1)}(j\gamma r)}.$$

For large values of r, the asymptotic forms may be inserted. Then
$$E_z = A\sqrt{\frac{2}{\pi j\gamma r}}\, e^{j\left(j\gamma r - \frac{\pi}{4}\right)},$$
$$H_\phi = \frac{-jA}{\zeta}\sqrt{\frac{2}{\pi j\gamma r}}\, e^{j\left(j\gamma r - \frac{3\pi}{4}\right)},$$
and
$$Z_w = j\zeta e^{j(\pi/2)} = -\zeta.$$

If γ is separated into its real and imaginary parts,
$$\gamma = \alpha + j\beta,$$

---
[1] See, for example, Eugene Jahnke and Fritz Emde, *Tables of Functions with Formulas and Curves*, Dover Publications, New York, 1943, p. 146.

then
$$E_z = C\sqrt{\frac{1}{r}}\, e^{-\alpha r} e^{-j\beta r},$$
$$H_\phi = -\frac{C}{\zeta}\sqrt{\frac{1}{r}}\, e^{-\alpha r} e^{-j\beta r},$$

where C is a new real constant. These are the equations of a cylindrical wave traveling outward and attenuated at the rate of α per unit length. The wave impedance is the same as for a plane wave, the negative sign being necessary because of the choice of the positive direction of the coordinates. For small r, both $\mathbf{E}$ and $\mathbf{H}$ become infinite and must be produced by an energy source at a finite radius.

The second solution is
$$E_z = BJ_0(jx) = BJ_0(j\gamma r),$$
$$H_\phi = \frac{B}{j\omega\mu}\frac{d}{dr}J_0(j\gamma r) = \frac{-jb}{\zeta}J_1(j\gamma r),$$

and the wave impedance is
$$Z_w = \zeta j\frac{J_0(j\gamma r)}{J_1(j\gamma r)}.$$

Again, for large values of r,
$$E_z = B\sqrt{\frac{2}{\pi j\gamma r}}\cos\left(j\gamma r - \frac{\pi}{4}\right),$$
$$H_\phi = \frac{-jB}{\zeta}\sqrt{\frac{2}{\pi j\gamma r}}\cos\left(j\gamma r - \frac{3\pi}{4}\right),$$
$$Z_w = j\zeta \cot\left(j\gamma r - \frac{\pi}{4}\right).$$

Thus this solution does not represent a propagating wave unless some loss is present. If γ is purely imaginary, Z_w is likewise purely imaginary and no propagation of energy takes place. It should be emphasized that for neither solution is Z_w independent of r, although for the outward-traveling wave it approaches a constant value. All the cases in which the electromagnetic vectors are purely transverse to the direction of propagation have now been described briefly. The longitudinal modes will be described in the portion of this chapter following Sec. 2·10.

2·10. Babinet's Principle.—Maxwell's equations have a symmetry in the electric and magnetic fields that is extremely useful in the discussion of electromagnetic problems. One aspect of this symmetry can be expressed as a generalization of Babinet's principle in optics. If $\mathbf{E}$ and $\mathbf{H}$ in Maxwell's equations are replaced by new fields $\mathbf{E}'$ and $\mathbf{H}'$, according to the relations

$$E = \pm \sqrt{\frac{\mu}{\epsilon}} H', \\ H = \pm \sqrt{\frac{\epsilon}{\mu}} E',$$
(22)

it is found that **E'** and **H'** also satisfy Maxwell's equations. The boundary conditions must be altered accordingly. On a metal wall, the tangential component of **E** vanishes; therefore the new boundary condition should be that the tangential component of **H'** vanish over this same surface. The electric wall in the unprimed system must be replaced by a magnetic wall in the primed system. Likewise magnetic walls in the unprimed case are to be replaced by electric walls. When the boundary conditions consist of the specification of a radiation condition, no change is required for the transformation. Poynting's vector is

$$S = E \times H = -H' \times E' = E' \times H' = S'$$

and is therefore invariant.

A simple example of the application of Babinet's principle is afforded by the reflection of a wave from a metal plate. At normal incidence, if the plate is perfectly conducting, a pure standing wave is set up, with planes of zero electric field parallel to the plate and spaced one-half wavelength apart. The magnetic field is a maximum at the metal plate and has planes of zero intensity wherever the electric field is a maximum. If **E** is replaced by **H'**, **H** by −**E'**, and the metal plate by a magnetic wall, another possible field configuration is obtained. The new solution is obviously identical with the old one if one imagines a magnetic wall placed one-quarter wavelength in front of the electric wall.

A second example can be found in the propagation of the *TEM*-mode between parallel plates. Since **H** is transverse to the direction of propagation, magnetic walls perpendicular to **H** can be inserted as indicated schematically in Fig. 2·5a.

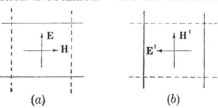

FIG. 2·5.—Babinet's principle applied to the *TEM*-mode between parallel plates. The electric walls are indicated by solid lines; the magnetic walls by dotted lines.

Application of Babinet's principle leads to the situation shown in Fig. 2·5b. This is obviously the same mode of propagation as the original one, but rotated 90° about the direction of propagation.

Although these two examples of the application of Babinet's principle are rather trivial, they serve to show that in order to apply the principle it is necessary to have a system with a high degree of symmetry. All the magnetic walls must exist by virtue of the symmetry. Many more fruitful examples of the application of this principle can be given for waves propagating in pipes.

ELECTROMAGNETIC WAVES WITH LONGITUDINAL COMPONENTS

2·11. General Procedure.—In the preceding sections it has been shown how the wave solutions of Maxwell's equations may be divided into three classes: solutions where both **E** and **H** are transverse to the direction of propagation, solutions where only **E** is longitudinal, and solutions where only **H** is longitudinal. Several cases of purely transverse propagation were also discussed. The longitudinal modes may now be considered. The general procedure will first be outlined and then will be applied to the specific cases that are most commonly met in practice.

Let us first investigate the solutions in which **E** is entirely transverse. These solutions represent transverse-electric waves (*TE*-waves), or *H*-waves. The conductivity of the medium will be omitted explicitly from the equations and will be assumed to be contained in the imaginary part of the dielectric constant ϵ. If E_z is taken equal to zero, Maxwell's equations, written in cartesian form, are

$$\frac{\partial E_y}{\partial z} = j\omega\mu H_x, \qquad \frac{\partial E_x}{\partial z} = -j\omega\mu H_y, \tag{23}$$

$$\frac{\partial H_z}{\partial y} - \frac{\partial H_y}{\partial z} = j\omega\epsilon E_x, \qquad \frac{\partial H_x}{\partial z} - \frac{\partial H_z}{\partial x} = j\omega\epsilon E_y, \tag{24}$$

$$\frac{\partial E_y}{\partial x} - \frac{\partial E_x}{\partial y} = -j\omega\mu H_z, \tag{25}$$

$$\frac{\partial H_y}{\partial x} - \frac{\partial H_x}{\partial y} = 0, \tag{26}$$

for the curl equations, and

$$\frac{\partial E_x}{\partial x} + \frac{\partial E_y}{\partial y} = 0, \tag{27}$$

$$\frac{\partial H_x}{\partial x} + \frac{\partial H_y}{\partial y} + \frac{\partial H_z}{\partial z} = 0. \tag{28}$$

for the divergence equations.

If plane waves are specified, and if it is assumed that the variation with z of the five components of **E** and **H** are given by $e^{-\gamma z}$, then the field components take the form $E_x(z) = e^{-\gamma z}E_x$, and similar expressions are written for the other components. This ambiguous notation should cause no confusion, since $E_x = E_x(z)$ will be used only in Eqs. (23) through (28); elsewhere E_x will be a function of x and y only. Substituting this, Eqs. (23) become

$$-\gamma E_y = j\omega\mu H_x, \qquad \gamma E_x = j\omega\mu H_y. \tag{29}$$

From these equations the first important result is obtained. The wave impedance Z_H is

$$Z_H = \frac{E_x}{H_y} = \frac{j\omega\mu}{\gamma}. \tag{30}$$

Also,

$$\frac{E_x}{E_y} = -\frac{H_y}{H_x}. \tag{31}$$

This result is significant. The set of lines in the transverse plane that give the direction of the transverse electric field E_t at any point have the slope $dy/dx = E_y/E_x$. A similar statement is true for the magnetic lines of force. Thus Eq. (31) is equivalent to

$$\left(\frac{dy}{dx}\right)_{\text{electric}} = -\frac{1}{\left(\frac{dy}{dx}\right)_{\text{magnetic}}},$$

and the lines of electric and magnetic force are therefore mutually perpendicular in the transverse plane.

If the assumed variation with z and the values of E_x and E_y given by Eqs. (29) are inserted in Eqs. (24), it is possible to solve for $\partial H_z/\partial x$ and $\partial H_z/\partial y$,

$$\begin{aligned}\frac{\partial H_z}{\partial x} &= -\frac{\gamma^2 + \omega^2\epsilon\mu}{\gamma} H_x, \\ \frac{\partial H_z}{\partial y} &= -\frac{\gamma^2 + \omega^2\epsilon\mu}{\gamma} H_y.\end{aligned} \tag{32}$$

Equations (25) and (28) become equivalent, because of Eqs. (29), and lead to

$$\frac{\partial H_x}{\partial x} + \frac{\partial H_y}{\partial y} - jH_z = 0.$$

Substituting for H_x and H_y from Eqs. (32),

$$\frac{\partial^2 H_z}{\partial x^2} + \frac{\partial^2 H_z}{\partial y^2} + (\gamma^2 + \omega^2\epsilon\mu)H_z = 0. \tag{33}$$

The remaining equations yield no new results. The procedure is thus straightforward. A solution of Eq. (33) for H_z is found; from Eqs. (32), H_x and H_y are determined; and from these values and Eqs. (29), E_x and E_y are obtained. Thus all the components of the field are determined from a single scalar quantity H_z. It should be noted that Eq. (33) is just the wave equation for H_z derived in Sec. 2·3.

To apply this to a waveguide of a particular shape, it is necessary to find an H_z that satisfies the proper boundary conditions at the walls of the guide. If the walls are perfect conductors, E_t must vanish over the surface. These boundary conditions will give rise to a relation connecting γ and ω and the dimensions of the waveguide. The form of this relation will depend upon the shape of the cross section of the guide.

If a quantity k_c is introduced, defined by

$$k_c^2 = \gamma^2 + \omega^2\epsilon\mu, \qquad (34)$$

then the numbers k_c^2 are the characteristic values, or eigenvalues, of Eq. (33). To each value of k_c^2 there will correspond a function H_z, a characteristic function, from which may be derived the other components of the fields. Some special cases will be treated in detail below.

The wave impedance Z_H can be written in a more useful form when the dielectric is not lossy. Since $\gamma = j(2\pi/\lambda_g)$, where λ_g is the wavelength in the waveguide,

$$Z_H = \frac{\lambda_g}{\lambda} \zeta, \qquad (35)$$

where λ is the wavelength in the dielectric medium for a TEM-wave.

The equations that H_z must satisfy are identical in form with those which apply to the velocity potential governing the propagation of sound waves. Sound waves are essentially simpler in nature than electromagnetic waves, since they can always be derived from a scalar field. Many electromagnetic problems require the vector representation.

Let us now consider the transverse-magnetic waves, or E-waves. The procedure of solution is entirely analogous to that of H-waves. It is assumed that $H_z = 0$ and that the remaining components are proportional to $e^{-\gamma z}$. A reduction of the equations results in expressions similar to Eqs. (33), (32), and (29). The results are

$$\frac{\partial^2 E_z}{\partial x^2} + \frac{\partial^2 E_z}{\partial y^2} + (\gamma^2 + \omega^2\epsilon\mu)E_z = 0, \qquad (36)$$

$$\left.\begin{aligned} \frac{\partial E_z}{\partial x} &= -\frac{\gamma^2 + \omega^2\epsilon\mu}{\gamma} E_x, \\ \frac{\partial E_z}{\partial y} &= -\frac{\gamma^2 + \omega^2\epsilon\mu}{\gamma} E_y, \end{aligned}\right\} \qquad (37)$$

$$Z_E = \frac{E_x}{H_y} = \frac{\gamma}{j\omega\mu} = -\frac{E_y}{H_x}. \qquad (38)$$

Thus E_t is again perpendicular to H_t, and four components of the fields are determined from one, E_z. The boundary conditions again determine a set of characteristic numbers k_c^2 with the associated functions E_z and hence a relation between γ and ω. The wave impedance may also be written in terms of λ_g as

$$Z_E = \frac{\lambda}{\lambda_g} \zeta, \qquad (39)$$

when no dielectric losses are present.

It is now obvious that any field can be represented as the sum of TE-, TM-, and TEM-modes. The TM-mode portion is given by the values of

E_x, E_y, H_x, and H_y determined from the value of E_z by Eqs. (37) and (38). In a similar way, by the use of H_z, the TE-mode portion can be separated out by means of Eqs. (29) and (32). The remainder of the field will then be purely transverse. It has already been shown that no TEM-mode is possible in a hollow pipe with no central conductor. Hence, any field in a hollow waveguide, however complicated, may be represented by a combination of TE- and TM-modes.

Once an expression for the fields has been found, the energy flow down the hollow pipe can be computed by integrating the value of Poynting's vector over the cross section of the waveguide. The power flow P is found to be

$$P = \frac{1}{2} \frac{1}{\mathrm{Re}\,(Z_w)} \int |E_t|^2\, dS = \frac{1}{2} \mathrm{Re}\,(Z_w) \int |H_t|^2\, dS,$$

where Z_w has been written for the E- or H-mode wave impedance, as the case may be. To maintain the fields in a hollow pipe, currents must flow in the walls, and the surface current density is equal to the tangential component of the magnetic field. For E-modes, the tangential component of H is equal to the total magnetic field at the guide walls. Since H is purely transverse, K is purely longitudinal. If a small slit is cut in a waveguide such that it is parallel to K, the field in the guide is not disturbed and there is no radiation from the slit.

2·12. The Normal Modes of Rectangular Pipes.—Let us consider a waveguide of rectangular cross section, which has dimensions b in the y-direction and a in the x-direction, as shown in Fig. 2·6, and which has

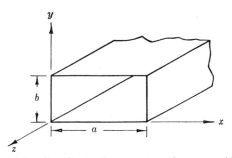

Fig. 2·6.—Coordinates for a rectangular waveguide.

perfectly conducting walls. Let us first consider the TE-modes. Equation (33) for H_z is obviously separable in rectangular coordinates and leads to simple sinusoidal solutions. Let

$$H_z = \cos k_x x \, \cos k_y y,$$

where the separation constants k_x and k_y are related by

$$k_c^2 = k_x^2 + k_y^2 = \gamma^2 + \omega^2 \epsilon \mu.$$

The quantities k_x and k_y are called the wave numbers in the x- and y-directions, respectively. Following the standard procedure, it is found that

$$H_x = \frac{\gamma k_x}{k_x^2 + k_y^2} \sin k_x x \cos k_y y,$$

$$H_y = \frac{\gamma k_y}{k_x^2 + k_y^2} \cos k_x x \sin k_y y,$$

$$E_x = \frac{j\omega\mu k_y}{k_x^2 + k_y^2} \cos k_x x \sin k_y y,$$

$$E_y = -\frac{j\omega\mu k_x}{k_x^2 + k_y^2} \sin k_x x \cos k_y y.$$

The boundary conditions must now be applied.

When $y = 0$ or $y = b$, $E_x = 0$ and $k_y = n\pi/b$, where n is an integer. Also $E_y = 0$ when $x = 0$ or $x = a$, and $k_x = m\pi/a$, where m is an integer. It is clear that n and m may take on any values, including zero, except that both n and m equal to zero is excluded. Thus γ is given by

$$\gamma^2 = \left(\frac{m\pi}{a}\right)^2 + \left(\frac{n\pi}{b}\right)^2 - \omega^2\epsilon\mu.$$

In order to have propagation down the pipe, γ^2 must be negative; hence *no waves are propagated below a certain frequency.* If no losses are present in the medium, that is, if ϵ is purely real, then there is a sharply defined critical frequency ω_c, which is given by

$$\omega_c^2 = \frac{1}{\epsilon\mu}\left[\left(\frac{m\pi}{a}\right)^2 + \left(\frac{n\pi}{b}\right)^2\right]. \quad (40)$$

The **existence of** this critical frequency, or cutoff frequency, is characteristic of all longitudinal waves in pipes, in both TE- and TM-modes. The value of the critical frequency depends upon the mode under consideration, that is, upon the values of n and m and the dimensions and shape of the hollow pipe. It is more customary to formulate this in terms of wavelength. If λ_g is the wavelength in the waveguide, then

$$\gamma = j\frac{2\pi}{\lambda_g},$$

and

$$\left(\frac{2\pi}{\lambda_g}\right)^2 = \left(\frac{2\pi}{\lambda}\right)^2 - \left(\frac{m\pi}{a}\right)^2 - \left(\frac{n\pi}{b}\right)^2,$$

or

$$\lambda_g = \frac{\lambda}{\sqrt{1 - \left(\frac{\lambda}{\lambda_c}\right)^2}}, \quad (41)$$

SEC. 2·12]   THE NORMAL MODES OF RECTANGULAR PIPES   35

where
$$\frac{1}{\lambda_c^2} = \left(\frac{n}{2a}\right)^2 + \left(\frac{m}{2b}\right)^2 \tag{42}$$

and λ_c is called the cutoff wavelength. A little consideration will show that Eq. (40) is universally applicable for both types of modes and all shapes of pipe; only λ_c differs in the different cases. The value of λ in Eq. (41) is the wavelength of a plane *TEM*-wave in the medium that fills the hollow pipe. In terms of λ_0, the wavelength of a *TEM*-wave in free space,

$$\lambda_0 = \lambda \sqrt{\frac{\epsilon\mu}{\epsilon_0\mu_0}},$$

Eq. (41) becomes

$$\lambda_g = \frac{\lambda_0}{\sqrt{\frac{\epsilon\mu}{\epsilon_0\mu_0} - \left(\frac{\lambda_0}{\lambda_c}\right)^2}}. \tag{43}$$

It will be noted that the cutoff wavelength defined in this way is *independent* of the dielectric material filling the waveguide; the critical frequency defined in Eq. (40) is not.

For frequencies below the critical value, γ becomes real and the waves are attenuated. For very low frequencies,

$$\gamma^2 = \left(\frac{m\pi}{a}\right)^2 + \left(\frac{n\pi}{b}\right)^2.$$

For $m = 1$ and $n = 0$, $\gamma = \pi/a$, which corresponds to an attenuation of 27.3 db in a distance equal to the width of the pipe.

A pair of values of n and m suffice to designate a particular mode that is called, according to the accepted notation, a TE_{mn}-mode. The mode that has the lowest critical frequency for propagation is the TE_{10}-mode, if $a > b$. The critical frequency is

$$\omega_c = \frac{\pi}{\epsilon\mu a}, \tag{44}$$

and the cutoff wavelength is

$$\lambda_c = 2a. \tag{45}$$

This lowest mode is called the dominant mode. Equations for the fields and diagrams showing the lines of electric and magnetic force for various modes may be found in Sec. 2·19.

It is of interest to examine in more detail the case of the lowest *H*-mode in rectangular guide, since this is by far the most important case. The fields in the guide have the values

$$H_y = E_x = 0,$$

$$H_x = \frac{\gamma}{k_x} \sin k_x x = j \frac{2a}{\lambda_g} \sin \frac{\pi x}{a},$$

$$E_y = -\frac{j\omega\mu}{k_x} \sin k_x x = -j \frac{2a}{\lambda} \sqrt{\frac{\mu}{\epsilon}} \sin \frac{\pi x}{a} = -j \frac{2a}{\lambda_g} Z_H \sin \frac{\pi x}{a}, \quad (46)$$

$$H_z = \cos k_x x = \cos \frac{\pi x}{a},$$

where it is assumed that there are no losses and hence γ is purely imaginary. The power flow is

$$P = \frac{1}{2} Z_H \int |H_x|^2 \, dS = \frac{a^3 b}{\lambda_g^2} Z_H, \quad (47)$$

for a unit amplitude of H_z. In terms of the maximum value of $|E_y|$, this becomes

$$P = \frac{ab}{4} \frac{|E_y|^2}{Z_H}.$$

Since $H_x = 0$ at the side walls of the guide, the current density on the side walls has only a y-component which is independent of y and which is of unit amplitude if H_z has unit amplitude. The current in the top and bottom of the guide has both longitudinal and transverse components

$$\left.\begin{array}{l} K_x = \cos \dfrac{\pi x}{a}, \\ K_z = j \dfrac{2a}{\lambda_g} \sin \dfrac{\pi x}{a}. \end{array}\right\} \quad (48)$$

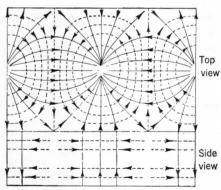

Fig. 2·7.—Lines of current flow (solid lines) for the lowest H-mode in rectangular waveguide. The dotted lines are the magnetic lines.

The transverse current K_x is thus zero at the center of the top and bottom walls, and a longitudinal slit can be cut here without disturbing the field. Figure 2·7 shows the lines of current flow on the top and side of the rectangular waveguide. It should be noticed that the flow is not divergenceless; the circuit made by following a current line can be completed only by including the displacement current on a portion of the path. The total longitudinal current is

$$I = \int_0^a K_z \, dx = j \frac{4a^2}{\pi \lambda_g}.$$

SEC. 2·12]   THE NORMAL MODES OF RECTANGULAR PIPES

An impedance may be defined on the current basis,

$$Z_I = \frac{2P}{I^2} = \frac{\pi^2}{8}\frac{b}{a} Z_H.$$

In a similar fashion, an impedance may be defined on a voltage basis. The voltage is defined as $V = bE$, and

$$Z_V = \frac{V^2}{2P} = 2\frac{b}{a} Z_H.$$

Thus impedances Z_w, Z_I, and Z_V are all different, and there is no unique way to define a useful quantity in the nature of an impedance for a single waveguide. This point will be discussed at length in later chapters.

The general solution of Maxwell's equations in terms of plane waves may be used with profit for the particularly simple case of the dominant H-mode in rectangular guide. Let us consider two plane wavefronts inclined at an angle θ to the z-axis, with the electric field in the direction of the y-axis, as shown in Fig. 2·8. If conducting plates are inserted at $y = 0$ and $y = b$, they will cause no distortion of the fields, since they are everywhere perpendicular to the electric field. Now, if vertical metal walls are placed at $x = 0$ and $x = a$, the plane waves will be successively reflected at a constant angle θ and will thus be

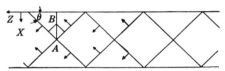

FIG. 2·8.—Propagation of the dominant H-mode in rectangular waveguide in terms of plane waves.

propagated down the guide. If the plane waves are taken to be in phase at the point A, then the electric fields of the two waves add to produce a maximum intensity at this point which we shall call E. The electric field, at a point such as B, is equal to the sum of the amplitudes of the two waves taken in the proper phase and is

$$\frac{E}{2} e^{j\omega(\overline{AB}\cos\theta)/c} + \frac{E}{2} e^{-j\omega(\overline{AB}\cos\theta)/c},$$

where c is the velocity of the plane waves and $\overline{AB}$ is the distance between the points A and B. Since $\overline{AB} = (a/2) - x$,

$$E_y = E \cos\left(\frac{a\pi}{\lambda}\cos\theta - x\frac{2\pi}{\lambda}\cos\theta\right).$$

Now by the choice of $\cos\theta = \lambda/2a$, E_y reduces to

$$E_y = E \sin\frac{\pi x}{a},$$

which is just the value for the H-mode field. In a similar manner the components of the magnetic field can be found, and they will agree with the H-mode values. Thus the H-mode field has been decomposed into that of two plane waves and a useful concept in describing some of the properties of the H-mode is gained. The point A moves with the velocity

$$\frac{c}{\sin \theta} = c \frac{\lambda_g}{\lambda},$$

which is just the value of the phase velocity of the H-mode waves.

The results for the rectangular waveguide may be applied to the higher modes between parallel plates. If the height of the guide, b, is allowed to become very large, the solution for parallel plates, when the electric field is parallel to the plates, is obtained. A series of modes exists, for all integral values of n excluding $n = 0$, corresponding to the TE_{0n}-modes in the rectangular guide.

The TM-modes may be treated in a very similar manner. The equation for E_z is again separable in rectangular coordinates, and the following values are found for the fields:

$$\left. \begin{aligned} E_z &= \sin k_x x \, \sin k_y y, \\ E_x &= -\frac{\gamma k_x}{k_x^2 + k_y^2} \cos k_x x \, \sin k_y y, \\ E_y &= -\frac{\gamma k_y}{k_x^2 + k_y^2} \sin k_x x \, \cos k_y y, \\ H_x &= j \frac{\omega \epsilon k_y}{k_x^2 + k_y^2} \sin k_x x \, \cos k_y y, \\ H_y &= -j \frac{\omega \epsilon k_x}{k_x^2 + k_y^2} \cos k_x x \, \sin k_y y, \end{aligned} \right\} \quad (49)$$

where as before

$$k_c^2 = k_x^2 + k_y^2 = \gamma^2 + \omega^2 \epsilon \mu,$$
$$k_y = \frac{n\pi}{a}, \qquad k_x = \frac{m\pi}{b}.$$

The modes are designated as TM_{mn}-modes, and it is evident that the lowest mode is the TM_{11}-mode, since the zero values of m and n are excluded. The cutoff frequencies are given by Eq. (40), and the cutoff wavelength of the lowest mode is

$$\lambda_c = \frac{2ab}{\sqrt{a^2 + b^2}}.$$

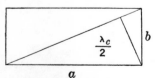

Fig. 2·9.—Cutoff wavelength of the TM_{11}-mode.

This wavelength has the simple geometrical interpretation shown in Fig. 2·9.

2·13. The Normal Modes in Round Pipes. To treat the case of waveguides of circular cross section, it is convenient to employ cylindrical coordinates r, θ, z and to choose the axis as the direction of propagation. For *TM*-modes, the wave equation for E_z is

$$\frac{\partial^2 E_z}{\partial r^2} + \frac{1}{r}\frac{\partial E_z}{\partial r} + \frac{1}{r^2}\frac{\partial^2 E_z}{\partial \theta^2} + (\gamma^2 + \omega^2\epsilon\mu)E_z = 0. \tag{50}$$

If the variables are separated by taking

$$E_z = R(r)\Theta(\theta),$$

the equation for Θ is

$$\frac{\partial^2 \Theta}{\partial \theta^2} + \chi^2 \Theta = 0,$$

where χ^2 is the separation constant. The solution is

$$\Theta = e^{j\chi\theta}, \tag{51}$$

and hence χ must be an integer or zero. The complex form of the function Θ is interpreted to mean that two solutions are possible: One is $\Theta = \cos \chi\theta$; the other is $\Theta = \sin \chi\theta$. Thus the modes are degenerate in pairs. The two modes may be thought of as two states of polarization of the field.

The equation for R is

$$\frac{\partial^2 R}{\partial r^2} + \frac{1}{r}\frac{\partial R}{\partial r} + \left(k_c^2 - \frac{\chi^2}{r^2}\right)R = 0, \tag{52}$$

where $k_c^2 = \gamma^2 + \omega^2\epsilon\mu$. This becomes Bessel's equation in the canonical form by the substitution $x = k_c r$. The solution for E_z is therefore

$$E_z = e^{j\chi\theta}J_\chi(k_c r), \tag{53}$$

where J_χ is the Bessel function of the first kind of order χ. The solution $N_\chi(k_c r)$, the Bessel function of the second kind, is excluded because of the singularity at $r = 0$. The boundary conditions may be applied immediately, since E_z must vanish at $r = a$, the radius of the tube. Thus k_c is determined by

$$\left. \begin{array}{r} J_\chi(k_c a) = 0, \\ \\ k_c a = t_{\chi n}, \end{array} \right\} \tag{54}$$

or

where $t_{\chi n}$ is the nth root of the Bessel function of order χ. The cutoff frequencies are then determined by

$$\omega_c^2 = \frac{1}{\epsilon\mu}\left(\frac{t_{\chi n}}{a}\right)^2. \tag{55}$$

The smallest value of $t_{\chi n}$ is $t_{01} = 2.405$. Other values are given in Table 2·2 at the end of this section.

The other components of the fields are determined in a manner analogous to that of the rectangular case. Thus,

$$\left.\begin{array}{l} H_r = \dfrac{j\omega\epsilon}{k_c^2}\dfrac{\partial E_z}{r\,\partial\theta} = \dfrac{\chi\omega\epsilon}{rk_c^2}e^{jx\theta}J_\chi(k_c r), \\[6pt] H_\theta = -\dfrac{j\omega\epsilon}{k_c^2}\dfrac{\partial E_z}{\partial r} = -j\dfrac{\omega\epsilon}{k_c}e^{jx\theta}J'_\chi(k_c r), \\[6pt] E_r = Z_E H_\theta, \qquad E_\theta = -Z_E H_r, \qquad Z_E = \dfrac{\gamma}{j\omega\epsilon}. \end{array}\right\} \quad (56)$$

The TM-modes are distinguished by the subscripts $TM_{\chi n}$, and the lowest TM-mode is the TM_{01}-mode. Equation (41) is valid for these modes if the values of λ_c given by

$$\lambda_c = \frac{2\pi a}{t_{\chi n}} \tag{57}$$

are used.

The TE-modes are treated in the same fashion. The solution of the wave equation for H_z leads to the same solution as that for E_z in the TM case:

$$H_z = e^{jx\theta}J_\chi(k_c r), \tag{58}$$

but the boundary conditions are determined from E_θ, which is

$$E_\theta = \frac{j\omega\mu}{k_c^2}\frac{\partial H_z}{\partial r} = \frac{j\omega\mu}{k_c}e^{jx\theta}J'_\chi(k_c r).$$

This component of the field must vanish when $r = a$. If $s_{\chi n}$ are the roots of J'_χ, then

$$k_c a = s_{\chi n}.$$

The cutoff frequency is

$$\omega_c^2 = \frac{1}{\epsilon\mu}\left(\frac{s_{\chi n}}{a}\right)^2, \tag{59}$$

and the cutoff wavelength is

$$\lambda_c = \frac{2\pi a}{s_{\chi n}}. \tag{60}$$

The modes are designated by $TE_{\chi n}$, and the dominant mode for round pipe is TE_{11}, for which $s_{11} = 1.841$. Table 2·2 gives other values of the roots of J'_χ. It should be noted that the TE-modes also have two states of polarization except when $\chi = 0$. The other components of the fields are

$$\left.\begin{array}{l} E_r = -\dfrac{j\omega\mu}{k_c^2}\dfrac{1}{r}\dfrac{\partial H_z}{\partial\theta} = \dfrac{\omega\mu\chi}{k_c^2 r}e^{jx\theta}J_\chi(k_c r), \\[6pt] H_r = -\dfrac{E_\theta}{Z_H}, \qquad H_\theta = \dfrac{E_r}{Z_H}. \end{array}\right\} \quad (61)$$

The principal formulas and pictures of the electric and magnetic lines are collected together in Sec. 2·19.

TABLE 2·2.—ROOTS OF THE BESSEL FUNCTIONS J_χ AND J'_χ

Mode	Root of	Root = $k_c a$
TE_{11}	J'_1	1.841
TM_{01}	J_0	2.405
TE_{21}	J'_2	3.054
TE_{01}, TM_{11}	J'_0, J_1	3.832
TE_{31}	J'_3	4.200
TM_{21}	J_2	5.135
TE_{41}	J'_4	5.30
TE_{12}	J'_1	5.330
TM_{02}	J_0	5.520
TM_{31}	J_3	6.379
TE_{51}	J'_5	6.41
TE_{22}	J'_2	6.71
TE_{02}, TM_{12}	J'_0, J_1	7.016

2·14. Higher Modes in Coaxial Cylinders.—In the earlier sections of this chapter, the principal, or transverse-electromagnetic, mode of propagation between coaxial cylinders was investigated. It was shown that this mode can exist for any frequency or, in other words, that the cutoff frequency is zero. There also exist higher modes of propagation which have nonvanishing frequencies of cutoff. These modes must satisfy Eq. (50) or the equivalent equation for H_z. The Bessel function of the second kind is here an admissible solution, since the origin is now within the center conductor and is excluded. Thus it is necessary to take a linear combination of the functions of the first and second kinds,

$$\left. \begin{array}{c} H_z \\ E_z \end{array} \right\} = e^{ix\theta}[AJ_\chi(k_c r) + BN_\chi(k_c r)]. \quad (62)$$

The boundary conditions for an E-mode are, then, that this quantity vanish at $r = a$ and $r = b$.

For the TM-modes, therefore,

$$-\frac{A}{B} = \frac{N_\chi(k_c a)}{J_\chi(k_c a)} = \frac{N_\chi(k_c b)}{J_\chi(k_c b)}. \quad (63)$$

The corresponding condition for TE-modes is

$$\frac{N'_\chi(k_c a)}{J'_\chi(k_c a)} = \frac{N'_\chi(k_c b)}{J'_\chi(k_c b)}. \quad (64)$$

The lowest mode is the TE_{11}-mode, where k_c is given by the first root of

$$J'_1(k_c b)N'_1(k_c a) - J'_1(k_c a)N'_1(k_c b) = 0.$$

The value of this root is represented in Fig. 2·10, where a quantity f,

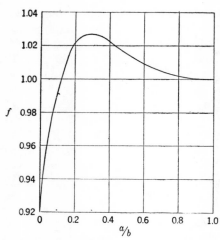

Fig. 2·10.—Cutoff wavelengths for the TE_{11}-mode between coaxial cylinders of radii a and b. The curve is a plot of f vs. a/b, where $f\lambda_c = (a + b)\pi$.

defined by the relation

$$f\lambda_c = (a + b)\pi, \qquad (65)$$

is plotted against a/b, where $(a + b)\pi$ is the mean circumference of the two coaxial cylinders. Thus as a/b becomes nearer to unity, the cutoff wavelength approaches this mean circumference. This is exactly as we should expect, since the effect of curvature must become small as a approaches b and the cutoff wavelength must approach the value for the TE_{10}-mode in a rectangular guide whose width is half the mean circumference. As the radius of the central conductor approaches zero, the TE_{11}-mode between coaxial cylinders goes smoothly over into the TE_{11}-mode within a pipe of circular cross section. The other components of the field of the TE_{11}-mode are derived as before. The values are

$$H_z = e^{j\theta}\left[J_1(k_c r)N_1'\left(k_c \frac{a}{b}\right) - J_1'\left(k_c \frac{a}{b}\right)N_1(k_c r)\right],$$

$$E_r = -\frac{j\omega\mu}{k_c^2}\frac{1}{r}\frac{\partial H_z}{\partial \theta} = \frac{\omega\mu}{k_c^2 r}\left[J_1(k_c r)N_1'\left(k_c \frac{a}{b}\right) - J_1'\left(k_c \frac{a}{b}\right)N_1(k_c r)\right], \qquad (66)$$

$$E_\theta = \frac{j\omega\mu}{k_c^2}\frac{\partial H_z}{\partial r} = \frac{j\omega\mu}{k_c}e^{j\theta}\left[J_1'(k_c r)N_1'\left(k_c \frac{a}{b}\right) - J_1'\left(k_c \frac{a}{b}\right)N_1'(k_c r)\right],$$

$$H_r = -\frac{E_\theta}{Z_H},$$

$$H_\theta = \frac{E_r}{Z_H}.$$

The next higher mode will be a TM-mode and will have a cutoff wavelength greater than the cutoff wavelength for the TE-mode by a factor of approximately $\pi/2$.

2·15. Normal Modes for Other Cross Sections.—There are several other cases in which it is possible to obtain a separation of the wave equation for H_z or E_z. If the cross section of the waveguide is elliptical, the fields are expressible in terms of Mathieu functions. The solutions have been investigated in detail by L. J. Chu.[1] The solution for a

[1] L. J. Chu, "Electromagnetic Waves in Elliptic Hollow Pipes of Metal," *J. Applied Phys.*, **9**, 583 (1938).

SEC. 2·15]   *NORMAL MODES FOR OTHER CROSS SECTIONS*   43

guide with parabolic walls has been obtained.[1] An exact solution is possible also when the guide cross section is an equilateral triangle.[2]

Many miscellaneous shapes can be treated by proceeding in the reverse order. Let us take for E_z any solution of the wave equation and plot its

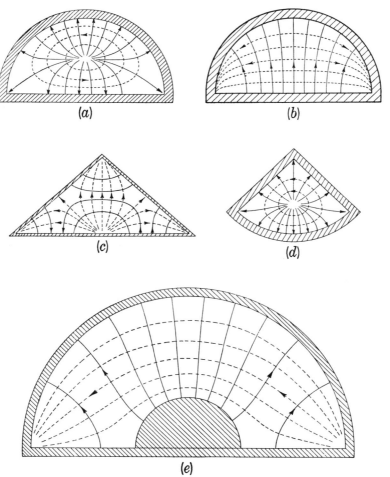

FIG. 2·11.—Modes derived by insertion of conducting surfaces perpendicular to lines of electric force.

contour lines, that is, the lines of $E_z = $ constant. Now the boundary condition for a TM-mode is that E_z vanish on the boundary. Hence the contour $E_z = 0$ may be chosen as the boundary of the cross section.

[1] R. D. Spence and C. P. Wells, *Phys. Rev.*, **62**, 58 (1942).
[2] S. A. Schelkunoff, *Electromagnetic Waves*, Van Nostrand, New York, 1943, Sec. 10·8, p. 393.

The contour lines now represent the magnetic lines in the transverse plane. The electric lines are then orthogonal to these magnetic lines. In a similar way, if H_z is assumed to be a solution of the wave equation, then the boundary of the waveguide for which this solution is valid is normal to the lines $H_z = $ constant. These contour lines represent the lines of electric force in the transverse plane, and the transverse lines of magnetic force are orthogonal to them. The cutoff wavelengths are prescribed when the functions H_z or E_z are specified, since they contain $k_c = 2\pi/\lambda_c$ as a parameter.

Moreover, if a solution for a simple case has been obtained, it is possible to derive other cases from it by inserting a conducting surface that is everywhere perpendicular to the lines of electric force. If such a surface includes a portion of the original boundary of the guide, the cutoff wavelength will remain unchanged. Figure 2·11 shows several examples of such derived modes.

It is always possible, of course, to solve the wave equation by employing all the well-known techniques of numerical integration, perturbation methods, and so forth. There is a general relation between the cutoff wavelength and the solution to the wave equation. The two-dimensional Green's theorem is

$$\int_S \left(\frac{\partial U}{\partial x} \frac{\partial V}{\partial x} + \frac{\partial U}{\partial y} \frac{\partial V}{\partial y} \right) dS = - \int_S U \left(\frac{\partial^2 V}{\partial x^2} + \frac{\partial^2 V}{\partial y^2} \right) dS + \oint U \frac{\partial V}{\partial n} \, dl, \tag{67}$$

where the surface integrals are taken over the guide cross section and the line integral around the boundary. Let us take U equal to V, and let it represent either E_z or H_z. Then

$$\int_S \left[\left(\frac{\partial U}{\partial x} \right)^2 + \left(\frac{\partial U}{\partial y} \right)^2 \right] dS = - \int_S U \left(\frac{\partial^2 U}{\partial x^2} + \frac{\partial^2 U}{\partial y^2} \right) dS + \oint U \frac{\partial U}{\partial n} \, dl, \tag{68}$$

but the first term on the right may be written $k_c^2 \int U^2 \, dS$, since U satisfies the wave equation, and the second term on the right vanishes, since either U or $\partial U/\partial n$ is zero. Therefore,

$$k_c^2 = \left(\frac{2\pi}{\lambda_c} \right)^2 = \frac{\int_S \left[\left(\frac{\partial U}{\partial x} \right)^2 + \left(\frac{\partial U}{\partial y} \right)^2 \right] dS}{\int_S U^2 \, dS}. \tag{69}$$

This quantity is always positive; therefore, for any arbitrary shape, some transmission mode is possible.

If some approximate form for U is known, λ_c can be calculated from this equation. It may be shown that this is a variational expression for

k_c^2; that is, the function U that results in the minimum value of k_c^2 is the correct value in the sense that it satisfies the wave equation and the boundary conditions. Therefore, if any function for U is used, the value of k_c^2 calculated from it will always be larger than the actual value that is the correct solution to the problem. It is also possible to establish a systematic method of successive approximations that converges on the correct values of k_c and U for the particular problem. This procedure has been discussed in detail by J. Schwinger. An example of the results of such calculations is shown in Fig. 2·12, which shows the cross sections of several waveguides with flat tops and bottoms and semicircular sides. The cutoff wavelengths are identical for all these shapes.

2·16. Transmission Losses.—Throughout the preceding sections it has been assumed that the walls of the waveguide are made of perfect conductors. A guide with real metal walls has a finite, although large, conductivity, and this must be taken into account. The alteration appears in the boundary conditions. It has been assumed that the tangential electric field vanishes on the surface of the conductor. In the case of a real metal, the tangential electric field does not quite vanish but has the small value determined by the product of the conductivity and the current density. The current density is equal to the tangential magnetic field at the metal surface. Thus the electric field is altered by

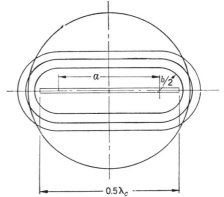

FIG. 2·12.—Various waveguide cross sections having the same cutoff wavelength.

the addition of a small component tangential to the metal. The magnetic field is also altered by the addition of a small field normal to the metal surface, and this normal component of the magnetic field is of the same order as the tangential component of the electric field. Therefore some energy flows into the metal. The rate at which power is lost into the metal per unit distance is

$$\frac{\partial P}{\partial z} = \frac{1}{2} \text{Re}\,(Z_m) \int_{\text{walls}} |H_{tan}|^2 \, ds,$$

where H_t is the tangential magnetic field and Z_m is the wave impedance of the metal. Since H_t varies as $e^{-\gamma z}$, $|H_{tan}|^2$ varies as $e^{-(\gamma+\gamma^*)z}$ which is $e^{-2\alpha z}$ Therefore

$$\frac{\partial P}{\partial z} = -2\alpha P.$$

The propagation constant is thus no longer purely imaginary but has a small real part. The change in the guide wavelength is, however, of the second order of small quantities and can be neglected. The wave impedance also becomes complex.

The amount of radiation into the metal is found by integrating the Poynting vector over the surface of the waveguide, choosing the component of S normal to the walls. This component is given by

$$S_n = \tfrac{1}{2} E_{\tan} H_{\tan}^*.$$

Since $E_{\tan}$ is small, it is permissible to use for $H_{\tan}$ the value that it would have in a guide of infinite conductivity. At the metal surface the tangential component of the electric field must be

$$E_{\tan} = Z_m H_{\tan}.$$

It will be recalled that

$$Z_m = \sqrt{\frac{j\omega\mu}{\sigma + j\omega\epsilon}}. \tag{70}$$

The attenuation constant α can then be calculated from the relation

$$\alpha = -\frac{1}{2P}\frac{dP}{dz} = \frac{1}{2P}\int_{\text{walls}} \text{Re}\,(S_n)\,ds = \frac{\text{Re}\,(Z_m)}{4P}\int_{\text{walls}} |H_{\tan}|^2\,ds,$$

where the element of area ds is a strip of unit length in the z-direction. For a good conductor, an approximate value of Z_m may be used, since $\omega\epsilon \ll \sigma$. By expansion, it is found that

$$Z_m = e^{j(\pi/4)}\sqrt{\frac{\omega\mu}{\sigma}}\left(1 - j\,\frac{\omega\epsilon}{2\sigma} - \frac{3}{8}\frac{\omega^2\epsilon^2}{\sigma^2} + \cdots\right). \tag{71}$$

For metals, σ is usually greater than 10^7 mhos per meter, ϵ is of the order of 10^{-11} farad per meter, and hence even for frequencies corresponding to ω equal to 10^{13} per second, $\omega\epsilon/\sigma$ is only 10^{-5}. Thus even the first power of $\omega\epsilon/\sigma$ may be neglected entirely, compared with unity, and

$$Z_m = \sqrt{\frac{\omega\mu}{2\sigma}}(1 + j). \tag{72}$$

Hence

$$\alpha = \frac{1}{4P}\sqrt{\frac{\omega\mu}{2\sigma}}\int_{\text{walls}} |H_{\tan}|^2\,ds. \tag{73}$$

The attenuation constant α can be calculated explicitly in terms of the dimensions of the guide and the mode, and the necessary expressions are given in Sec. 2·19. A convenient way to express the metal losses is in terms of a quantity δ called the "skin depth," defined by the expression

$$\text{Re}\,(S) = \tfrac{1}{2}\text{Re}\,(Z_m)|H_{\tan}|^2 = \tfrac{1}{2}\text{Re}\,(Z_m)K^2 = \frac{1}{2}\frac{1}{\delta\sigma}K^2.$$

Thus the metal losses are equal to those which would be produced by a uniform current K flowing through a surface layer of conductivity σ and thickness δ. Therefore

$$\delta = \sqrt{\frac{2}{\omega\mu\sigma}}, \qquad (74)$$

and hence δ is characteristic of the metal and of the frequency. Table 2·3 shows values of σ for various metals, values of δ for a frequency of 10^{10} cps, assuming $\mu = \mu_0$, and the relative losses per meter in waveguides constructed of the various metals. The propagation constant in the metal is $\gamma = \sqrt{j\omega\mu\sigma - \omega^2\epsilon\mu}$, which, for large σ, becomes

$$\gamma = \sqrt{\frac{\omega\mu\sigma}{2}}(1+j) = \frac{1}{\delta}(1+j). \qquad (75)$$

Hence the fields within the metal fall off to $1/e$ of their value at the surface at a depth equal to δ.

TABLE 2·3.—SKIN DEPTH AND RELATIVE LOSS OF VARIOUS METALS

Metal	Conductivity σ, mhos/meter	Skin depth δ for 10^{10} cps, meters	Relative loss per meter
Ag	6.17×10^7	6.42×10^{-7}	0.97
Cu	5.80	6.60	1.00
Au	4.10	7.85	1.19
Cr	3.84	8.11	1.23
Al	3.72	8.26	1.25
70-30 brass	1.57	12.7	1.92
P	0.9	17.0	2.5
Solder	0.71	18.5	2.8

For the dominant mode in rectangular waveguide (H_{10}-mode), the value of α is easily determined. From Eqs. (46),

$$|H_{\tan}|^2 = |H_x|^2 + |H_z|^2$$
$$= \frac{4a^2}{\lambda_g^2}\sin^2\frac{\pi x}{a} + \cos^2\frac{\pi x}{a}.$$

Equation (47) gives the value of P for this mode. The attenuation α may therefore be written

$$\alpha = \sqrt{\frac{\epsilon\omega}{2\sigma}}\frac{1}{b}\frac{\left(1 + \frac{2b}{a}\left(\frac{\lambda}{2a}\right)^2\right)}{\sqrt{1 - \left(\frac{\lambda}{2a}\right)^2}}. \qquad (76)$$

Thus α is infinite for $\lambda = 2a$ and, for smaller values of λ, decreases, passes through a minimum, and increases again, approaching infinity as λ

approaches zero. Figure 2·13 shows the calculated values of the attenuation in a rectangular copper waveguide for four modes.

2·17. Cylindrical Cavities.—Suppose that a piece of waveguide of length l is closed off by metal walls perpendicular to the axis of the guide. If there are electromagnetic waves in the cavity so formed, they will be reflected from the ends and will travel back and forth until they are all dissipated in heating the metal. For certain frequencies, a cavity of this

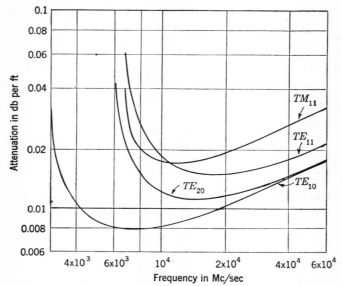

Fig. 2·13.—Attenuation in rectangular copper waveguide for several modes; $a = 2$ in., $b = 1$ in.

kind is said to be resonant; and in exact analogy with the vibrations of a taut string fixed at both ends, these frequencies are given by the condition that

$$l = \frac{n\lambda_g}{2}, \qquad (77)$$

where n is an integer and λ_g is the wavelength in the guide. The discussion of most of the properties of resonators will be found in Chap. 7. Only the losses in the cavity will be discussed here. These losses are most conveniently expressed in terms of a quantity called the Q of the cavity. This quantity Q is defined as the energy stored in the cavity divided by the energy lost per radian. If the losses occur only in the cavity itself and not by transfer of energy to other systems, the pertinent quantity is the "unloaded" Q, which is denoted by Q_0. This concept is a natural extension of low-frequency terminology and is useful in very much the same way, as will be shown in more detail later.

SEC. 2·17]   CYLINDRICAL CAVITIES   49

If the cavity is of resonant length, the field pattern is in the form of a standing wave having nodes at the two ends and $(n - 1)$ nodes along the length of the cavity. The stored energy could be calculated by integrating, over the cavity, the quantities $\frac{1}{2}\epsilon E^2$ and $\frac{1}{2}\mu H^2$. Likewise the losses could be found by integrating the square of the tangential magnetic fields over the walls and the ends of the cavity. This calculation has already been performed in effect, however, and a value for Q_0 may be derived from the previous results. The standing-wave pattern of the fields may be decomposed into two waves of equal amplitude traveling in opposite directions. It will be shown in Sec. 2·18 that the waves carry energy with the group velocity v_g. If this is assumed to be true, the energy stored, W, is seen to be

$$W = \frac{2Pl}{v_g}, \tag{78}$$

where P is, as before, the power flow in the waveguide. Since

$$v_g = \frac{\lambda}{\lambda_g} c,$$

where λ and c are the wavelength and the velocity in the dielectric medium, the expression for the power flow may be written

$$W = \frac{2\pi n P}{\omega} \left(\frac{\lambda_g}{\lambda}\right)^2. \tag{79}$$

The energy lost in the walls, W_1, can be written

$$W_1 = \frac{4\alpha Pl}{\omega} = \frac{2n\alpha P \lambda_g}{\omega}. \tag{80}$$

The energy lost in the end plates, W_2, is given by

$$W_2 = \frac{1}{2\omega} \int_{\text{ends}} \frac{1}{\delta\sigma} |H_t|^2 \, dS,$$

where the integral is taken over the two ends. This is directly related to the quantity P by

$$\frac{1}{2} \int_{\text{ends}} |H_t|^2 \, dS = \frac{8P}{Z_w},$$

where Z_w is the wave impedance. The factor 8 is a combination of two effects: a factor of 2 because the cavity has two ends, and a factor of 4 that arises because at each end the magnetic field is twice that for a traveling wave and the second power of this field appears in the equation. Therefore

$$W_2 = \frac{8P}{\omega \delta\sigma Z_w}. \tag{81}$$

The combination of these results gives

$$\frac{1}{Q_0} = \frac{W_1}{W} + \frac{W_2}{W},\tag{82}$$

or

$$\frac{1}{Q_0} = \frac{\alpha\lambda^2}{\pi\lambda_g} + \frac{4\lambda^2}{n\pi\lambda_g^2 Z_w \delta\sigma}.\tag{83}$$

For a longitudinal H-mode (TE-mode), $Z_w = (\lambda_g/\lambda)\zeta$; and for a longitudinal E-mode (TM-mode), $Z_w = (\lambda/\lambda_g)\zeta$. Thus Q_0 is expressed in terms of quantities already calculated. Values of Q_0 for the various cases are included in Sec. 2·19. It has been assumed here that the losses in the dielectric material in the cavity may be neglected. In Chap. 11 the dielectric losses will be taken into account.

2·18. Energy Density and Power Flow in Waveguides.—To complete this survey of longitudinal electromagnetic waves, it remains to prove some general theorems regarding the normal modes and to calculate the power flow and stored energy in waveguides. It has been shown that the fields for both E- and H-modes are completely determined once a single component of the field is known. If either the longitudinal component of the magnetic field for H-modes or the electric field for E-modes is designated by V_z, this quantity is determined by

$$(\nabla_t^2 + k_c^2)V_z = 0,\tag{84}$$

where $k_c^2 = \gamma^2 + \omega^2\epsilon\mu$, and ∇_t^2 is the Laplacian operator in the transverse coordinates [Eqs. (33) and (36)] with the boundary condition that on the guide walls,

$$\left.\begin{array}{ll} V_z = 0 & \text{for } E\text{-modes,} \\ \dfrac{\partial V_z}{\partial n} = 0 & \text{for } H\text{-modes,} \end{array}\right\}\tag{85}$$

where $\partial/\partial n$ is the derivative in the direction normal to the guide walls. The values of k_c^2 are the characteristic numbers of the problem which determine the cutoff wavelength

$$\lambda_c = \frac{2\pi}{k_c}.\tag{86}$$

The cutoff frequency is

$$\omega_c = \frac{k_c}{\sqrt{\epsilon\mu}},\tag{87}$$

and the guide wavelength is given by

$$\frac{1}{\lambda_g^2} = \frac{1}{\lambda^2} - \frac{1}{\lambda_c^2}.\tag{88}$$

The transverse components of $\mathbf{V}$ are

$$\mathbf{V}_t = -\frac{\gamma}{k_c^2}\operatorname{grad}_t V_z,\tag{89}$$

where grad_t is the gradient in the transverse coordinates. The remaining relation necessary to determine the fields completely is

$$|\mathbf{E}_t| = Z_w |\mathbf{H}_t|, \tag{90}$$

where the wave impedance Z_w is

$$\left.\begin{array}{l} Z_H = \dfrac{j\omega\mu}{\gamma} \quad \text{for } H\text{-modes,} \\[6pt] Z_E = \dfrac{\gamma}{j\omega\epsilon} \quad \text{for } E\text{-modes.} \end{array}\right\} \tag{91}$$

It may now be shown that the characteristic functions found in this way are othogonal. First, if the longitudinal components of two H-modes are considered, it is to be shown that

$$\int H_{za} H_{zb}\, dS = 0, \tag{92}$$

where a and b refer to two distinct modes. By Eq. (84),

$$\int H_{za} H_{zb}\, dS = \frac{1}{k_{ca}^2 - k_{cb}^2} \int (H_{za} \nabla^2 H_{zb} - H_{zb} \nabla^2 H_{za})\, dS.$$

By Green's second theorem, this integral becomes

$$\int \left(H_{za} \frac{\partial H_{zb}}{\partial n} - H_{zb} \frac{\partial H_{za}}{\partial n} \right) dl, \tag{93}$$

where the integral is taken over the curve bounding the guide and vanishes by virtue of the boundary conditions [Eqs. (85)]. For the transverse components,

$$\int \mathbf{H}_{ta} \cdot \mathbf{H}_{tb}\, dS = \frac{\gamma_a \gamma_b}{k_{ca}^2 k_{cb}^2} \int \text{grad}_t H_{za} \cdot \text{grad}_t H_{zb}\, dS,$$

by Eq. (89). The integral on the right can be transformed, by Green's first theorem, to

$$\int H_{za} \frac{\partial H_{zb}}{\partial n}\, dl - \int H_{za} \nabla^2 H_{zb}\, dS = \int H_{za} \frac{\partial H_{zb}}{\partial n}\, dl \\ + k_{cb}^2 \int H_{za} H_{zb}\, dS. \tag{94}$$

The first integral vanishes because of the boundary conditions, and it has just been proved that the second integral vanishes. Because the transverse electric fields are proportional to the transverse magnetic fields, the orthogonality of the electric fields also is proved.

The proof for two E-modes is exactly analogous. For the longitudinal fields the integral in Eq. (93) again applies, with E_z written for H_z. The integral again vanishes because of the boundary condition $E_z = 0$. For the transverse components there is an expression similar to Eq. (94) which vanishes in the same manner. For one H-mode (a) and one E-mode (b), the longitudinal components are orthogonal;

$$\int H_{za} H_{zb} \, dS = \int E_{za} E_{zb} \, dS = 0, \qquad (95)$$

since

$$E_{za} = H_{zb} = 0.$$

For the transverse magnetic fields,

$$\int \mathbf{H}_{ta} \cdot \mathbf{H}_{tb} \, dS = \frac{\gamma_a \gamma_b}{k_{ca}^2 k_{cb}^2} \int \mathrm{grad}_t \, H_{za} \cdot Z_w \mathbf{n} \times \mathrm{grad}_t \, E_{zb} \, dS, \qquad (96)$$

where $\mathbf{n}$ is a unit vector in the z-direction. The integral may be transformed into

$$\int \mathrm{div}_t \, (H_{za} \mathbf{n} \times \mathrm{grad}_t \, E_{zb}) \, dS - \int H_{za} \, \mathrm{div}_t \, (\mathbf{n} \times \mathrm{grad}_t \, E_{zb}) \, dS.$$

The first integral may be changed to a line integral which vanishes, since $\mathbf{n} \times \mathrm{grad}_t \, E_{zb}$ is tangential to the guide walls. In the second integral,

$$\mathrm{div}_t \, (\mathbf{n} \times \mathrm{grad}_t \, E_{zb}) = -\mathbf{n} \cdot \mathrm{curl} \, \mathrm{grad} \, E_{zb} = 0,$$

and hence the transverse magnetic fields are orthogonal. The proof for the transverse electric fields is almost identical. Thus it is clear that the longitudinal components of the electric and magnetic fields and the transverse components are all separately orthogonal for any two different modes.

It remains now to show that the energy flow for two modes contains no mixed terms. If two H-modes are considered, the power flow contains terms such as

$$P = \int \mathbf{E}_{ta} \times \mathbf{H}_{tb}^* \cdot \mathbf{n} \, dS = \int \mathbf{H}_{ta} \times \mathbf{n} \times \mathbf{H}_{tb}^* \cdot \mathbf{n} \, dS$$

$$= \int \mathbf{H}_{ta} \cdot \mathbf{H}_{tb}^* \, dS = 0, \qquad (97)$$

as has already been shown. The argument is identical for two E-modes, and for one H-mode and one E-mode. Thus when several modes exist at the same time in a waveguide, the flux of power of two modes can be computed independently and added. It should be noted that this is not true for the loss in a waveguide, since there can be a mixed term of the form $H_{ta} H_{zb}$ the integral of which does not vanish.

Expressions for the stored electric and magnetic energy in a waveguide

ENERGY DENSITY AND POWER FLOW

will now be given. First, for H-modes, the electric energy is

$$W^e = \frac{1}{4}\epsilon \int |E_t|^2 \, dS = \frac{1}{4}\epsilon \frac{\omega^2\mu^2}{k_c^4} \int |\text{grad}_t \, H_z|^2 \, dS,$$

the factor being $\frac{1}{4}$ because an average has been taken over the z-direction. Using Eq. (84), this becomes

$$W^e = \frac{1}{4}\epsilon \frac{\omega^2\mu^2}{k_c^2} \int |H_z|^2 \, dS. \tag{98}$$

The energy associated with the longitudinal magnetic field is

$$W_l^m = \tfrac{1}{4}\mu \int |H_z|^2 \, dS, \tag{99}$$

and the transverse magnetic energy density is

$$W_t^m = \frac{1}{4}\mu \int |H_t|^2 \, dS = -\frac{1}{4}\mu \frac{\gamma^2}{k_c^2} \int |H_z|^2 \, dS. \tag{100}$$

The total magnetic energy is

$$W_l^m + W_t^m = \frac{1}{4}\mu\left(1 - \frac{\gamma^2}{k_c^2}\right)\int |H_z|^2 \, dS = \frac{1}{4}\frac{\epsilon\mu^2\omega^2}{k_c^2}\int |H_z|^2 \, dS, \tag{101}$$

which is equal to the total electric energy. The rate of flow of energy is

$$P = \frac{1}{2}\text{Re}\left(\int E_t H_t^* \, dS\right) = -\frac{1}{2}Z_H \frac{\gamma^2}{k_c^2} \int |H_z|^2 \, dS.$$

The velocity v by which energy is transported is the rate of flow of energy divided by the energy density, or

$$v = -\frac{1}{2}Z_H \frac{\gamma^2}{k_c^2} \frac{2k_c^2}{\epsilon\mu^2\omega^2} = \frac{\lambda}{\lambda_g}\frac{1}{\sqrt{\epsilon\mu}}. \tag{102}$$

Now the phase velocity of H-waves is

$$v_{\text{ph}} = \frac{\omega}{\beta} = \frac{\lambda_g}{\lambda}\frac{1}{\sqrt{\epsilon\mu}}, \tag{103}$$

while the group velocity is, using Eq. (88),

$$v_g = \frac{d\omega}{d\beta} = \frac{1}{\sqrt{\epsilon\mu}}\frac{\beta}{\sqrt{k_c^2 + \beta^2}} = \frac{1}{\sqrt{\epsilon\mu}}\frac{\lambda}{\lambda_g}. \tag{104}$$

Thus the velocity of transport of energy is equal to the group velocity.[1]

---

[1] For a precise discussion of the five velocities (front, phase, signal, group, and energy-transport velocity) that are associated with wave motion, the reader is referred to an excellent article by L. Brillouin, *Congrès International d'Électricité*, Vol. II, 1re Section, Paris, 1932.

There are analogous expressions for E-modes

$$\left.\begin{aligned}
W^m &= \frac{1}{4}\mu \int |H_t|^2\,dS = \frac{1}{4}\mu\frac{\omega^2\epsilon^2}{k_c^2}\int |E_z|^2\,dS, \\
W_t^e &= \frac{1}{4}\epsilon \int |E_t|^2\,dS = -\frac{1}{4}\frac{\epsilon\gamma^2}{k_c^2}\int |E_z|^2\,dS, \\
W_l^e &= \tfrac{1}{4}\epsilon\int |E_z|^2\,dS, \\
P &= \frac{1}{2}Z_E\int |H_t|^2\,dS = \frac{1}{2}Z_E\frac{\omega^2\epsilon^2}{k_c^2}\int |E_z|^2\,dS.
\end{aligned}\right\} \quad (105)$$

Here again, therefore, the total stored energy is equally divided between electric and magnetic energy. It can be shown as above that the same expressions for $v_{\rm ph}$, v_g, and v hold for E-modes as for H-modes.

2·19. Summary of Results.—The survey of the classical electromagnetic theory of both transverse and longitudinal waves has been completed. It remains only to summarize the results in a manner that will be convenient for ready reference. For each of the modes that are of practical importance, the specific form of the fields will be given, together with the cutoff wavelength, formulas for the power flow and the attenuation, and the expressions for the unloaded Q of a cavity n half-wavelengths long.

Coaxial TEM-mode.—The transverse cross section of a coaxial transmission line operating in this mode is shown in Fig. 2·14. The fields are given by

$$E_z = H_z = E_\phi = H_r = 0,$$
$$E_r = \zeta\frac{I}{2\pi r}, \qquad H_\phi = \frac{I}{2\pi r},$$

where I is the total current in either the inner or outer conductor. The power relations are

$$P = \frac{I^2}{4\pi}\zeta\ln\frac{b}{a},$$
$$\alpha = \frac{1}{2\delta\sigma}\eta\frac{1}{\ln\frac{b}{a}}\left(\frac{1}{a}+\frac{1}{b}\right),$$
$$\frac{1}{Q_0} = \frac{2}{\pi\delta\sigma}\eta\left[\frac{\lambda}{4\ln\frac{b}{a}}\left(\frac{1}{a}+\frac{1}{b}\right)+\frac{2}{n}\right].$$

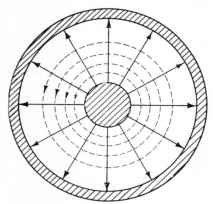

Fig. 2·14.—Transverse cross section of coaxial transmission line operating in the TEM-mode.

The H_{10}-mode (TE_{10}-mode) in Rectangular Waveguide.—The field configuration for this mode is shown in Fig. 2·15. The equations for the fields and the power relations are

$$H_z = \cos\frac{\pi x}{a},$$

$$E_x = 0, \qquad H_x = j\frac{2a}{\lambda_g}\sin\frac{\pi x}{a},$$

$$E_y = -j\frac{2a}{\lambda}\zeta\sin\frac{\pi x}{a}, \qquad H_y = 0,$$

$$\lambda_c = 2a, \qquad P = \frac{a^3 b}{\lambda_g^2}\zeta,$$

$$\alpha = \frac{1}{\delta\sigma}\eta\frac{1}{b}\frac{1 + \frac{2b}{a}\left(\frac{\lambda}{2a}\right)^2}{\sqrt{1 - \left(\frac{\lambda}{2a}\right)^2}},$$

$$\frac{1}{Q_0} = \frac{1}{\pi\delta\sigma}\eta\left\{\frac{\lambda}{b}\left[1 + \frac{2b}{a}\left(\frac{\lambda}{2a}\right)^2\right] + \frac{4}{n}\left[1 - \left(\frac{\lambda}{2a}\right)^2\right]^{3/2}\right\}.$$

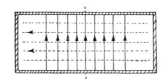

Fig. 2·15.—Field configuration for H_{10}-mode (TE_{10}-mode) in rectangular waveguide.

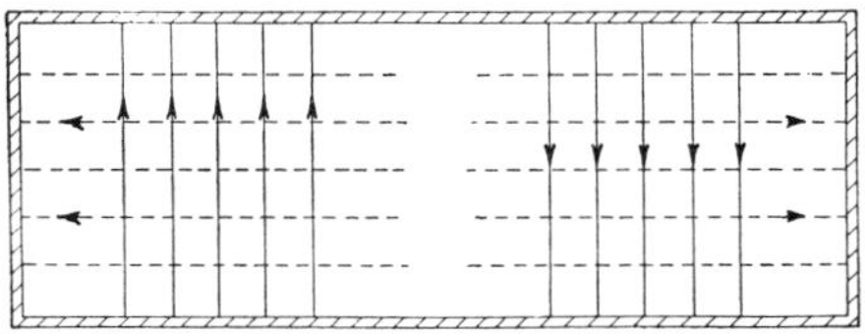

Fig. 2·16.—Cross section of waveguide for TE_{20}-mode.

The H_{20}-mode (TE_{20}-mode) in Rectangular Waveguide.—The cross section of a waveguide carrying the TE_{20}-mode is shown in Fig. 2·16. The equations are

$$H_z = \cos\frac{2\pi x}{a},$$

$$E_x = 0, \qquad H_x = j\frac{a}{\lambda_g}\sin\frac{2\pi x}{a},$$

$$E_y = -j\frac{a}{\lambda}\zeta\sin\frac{2\pi x}{a}, \qquad H_y = 0,$$

$$\lambda_c = a, \qquad P = \frac{1}{4}\frac{a^3 b}{\lambda_g^2}\zeta,$$

$$\alpha = \frac{1}{\delta\sigma}\eta\frac{1}{b}\frac{1 + \frac{2b}{a}\left(\frac{\lambda}{a}\right)^2}{1 - \left(\frac{\lambda}{a}\right)^2},$$

$$\frac{1}{Q_0} = \frac{1}{\pi\delta\sigma}\eta\left\{\frac{\lambda}{b}\frac{1 + \frac{2b}{a}\left(\frac{\lambda}{a}\right)^2}{\left[1 - \left(\frac{\lambda}{a}\right)^2\right]^{1/2}} + \frac{4}{n}\left[1 - \left(\frac{\lambda}{a}\right)^2\right]^{3/2}\right\}.$$

The E_{11}-mode (TM_{11}-mode) in Rectangular Waveguide.—The next mode that can be propagated in a rectangular waveguide, as the frequency is increased, is the H_{01}-mode, provided that the dimensions are such that $2b < a$. The relevant equations can be easily obtained by setting $x = y$ and $a = b$ in the H_{10}-mode equations. A further increase in frequency allows the E_{11}-mode to propagate. The transverse fields for this mode are shown in Fig. 2·17. The fields are given by

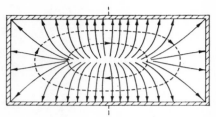

Fig. 2·17.—Transverse fields for the E_{11}-mode.

$$E_z = \sin\frac{\pi x}{a} \sin\frac{\pi y}{b},$$

$$E_x = -j\frac{\lambda_c^2}{2a\lambda_g}\cos\frac{\pi x}{a}\sin\frac{\pi y}{b}, \qquad H_x = j\frac{\lambda_c^2}{2b\lambda}\eta\sin\frac{\pi x}{a}\cos\frac{\pi y}{b},$$

$$E_y = -j\frac{\lambda_c^2}{2b\lambda_g}\sin\frac{\pi x}{a}\cos\frac{\pi y}{b}, \qquad H_y = -j\frac{\lambda_c^2}{2a\lambda}\eta\cos\frac{\pi x}{a}\sin\frac{\pi y}{b},$$

$$\lambda_c = \frac{2ab}{\sqrt{a^2+b^2}}, \qquad P = \frac{1}{8}\eta\frac{ab\lambda_c^2}{\lambda\lambda_g},$$

$$\alpha = \frac{2}{\delta\sigma}\eta\frac{a^3+b^3}{ab(a^2+b^2)}\left[1-\left(\frac{\lambda}{\lambda_c}\right)^2\right]^{-\frac{1}{2}},$$

$$\frac{1}{Q_0} = \frac{2}{\pi\delta\sigma}\eta\left\{\frac{a^3+b^3}{ab(a^2+b^2)}\lambda + \frac{2}{n}\left[1-\left(\frac{\lambda}{\lambda_c}\right)^2\right]^{\frac{1}{2}}\right\}.$$

The H_{11}-mode (TE_{11}-mode) in Round Waveguide.—The TE_{11}-mode is the dominant mode in waveguide of circular cross section. The fields are shown in Fig. 2·18. The field and power relations are

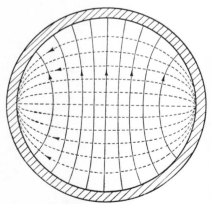 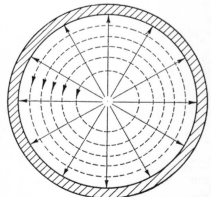

Fig. 2·18.—Fields for the TE_{11}-mode in round waveguide.

Fig. 2·19.—Fields for the TM_{01}-mode in round waveguide.

SEC. 2·19] SUMMARY OF RESULTS 57

$$H_z = e^{j\theta} J_1\left(s_{11}\frac{r}{a}\right),$$

$$E_r = \frac{\lambda_c^2}{2\pi\lambda r}\zeta\, e^{j\theta} J_1\left(s_{11}\frac{r}{a}\right), \qquad H_r = -j\frac{\lambda_c}{\lambda_g} e^{j\theta} J_1'\left(s_{11}\frac{r}{a}\right),$$

$$E_\theta = j\frac{\lambda_c}{\lambda}\zeta\, e^{j\theta} J_1'\left(s_{11}\frac{r}{a}\right), \qquad H_\theta = \frac{\lambda_c^2}{2\pi\lambda_g r} e^{j\theta} J_1\left(s_{11}\frac{r}{a}\right),$$

where s_{11} is the first root of the Bessel function J_1';

$$s_{11} = 1.841, \qquad \lambda_c = 1.706(2a),$$

$$P = \frac{1}{4\pi}\frac{\lambda_c^4}{\lambda\lambda_g}\zeta\,\frac{s_{11}^2 - 1}{2} J_1^2(s_{11}) = 0.0322\frac{\lambda_c^4}{\lambda\lambda_g}\zeta,$$

$$\alpha = \frac{1}{\delta\sigma}\eta\frac{1}{a}\left[\frac{1}{s_{11}^2 - 1} + \left(\frac{\lambda}{\lambda_c}\right)^2\right]\left[1 - \left(\frac{\lambda}{\lambda_c}\right)^2\right]^{-\frac{1}{2}},$$

$$\frac{1}{Q_0} = \frac{1}{\pi\delta\sigma}\eta\left\{\frac{\lambda}{a}\left[\frac{1}{s_{11}^2 - 1} + \left(\frac{\lambda}{\lambda_c}\right)^2\right] + \frac{4}{n}\left[1 - \left(\frac{\lambda}{\lambda_c}\right)^2\right]^{\frac{3}{2}}\right\}.$$

The E_{01}-mode (TM_{01}-mode) in Round Waveguide.—The lowest mode with circular symmetry is the E_{01}-mode which is of considerable practical importance. The fields are shown in Fig. 2·19. The relevant equations are

$$E_z = J_0\left(t_{01}\frac{r}{a}\right),$$

$$E_r = -j\frac{2\pi}{\lambda_g} J_0'\left(t_{01}\frac{r}{a}\right), \qquad H_r = 0,$$

$$E_\theta = 0, \qquad H_\theta = -j\frac{2\pi}{\lambda} J_0'\left(t_{01}\frac{r}{a}\right),$$

where t_{01} is the first root of J_0; it has the value

$$t_{01} = 2.405, \qquad \lambda_c = 1.306(2a).$$

The power relations are

$$P = \frac{1}{8\pi}\frac{\lambda_c^4}{\lambda\lambda_g}\eta[t_{01}J_0'(t_{01})]^2 = 0.0620\frac{\lambda_c^4}{\lambda\lambda_g}\eta,$$

$$\alpha = \frac{1}{\delta\sigma}\eta\frac{1}{a}\left[1 - \left(\frac{\lambda}{\lambda_c}\right)^2\right]^{-\frac{1}{2}},$$

$$\frac{1}{Q_0} = \frac{\delta}{a} + \frac{4a}{n\lambda}\left[1 - \left(\frac{\lambda}{\lambda_c}\right)^2\right]^{-\frac{1}{2}}.$$

The H_{21}-mode (TE_{21}-mode) in Round Waveguide.—The next mode in round waveguide, in order of decreasing cutoff wavelengths, is the H_{21}-mode. The fields are shown in Fig. 2·20. The equations for this mode are

$$H_z = e^{2j\theta} J_2\left(s_{21}\frac{r}{a}\right),$$

$$E_r = \frac{\lambda_c^2}{\pi\lambda r}\zeta e^{2j\theta} J_2\left(s_{21}\frac{r}{a}\right), \qquad H_r = -j\frac{\lambda_c}{\lambda_g} e^{2j\theta} J_2'\left(s_{21}\frac{r}{a}\right),$$

$$E_\theta = j\frac{\lambda_c}{\lambda}\zeta e^{2j\theta} J_2'\left(s_{21}\frac{r}{a}\right), \qquad H_\theta = \frac{\lambda_c^2}{\pi\lambda_g r} e^{2j\theta} J_2\left(s_{21}\frac{r}{a}\right),$$

$$s_{21} = 3.054, \qquad \lambda_c = 1.029(2a),$$

$$P = \frac{1}{4\pi}\frac{\lambda_c^4}{\lambda\lambda_g}\zeta\frac{s_{21}^2 - 4}{2} J_2^2(s_{21}) = 0.0500 \frac{\lambda_c^4}{\lambda\lambda_g}\zeta,$$

$$\alpha = \frac{1}{\delta\sigma}\eta\frac{1}{a}\left[\frac{4}{s_{21}^2 - 4} + \left(\frac{\lambda}{\lambda_c}\right)^2\right]\left[1 - \left(\frac{\lambda}{\lambda_c}\right)^2\right]^{-\frac{1}{2}},$$

$$\frac{1}{Q_0} = \frac{1}{\pi\delta\sigma}\eta\left\{\frac{\lambda}{a}\left[\frac{4}{s_{21}^2 - 4} + \left(\frac{\lambda}{\lambda_c}\right)^2\right] + \frac{4}{n}\left[1 - \left(\frac{\lambda}{\lambda_c}\right)^2\right]^{\frac{3}{2}}\right\}.$$

The H_{01}-mode (TE_{01}-mode) in Round Waveguide.—The first H-mode with circular symmetry is the TE_{01}-mode. The fields are shown in Fig.

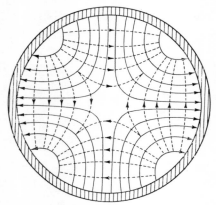

Fig. 2·20.—Fields for the TE_{21}-mode in round waveguide.

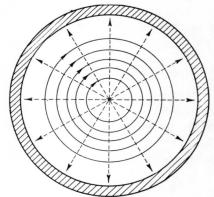

Fig. 2·21.—Fields for the TE_{01}-mode in round waveguide.

2·21. The field equations are

$$H_z = J_0\left(s_{01}\frac{r}{a}\right),$$

$$E_r = 0, \qquad H_r = -j\frac{\lambda_c}{\lambda_g} J_0'\left(s_{01}\frac{r}{a}\right),$$

$$E_\theta = j\frac{\lambda_c}{\lambda}\zeta J_0'\left(s_{01}\frac{r}{a}\right), \qquad H_\theta = 0,$$

$$s_{01} = 3.832, \qquad \lambda_c = 0.820\,(2a).$$

The power relations are

SUMMARY OF RESULTS

$$P = \frac{1}{8\pi} \frac{\lambda_c^4}{\lambda\lambda_g} \zeta s_{01}^2 J_0^2(s_{01}) = 0.0948 \frac{\lambda_c^4}{\lambda\lambda_g} \zeta,$$

$$\alpha = \frac{1}{\delta\sigma} \eta \frac{1}{a} \left(\frac{\lambda}{\lambda_c}\right)^2 \left[1 - \left(\frac{\lambda}{\lambda_c}\right)^2\right]^{-\frac{1}{2}},$$

$$\frac{1}{Q_0} = \frac{1}{\pi\delta\sigma} \eta \left\{ \frac{\lambda}{a}\left(\frac{\lambda}{\lambda_c}\right)^2 + \frac{4}{n}\left[1 - \left(\frac{\lambda}{\lambda_c}\right)^2\right]^{\frac{3}{2}} \right\}.$$

It should be noted that the attenuation for this mode has the unique property that it decreases continuously with decreasing wavelength.

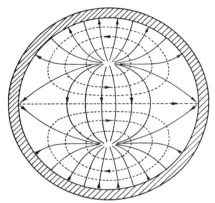

FIG. 2·22.—Field configurations for TM_{11}-mode in round waveguide.

The E_{11}-mode (TM_{11}-mode) in Round Waveguide.—The chief importance of the E_{11}-mode lies in the circumstance that it is a degenerate mode, having the same value of cutoff wavelength as the H_{01}-mode. The field configurations are shown in Fig. 2·22. The equations for the field components are

$$E_z = e^{j\theta} J_1\left(t_{11}\frac{r}{a}\right),$$

$$E_r = -j\frac{\lambda_c}{\lambda_g} e^{j\theta} J_1'\left(t_{11}\frac{r}{a}\right), \qquad H_r = \frac{\lambda_c^2}{2\pi\lambda r} \eta e^{j\theta} J_1\left(t_{11}\frac{r}{c}\right),$$

$$E_\theta = -\frac{\lambda_c^2}{2\pi\lambda_g r} e^{j\theta} J_1\left(t_{11}\frac{r}{c}\right), \qquad H_\theta = -j\frac{\lambda_c}{\lambda} \eta e^{j\theta} J_1'\left(t_{11}\frac{r}{a}\right),$$

$$t_{11} = 3.832, \qquad \lambda_c = 0.820(2a),$$

$$P = \frac{1}{8\pi} \frac{\lambda_c^4}{\lambda\lambda_g} \eta [t_{11} J_1'(t_{11})]^2 = 0.0518 \frac{\lambda_c^4}{\lambda\lambda_g} \eta,$$

$$\alpha = \frac{1}{\delta\sigma} \eta \frac{1}{a} \left[1 - \left(\frac{\lambda}{\lambda_c}\right)^2\right]^{-\frac{1}{2}},$$

$$\frac{1}{Q_0} = \frac{\delta}{a} + \frac{4a}{n\lambda}\left[1 - \left(\frac{\lambda}{\lambda_c}\right)^2\right]^{-\frac{1}{2}}.$$

CHAPTER 3

WAVEGUIDES AS TRANSMISSION LINES

By C. G. Montgomery

3·1. Some General Properties of Guided Waves.—In the previous chapter, it was shown that waves may travel in hollow pipes in many modes of transmission and that for each of these modes there is a corresponding cutoff frequency. For frequencies below the cutoff frequency the energy in the mode is quickly attenuated; above the cutoff frequency it is freely transmitted. The most important condition in practice is that in which the frequency lies above the cutoff frequency for the lowest mode but below the cutoff frequency for the next higher mode. The lowest, or dominant, mode will then propagate energy in the waveguide. Let us take as the direction of propagation the positive z-direction. Let us suppose that this traveling wave encounters an obstacle in the waveguide, such as a conducting wire placed across the guide. In the neighborhood of this wire, the solution of Maxwell's equations that corresponds to the dominant mode will no longer suffice to satisfy the boundary conditions. There must be, in fact, other modes present that are excited by the currents flowing in the wire. These higher modes, however, are not propagated, and their amplitudes die out rapidly in both positive and negative z-directions away from the wire. The actual fields near the obstacle are determined, of course, by a solution of the electromagnetic equations that satisfies the particular boundary conditions imposed by the geometrical configuration. The higher modes can be regarded as representing a kind of Fourier expansion of the fields near the obstacle. For rectangular waveguide in which the TE_{10}-mode is the dominant one, the expansion is an actual Fourier series in terms of sines and cosines; for more complicated cases it is a generalized expansion in terms of other functions. One member of this expansion will be the dominant-mode term representing a wave progressing in the direction opposite the incident wave, that is, a reflected wave. On the far side of the obstacle and some distance away from it only the dominant mode exists progressing in the positive z-direction but having a reduced amplitude. Some of the energy has been reflected; some may have been absorbed at the obstacle. The resulting situation is illustrated schematically in Fig. 3·1.

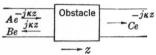

Fig. 3·1.—Electromagnetic waves incident upon and reflected from an obstacle in waveguide.

To the left of the obstacle there is an incident wave of amplitude A and a reflected wave of amplitude B. To the right there is a transmitted wave of amplitude C. Each of these waves varies sinusoidally with time, the transmitted wave being

$$Ce^{j(\omega t - \kappa z)},$$

where the amplitude C is to be regarded as complex and hence contains the phase of the wave. Only the case in which the losses along the waveguide can be neglected, and hence κ is real and $\kappa = 2\pi/\lambda_g$, will be considered. At some distance from the obstacle, the field is completely characterized by the amplitudes of the reflected and transmitted waves, and for many purposes it is not necessary to enquire further into its nature. It may be assumed without loss of generality that the amplitude A is real.

To the left of the obstacle there are two waves, of amplitudes A and B, traveling in opposite directions. The amplitudes A and B will be taken proportional to the *transverse electric* field at some point in the transverse plane. This is purely a convention; the transverse magnetic field could have been chosen instead, and the only differences would be those of sign in certain expressions which will be derived. The significance of this will be seen later. These waves may be thought of as being represented by two radius vectors in the complex plane, as in Fig. 3·2. Let ϕ be the argument of B. The resultant of the two vectors is then proportional to the total transverse electric field. The incident-wave vector lies along the real axis at $t = 0$, $z = 0$. For a constant value of z, both vectors rotate together counterclockwise at a constant angular velocity ω as t increases. If t is constant, motion along the guide in the direction of increasing z corresponds to a clockwise rotation of the incident-wave vector about the origin and a counterclockwise rotation of the reflected-wave vector.

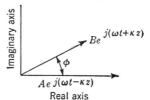

Fig. 3·2.—Vector representation of waves in the complex plane.

The amplitude of the resultant wave will thus pass through a minimum value when the vectors are oppositely directed, increase to a maximum, decrease to a minimum, and so on, as the obstacle is approached. The incident and reflected waves may therefore be resolved into a traveling wave and a standing wave. The distance toward the obstacle from the first minimum in the standing-wave pattern is given by

$$z = \frac{\pi - \phi}{2\kappa}. \quad (1)$$

This description can be expressed more precisely. The resultant transverse field F is the sum of the contribution of two waves

$$F = Ae^{j(\omega t-\kappa z)} + Be^{j(\omega t+\kappa z)}$$
$$= e^{j\omega t}[Ae^{-j\kappa z} + B_1 e^{j(\kappa z+\phi)}],$$

where
$$B = B_1 e^{j\phi}.$$

The addition and subtraction of $B_1 e^{-j(\kappa z+\phi)}$ results in

$$Fe^{-j\omega t} = De^{-j\kappa z+j\theta} + 2B_1 \cos(\kappa z + \phi),$$

where
$$D = |A - B_1 e^{-j\phi}| = (A^2 + B_1^2 - 2AB_1 \cos \phi)^{\frac{1}{2}}$$

is real and
$$\theta = \tan^{-1}\frac{B_1 \sin \phi}{A - B_1 \cos \phi}.$$

The first term is the traveling wave of amplitude D, and the second term is the standing wave of amplitude $2B_1$.

It is easy to measure a quantity proportional to the amplitude of F by inserting a small probe in the waveguide, and in this way the standing-wave pattern can be measured. The methods for making such measurements are discussed in detail in Vol. 11 of this series. The standing-wave ratio will be denoted by r. It is given by

$$r = \frac{|F|_{\max}}{|F|_{\min}} = \frac{|A| + |B|}{|A| - |B|} = \frac{A + B_1}{A - B_1}.$$

Commonly, r is expressed as a ratio of fields at points along the waveguide, but it may also be expressed as a power ratio P, in decibels. The connection is

$$P = 10 \log_{10} r^2 \quad \text{db}, \tag{2}$$

since power is proportional to the square of the transverse electric field. The quantity P is not directly related to the ratio of power transfer at two points but merely furnishes a convenient manner of expressing a ratio, particularly when the ratio is large. The quantity r is always greater than unity except when there is no reflected wave, in which case it has the value 1. It is clear that since only the amplitude of the field is measured, the standing-wave pattern has a period, with respect to z, twice that of the incident wave, and that the distance between successive minima is only one-half wavelength. Moreover, the time phase of the standing-wave portion of the field is constant for a half wavelength and then changes abruptly by π for the next half wavelength. One other parameter is necessary to specify the standing-wave pattern. This other parameter is conveniently expressed as the distance, measured in wavelengths in the waveguide, from some reference plane in the obstacle to the first minimum, and will be denoted by $d/\lambda_g = (\kappa/2\pi)d$. This quantity is dimensionless and is therefore sometimes called the "phase" of the standing-wave pattern.

SEC. 3·1] SOME GENERAL PROPERTIES OF GUIDED WAVES 63

Another parameter that is convenient for characterizing an obstacle is the "reflection coefficient." This is defined as the ratio of the amplitude of the reflected wave to the amplitude of the incident wave and is denoted by Γ. It is given by

$$\Gamma = \frac{Be^{j\kappa z}}{Ae^{-j\kappa z}} = \frac{B_1}{A} e^{j(2\kappa z + \phi)}. \tag{3}$$

Thus Γ is a complex number whose magnitude B_1/A varies from zero to unity and whose argument is a function of position along the guide and of the phase of the reflected wave. It is possible, of course, to relate Γ to the standing-wave ratio;

$$r = \frac{A + B_1}{A - B_1} = \frac{1 + \dfrac{B_1}{A}}{1 - \dfrac{B_1}{A}} = \frac{1 + |\Gamma|}{1 - |\Gamma|}, \tag{4}$$

and

$$|\Gamma| = \frac{r - 1}{r + 1}. \tag{5}$$

At the position of the minimum in the standing-wave pattern, the phase of Γ is π, and

$$\Gamma = -\frac{B_1}{A}.$$

At the maximum, the phase is zero and $\Gamma = +B_1/A$. If the phase of Γ at the reference plane of the obstacle is θ, then

$$\theta = \pi + 2\kappa d + \phi. \tag{6}$$

The period of Γ with respect to z is π/κ, or a half wavelength.

It has so far been tacitly assumed that it is unnecessary to consider what happens to the reflected wave as it travels away from the obstacle to the left. If it is absorbed in the generator that produces the incident waves, there is no difficulty. Suppose, however, that the generator reflects the wave with a reflection coefficient Γ_1. The wave reflected from both the generator and the obstacle is traveling to the right with an amplitude $\Gamma\Gamma_1 A$. This wave will be reflected from the obstacle a second time and then again from the generator. The total amplitude A' of the waves traveling to the right will be

$$A' = A + \Gamma\Gamma_1 A + \Gamma^2\Gamma_1^2 A + \cdots,$$

which converges, since $|\Gamma\Gamma_1| < 1$. The total wave amplitude traveling to the left will be

$$\Gamma A + \Gamma^2\Gamma_1 A + \Gamma^3\Gamma_1 A + \cdots = \Gamma A'.$$

Thus the whole effect of a reflection from the generator is to change the amplitude of the incident wave from A to A', and the situation is otherwise unaltered.

Some simple relations between the transmitted power and the reflected power, in terms of Γ and r, may now be written. The fraction of the power reflected, P_r, is

$$P_r = \Gamma\Gamma^* = |\Gamma|^2 = \left(\frac{r-1}{r+1}\right)^2.$$

The fraction of the power transmitted by the obstacle, P_t, is

$$P_t = \left|\frac{C}{A}\right|^2.$$

The insertion loss of the obstacle is defined as $L = 1 - P_t$ and is made up of the reflection loss P_r plus the power absorbed in the obstacle. It is not possible to measure A directly because of the standing waves to the left of the obstacle. It is possible, however, to measure the ratio of the field amplitude on the right to the field amplitude at the maximum or minimum of the standing-wave pattern; that is, the quantities

$$\frac{|C|}{A + B_1} \quad \text{or} \quad \frac{|C|}{A - B_1}$$

can be measured. Therefore,

$$L = 1 - P_t = 1 - \frac{|C|^2}{(A + B_1)^2}(1 + |\Gamma|)^2$$

$$= 1 - \frac{|C|^2}{(A - B_1)^2}(1 - |\Gamma|)^2, \quad (7)$$

or

$$L = 1 - \frac{|C|^2}{(A + B_1)^2}\frac{4r^2}{(r + 1)^2} = 1 - \frac{|C|^2}{(A - B_1)^2}\frac{4}{(r + 1)^2}. \quad (8)$$

The input power to the obstacle, P_{in}, is

$$P_{\text{in}} = A^2 - B_1^2 = (A + B_1)(A - B_1) = |F|_{\max}|F|_{\min}.$$

Let us consider the simple case where the obstacle is a transverse plate of metal entirely across the waveguide. The transverse electric field is zero over such a plate, and the incident- and reflected-wave vectors must be equal and opposite. Thus $\phi = \pi$, $\Gamma = -1$, $\theta = 0$, $r = \infty$, and d/λ_g is zero or $\frac{1}{2}$. If the obstacle consists of a transverse magnetic wall over which E_t is a maximum but $H_t = 0$, then $\phi = 0$, $\Gamma = 1$, $\theta = \pi$, $r = \infty$, and d/λ_g is $\frac{1}{4}$. These two cases are called the "short-circuit" and the "open-circuit" cases respectively.

3·2. Low-frequency Transmission Lines.—The reader familiar with conventional transmission-line theory may by now have become suspicious. Standing waves, short and open circuits, and other things usually associated with ordinary low-frequency circuits where currents and voltages and not electric and magnetic fields are taken to be the fundamental quantities have been mentioned. These suspicions may be lulled by establishing more explicitly the connections with low-frequency circuits. It should be emphasized, however, that up to this point only the fact that there are waves traveling down a waveguide and being reflected or transmitted by obstacles has been utilized, and therefore the results are completely general. But one restriction has been made, namely, that only the dominant mode can be propagated in the guide.

In a coaxial transmission line, energy is propagated in the principal or TEM-mode. In Sec. 2·7 the expression for the fields and the equations that they satisfy have already been derived. It was found that if losses are neglected,

$$\frac{\partial E_r}{\partial z} = -j\omega\mu H_\phi, \qquad \frac{\partial H_\phi}{\partial z} = -j\omega\epsilon E_r.$$

The solution of these equations is

$$E_r = \zeta \frac{I}{2\pi r}, \qquad H_\phi = \frac{I}{2\pi r},$$

where I is the total current flowing in the walls of either the inner or outer conductor. These equations can be put into a slightly different form. If the voltage across the line is defined as

$$V = \int_a^b E_r \, dr,$$

this value of V is independent of the path of integration from the inner to the outer cylinder provided only that the path be restricted to a transverse plane, since H is purely transverse. If the equations are integrated with respect to r over such a path, and if I is substituted for $2\pi r H_\phi$, then

$$\frac{\partial V}{\partial z} = -j\omega\mu \frac{\ln \frac{b}{a}}{2\pi} I, \qquad \frac{\partial I}{\partial z} = -j\omega\epsilon \frac{2\pi}{\ln \frac{b}{a}} V.$$

These equations are a special case of the general transmission-line equations

$$\frac{\partial V}{\partial z} = -ZI, \qquad \frac{\partial I}{\partial z} = -YV, \tag{9}$$

where Z and Y are the series impedance and shunt admittance per unit length of the line. These equations are rigorously true for the coaxial line if

$$Z = j\omega\mu \frac{\ln\frac{b}{a}}{2\pi}, \quad \text{and} \quad Y = j\omega\epsilon \frac{2\pi}{\ln\frac{b}{a}}. \tag{10}$$

These values could have been found, not only from Maxwell's equations directly as has been done here, but also from a calculation of the inductance and capacitance per unit length between coaxial cylinders. For transmission lines of other shapes, such as parallel wires, Eq. (9) is valid if the usual low-frequency approximations are made. The values of Z and Y will, of course, be different; they will be those characteristic of the particular line under consideration. The solution of Eq. (9) may now be written as the sum of waves traveling to the right and to the left of the point of observation

$$\left. \begin{array}{l} V = Ae^{-\gamma z} + Be^{\gamma z}, \\ I = \dfrac{1}{Z_0} Ae^{-\gamma z} - \dfrac{1}{Z_0} Be^{\gamma z}, \end{array} \right\} \tag{11}$$

where Z_0 is the characteristic impedance of the transmission line and γ is the propagation constant; thus

$$Z_0 = \sqrt{\frac{Z}{Y}} \quad \text{and} \quad \gamma = \sqrt{ZY}. \tag{12}$$

The transmission-line equations may be conveniently rewritten in terms of these parameters, since

$$\left. \begin{array}{l} Z = \gamma Z_0, \\ Y = \dfrac{\gamma}{Z_0} = \gamma Y_0. \end{array} \right\} \tag{13}$$

Equations (9) become

$$\frac{\partial V}{\partial z} = -\gamma Z_0 I, \quad \frac{\partial I}{\partial z} = -\gamma Y_0 V. \tag{14}$$

These equations represent waves of voltage and current, instead of electric and magnetic fields, but the discussion of reflection coefficients, standing waves, and so forth, of the previous section is valid here also. A new concept, however, has been introduced: the "impedance" at any point on the line, which is the ratio V/I. This quantity is uniquely defined. This is true for a coaxial line at any frequency for the principal mode or for a more general type of transmission line at low frequencies where the ordinary ideas of circuits and circuit elements are valid. No unique definition of V/I in a waveguide can be made, since (1) the current for a given value of z is a function of the coordinates in the xy-plane and (2) the line integral of the electric field is not independent of the end points, even though the path of integration does lie in the transverse plane. Impedance is such a useful concept, however, that it is desired to retain its use for waveguides, and some generalization must be made. Before proceeding with the generalization, the results that are valid at low frequencies will be reviewed.

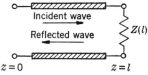

FIG. 3·3.—A transmission line terminated by an impedance $Z(l)$.

3·3. The Transformation of Impedances. Let us consider a line terminated at $z = l$ by an impedance $Z = Z(l)$, as in Fig. 3·3. Equations (10) are subject to the boundary condition that

$$\frac{V(l)}{I(l)} = Z(l).$$

Thus,

$$Z(l) = \frac{V}{I} = Z_0 \frac{Ae^{-\gamma l} + Be^{\gamma l}}{Ae^{-\gamma l} - Be^{\gamma l}}. \tag{15}$$

The voltage reflection coefficient is

$$\Gamma_V = \frac{B}{A} e^{2\gamma l} = \frac{Z(l) - Z_0}{Z(l) + Z_0}. \tag{16}$$

The current reflection coefficient can be defined as the ratio of the reflected current wave amplitude to the incident current wave amplitude. Hence

$$\Gamma_I = -\frac{B}{A} e^{2\gamma l} = \frac{Z_0 - Z(l)}{Z_0 + Z(l)} = -\Gamma_V.$$

Some authors, in the discussion of transmission lines, prefer to use the current reflection coefficient instead of the voltage reflection coefficient. In this chapter the voltage reflection coefficient will be used, and the symbol Γ will be understood to be equivalent to Γ_V.

Equation (16) is the transformation equation of Γ along the line; now the transformation equation for Z must be found. At $z = 0$, the input impedance is

$$Z_{in} = Z_0 \frac{A + B}{A - B} = Z_0 \frac{1 + \frac{B}{A}}{1 - \frac{B}{A}}.$$

The elimination of the ratio B/A, by means of Eq. (16), has the result

$$Z_{in} = Z_0 \frac{Z(l) + Z_0 \tanh \gamma l}{Z_0 + Z(l) \tanh \gamma l}. \tag{17}$$

It should be emphasized that Eq. (17) is a direct result of the boundary conditions imposed on A and B by specifying a termination $Z(l)$ on the end of the line. It is worth while to restate more explicitly what has been done. It has been shown that the total or transmitted voltage across the load $Z(l)$ is the sum of an incident and a reflected voltage wave, and that the transmitted current is the sum of an incident and a reflected current wave; that is

$$V_i + V_r = V_t, \qquad I_i + I_r = I_t.$$

These currents and voltages are separately related by means of the impedance; thus

$$V_i = Z_0 I_i, \qquad V_r = -Z_0 I_r, \qquad V_t = Z(l) I_t.$$

These two sets of equations state the boundary conditions in a form that makes obvious the circumstance that the currents and voltages combine additively at a boundary. It is possible to solve for

$$\Gamma_V = \frac{V_r}{V_i} = \frac{Z(l) - Z_0}{Z(l) + Z_0} = -\frac{I_r}{I_i} = -\Gamma_I.$$

The other ratio,

$$T_V = \frac{V_t}{V_i} = \frac{2Z(l)}{Z(l) + Z_0} = 1 + \Gamma_V, \tag{18}$$

can be defined as the voltage transmission coefficient. When these relations are combined with the fact that V and I transform along the lines as waves with a propagation constant γ, Eqs. (16) and (17) can again be derived.

It is also possible to express the relationships between these quantities in terms of admittances, which are simply the reciprocals of the corresponding impedances. Thus

$$Y_0 = \frac{1}{Z_0}, \qquad Y(l) = \frac{1}{Z(l)}, \qquad Y_{\text{in}} = \frac{1}{Z_{\text{in}}}.$$

The formulas become

$$\Gamma_V = \frac{Y_0 - Y(l)}{Y_0 + Y(l)} = |\Gamma_V| e^{2\gamma l}, \tag{19}$$

$$Y_{\text{in}} = Y_0 \frac{Y(l) + Y_0 \tanh \gamma l}{Y_0 + Y(l) \tanh \gamma l}, \tag{20}$$

$$T_V = \frac{2Y_0}{Y_0 + Y(l)} = 1 + \Gamma_V. \tag{21}$$

When it is possible to neglect the attenuation in the line, γ is purely imaginary; and if $\gamma = j\kappa$, the expressions become

$$Z_{\text{in}} = Z_0 \frac{Z(l) + jZ_0 \tan \kappa l}{Z_0 + jZ(l) \tan \kappa l}, \tag{22}$$

$$Y_{\text{in}} = Y_0 \frac{Y(l) + jY_0 \tan \kappa l}{Y_0 + jY(l) \tan \kappa l}. \tag{23}$$

When the line is terminated by a short circuit, $Z(l) = 0$, $Y(l) = \infty$,

$$Z_{\text{short}} = jZ_0 \tan \kappa l,$$
$$Y_{\text{short}} = -jY_0 \cot \kappa l.$$

For an open-circuited line, $Z(l) = \infty$, $Y(l) = 0$,

$$Z_{\text{open}} = -jZ_0 \cot \kappa l,$$
$$Y_{\text{open}} = jY_0 \tan \kappa l.$$

If $l = \lambda_g/4$, $\tan \kappa l = \tan \pi/2 = \infty$ and the input impedance and admittance become

$$Z_{\text{in}} = \frac{Z_0^2}{Z(l)},$$
$$Y_{\text{in}} = \frac{Y_0^2}{Y(l)}. \tag{24}$$

Thus a quarter-wavelength line inverts the impedance and admittance with respect to the characteristic impedance or admittance.

One notable characteristic of all the expressions for Γ, T, Z_{in}, and Y_{in} is that they can all be written in terms of relative impedances. For example, Eq. (17) may be written

$$\frac{Z_{\text{in}}}{Z_0} = \frac{\dfrac{Z(l)}{Z_0} + \tanh \gamma l}{1 + \dfrac{Z(l)}{Z_0} \tanh \gamma l}.$$

It is just this circumstance which makes these expressions valuable for use with waveguides, where Z_0 cannot be uniquely defined. This characteristic is really only a result of the wave nature of the solutions to Maxwell's equations, not of any special assumptions that have been made. This point will receive further consideration later.

3·4. Power Flow.—The power flow into a line is, of course, given by the real part of $\frac{1}{2} V_{\text{in}} I_{\text{in}}^*$, or

$$P = \tfrac{1}{2} \text{Re } (V_{\text{in}} I_{\text{in}}^*) = \tfrac{1}{2} \text{Re } (Z_{\text{in}}) |I_{\text{in}}|^2 = \tfrac{1}{2} \text{Re } (Y_{\text{in}}) |V_{\text{in}}|^2. \tag{25}$$

This result is not to be derived from the transmission-line equations but represents a second physical property of the quantities V and I. It can be proved in the low-frequency approximation for the ordinary circuit equations. This will not be done, however. The result is fundamentally a consequence of Poynting's theorem. In the case of a coaxial line it is possible to verify Eq. (25) by substituting for V and I the values of the field strengths and integrating over the area of the line. It is more helpful to think of the equation as an expression of a physical characteristic of the quantities V and I that they must satisfy in order to maintain consistency with the fundamental electromagnetic equations. All the remarks made in an earlier section relative to the power carried by the incident and reflected waves in relation to the reflection coefficient and the standing-wave ratio are still valid here.

3·5. The Combination of Admittances.—Let us suppose that an admittance Y is shunted across an infinite line of characteristic admittance Y_0, as shown in Fig. 3·4a, and that there is a wave incident from the

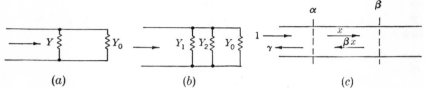

Fig. 3·4.—The combination of admittances and reflection coefficients.

left to the right. The admittance to the right from a point just to the left of Y is $Y + Y_0$, and the voltage reflection coefficient is

$$\Gamma = \frac{-Y}{Y + 2Y_0}. \tag{26}$$

Let us now regard Y as made up of two admittances Y_1 and Y_2 in parallel. The total admittance is $Y_1 + Y_2 + Y_0$, and the reflection coefficient is

$$\Gamma = -\frac{Y_1 + Y_2}{Y_1 + Y_2 + 2Y_0}.$$

Thus the law of combination of admittances is simply additive. It is now of interest to know the law of combination of reflection coefficients. Suppose that the Y_1 mentioned above is shunted across the line. Let the reflection coefficient be α. Let β be the reflection coefficient when Y_2 is shunted across the line. The problem of how α and β should be combined to give the value of Γ above may be treated by considering successive reflections of the waves.

Let us assume that the wave is incident first upon Y_1. There will be an incident wave whose amplitude may be taken as unity, a transmitted wave of amplitude x, and a reflected wave whose amplitude it is desired to find. Let the amplitude of this reflected wave be γ as indicated in Fig. 3·4c. The wave of amplitude x is made up, of course, of all the waves traveling to the right resulting from the successive reflections between the two admittances. It may be regarded, however, as being made up of the transmitted wave $1 + \alpha$ from the first admittance plus the sum of all the reflected waves from the second admittance which are again reflected from the first admittance; thus

$$x = 1 + \alpha + \alpha\beta x.$$

Likewise

$$\gamma = \alpha + \beta x(1 + \alpha).$$

If x is eliminated,

$$\gamma = \alpha + \beta \frac{(1 + \alpha)^2}{1 - \alpha\beta}. \tag{27}$$

Thus the law of addition is a very complicated one indeed. To verify this result, the expression $-2\gamma/(1 + \gamma)$ may be formed, which is

$$-\frac{2\gamma}{1 + \gamma} = -\frac{2\alpha}{1 + \alpha} - \frac{2\beta}{1 + \beta}.$$

From Eq. (26),

$$\frac{Y}{Y_0} = \frac{-2\Gamma}{1 + \Gamma}.$$

Thus the law of additivity of shunt admittances has been verified from the wave picture.

The argument just stated could have been carried through using the concept of an equivalent series impedance that combined simply with another series impedance. Again the reflection coefficients do not combine simply. This is another aspect of the importance of the admittance or impedance concept for use in waveguides, where neither currents nor voltages may be uniquely defined.

3·6. Transmission-line Charts.—It has been shown, in the preceding section, that a reflection in a transmission line can be described in several alternative ways. Each of these ways is convenient for certain problems; all are in common use. A reflection can be described by any of four pairs of variables:

1. The standing-wave ratio and the position of the minimum, r and d.
2. The real and imaginary parts of an equivalent shunt admittance $Y/Y_0 = G/Y_0 + jB/Y_0$.
3. The real and imaginary parts of an equivalent series impedance, $Z/Z_0 = R/Z_0 + jX/Z_0$.
4. The modulus and phase of the voltage reflection coefficient, $\Gamma = |\Gamma|e^{j\phi}$.

A fifth pair of parameters is sometimes used, namely, the modulus and phase of the current reflection coefficient. Since it has already been shown, however, that $\Gamma_I = -\Gamma_V$, this represents a more or less trivial addition. It must be remembered that impedances and admittances occur only as the ratio to the characteristic impedance and admittance of the transmission line, and therefore all of these parameters are dimensionless quantities. The four pairs of parameters listed above are related to one another by equations that have been stated in the preceding sections. Each of the complex quantities Γ, Z/Z_0, and Y/Y_0 may be regarded as a function of a complex variable, but this is not true of the quantity $re^{i\kappa d}$. The fundamental relations that have been derived above are

$$\Gamma = |\Gamma|e^{i\phi} = \frac{Z - Z_0}{Z + Z_0} = \frac{Y_0 - Y}{Y_0 + Y},$$

$$r = \frac{1 + |\Gamma|}{1 - |\Gamma|}, \qquad \kappa d = \frac{\phi - \pi}{2}.$$

These relations may be separated into their real and imaginary parts. Thus,

$$r = \frac{1 + |\Gamma|}{1 - |\Gamma|} = \frac{\sqrt{(R+1)^2 + X^2} + \sqrt{(R-1)^2 + X^2}}{\sqrt{(R+1)^2 + X^2} - \sqrt{(R-1)^2 + X^2}}$$
$$= \frac{\sqrt{(G+1)^2 + B^2} + \sqrt{(G-1)^2 + B^2}}{\sqrt{(G+1)^2 + B^2} - \sqrt{(G-1)^2 + B^2}}, \quad (28)$$

$$\kappa d = \frac{\phi - \pi}{2} = \frac{1}{2}\tan^{-1}\frac{2X}{R^2 + X^2 - 1} - \frac{\pi}{2}$$
$$= \frac{1}{2}\tan^{-1}\frac{2B}{G^2 + B^2 - 1} - \frac{\pi}{2}, \quad (29)$$

$$|\Gamma| = \frac{r-1}{r+1} = \sqrt{\frac{(R-1)^2 + X^2}{(R+1)^2 + X^2}} = \sqrt{\frac{(G-1)^2 + B^2}{(G+1)^2 + B^2}}, \quad (30)$$

$$\phi = 2\kappa d + \pi = \tan^{-1}\frac{2X}{R^2 + X^2 - 1} = \tan^{-1}\frac{2B}{G^2 + B^2 - 1}, \quad (31)$$

$$R = \frac{r}{r^2 \cos^2 \kappa d + \sin^2 \kappa d} = \frac{1 - |\Gamma|^2}{1 - 2|\Gamma|\cos\phi + |\Gamma|^2} = \frac{G}{G^2 + B^2}, \quad (32)$$

$$X = \frac{(1 - r^2)\sin \kappa d \cos \kappa d}{r^2 \cos^2 \kappa d + \sin^2 \kappa d} = \frac{2|\Gamma|\sin\phi}{1 - 2|\Gamma|\cos\phi + |\Gamma|^2} = \frac{-B}{G^2 + B^2} \quad (33)$$

$$G = \frac{r}{r^2 \sin^2 \kappa d + \cos^2 \kappa d} = \frac{1 - |\Gamma|^2}{1 + 2|\Gamma|\cos\phi + |\Gamma|^2} = \frac{R}{R^2 + X^2}, \quad (34)$$

$$B = \frac{(r^2 - 1)\sin \kappa d \cos \kappa d}{r^2 \sin^2 \kappa d \cos^2 \kappa d} = \frac{-2|\Gamma|\sin\phi}{1 + 2|\Gamma|\cos\phi + |\Gamma|^2} = \frac{-X}{R^2 + X^2}. \quad (35)$$

For convenience of notation Y_0 and Z_0 have, in the above equations, been set equal to unity. The values of R, X, G, and B above are thus measured relative to the characteristic impedance or admittance of the line.

The transformations represented by the above equations are conformal—the true values of angles are preserved in the transformation. They are also bilinear transformations, that is, of the form

$$z = \frac{aw + b}{cw + d}.$$

Thus it is possible to apply many general theorems which are well known for transformations of this type.

These equations are sufficiently numerous and complicated that some graphical method of handling them is almost essential. Fortunately, a method exists that is convenient and easy to use, whereby these 24 relations can be represented by a single chart. This chart, designed by

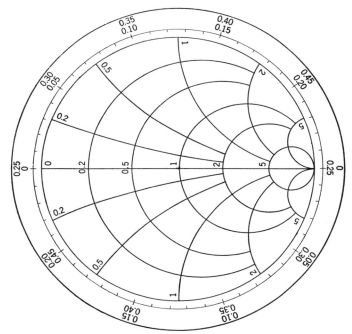

Fig. 3·5.—The Smith impedance chart.

P. H. Smith,[1] is illustrated in Fig. 3·5. The quantities $|\Gamma|$ and ϕ are chosen as polar coordinates, and lines of constant R and constant X are plotted. The region of interest is within the circle of unit radius, $|\Gamma| = 1$. The family of curves R = constant, X = constant consists of orthogonal circles. In terms of rectangular coordinates u and v in the Γ-plane, these circles are given by

$$\left(u - \frac{R}{R+1}\right)^2 + v^2 = \frac{1}{(R+1)^2}$$

$$(u - 1)^2 + \left(v - \frac{1}{X}\right)^2 = \frac{1}{X^2}.$$

The R-circles all have their centers on the u-axis and all pass through the point $u = 1$, $v = 0$. The X-circles all have their centers on the line $u = 1$, and all pass through the point $u = 1$, $v = 0$. All values of R

[1] P. H. Smith, *Electronics*, January 1939, January 1944.

from zero to plus infinity and all values of X from minus infinity to plus infinity are included within the unit circle. Thus there is a convenient means of transformation from $|\Gamma|$ and ϕ to R and X and inversely. If the reference plane is moved nearer to the generator, that is, in the negative z-direction, the vector Γ rotates clockwise, making one revolution in

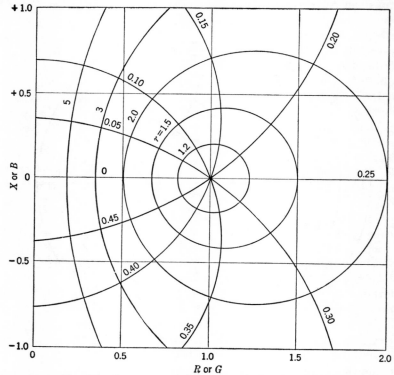

Fig. 3·6.—Impedance chart with rectangular coordinates.

half a wavelength. An auxiliary scale outside the unit circle, running from 0 to 0.5 around the circumference, facilitates this transformation. Curves of constant standing-wave ratio are concentric circles about the origin which pass through the points $r = R$. The parameter d/λ_g is read on the external circular scale. The relation between impedance and admittance is obtained in the following manner. A shift of reference plane of one-quarter wavelength inverts the value of the relative impedance; the shunt admittance equivalent to a series impedance is given, therefore, by the point diametrically opposite the origin from the impedance point at the same radius. Moreover, it is apparent from the transformation equations that if $|\Gamma|$ is replaced by $-|\Gamma|$, then R must be replaced by G and X by B. Thus the same chart may be used for admittances provided the value of ϕ is increased by π.

The use of a Smith chart is very similar to the use of a slide rule; many tricks and short cuts are possible that are hard to describe but greatly facilitate computations. The Emeloid Company, of Arlington, N. J., makes a chart of this kind, of celluloid, which is called the "Radio Transmission Line Calculator."

Impedance charts of other varieties have been made and used, but only one other is commonly encountered. In this version, R and X are used as rectangular coordinates, and the lines of constant r and d/λ_g are plotted. The chart has the same form when G and B are used as coordinates. The reflection coefficient cannot be read easily from the diagram. The lines of constant r are a family of circles with centers on the real axis, and the lines of constant d/λ_g are circles centered on the imaginary axis and orthogonal to the r-circles. An outstanding difficulty with a chart of this type is that the points of infinite R and X are not accessible. This rectangular form of impedance chart is illustrated in Fig. 3·6.

3·7. The Impedance Concept in Waveguide Problems.—It has been shown in preceding sections that the properties of both waveguides and low-frequency transmission lines can be described in terms of incident and reflected waves. The state of the line or waveguide can be expressed by means of reflection coefficients that are, with the exception of a constant factor, sufficient to specify this state completely. In addition, it has been seen that the rule of combination of reflection coefficients is complicated even in the simplest cases.

On the other hand, the state of a low-frequency transmission line may be expressed equally well in terms of a relative impedance or admittance, that is, the ratio of the impedance or admittance to the characteristic impedance or admittance of the transmission line. The impedance or admittance combines simply with other impedances, and it is this property which leads to a demand for an equivalent concept for the characteristic impedance of a waveguide. It has been seen that the reflection coefficient in a waveguide can be replaced, at least formally, by a relative impedance that is completely equivalent and that expresses the state of the fields to within an unknown factor. In any configuration of waveguides of a single kind, relative impedances or admittances may be defined in terms of r and ϕ and combined according to the usual low-frequency rules. It is not necessary to specify exactly what is meant by the characteristic impedance of the guide.

Let us now consider the junction of two waveguides as illustrated in Fig. 3·7. If radiation is incident upon the junction from guide 1, there will be, in general, a reflected wave in guide 1. This reflected wave may be described in terms of the reflection coefficient or in terms of an equivalent relative shunt admittance or series impedance that terminates guide 1 at the junction. Provided the losses in the neighborhood of the junction may be neglected, the power flowing in guide 2 must be equal to the difference between the incident and reflected powers in guide 1. The amount of reflected power will be determined by the actual electric and magnetic fields in the aperture, which, of course, satisfy Maxwell's equations and the appropriate boundary conditions. In particular, across any transverse plane, the tangential electric and magnetic fields must be continuous. To complete the analogy with low-frequency transmission lines, quantities analogous to the current and voltage must be defined for waveguides, since it is in terms of the values of current and voltage that the terminal conditions must be specified. A few possibilities will be discussed.

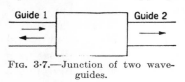

Fig. 3·7.—Junction of two waveguides.

The voltage and current should be linear in the magnetic and electric fields, since it is desired that their product be a measure of the power. Thus let
$$V = aE_t + bH_t,$$
and
$$I = cE_t + dH_t,$$
where E_t and H_t are some mean values of the transverse fields. The complex power is then
$$P = \tfrac{1}{2}VI^* = \tfrac{1}{2}(ac^*|E_t|^2 + bc^*H_tE_t^* + ad^*E_tH_t^* + bd^*|H_t|^2).$$

The Poynting theorem states, however, that P is proportional to $E_tH_t^*$ and independent of $|E_t|^2$ and $|H_t|^2$. Therefore
$$ac^* = bd^* = 0;$$
hence either
$$a = d = 0, \quad \text{or} \quad b = c = 0.$$

Now if it is required that the voltage be zero at a short circuit where $E_t = 0$ and that the current be zero at an open circuit where $H_t = 0$, then it is necessary that $b = c = 0$, and therefore
$$V = aE_t, \quad I = dH_t.$$

The impedance at any point is then
$$\frac{V}{I} = \frac{a}{d}\frac{E_t}{H_t}, \tag{36}$$
and the power flow
$$P = \tfrac{1}{2}ad^*E_tH_t^*. \tag{37}$$

It has already been shown that all properties of reflected waves can be expressed in terms of a relative impedance, and no condition is imposed on the proportionality factor of Eq. (36). The second condition [Eq.

(37)] when applied to two guides such as indicated in Fig. 3·7 does represent, however, a new condition. Any mean value may be chosen for E_t and H_t, and in fact different values may be taken for two different waveguides, provided only that the conservation of power at a junction between two guides in ensured. This condition cannot be written explicitly, since it depends upon the nature of the junction. It can be seen, however, that if the current is identified, for example, with the transverse magnetic field at the center of the waveguide, then the constant of proportionality between the voltage and the electric field is definite and is determined so that $P = \frac{1}{2}VI^*$ represents the true power flow. It may be pointed out that the ratio between I and H_t need never be specified and may be chosen at will. If particular values of E_t and H_t are chosen, then only the product ad^* is determined, but neither a nor d separately.

The situation is somewhat analogous to that arising from the insertion of ideal transformers in a network. If an ideal transformer were connected to each voltmeter and ammeter in a network in such a manner that the product of the readings remained the same, the result would be an effective change in the definition of impedance, all the power relations being conserved.

3·8. Equivalent T-network of a Length of Waveguide.—If the concept of impedance in a waveguide is to be useful, it is important to determine whether or not it can be used in the same manner as the impedance in low-frequency circuits. It has already been seen that the reflections in a long line are correctly described in terms of an equivalent shunt admittance or series impedance. Now the question is whether or not more complicated structures can be represented by equivalent circuits. If a straight piece of waveguide is terminated in such a way that the reflection is described by an impedance Z_l at the end of the line, then the equivalent impedance at the input terminals of the line is

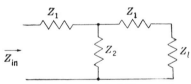

Fig. 3·8.—Symmetrical T-network.

$$Z_{\text{in}} = Z_0 \frac{Z_l + jZ_0 \tan \kappa l}{Z_0 + jZ_l \tan \kappa l}. \tag{3.17}$$

Now to find a simple network equivalent, the symmetrical T-network shown in Fig. 3·8 may be tried. This network has two parameters Z_1 and Z_2 in terms of which the line parameters Z_0 and κl might be expressed. For the network

$$Z_{\text{in}} = Z_1 + \frac{Z_2(Z_1 + Z_l)}{Z_2 + Z_1 + Z_l},$$

or
$$Z_{in} = \frac{Z_1 \dfrac{Z_1 + 2Z_2}{Z_1 + Z_2} + Z_l}{1 + \dfrac{Z_l}{Z_1 + Z_2}}.$$

By comparison with Eq. (3·17), the following identification can be made:
$$Z_1 \frac{Z_1 + 2Z_2}{Z_1 + Z_2} = jZ_0 \tan \kappa l,$$
$$\frac{1}{Z_1 + Z_2} = j \frac{\tan \kappa l}{Z_0}.$$

If these equations are solved for Z_1 and Z_2, the result is, with some trigonometric transformations,
$$Z_1 = jZ_0 \tan \frac{\kappa l}{2},$$
$$Z_2 = -jZ_0 \csc \kappa l. \tag{38}$$

These values are independent of Z_l. Hence this equivalent circuit is in all respects similar to the waveguide, and it can be used with confidence.

For easy reference, Table 3·1 presents the values of the circuit parameters for lines of commonly used lengths. The parameters refer either to the T-form or to the π-form of the network. The equivalence of these forms is discussed in Chap. 4.

TABLE 3·1.—NETWORK PARAMETERS OF LINES OF VARIOUS LENGTHS

Length of line	Series impedance of T-network or shunt admittance of π-network	Shunt impedance of T-network or series admittance of π-network
0	0	∞
$\lambda_g/8$	$j(\sqrt{2} - 1)$	$-j\sqrt{2}$
$\lambda_g/4$	j	$-j$
$3\lambda_g/8$	$j(\sqrt{2} + 1)$	$-j\sqrt{2}$
$\lambda_g/2$	∞	∞
$5\lambda_g/8$	$-j(\sqrt{2} + 1)$	$j\sqrt{2}$
$3\lambda_g/4$	$-j$	j
$7\lambda_g/8$	$-j(\sqrt{2} - 1)$	$j\sqrt{2}$
λ_g	0	∞

In a similar manner, it would be expected that any configuration of metal in a waveguide would have some equivalent-circuit representation. If this equivalent circuit is known, all the techniques known at low frequencies can then be applied to investigate the behavior of the waveguide configuration, both by itself and in combination with other configurations. The equivalent T-network that has been found for a straight length of waveguide has been shown to be equivalent only in a mathematical sense. It has not been proved that the equivalence has any physical basis. However, this physical basis will be established by arguments that are given in Chap. 5. Chapter 4 will be devoted to a review of those elements of conventional circuit theory which will be most useful.

3·9. Transmission-line Equations for the H_{10}-mode.—It is possible to express the field equations for a single mode of propagation in a waveguide in such a form that the correspondence to a transmission line is made obvious. The dominant mode only will be considered, but the extension to higher modes may be easily accomplished. The electric field in a rectangular waveguide in the dominant H_{10}-mode may be written

$$E_y = A \sin \frac{\pi x}{a} V(z),$$

where $V(z)$ expresses the field variation along the line and may be called the voltage. Likewise the transverse magnetic field is

$$H_x = -A \sin \frac{\pi x}{a} I(z),$$

where $I(z)$ may be called the current. Then from Eq. (2·46) the longitudinal field is

$$H_z = A \frac{j\pi}{\omega \mu a} \cos \frac{\pi x}{a} V(z).$$

The relevant Maxwell equations are

$$\frac{\partial E_y}{\partial z} = j\omega\mu H_x,$$

$$\frac{\partial H_x}{\partial z} - \frac{\partial H_z}{\partial x} = j\omega\epsilon E_y.$$

If the values of the fields are substituted in the first equation to find the voltage and current, the result is

$$\frac{\partial V(z)}{\partial z} = -j\omega\mu I(z).$$

This is one equation for a transmission line, upon identification of

$$j\omega\mu = \gamma Z_0 = Z.$$

The second Maxwell equation results in

$$\frac{\partial I(z)}{\partial z} = -j\left[\omega\epsilon - \frac{1}{\omega\mu}\left(\frac{\pi}{a}\right)^2\right] V(z),$$

which is the second transmission-line equation, where

$$j\left[\omega\epsilon - \frac{1}{\omega\mu}\left(\frac{\pi}{a}\right)^2\right] = \gamma Y_0 = Y.$$

The propagation constant γ of the transmission line has the same value as that of the waveguide

$$\gamma^2 = -\omega^2\epsilon\mu + \left(\frac{\pi}{a}\right)^2,$$

and the characteristic impedance is

$$Z_0 = j\frac{\omega\mu}{\gamma} = Z_H.$$

The constant A may be chosen to obtain the correct power-transfer relation if

$$\int S_z\, dx\, dy = -\frac{b}{2}\int_0^a E_y H_x^*\, dx = \frac{1}{2} VI^*.$$

Therefore

$$A = \sqrt{\frac{2}{ab}}.$$

It is important to emphasize the arbitrariness in the choice of constants. The value of A was chosen so that the complex power is $\frac{1}{2}VI^*$.

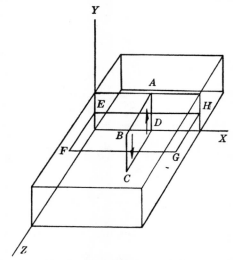

Fig. 3·9.—Dominant mode in rectangular waveguide.

The use of the same multiplying constants in the expressions for E_y and H_x results in the characteristic impedance of the line being equal to the wave impedance. Any other choice that preserved the power relation would have been equally acceptable. A different value of the impedance would have been obtained.

EQUATIONS FOR THE H_{10}-MODE

The transmission-line equations can be derived in another manner. Consider the waveguide in Fig. 3·9 operating in the dominant mode. Let us choose for the voltage $V(z)$ the integral of the electric field at the center of the guide,

$$V(z) = \int_0^b E_y \, dy,$$

and for the line current a quantity proportional to the longitudinal current flowing at the center of the broad face of the waveguide. This longitudinal current is equal to the maximum value of the transverse magnetic field

$$I(z) = \alpha H_x(z).$$

Let us consider the line integral of the electric field around the rectangular path $ABCD$ in Fig. 3·9. As the distances $AB = CD$ become infinitesimal, the line integral approaches dV/dz. By Faraday's law (the curl **E** equation)

$$\frac{dV}{dz} = \frac{j\omega\mu b I}{\alpha}.$$

If likewise the limit of the line integral of H_x is taken around the path $EFGH$ the result is

$$\frac{1}{\alpha}\frac{dI}{dz}\int_0^a \sin\frac{\pi x}{a}\,dx = j\omega\epsilon\frac{V}{b}\int_0^a \sin\frac{\pi x}{a}\,dx - 2K_t,$$

where K_t is the transverse current density across EF or GH. Its value is

$$K_t = H_z = \frac{j\pi}{\omega\mu ab}V;$$

hence

$$\frac{dI}{dz} = j\alpha\left(\frac{\omega\epsilon}{b} - \frac{\pi^2}{\omega\mu a^2 b}\right)V.$$

The impedance and admittance per unit length of the transmission line are therefore

$$Z = \frac{-j\omega\mu b}{\alpha},$$

$$Y = -j\alpha\left(\frac{\omega\epsilon}{b} - \frac{\pi^2}{\omega\mu a^2 b}\right).$$

The propagation constant of the line is given by

$$\gamma^2 = ZY = -\omega^2\epsilon\mu + \frac{\pi^2}{a^2},$$

and the characteristic impedance

$$Z_0^2 = \frac{Z}{Y} = \frac{1}{\alpha^2}\frac{(\omega\mu b)^2}{\omega^2\epsilon\mu - \frac{\pi^2}{a^2}} = -\frac{1}{\alpha^2}\left(\frac{\omega\mu b}{\gamma}\right)^2.$$

The choice of α may now be made such that $\frac{1}{2}VI^*$ is equal to the complex power.

J. Schwinger has shown that it is possible to proceed in an entirely general way and transform Maxwell's equations directly into the transmission-line equations whenever the boundary conditions are independent of the z-coordinate. Such a general case need not be considered here. The procedure that will be uniformly adopted here is as follows:

1. The voltage V is chosen proportional to the transverse electric field in the waveguide.
2. The current I is chosen proportional to the transverse magnetic field.
3. The constants of proportionality are normalized in such a way that $\frac{1}{2}VI^*$ is equal to the complex power flow.

Thus, it is assumed that

$$\mathbf{E}_t(x,y,z) = V(z)\mathbf{f}(x,y), \tag{39}$$
$$\mathbf{H}_t(x,y,z) = I(z)\mathbf{g}(x,y), \tag{40}$$

where $\mathbf{f}$ and $\mathbf{g}$ are real and so normalized that

$$\int \mathbf{f} \times \mathbf{g} \cdot d\mathbf{S} = 1. \tag{41}$$

CHAPTER 4

ELEMENTS OF NETWORK THEORY

By C. G. Montgomery

4·1. Elementary Considerations.—In this chapter will be presented some of the elementary results of network theory that are useful in the study of microwave circuits. The approach will be in terms of what may be called the low-frequency approximation to electromagnetic theory. This approximation is the one usually employed in conventional circuit theory, and the results are well known and available in many standard textbooks. For the convenience of the reader and also to aid in a more orderly presentation of the properties of high-frequency circuits, some of the more useful material has been collected. This material is offered, in general, without detailed proof of its correctness. Many of the results are proved in Chap. 5 as special cases of more general theorems. In other cases only the method of proof is outlined. The reader will find himself already acquainted with a large part of this discussion.

In this chapter, the concept of an impedance element, or impedor, will be considered as fundamental. An impedance element is a device that has two accessible terminals. It may be a simple device, such as a piece of poorly conducting material (a resistor), or it may be a very complicated structure. It is required, however, that it be passive, that is, that no energy is generated within the element. Charge may be transferred to the element only by means of the terminals; and if a current flows into one terminal, an equal current must flow out of the other. This is the first portion of the low-frequency approximation mentioned above. Thus a conducting sphere is not an impedor, since it has only a single terminal, but the equivalent impedance element can be supposed to have one terminal at the sphere and the other terminal at the point of zero potential or ground, perhaps at infinity. In the region between the terminals of the impedance element there exists an electric field. The potential difference, or voltage, between the terminals is defined as the line integral of the electric field from one terminal to the other. The second portion of the low-frequency approximation under which network theory is here treated requires that this line integral be independent of the path between the two terminals. The difference in voltage, for any two paths, will be proportional to the magnetic field integrated over the area enclosed between the two paths and to the frequency, and can be made as small as desired by the choice of a sufficiently low frequency. The ratio of the voltage across the terminals to the current entering and leaving the terminals is the *impedance* Z of the element. The reciprocal of the impedance is the *admittance* Y of the element. Only the cases where Z is independent of V or I will be considered, and for these cases the impedance element is said to be *linear*. At low frequencies an impedance element has a variation with frequency of the form

$$Z = -\frac{j}{\omega C} + R + j\omega L, \qquad (1)$$

where R, L, and C are positive constant parameters. The real part of the impedance, R, is called the resistive part, or *resistance;* the imaginary part of Z, $[\omega L - (1/\omega C)]$, is called the *reactance*. The parameters L and C are the *inductance* and the *capacitance* of the element. The capacitance C may be infinite, but in physical elements neither L nor R is truly zero, although, of course, in many cases they may have negligible values. The reactance is often denoted by the symbol X. At higher frequencies, when the impedance elements of waveguide structures are considered, this simple form of frequency dependence is no longer valid.

In a similar manner, the admittance Y can be broken up into its real and imaginary parts,

$$Y = G + jB,$$

where G is called the *conductance* and B the *susceptance* of the element. An impedance element whose frequency dependence is given by Eq. (1) is often broken up, for the convenience of the mathematical symbolism, into two or three elements in series, one for the real and one for the imaginary part, or one for each term with a characteristic frequency dependence. Since the admittance is the ratio of the current to the voltage, if an admittance is split into parts, the component admittances must be combined in parallel. The currents through the separate elements then add, and the voltages across them are equal. Thus if an inductance, a resistance, and a capacitance are combined in parallel, the admittance obtained is

$$Y = \frac{1}{R} + j\omega C - \frac{j}{\omega L}.$$

The impedance of Eq. (1) has, on the other hand, an admittance made up of the conductance

$$G = \frac{R}{R^2 + \left(\omega L - \dfrac{1}{\omega C}\right)^2} = \frac{R}{R^2 + X^2}$$

and the susceptance

$$B = \frac{-\left(\omega L - \dfrac{1}{\omega C}\right)}{R^2 + \left(\omega L - \dfrac{1}{\omega C}\right)^2} = \frac{-X}{R^2 + X^2}. \qquad (2)$$

and the variation with ω is characteristic of R, L, and C in series.

Here is found the first illustration of the *duality principle,* which is of great convenience in network theory. An impedance may be regarded as the sum of several impedances in series, while an admittance is the sum of other admittances in parallel. One quantity is said to be the "dual" of another if, in a statement or equation, the two quantities can be interchanged without invalidation of that statement or equation. Thus it is seen that

Impedance	is the dual of	admittance.
Series	is the dual of	parallel.
Voltage	is the dual of	current.
Resistance	is the dual of	conductance.
Reactance	is the dual of	susceptance.

The simple relation $Z = V/I$ remains true if the quantities are all replaced by their duals; that is, $Y = I/V$. Likewise the statement "impedances are added in series" becomes "admittances are added in parallel." If the duality replacement is made in Eqs. (2), they become

$$\left. \begin{array}{l} R = \dfrac{G}{G^2 + B^2}, \\ X = \dfrac{-B}{G^2 + B^2}. \end{array} \right\} \quad (3)$$

The duality principle results entirely from the fundamental symmetries of Maxwell's equations. An equivalent formulation was discussed in Sec. 2·10 under the name of Babinet's principle.

Several impedance elements and voltage sources may be connected together to form a network such as the one shown in Fig. 4·1. A network is composed of *branches* that may be individual

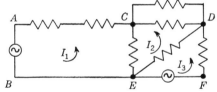

Fig. 4·1.—An arbitrary network.

impedance elements, such as CE, or may consist of several impedance elements in series, as the branch AC, or in parallel, as CD. The branches are connected together at *branch points,* or *nodes.* They form in this way a collection of individual *circuits* or *meshes* such as $CDEC$ or $ACDFEBA$.

If the impedances in the branches are known, it is possible to find the currents in the network in terms of the sources of electromotive force or applied voltages. The necessary relations are Kirchhoff's laws. The first law states that the algebraic sum of the voltage changes around any circuit must be zero. This law may be regarded as an expression of the law of conservation of energy for a charge that is carried completely around the circuit. Explicit use is made here of the low-frequency approximation that the line integral of the electric field is independent of the path. Kirchoff's second law states that the algebraic sum of the currents flowing into each branch point must be zero. This law follows from the conservation of charge, since charge cannot accumulate at the branch point. A sufficient number of linear relations may be set up by means of these two laws to make possible the determination of all currents.

A simpler set of equations is obtained if the currents in the meshes are used as the unknown variables. These currents, which are indicated in Fig. 4·1, are sometimes called the circulating currents. When circulating currents are so chosen, Kirchhoff's second law is automatically satisfied. Thus there is a set of equations of the from

$$\left. \begin{aligned} v_1 &= Z_{11}i_1 + Z_{12}i_2 + Z_{13}i_3 + \cdots + Z_{1m}i_m, \\ v_2 &= Z_{21}i_1 + Z_{22}i_2 + \quad\quad\quad \cdots + Z_{2m}i_m, \\ & \cdot \quad \cdot \quad \cdot \quad\quad \cdot \quad \cdot \\ v_m &= Z_{m1}i_1 + Z_{m2}i_2 + \quad\quad\quad \cdots + Z_{mm}i_m, \end{aligned} \right\} \quad (4)$$

where $i_1, i_2, \ldots, i_m$ are the mesh currents and $v_1, v_2, \ldots, v_m$ are the applied voltages in each mesh. The coefficients Z_{ii} are called the *self-impedances* of the mesh, and a coefficient Z_{ij} is the *mutual impedance* of mesh i to mesh j. It is to be noticed that the directions of the currents and the voltages in the meshes may be chosen in an arbitrary manner. The convention usually adopted is that the relation between the voltage and the current in any one mesh is such that the product $i_j v_j$ represents the power dissipated in the positive real part of Z_{jj}. The signs of the currents are, however, completely arbitrary. This entails a corresponding arbitrariness in the sign of the mutual elements of the network. For certain cases uniform conventions will be adopted for the positive direction of the currents. It should be emphasized that although the signs of the coefficients of the currents are indefinite, the network itself is not, and the results of the calculations are independent of the choice of the positive directions of the currents. The number of equations is, of course, equal to the number of independent meshes, m, and can be determined from the relation

$$m + n = b + 1, \quad (5)$$

where n is the number of nodes and b the number of branches in the network. This relation can be easily proved by a process of mathematical induction. It is observed that this relation is valid regardless of how the number of branches is chosen. Thus, in Fig. 4·1, $EBAC$ may be called one branch ending in the nodes E and C or two branches EBA and AC with nodes at E, A, and C.

Suppose that there are two sets of applied voltages $v_i^{(1)}$ and $v_i^{(2)}$. These voltages may be of the same frequency or of different frequencies. Let $i_i^{(1)}$ and $i_i^{(2)}$ be the corresponding sets of currents. Then, since Eqs. (4) are linear, the currents $i_i^{(1)} + i_i^{(2)}$ are the mesh currents when the applied voltages are $v_s^{(1)} + v_s^{(2)}$. This is known as the *superposition theorem* for linear networks.

The fundamental set of network equations may be established on a node basis rather than on a loop basis as are Eqs. (4). The independent variables are the voltages of the nodes, and the dependent variables the currents flowing in and out of the nodes. The equations thus obtained,

$$\left.\begin{aligned} i_1 &= Y_{11}v_1 + Y_{12}v_2 + \cdots + Y_{1m}v_m, \\ &\quad \vdots \\ i_m &= Y_{m1}v_1 + \phantom{Y_{12}v_2} \cdots + Y_{mm}v_m, \end{aligned}\right\} \quad (6)$$

are the duals of Eqs. (4). The coefficients Y_{ii} are the self-admittances of the network, and Y_{ij} are the mutual admittances.

4·2. The Use of Matrices in Network Theory.—Many of the results of network theory, both in the low- and high-frequency approximations, can be written most conveniently and concisely by the use of matrix notation. For the convenience of the reader, a summary of the rules of matrix manipulation is presented in this section. Eqs. (4) are written as

$$\mathsf{v} = \mathsf{Z}\mathsf{i} \qquad (7)$$

where Z is the impedance matrix of the system. It is a square array of the coefficients of Eqs. (4),

$$\begin{pmatrix} Z_{11} & Z_{12} & Z_{13} & \cdots & Z_{1m} \\ Z_{21} & Z_{22} & \cdot & \cdots & Z_{2m} \\ \cdot & \cdot & \cdot & \cdots & \cdot \\ Z_{m1} & Z_{m2} & \cdot & \cdots & Z_{mm} \end{pmatrix}.$$

The quantities v and i are also matrices consisting of one column only and are more often called "column vectors." They are

$$\mathsf{v} = \begin{pmatrix} v_1 \\ v_2 \\ \cdot \\ \cdot \\ \cdot \\ v_m \end{pmatrix}, \qquad \mathsf{i} = \begin{pmatrix} i_1 \\ i_2 \\ \cdot \\ \cdot \\ \cdot \\ i_m \end{pmatrix}.$$

All matrices are distinguished by the characteristic sans serif type used above; their components are printed in the usual italic type, since they are ordinary scalar quantities.

The operation of combining Z and i is called multiplication. It is defined by the following equation which holds for the multiplication of any two matrices provided the number of columns of the first matrix is equal to the number of rows of the second matrix.

$$A = BC, \qquad A_{ij} = \sum_{n=1}^{m} B_{in}C_{nj}. \tag{8}$$

It should be noted that $CB \neq BC$, in general. When the product is independent of the order of multiplication, the matrices are said to commute. Matrices are equal only if each element of one is equal to the corresponding element of the other. It is apparent that Eq. (7) is the same as Eqs. (4) when the above rules are applied.

Addition, too, may be defined for matrices. The sum of two matrices is the matrix whose elements are the sums of the elements of the two matrices; thus

$$A = B + C,$$
$$A_{ij} = B_{ij} + C_{ij}.$$

It is possible also to define a zero matrix O whose elements are all zero, and the equation

$$A - B = O$$

means that

$$A_{ij} = B_{ij}, \qquad \text{for all } i \text{ and } j.$$

Multiplication of a matrix by a scalar is therefore also defined as

$$cA = (cA_{ij}) = Ac.$$

Likewise, the unit matrix I may be defined, whose elements along the diagonal running from the upper left-hand corner to the lower right-hand corner are unity and whose other elements are zero,

$$I = \begin{pmatrix} 1 & 0 & 0 & \cdots & 0 \\ 0 & 1 & \cdot & \cdots & \cdot \\ 0 & 0 & 1 & \cdots & \cdot \\ \cdot & \cdot & & & \\ \cdot & \cdot & & & \\ 0 & \cdot & \cdot & \cdots & 1 \end{pmatrix}, \qquad \text{or} \qquad I_{ij} = \delta_{ij},$$

where δ_{ij} is the Kronecker delta. For any matrix

$$AI = IA = A.$$

A matrix that has elements only along the diagonal is called a diagonal matrix. Two diagonal matrices always commute with each other.

If Eqs. (4) are solved for the $i_1, i_2, \ldots, i_m$ in terms of $v_1, v_2, \ldots, v_m$, the resulting equations can be written as a matrix equation

$$i = Yv,$$

where Y is the admittance matrix. If this equation is multiplied by Z,

$$Zi = ZYv = v.$$

Thus Y is the reciprocal of Z and can be written $Y = Z^{-1}$. The solutions of Eqs. (4) for i_j in terms of $v_1, v_2, \ldots, v_m$ can be written by Cramer's rule as

$$i_j = \frac{1}{\det(Z)}(v_1 Z^{1j} + v_2 Z^{2j} \cdots v_m Z^{mj}),$$

where det (Z) is the determinant formed from the elements of Z and Z^{ij} is the cofactor of the element Z_{ij} in det (Z). Therefore it is evident that the reciprocal of a matrix can be defined as

$$Z^{-1} = \frac{(Z^{ji})}{\det(Z)}. \qquad (9)$$

The matrix (Z^{ji}) is formed by arranging the cofactors of the elements Z_{ij} in a matrix array and then interchanging the rows and columns. This operation is called transposition and is described by the equation

$$\tilde{Z} = (Z_{ji}), \quad \text{if} \quad Z = (Z_{ij}).$$

Reciprocal matrices have the property

$$AA^{-1} = A^{-1}A = I.$$

Furthermore, if

$$AB = C,$$

then

$$C^{-1} = B^{-1}A^{-1}.$$

Also

$$\tilde{C} = \tilde{B}\tilde{A}.$$

A column vector is also subject to the operation of transposition and the resulting matrix is a row vector. Thus,

$$i = \begin{pmatrix} i_1 \\ i_2 \\ i_3 \\ \cdot \\ \cdot \end{pmatrix}, \qquad \tilde{i} = (i_1, i_2, i_3, \cdots).$$

The product of a column vector and a row vector, taken in that order, is a square matrix

$$a\tilde{b} = A,$$

where

$$A_{ij} = a_i b_j.$$

If the product is taken in the reverse order, the result is a scalar

$$\tilde{a}b = c = \sum_j a_j b_j.$$

4·3. Fundamental Network Theorems.—The fundamental physical principles that form the basis of network theory are embodied in Maxwell's equations for the electromagnetic field. These equations, together with the force equation, the appropriate boundary conditions, and Ohm's law, would be sufficient for all further developments. Network theory is, however, limited to some rather special cases of all those to which the general electromagnetic equations may be applied. It would thus be possible to formulate network theory from several specialized and rather simple theorems or postulates which, in turn, are derivable from the general equations. It will not be attempted here to erect this logical structure, because the problems considered in waveguide networks are more general than those treated by ordinary network theory. The discussion will be confined to a mere statement of the network theorems without a rigorous justification for them. The more general point of view will be adopted in Chap. 5. The choice of theorems that are to be regarded as the primary ones, from which all the others can be derived, and those which are corollaries to the primary theorems is, of course, to some extent arbitrary.

The first network theorem, the *superposition theorem*, was stated in the first section of this chapter. This theorem follows directly from the linearity of Maxwell's equations. The second important theorem is called the *reciprocity theorem*. This theorem is most concisely expressed by the statement that the impedance matrix of a network is symmetrical; the element Z_{ij} is equal to the element Z_{ji}. This theorem follows from the symmetry of Maxwell's equations and will be proved in Chap. 5. Since $Y = Z^{-1}$, it follows that Y is also a symmetric matrix and $Y_{ij} = Y_{ji}$. The reciprocity theorem is often expressed by the rather ambiguous statement that it is possible to interchange the position of a generator and an ammeter without changing the ammeter reading. An inspection of Eqs. (4) will convince the reader that the statement is vague but correct.

The third important network theorem is called *Thévenin's theorem*. This theorem states that a network having two accessible terminals and containing sources of electromotive force may be replaced by an electromotive force in series with an impedance. The magnitude of this electromotive force is that which would exist across the two terminals if they were open-circuited, and the impedance is that presented between the two terminals when all the voltage sources within the network are replaced by their internal impedances. This equivalence is illustrated in

Fig. 4·2. Let V be the open-circuit voltage of the network and Z the impedance looking back into the network when the internal electromotive forces are zero. Let Z_L be the load impedance placed across the network terminals. Then Thévenin's theorem states that the two networks shown in Fig. 4·2 are equivalent. To make this evident, let us imagine that a source of voltage $-V$ is placed in series with the load Z_L. No current will then flow in Z_L. Now, invoking the superposition theorem, this zero current may be considered as composed of two equal and opposite cur-

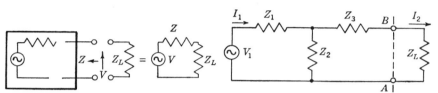

FIG. 4·2.—Equivalent networks demonstrating Thévenin's theorem.

FIG. 4·3.—Circuit to illustrate Thévenin's theorem.

rents, one excited by the external source and the other by the source within the network. However, the value of the former current is $-V/(Z + Z_L)$; therefore, the current through Z_L from the sources within the network is $V/(Z + Z_L)$, which proves the equivalence.

As an example of the application of this theorem, let us consider the circuit shown in Fig. 4·3. The impedance Z looking to the left of AB when V_1 is short-circuited is

$$Z = Z_3 + \frac{Z_1 Z_2}{Z_1 + Z_2}.$$

The open-circuit voltage at AB is

$$V = -V_1 \frac{Z_2}{Z_1 + Z_2}.$$

Hence

$$I_2 = -\frac{V_1 \dfrac{Z_2}{Z_1 + Z_2}}{Z_3 + \dfrac{Z_1 Z_2}{Z_1 + Z_2} + Z_L} = -\frac{V_1 Z_2}{Z_1 Z_2 + Z_2 Z_3 + Z_1 Z_3 + Z_L(Z_1 + Z_2)}.$$

Thévenin's theorem is one of a class of similar relations, each of which is particularly useful in certain applications. Let us consider, for example, a network with two pairs of accessible terminals with the voltages and currents as indicated in Fig. 4·4. The output voltage is

$$V_2 = Z_{12} I_1 + Z_{22} I_2,$$

but

$$Z_L = -\frac{V_2}{I_2},$$

and hence

$$I_2 = -\frac{Z_{12}I_1}{Z_{22} + Z_L}.$$

An equivalent output circuit is, therefore, that shown in Fig. 4·5a. In a similar manner, it is found from the admittance equation that

$$V_2 = -\frac{V_{12}V_1}{Y_{22} + Y_L};$$

therefore, an equivalent circuit can be drawn as in Fig. 4·5b.

Another theorem that is useful in the study of the behavior of networks is the *compensation theorem*. If a network is modified by making a change ΔZ in the impedance of one of its branches, the change in the current at any point is equal to the current that would be produced by an electromotive force in series with the modified branch of $-i\Delta Z$, where i is the current in the branch. This is immediately evident from the superposition theorem, since, if the network is altered by both changing the impedance and inserting a series electromotive force $-i\Delta Z$, no alteration is caused in any of the network currents. Consequently the two alterations have equal and opposite effects. This is a statement of the compensation theorem. It is necessary, however, to consider the special case for which ΔZ is infinite, that is, when the branch is open-circuited. Let the impedance in the Kth branch be Z_K and the current through it I_K. Now let us suppose that Z_K becomes infinite.

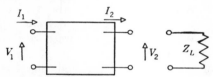

Fig. 4·4.—Two-terminal-pair network.

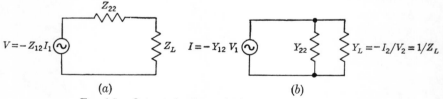

(a)   (b)

Fig. 4·5.—Output circuits equivalent to circuit of Fig. 4·4.

Let Z' be the input impedance of the network at the terminals of Z_K. Then by Thévenin's theorem,

$$I_K = \frac{V'}{Z_K + Z'},$$

where V' is the voltage across the terminals of Z_K when Z_K becomes infinite. If an electromotive force equal to $-V'$ is introduced, the branch may be open-circuited without disturbing the currents in the remainder of the network. Hence the change in the network currents caused by opening a branch is equal to the currents that would be produced by the electromotive force

$$-V' = -I_K(Z_K + Z')$$

acting in series with Z_K.

The several theorems that have been considered are based on the elementary properties of Maxwell's equations. The reciprocity theorem, for example, depends essentially on the symmetry of the electromagnetic equations. On the other hand, the compensation theorem depends on the linearity of the equations. Other fundamental properties of networks, or network theorems, are based fundamentally on the physical principles of energy flow and energy dissipation. If the force equation for the electron is to be regarded as fundamental, these properties can be regarded as derived from that. If a current i flows into the terminals of an impedance Z, across the terminals of which there exists a voltage V, then the complex power associated with the impedance is given by $P = \frac{1}{2}Vi^*$, where the asterisk denotes the complex conjugate. The real part of the complex power is the energy dissipated per second in the impedance; the imaginary part is the difference between the stored magnetic and the stored electric energy. Thus, if Z is purely real,

$$P = \frac{1}{2} V \frac{V^*}{Z} = \frac{1}{2} Z i i^*$$

is purely real and no energy is stored. If Z is purely imaginary, the real part of P is zero and no energy is dissipated. The factor $\frac{1}{2}$ in the expression for complex power comes from the fact that the amplitudes of the currents and voltage are used; that is, the peak values are employed rather than the root-mean-square, rms, values. If the rms values were used, the factor $\frac{1}{2}$ would be absent.

As an example, let us consider a simple circuit consisting of a resistance, an inductance, and a capacitance in series with a generator. The impedance Z is

$$Z = R + j\left(\omega L - \frac{1}{\omega C}\right),$$

and

$$P = \frac{1}{2} Vi^* = \frac{|V|^2}{2Z^*} = \frac{\frac{|V|^2}{2}}{R - j\left(\omega L - \frac{1}{\omega C}\right)}.$$

Rationalized, this becomes

$$P = \frac{1}{2}\left[\frac{|V|^2 R}{R^2 + \left(\omega L - \frac{1}{\omega C}\right)^2} + j \frac{|V|^2 \left(\omega L - \frac{1}{\omega C}\right)}{R^2 + \left(\omega L - \frac{1}{\omega C}\right)^2}\right].$$

The real part is recognized as the energy dissipated in the resistance. The energy stored in the inductance is, from elementary considerations,

$$W_L = \tfrac{1}{2} L |i|^2 \omega,$$

and the electric energy stored in the capacitance is, similarly,

$$W_C = \tfrac{1}{2} C |V_C|^2 \omega,$$

where V_C is the voltage across the capacitance. By use of the relation

$$\frac{V_C}{V} = \frac{\dfrac{-j}{\omega C}}{Z},$$

some algebraic manipulation will show that

$$P = \frac{1}{2} \frac{|V|^2 R}{|Z|^2} + j(W_L - W_C).$$

The theorem on complex power is quite analogous to the complex Poynting energy theorem which was derived in Chap. 2. This is an extremely important point, and in Chap. 5 attention will be devoted to its elaboration in the more general form suitable for application to waveguide circuits as well as to the low-frequency approximation considered here. As might be expected, there is a theorem, analagous to Poynting's energy theorem for real fields, in terms of currents and voltages in which the *sum* of the stored electric and magnetic energies enters. A discussion of this theorem will be found in Chap. 5.

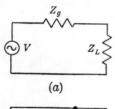

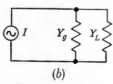

Fig. 4·6.—Circuits for power-transfer relationships.

It is now desired to state the relations for the maximum power transfer from a generator to a load. From Thévenin's theorem it is clear that the generator, however complex its nature, may be represented by an impedance Z_g in series with an ideal generator of zero impedance. The equivalent circuit of the generator with the load Z_L attached is thus simply shown in Fig. 4·6a. The power absorbed in the load is

$$P = \tfrac{1}{2} \text{ Re } (Z_L) |i|^2$$

or

$$P = \frac{1}{2} R_L \frac{V^2}{|Z_g + Z_L|^2} = \frac{1}{2} \frac{R_L V^2}{(R_g + R_L)^2 + (X_g + X_L)^2}.$$

It is easy to see that if the load impedance is varied, the conditions for P to be a maximum are that

$$R_L = R_g$$

and
$$X_L = -X_g.$$
This maximum value is
$$P_{max} = \frac{1}{8}\frac{V^2}{R_L}.$$

It should be noted that here again the factor $\frac{1}{8}$ would become $\frac{1}{4}$ if the rms value of V were used. The factor would, of course, be $\frac{1}{4}$ in the d-c case also.

For current generators, the dual relation is in terms of admittances. The circuit becomes that of Fig. 4·6b. For this case,
$$P = \tfrac{1}{2}\,\mathrm{Re}\,(Y_L)|V|^2,$$
$$P_{max} = \frac{1}{8}\frac{I^2}{G_L},$$
where
$$G_L = G_g, \qquad B_L = -B_g.$$

When the conditions for maximum power transfer are satisfied, the load impedance is said to be the conjugate image impedance. The quantity P_{max} is often called the "available power" of the generator.

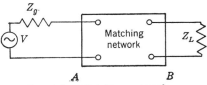

Fig. 4·7.—Matching network.

If the load impedance does not satisfy the conditions for maximum power transfer, a matching network is often inserted between the generator and the load, as shown in Fig. 4·7. If the network is lossless, the condition for maximum power transfer may be applied, with the same result, at either A or B or indeed at some point at the interior of the matching network. If the network is lossy, the two conditions are different, and the proper procedure depends upon considerations of design.

4·4. The Synthesis Problem and Networks with One Terminal Pair.—The problem of finding the properties of a network when its structure and the behavior of its component inductances, condensers, and resistors are known has been considered. The problem inverse to this, and more often encountered in practice, is that of constructing a network having certain specified properties. It will be seen that there are severe limitations on the possible behavior of networks; these limitations are fundamental to network design. Networks have been considered as composed of a number of branches forming complete circuits and containing sources of electromotive force and resistors in which power is dissipated. In general, the purpose of a network is to transfer power from a generator to one or more impedances which absorb the power. It is convenient then to remove the generators and the loads from the network and to consider the properties of the network with a number of accessible terminals separate from any loads and generators that may be connected to the terminals. Networks may thus be classified according to the number of accessible terminals. In the simplest case, there are only two accessible terminals, and the network is described by a single quantity Z_{11}, the self-impedance, or more simply, the impedance. It must be remembered, however, that Z_{11} is in general complex and, so far, there are no restrictions on the frequency dependence of the real and imaginary parts. Because the problem is still restricted to passive networks, $\text{Re}(Z_{11})$ must be greater than or equal to zero. Power is therefore transferred from a generator connected to the terminals only if Z_{11} has a real component.

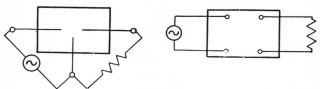

Fig. 4·8.—Network having three accessible terminals.

For networks having more than two accessible terminals, only the transfer of power from a generator connected to one pair of terminals to various loads connected to other terminal pairs is usually of interest. The potential difference between one terminal that is connected to the generator and another that is connected to the load is of no importance. Thus a network with three accessible terminals may be regarded as a two-terminal-pair network, one of the terminals being common to the input and output circuits, as indicated in Fig. 4·8.

Whether or not the two lower terminals shown in the right-hand figure are connected is usually immaterial. Thus a network with n accessible terminals may be regarded as possessing $n - 1$ terminal pairs. The general network equations [Eqs. (4)] may thus be reinterpreted as expressing the linear relations between the currents that flow in and out of each terminal pair and the voltages across the pairs of terminals. The order of the impedance or admittance matrix is thus an index of the complexity of the network under consideration. Networks with one, two, and more terminal pairs will be treated in succession.

The problem is essentially one of synthesis. Given a junction that has N pairs of terminals, the contents of the junction being specified by the elements of the admittance or impedance matrix, it is required to find the possible meshes inside the junctions and the values of the individual elements in the meshes. The problem can be solved in two stages having different degrees of complexity. First, ways must be found to connect individual elements and the impedances of these elements so that the impedance matrix has the proper value at a given frequency. It is relatively easy to do this. Second, the meshes and the elements must be chosen in such a way that the frequency dependence of the impedance matrix is reproduced. The frequency dependence of the elements is usually chosen to be that specified in Eq. (1), since at low frequencies such elements are easily obtainable with ordinary coils and condensers. Such networks are certainly physically realizable. At the higher frequencies, where guided waves are the primary concern, it is not at all certain whether it is advisable or even possible to form equivalent circuits of individual elements whose frequency dependence is given by Eq. (1). In general, therefore, the discussion will be limited to a consideration of the first part of the synthesis problem, namely, that of finding the meshes and the elements at a single frequency. Some important things may, however, be said about the possible dependence on frequency of the equivalent circuits for guided waves.

Let us consider the simplest case of a network with only one terminal pair. As has been already stated, one equivalent network for a first-order impedance matrix is a single impedance having the specified value at the given frequency. However, there may also be several impedances in series or in parallel that together have the required impedance. Thus it is obvious that there is not a unique solution to the problem. An attempt must be made, then, to find the simplest and most convenient solutions. The simplest solution, in this case, is obviously a single impedance element. A more complicated solution is necessary if it is desired to proceed to the second stage in the synthesis problem. If the network must have a complicated frequency dependence, then its composition out of simple elements will be correspondingly complicated. Indeed, it may be that there exists no network, however complicated, composed of simple elements such as R, ωL, $-1/\omega C$, which will give the frequency dependence that is desired. Only certain variations with frequency can be obtained with physically realizable elements. A discussion of this question resolves itself into two parts, and it is convenient to treat first the case where the impedance Z_{11} is purely imaginary.

The restriction on the variation with frequency for this case is known as *Foster's theorem*, and this theorem may well be classed with the other network theorems which have been called fundamental. Foster's theorem states that the impedance, or admittance, of a lossless network with two accessible terminals must always increase with increasing frequency. This is obviously true for the simple element given in Eq. (1). It is also true, as is shown in Chap. 5, for the general case of guided waves. A simple proof for the low-frequency case[1] will be presented here. If

[1] A. T. Starr, *Electric Circuits and Wave Filters*, 2d ed., Pitman, London, 1944, Appendix X, p. 463.

the value of the impedance Z_r in Fig. 4·9a is changed by an amount ΔZ_r, new currents will flow in the network. By the compensation theorem, these new currents will be equal to those which would be caused by the generator shown in the equivalent circuit of Fig. 4·9b. The original admittance of the network was $Y = I/V$, and the electromotive force V

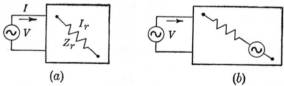

FIG. 4·9.—Equivalent circuit demonstrating Foster's theorem.

produced a current I_r in the branch r. Hence, by the reciprocity theorem, an electromotive force of $-I_r\Delta Z_r$ in the branch r will produce an input current of

$$\frac{-I_r \Delta Z_r}{V} I_r.$$

The change in the admittance is therefore

$$\Delta Y = -\sum_r \frac{I_r^2 \Delta Z_r}{V^2}.$$

If the network is purely reactive, I_r will be of the form $j\alpha_r V$, where α_r is real, and ΔZ_r must be $jk_r\Delta\omega$, where k_r is real and positive. Hence

$$\frac{\Delta Y}{\Delta \omega} = \sum_r j\alpha_r^2 k_r.$$

Since this is greater than zero, the theorem is proved. At zero frequency, the network must be either a pure inductance or a capacitance, and the admittance must therefore be either zero or minus infinity. Since the admittance must increase with frequency, it will increase to infinity,

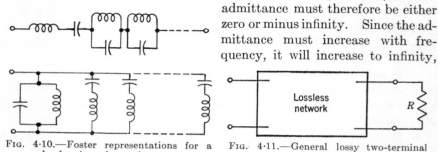

FIG. 4·10.—Foster representations for a lossless impedance element.

FIG. 4·11.—General lossy two-terminal network.

then change to minus infinity, and increase through zero to plus infinity again, repeating the process perhaps many times. From the duality principle it is clear that this must be true also for the impedance. This dual relationship is illustrated by the two possible equivalent circuits shown in Fig. 4·10. It should be emphasized that by means of networks of this form, the most general frequency dependence that is possible with elements in the simple form of inductances and condensers can be realized.

If the network contains resistances, the above statements are no longer true. Darlington[1] has shown that the most general lossy two-terminal network can be represented by the circuit shown in Fig. 4·11. Here the resistance R is independent of frequency, and the four-terminal network contains no lossy elements. For a further discussion of this important and interesting aspect of circuit theory the reader is referred to Vol. 9, Chap. 9.

4·5. The Circuit Parameters of Two-terminal-pair Networks.—Perhaps the most important type of network for microwave applications is that for which the impedance matrix is of the second order. The network has two pairs of terminals, and three independent parameters are necessary to describe its properties. The importance of such networks lies chiefly in the fact that they are readily connected in cascade or with lengths of transmission lines to form a transmission system. The whole assembly can then be reduced to an equivalent linear device represented, again, by three parameters. A common name for a two-terminal-pair network is "transducer." The impedance matrix is of the form

$$\mathbf{Z} = \begin{pmatrix} Z_{11} & Z_{12} \\ Z_{12} & Z_{22} \end{pmatrix}. \tag{10}$$

The simplest equivalent circuit for this is the familiar T-network shown in Fig. 4·12. The positive directions of the currents and voltages that

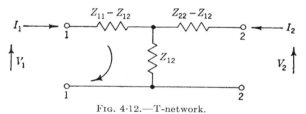

Fig. 4·12.—T-network.

will be adopted are indicated on the figure. It should be noticed that if the currents and voltages have positive values, power is flowing into the network at both pairs of terminals. The values of the impedances shown on the figure are easily derived. For example, consider the mesh starting at terminals (1) and indicated by the arrow in Fig. 4·12. If the voltage drops around this mesh are added, it is found that

$$V_1 = (Z_{11} - Z_{12})I_1 + Z_{12}I_1 + Z_{12}I_2, \tag{11}$$

[1] Sidney Darlington, "Synthesis of Reactance 4-Poles," *J. Math. Phys.*, **18**, (1939).

and V_2 may be found in a similar fashion. The current in the first mesh flows through the impedance Z_{12} in the same direction as the current in the second mesh. Thus if the shunt impedance in the network is positive, the mutual impedance element in the matrix is also positive. This is the most cogent reason for the choice that has been made of the positive directions of the current. The choice of the positive directions will be made, whenever possible, so that if power is flowing into all terminals of a complicated network, all the impedance or admittance matrix elements are positive definite if the corresponding network elements are also positive definite.

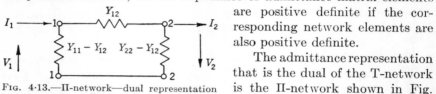

Fig. 4·13.—Π-network—dual representation of T-network.

The admittance representation that is the dual of the T-network is the Π-network shown in Fig. 4·13. It will be noticed that the direction of I_2 and also the direction of V_2 have been reversed in this case. In this way the requirements for positive matrix elements are satisfied. It is easy to verify the fact that the correct values have been assigned to the circuit elements by considering the currents flowing in and out of the node at the upper terminal (1). Thus

$$I_1 = (Y_{11} - Y_{12})V_1 + Y_{12}(V_1 + V_2), \tag{12}$$

and a similar expression exists for I_2.

It is possible also to find the admittance matrix corresponding to the T-network by the relation

$$\mathsf{Y} = \mathsf{Z}^{-1}. \tag{13}$$

The elements of Y may be easily found from the definition of a reciprocal matrix given in Sec. 4·2,

$$\left. \begin{array}{l} Y_{11} = \dfrac{Z_{22}}{D}, \\[4pt] Y_{12} = -\dfrac{Z_{12}}{D}, \\[4pt] Y_{22} = \dfrac{Z_{11}}{D}, \end{array} \right\} \tag{14}$$

where

$$D = Z_{11}Z_{22} - Z_{12}^2. \tag{15}$$

It will be noticed from the second of Eqs. (14) that if Z_{12} is positive, Y_{12} is negative. This arises because the positive direction of the currents shown by the network of Fig. 4·12 would not be correct for an admittance representation. In a similar way, it is possible to write

$$Z_{11} = \frac{Y_{22}}{\Delta},$$
$$Z_{12} = -\frac{Y_{12}}{\Delta}, \qquad (16)$$
$$Z_{22} = \frac{Y_{11}}{\Delta},$$
$$\Delta = Y_{11}Y_{22} - Y_{12}^2.$$

Moreover,
$$D\Delta = 1. \qquad (17)$$

These relations will be found useful in many ways.

A most important relation is the one giving the value of the input impedance of the network when the network is terminated by an arbitrary load Z_L. The second relation between the voltages and currents is given by
$$V_2 = Z_{12}I_1 + Z_{22}I_2, \qquad (18)$$
and it is required that
$$\frac{V_2}{I_2} = -Z_L \qquad (19)$$
or
$$I_2 = -\frac{Z_{12}}{Z_{22} + Z_L} I_1. \qquad (20)$$

Hence the input impedance is given by
$$Z_{\text{in}} = \frac{V_1}{I_1} = Z_{11} - \frac{Z_{12}^2}{Z_{22} + Z_L}. \qquad (21)$$

By the duality principle, therefore,
$$Y_{\text{in}} = Y_{11} - \frac{Y_{12}^2}{Y_{22} + Y_L}. \qquad (22)$$

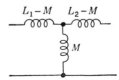

Fig. 4·14.—T-network equivalent of loosely coupled coils.

It should be noticed that the value of the input impedance (or admittance) is independent of the directions of the currents; this is guaranteed by the fact that it depends on Z_{12}^2 (or Y_{12}^2).

The parameter that is sometimes used to express the degree of coupling between the circuit connected to terminals (1) and that connected to terminals (2) is the coefficient of coupling k which is defined by
$$k = \frac{Z_{12}}{\sqrt{Z_{11}Z_{22}}}. \qquad (23)$$

This parameter is most commonly used in connection with loosely coupled coils. The equivalent circuit of this device in the T-network representation is shown in Fig. 4·14. The value for k in this case is

$$k = \frac{M}{\sqrt{L_1 L_2}}. \tag{24}$$

The parameter k thus has some value between zero and unity.

For many cases, the most useful circuit parameters are the input impedances of the network when the output end is open- or short-circuited. The impedance looking into terminals (1) when terminals (2) are short-circuited will be denoted by $Z_{\text{sc}}^{(1)}$; correspondingly, when terminals (2) are open-circuited, the input impedance is $Z_{\text{oc}}^{(1)}$. The superscript (2) will be used to denote the corresponding quantities for terminals (2). From Eq. (21) it is clear that

$$\left. \begin{array}{l} Z_{\text{oc}}^{(1)} = Z_{11}, \\[4pt] Z_{\text{sc}}^{(1)} = Z_{11} - \dfrac{Z_{12}^2}{Z_{22}}, \\[8pt] Z_{\text{oc}}^{(2)} = Z_{22}, \\[4pt] Z_{\text{sc}}^{(2)} = Z_{22} - \dfrac{Z_{12}^2}{Z_{11}}. \end{array} \right\} \tag{25}$$

Also
$$\frac{1}{Y_{11}} = \frac{D}{Z_{22}} = Z_{\text{sc}}^{(1)}. \tag{26}$$

Then from Eq. (14) it is noted that

$$Y_{11} Z_{11} = Y_{22} Z_{22}. \tag{27}$$

In terms of the short- and open-circuit parameters, this relation becomes

$$\frac{Z_{\text{sc}}^{(1)}}{Z_{\text{oc}}^{(1)}} = \frac{Z_{\text{sc}}^{(2)}}{Z_{\text{oc}}^{(2)}}, \tag{28}$$

and by the duality principle

$$\frac{Y_{\text{sc}}^{(1)}}{Y_{\text{oc}}^{(1)}} = \frac{Y_{\text{sc}}^{(2)}}{Y_{\text{oc}}^{(2)}}. \tag{29}$$

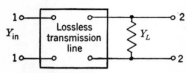

Fig. 4·15.—Transformation of an admittance by adding length of transmission line.

A good example of the usefulness of these relations is afforded by the following. It is easy to calculate the expression for the transformation of an admittance that occurs when a length of transmission line is added to it. Let terminals (2) be chosen as indicated in Fig. 4·15. Application of Eq. (29) to the combination of the admittances of the line and of the shunt element results in

$$\frac{Y_{\text{in}}}{-jY_0 \coth x} = \frac{jY_0 \tanh x + Y_L}{-jY_0 \coth x + Y_L}, \tag{30}$$

which can be reduced to the usual form

$$Y_{in} = Y_0 \frac{Y_L + jY_0 \tanh x}{Y_0 + jY_L \tanh x}. \tag{31}$$

Other useful parameters are the coefficients in the set of linear equations that relate the input current and voltage to the output current and voltage. They are given by

$$\left.\begin{array}{l} V_1 = \mathfrak{A} V_2 - \mathfrak{B} I_2, \\ I_1 = \mathfrak{C} V_2 - \mathfrak{D} I_2. \end{array}\right\} \tag{32}$$

The parameters are related to the impedance matrix elements by

$$\left.\begin{array}{ll} \mathfrak{A} = \dfrac{Z_{11}}{Z_{12}}, & \mathfrak{B} = \dfrac{D}{Z_{12}}, \\ \mathfrak{C} = \dfrac{1}{Z_{12}}, & \mathfrak{D} = \dfrac{Z_{22}}{Z_{12}}. \end{array}\right\} \tag{33}$$

There must be a relation that corresponds to Eq. (28) and that also expresses the reciprocity condition. This has the form

$$\begin{vmatrix} \mathfrak{A} & \mathfrak{B} \\ \mathfrak{C} & \mathfrak{D} \end{vmatrix} = 1. \tag{34}$$

If $\begin{pmatrix} V_1 \\ I_1 \end{pmatrix}$ and $\begin{pmatrix} V_2 \\ -I_2 \end{pmatrix}$ are chosen as the column vectors, then Eq. (34) is the determinate of the matrix. If Eq. (32) is solved for V_2 and I_2, it can be written, in matrix form, as

$$\begin{pmatrix} V_2 \\ I_2 \end{pmatrix} = \begin{pmatrix} \mathfrak{D} & \mathfrak{B} \\ \mathfrak{C} & \mathfrak{A} \end{pmatrix} \begin{pmatrix} V_1 \\ -I_1 \end{pmatrix}. \tag{35}$$

The elements of the matrix are the same as those of Eqs. (32), but in a different order, and the determinant of the matrix of Eq. (35) is equal to unity. This set of parameters has been in use for a long time, particularly for applications involving power transmission lines. From Eqs. (33) it is immediately evident that if the network is lossless and the Z's are all pure imaginary, the elements $\mathfrak{A}$ and $\mathfrak{D}$ are pure real and $\mathfrak{B}$ and $\mathfrak{C}$ are pure imaginary. If the network is symmetrical, $Z_{11} = Z_{22}$ and $\mathfrak{A} = \mathfrak{D}$.

The utility of this set of parameters becomes more obvious when networks connected in cascade are to be considered. This situation will be treated in more detail in the following section. An important example of the use of these parameters is afforded by the case of the ideal transformer. The matrix of a transformer takes the form

$$\begin{pmatrix} \dfrac{1}{n} & 0 \\ 0 & n \end{pmatrix}, \tag{36}$$

where n is the turn ratio of the transformer, the direction of voltage stepup being from terminals (1) to terminals (2). No impedance or admittance matrix exists for the ideal transformer. The elements all become infinitely large, and therefore the series impedances of the equivalent T-network are indeterminate.

Three important sets of parameters and their duals have just been presented which may be used to specify completely the properties of a general, passive, linear network with two pairs of terminals. The set that is most convenient to use depends on the particular application in question.

4·6. Equivalent Circuits of Two-terminal-pair Networks.—Another method of describing the behavior of a network is by means of an equivalent circuit. It has been shown that three parameters are necessary for the complete specification of a network with two pairs of terminals. The equivalent circuit must therefore contain at least three circuit elements. There is, of course, no unique equivalent circuit but an infinite number of them. Moreover, they may contain more than three circuit elements. Two examples have already been given—the familiar T- and Π-representations. For microwave applications other representations are also useful. Portions of a transmission line have been introduced as circuit elements. In Chap. 3, lines of this type were discussed, and the Π- and T-equivalents for such lines were given there. These lines will now be considered as convenient circuit elements, and the electrical length and characteristic impedance to define their properties will be specified. Although this could be done for the general case of lossy transmission lines, the discussion will be confined to the case where the lines are lossless, since these are by far the most important cases for microwave applications.

A simple case of such equivalent circuits is demonstrated by the circuits shown in Fig. 4·16. These circuits consist of transmission lines, one shunted by an arbitrary admittance and the other having an arbitrary impedance in series. The three parameters are thus the value of this admittance Y, the length of the line l, and its characteristic admittance Y_0. The line with the series impedance might be termed the dual representation. These two circuits are duals of each other in the sense that the relation between Y and the elements of the *admittance* matrix corresponding to one circuit is identical with the relation between Z and the *impedance* matrix elements that describe the other circuit. Not all of the important circuits of this type will be discussed in detail, but several are shown in Figs. 4·16 to 4·23. The relations between the cir-

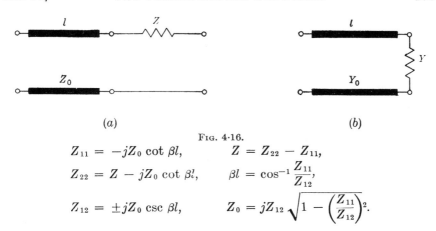

Fig. 4·16.

$$Z_{11} = -jZ_0 \cot \beta l, \qquad Z = Z_{22} - Z_{11},$$
$$Z_{22} = Z - jZ_0 \cot \beta l, \qquad \beta l = \cos^{-1}\frac{Z_{11}}{Z_{12}},$$
$$Z_{12} = \pm jZ_0 \csc \beta l, \qquad Z_0 = jZ_{12}\sqrt{1 - \left(\frac{Z_{11}}{Z_{12}}\right)^2}.$$

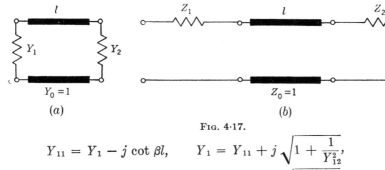

Fig. 4·17.

$$Y_{11} = Y_1 - j \cot \beta l, \qquad Y_1 = Y_{11} + j\sqrt{1 + \frac{1}{Y_{12}^2}},$$
$$Y_{22} = Y_2 - j \cot \beta l, \qquad Y_2 = Y_{22} + j\sqrt{1 + \frac{1}{Y_{12}^2}},$$
$$Y_{12} = \pm j \csc \beta l, \qquad \beta l = \csc^{-1}(\mp jY_{12}) = \sin^{-1}\left(\frac{\pm j}{Y_{12}}\right).$$

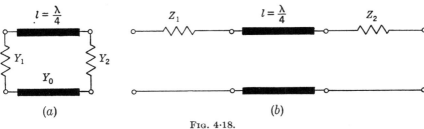

Fig. 4·18.

$$Y_{11} = Y_1, \qquad Y_1 = Y_{11},$$
$$Y_{22} = Y_2, \qquad Y_2 = Y_{22},$$
$$Y_{12} = jY_0, \qquad Y_0 = -jY_{12}.$$

cuit elements and the impedance, or admittance, matrix elements are given in the legends. Each set of equations given in the figure legend refers to the (a) circuit, which is shown on the left of each figure; the

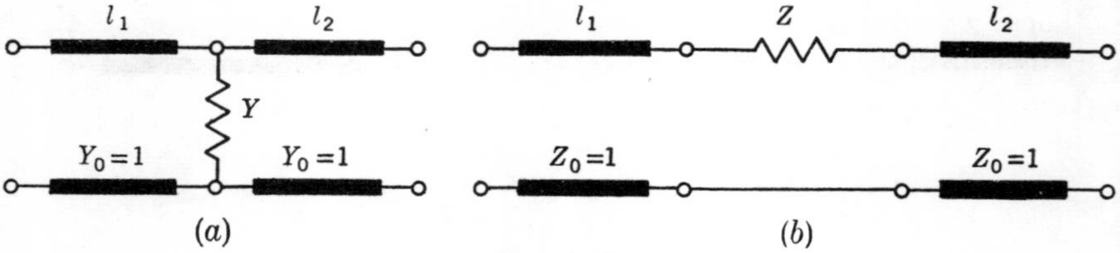

FIG. 4·19.

$$Y_{11} = j\frac{1 - \cot\beta l_1 \cot\beta l_2 - jY \cot\beta l_1}{\cot\beta l_1 + \cot\beta l_2 + jY},$$

$$Y_{22} = j\frac{1 - \cot\beta l_1 \cot\beta l_2 - jY \cot\beta l_2}{\cot\beta l_1 + \cot\beta l_2 + jY},$$

$$Y_{12} = \frac{j \csc\beta l_1 \csc\beta l_2}{\cot\beta l_1 + \cot\beta l_2 + jY}.$$

Symmetrical Case Only

$$Y = \frac{1}{Y_{12}[1 - (Y_{11} + Y_{12})^2]} + 2(Y_{11} + Y_{12}),$$

$$\beta l = \cot^{-1}(-jY_{11} - jY_{12}).$$

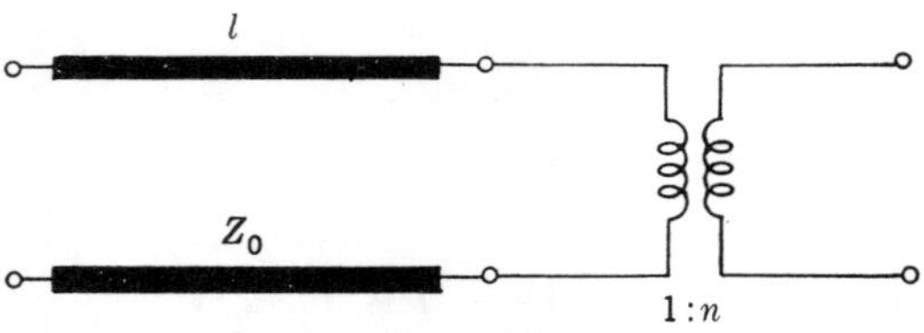

FIG. 4·20.

$$Z_{11} = -jZ_0 \cot\beta l, \qquad \beta l = \cos^{-1}\frac{\sqrt{Z_{11}Z_{22}}}{Z_{12}},$$

$$Z_{22} = -jn^2Z_0 \cot\beta l, \qquad Z_0 = -jZ_{11}\sqrt{\frac{Z_{12}^2}{Z_{11}Z_{22}} - 1},$$

$$Z_{12} = jnZ_0 \csc\beta l, \qquad n = \sqrt{\frac{Z_{22}}{Z_{11}}}.$$

circuit (b), on the right, is the dual circuit. Thus in the legend of Fig. 4·17 the relation

$$Y_{11} = Y_1 - j\cot\beta l$$

is given. The corresponding equation for the dual circuit is

$$Z_{11} = Z_1 - j\cot\beta l,$$

which is seen to be identical in form with the equation for Y_{11}. In a

similar way,
$$Z_{12} = j \csc \beta l, \qquad (37)$$
$$Z_{22} = Z_2 - j \cot \beta l.$$

These circuits are of importance for microwave applications principally because they can be reduced to very simple circuits by adding portions of transmission lines. This corresponds to shifting the planes

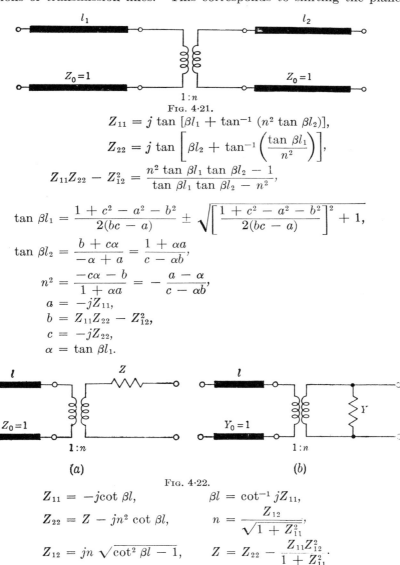

Fig. 4·21.

$$Z_{11} = j \tan [\beta l_1 + \tan^{-1}(n^2 \tan \beta l_2)],$$
$$Z_{22} = j \tan \left[\beta l_2 + \tan^{-1}\left(\frac{\tan \beta l_1}{n^2}\right) \right],$$
$$Z_{11}Z_{22} - Z_{12}^2 = \frac{n^2 \tan \beta l_1 \tan \beta l_2 - 1}{\tan \beta l_1 \tan \beta l_2 - n^2},$$
$$\tan \beta l_1 = \frac{1 + c^2 - a^2 - b^2}{2(bc - a)} \pm \sqrt{\left[\frac{1 + c^2 - a^2 - b^2}{2(bc - a)}\right]^2 + 1},$$
$$\tan \beta l_2 = \frac{b + c\alpha}{-\alpha + a} = \frac{1 + \alpha a}{c - \alpha b},$$
$$n^2 = \frac{-c\alpha - b}{1 + \alpha a} = -\frac{a - \alpha}{c - \alpha b},$$
$$a = -jZ_{11},$$
$$b = Z_{11}Z_{22} - Z_{12}^2,$$
$$c = -jZ_{22},$$
$$\alpha = \tan \beta l_1.$$

(a)   (b)

Fig. 4·22.

$$Z_{11} = -j\cot \beta l, \qquad\qquad \beta l = \cot^{-1} jZ_{11},$$
$$Z_{22} = Z - jn^2 \cot \beta l, \qquad n = \frac{Z_{12}}{\sqrt{1 + Z_{11}^2}},$$
$$Z_{12} = jn \sqrt{\cot^2 \beta l - 1}, \qquad Z = Z_{22} - \frac{Z_{11}Z_{12}^2}{1 + Z_{11}^2}.$$

of reference in a waveguide circuit to those points which are most convenient for the purpose at hand. Thus in Fig. 4·21, if an additional line length is added to each end, so that the total length of each of the transmission lines connected to the transformer is an integral multiple of a half wavelength, then the transmission lines may be omitted entirely and the circuit reduces to that of an ideal transformer alone. The circuits shown in Fig. 4·18 may be useful in finding the arrangement of waveguide or coaxial line necessary to reproduce a given T-network representation. The impedance of a coaxial line can be easily adjusted, and the shunt susceptance may be introduced by diaphragms. This

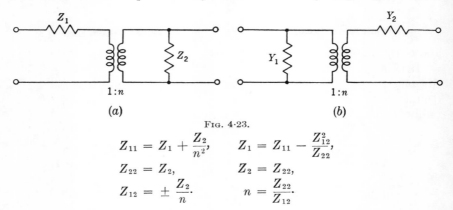

Fig. 4·23.

$$Z_{11} = Z_1 + \frac{Z_2}{n^2}, \qquad Z_1 = Z_{11} - \frac{Z_{12}^2}{Z_{22}},$$

$$Z_{22} = Z_2, \qquad Z_2 = Z_{22},$$

$$Z_{12} = \pm \frac{Z_2}{n}. \qquad n = \frac{Z_{22}}{Z_{12}}.$$

question can be discussed more completely after the equivalence of certain waveguide configurations to their networks has been shown. The circuit shown in Fig. 4·19 depicts another case in which an extremely simple circuit can result from the addition of lengths of transmission lines. It is evident, therefore, that a general two-terminal-pair network can be reduced either to a simple shunt element or to a simple series element. Thus the concepts of "shunt" and "series" lose much of their significance in transmission-line applications. As a corollary to this, it is easily seen that a pure shunt element is equivalent to a pure series element plus a transmission line one-quarter wavelength long. Likewise, a pure series element is equivalent to a pure shunt element plus a quarter-wavelength line.

The existence of the circuit shown in Fig. 4·21 is a sound justification for a terminology that was introduced as slang. Any device, such as a diaphragm or a screw, introduced into a length of waveguide was spoken of as a "transformer." The meaning that this phrase was intended to convey was merely that the diaphragm could *change* the amount of reflected and transmitted energy in the line. Figure 4·21 shows that this expression can be interpreted quite literally, and the turn ratio of the equivalent ideal transformer can be calculated in any given case. The transformer is not located at the position of the diaphragm, but at some other place along the line.

The equivalent circuit of Fig. 4·21 is particularly useful for interpreting the measured properties of a waveguide junction.[1] The turn ratio of the transformer is numerically equal to the voltage standing-wave ratio at the input terminals when the output terminals are connected to a matched transmission line. There is a simple relation between the position of a short circuit in the output-terminal line and the equivalent short circuit in the line connected to the input terminals. It is

$$\tan \beta(l_3 - l_1) = n \tan \beta(l_4 - l_2), \tag{38}$$

where l_3 and l_4 represent the distances from the reference planes and l_1 and l_2 are the network parameters.

It is important to notice one fact about all these equivalent circuits. Although, at a given frequency, the elements of the circuit are perfectly definite and can be represented by circuit elements familiar to low-frequency practice, these elements do not have the proper variation with

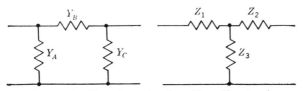

Fig. 4·24.—Transformation from Π- to T-network.

frequency. A circuit containing negative susceptances may be considered, at a single frequency, to be made up of inductances. These inductances at low frequencies have susceptances inversely proportional to the frequency. In the waveguide equivalent of this circuit, the inductances may have an arbitrary frequency variation. This serves merely to emphasize the fact that equivalent circuits are purely artificial devices and do not completely correspond to reality. In particular cases it is possible to find equivalent circuits that represent the waveguide configurations not only at one frequency but, to a good approximation, at a whole range of frequencies, provided that this range is less than one octave.

Perhaps the most useful transformation from one equivalent circuit to another is the familiar one from a Π- to a T-network. Let the circuit elements be designated as indicated in Fig. 4·24. The relations between them are given by

---

[1] N. Marcuvitz, "Waveguide Handbook Supplement," RL Report No. 41, Jan. 23, 1945, p. 2.

$$Z_1 = \frac{Y_C}{D'},$$
$$Z_2 = \frac{Y_A}{D'},$$
$$Z_{12} = \frac{Y_B}{D'},$$
(39)

where
$$D' = Y_A Y_B + Y_A Y_C + Y_B Y_C. \qquad (40)$$

Since these circuits are duals of one another, the inverse relationships are identical in form. For example,
$$Y_B = \frac{Z_3}{Z_1 Z_2 + Z_1 Z_3 + Z_2 Z_3}. \qquad (41)$$

4·7. Symmetrical Two-terminal-pair Networks.—Many waveguide configurations are symmetrical about some plane perpendicular to the axis of the transmission line. If this is the case, the input and output terminals are indistinguishable, and the number of independent parameters needed to specify the network is reduced from three to two. In the matrix representation, Z_{11} becomes identical with Z_{22}. The circuits shown in Figs. 4·16 and 4·20 reduce simply to a transmission line with the usual parameters: length and characteristic impedance. The relationship between a symmetrical two-terminal-pair network and a line is well known, and it will not be considered further here. We can state, however, a useful theorem known as the "bisection theorem."[1] This theorem can be formulated in a somewhat simpler form than that in which it was originally stated. If equal voltages are applied to the terminal pairs (1) and (2) of a symmetrical network, equal currents will flow into the two pairs of terminals and no current will flow across the plane of symmetry. The input impedance is then simply $(Z_{11} + Z_{12})$. This may be called the open-circuit impedance of half the network, $Z_{oc}^{(\frac{1}{2})}$. If equal voltages are applied to the two pairs of terminals but in opposite directions, the voltage across the center line of the network must be zero and the currents entering the terminals equal and opposite. The input impedance under these conditions is $(Z_{11} - Z_{12})$. This impedance is written as $Z_{sc}^{(\frac{1}{2})}$. These two values of input impedance are convenient ones to use, in some cases, to specify a symmetrical network.

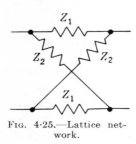

Fig. 4·25.—Lattice network.

A good example of the application of this theorem is the lattice form of network shown in Fig. 4·25. If equal voltages are applied to the two ends of the lattice, no current will flow in the impedance Z_1. Hence
$$Z_{oc}^{(\frac{1}{2})} = Z_2 = Z_{11} + Z_{12}. \qquad (42)$$

[1] A. C. Bartlett, *The Theory of Electrical Artificial Lines and Filters*, Wiley, New York, 1930, p. 28.

Likewise, if equal but opposite voltages are applied to the terminals, no current will flow in the impedance Z_2 and

$$Z_{sc}^{(1/2)} = Z_1 = Z_{11} - Z_{12}. \tag{43}$$

From these equations, the matrix parameters corresponding to the lattice case may be found,

$$\left.\begin{aligned} Z_{11} &= \frac{Z_1 + Z_2}{2}, \\ Z_{12} &= \frac{Z_2 - Z_1}{2}. \end{aligned}\right\} \tag{44}$$

The bisection theorem as originally stated by Bartlett was phrased in terms of cutting the symmetrical network into two equal parts. The theorem stated the values of the input impedance of half of the network when the terminals exposed by this bisection were either open- or short-circuited. The example of the lattice network has been given because, for this case, it is difficult to see just how the network should be divided. The derivation that involves the application of two sources of potential avoids this difficulty. The lattice network is particularly suitable for theoretical investigations of the properties of low-frequency networks and has been much used for this purpose. It can be shown that the lattice equivalent of any four-terminal network is physically realizable in the lattice form. "Physically realizable" means, in this case, that it is unnecessary to use any negative inductances or capacitances to construct the lattice. The lattice form, on the other hand, is quite unsuitable for the construction of practical networks at low frequencies, since no portion is grounded and the inevitable interaction between the elements of the network destroys its usefulness. This is not true of microwave applications. A configuration of conductors that can be reduced to the lattice form may well be a practical microwave circuit. As will be shown later, a magic T with appropriate impedances connected to two of the arms is a lattice circuit having the other two arms as the input and output terminals.

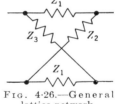

FIG. 4·26.—General lattice network.

For a network to be symmetrical it is, of course, not necessary that the arrangement of components be symmetrical. Thus, the more general lattice shown in Fig. 4·26 has a symmetrical T-network whose matrix elements are given by

$$\left.\begin{aligned} Z_{11} = Z_{22} &= \frac{(Z_1 + Z_2)(Z_1 + Z_3)}{(Z_1 + Z_2) + (Z_1 + Z_3)}, \\ Z_{12} &= \frac{Z_1^2 - Z_2 Z_3}{(Z_1 + Z_2) + (Z_1 + Z_3)}. \end{aligned}\right\} \tag{45}$$

The circuit shown in Fig. 4·21 also has no obvious symmetry when the input and output terminals are identical. The matrix elements reduce to

$$\left.\begin{aligned}Z_{11} &= -\frac{(n^2 + 1)\tan \beta l_1}{n^2 + \tan^2 \beta l_1}, \\ Z_{11}^2 - Z_{12}^2 &= \frac{n^2 + 1}{n^2 - 1},\end{aligned}\right\} \quad (46)$$

for this case. The three network parameters given in Fig. 4·21 must be subject to one condition. This condition is that

$$\tan \beta l_1 = -\cot \beta l_2. \quad (47)$$

There are many other useful equivalent circuits of more complicated forms which will not be discussed here in detail.[1]

4·8. Chains of Four-terminal Networks.—The great utility of the theory of the two-terminal-pair network lies in the fact that complicated transmission lines can be regarded as composed of a number of such networks connected in cascade. The transmission line can then be treated as a whole, or a small part of it can be reduced to a new T-network with the proper values of the network parameters. For two

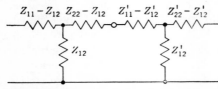

FIG. 4·27.—Two-terminal networks in cascade.

T-networks in cascade, as shown in Fig. 4·27, the matrix elements of the combination are given by

$$\left.\begin{aligned}Z_{11}^{(t)} &= Z_{11} - \frac{Z_{12}^2}{Z_{12} + Z_{11}'}, \\ Z_{12}^{(t)} &= \frac{Z_{12}Z_{12}'}{Z_{22} + Z_{11}'}, \\ Z_{22}^{(t)} &= Z_{22} - \frac{Z_{12}'^2}{Z_{22} + Z_{11}'},\end{aligned}\right\} \quad (48)$$

where the superscript t refers to the combination. The process of combination can be continued to any extent, and the whole transmission line reduced to an equivalent network with three parameters. The utility of the $\mathcal{ABCD}$ matrix is evident when it is applied to this problem. If the constants of the first network are denoted by unprimed letters and those of the second by primed ones, then by direct substitution it is found that

$$\begin{pmatrix} V_1 \\ I_1 \end{pmatrix} = \begin{pmatrix} \mathcal{A} & \mathcal{B} \\ \mathcal{C} & \mathcal{D} \end{pmatrix} \times \begin{pmatrix} \mathcal{A}' & \mathcal{B}' \\ \mathcal{C}' & \mathcal{D}' \end{pmatrix} \times \begin{pmatrix} V_2 \\ -I_2 \end{pmatrix}. \quad (49)$$

---

[1] The reader is referred to the very useful appendices in K. S. Johnson, *Transmission Circuits for Telephonic Communication*, Van Nostrand, New York, 1943.

It is evident from this that the matrix of the combination is equal to the product of the matrices of the components.

Let us consider an infinite chain of identical networks. At any pair of terminals the impedance seen looking in either direction must be independent of the particular pair of terminals chosen. It must be given by

$$Z = Z_{11} - \frac{Z_{12}^2}{Z_{22} + Z}. \tag{50}$$

The solution of this expression is

$$Z = \tfrac{1}{2}(Z_{11} - Z_{22}) \pm \sqrt{\tfrac{1}{4}(Z_{11} + Z_{22})^2 - Z_{12}^2}. \tag{51}$$

Since the networks are identical, this impedance is commonly called the *iterative impedance*. The two signs before the square root refer to the two values of the impedance seen in opposite directions from the pair of terminals. These impedances are alternative parameters with which to describe the network behavior. Moreover, the ratio of input to output currents in any network in the chain is given by

$$Z_{12}I_n - (Z_{22} + Z)I_{n+1} = 0, \tag{52}$$

where the negative sign arises from the convention, earlier established, that the currents always flow into the network at the upper terminals. The third network parameter is then defined by

$$\frac{I_{n+1}}{I_n} = e^{-\Gamma} \tag{53}$$

where Γ is called the *iterative propagation constant* or the *transfer constant* of the network. It is given by

$$e^{-\Gamma} = \frac{-(Z_{11} + Z_{22}) \pm \sqrt{(Z_{11} + Z_{22})^2 - 4Z_{12}^2}}{2Z_{12}}. \tag{54}$$

This result may be rewritten in a much neater form as

$$\cosh \Gamma = \frac{Z_{11} + Z_{22}}{2Z_{12}}. \tag{55}$$

A chain of networks of this sort is thus somewhat analogous to a transmission line but is an unsymmetrical case. The voltage across any pair of terminals either decreases or increases in a constant ratio along the network, corresponding in a way to waves propagating either to the right or to the left.

The iterative impedance has another significance that is sometimes useful. A two-terminal-pair network may be regarded as a transformation in the complex plane. If the load impedance is represented as a point on the complex Z_L-plane, the input impedance is a point on the

Z_{in}-plane that is related to the first by the transformation given by

$$Z_{\text{in}} = \frac{aZ_L + b}{cZ_L + d}.$$

Thus, the output plane may be said to be mapped onto the input plane by this transformation. Transformations of this form are called bilinear transformations or linear-fractional transformations. They have the important property that they are conformal; that is, angles in one plane are transformed to equal angles in the other plane. Thus a grid of perpendicular intersecting lines is transformed to two sets of circles that are mutually orthogonal. The iterative impedance, as is evident directly from Eq. (50), is represented by the point whose coordinates are unchanged by the transformation. It is thus sometimes referred to as the *fixed point*.

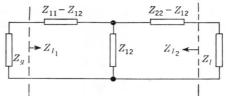

Fig. 4·28.—Networks connected in cascade on the image basis.

Another very common method of connecting networks in cascade is shown in Fig. 4·28. If the impedances connected to the network satisfy the relations

$$\left.\begin{array}{l} Z_G = Z_{I_1}, \\ Z_L = Z_{I_2}, \end{array}\right\} \tag{56}$$

then the network is said to be connected on an image-impedance basis. If Eqs. (56) are evaluated in terms of the network parameters, Z_{I_1} and Z_{I_2} are given by

$$\left.\begin{array}{l} Z_{I_1} = \sqrt{\dfrac{Z_{11}}{Z_{22}}(Z_{11}Z_{22} - Z_{12}^2)}, \\ Z_{I_2} = \sqrt{\dfrac{Z_{22}}{Z_{11}}(Z_{11}Z_{22} - Z_{12}^2)}. \end{array}\right\} \tag{57}$$

From Eqs. (25) it is evident that these impedances can also be expressed as

$$\left.\begin{array}{l} Z_{I_1} = \sqrt{Z_{\text{oc}}^{(1)} Z_{\text{sc}}^{(1)}}, \\ Z_{I_2} = \sqrt{Z_{\text{oc}}^{(2)} Z_{\text{sc}}^{(2)}}. \end{array}\right\} \tag{58}$$

The third network parameter is again defined in terms of a voltage or current ratio as

$$\frac{V_2}{V_1} = \sqrt{\frac{Z_{I_2}}{Z_{I_1}}} e^{-\theta}. \tag{59}$$

The quantity θ is called the image transfer constant. As may be easily verified, it is given by

$$\tanh \theta = \sqrt{\frac{Z_{sc}^{(1)}}{Z_{oc}^{(1)}}} = \sqrt{\frac{Z_{sc}^{(2)}}{Z_{oc}^{(2)}}}. \tag{60}$$

$$\cosh \theta = \frac{\sqrt{Z_{11}Z_{22}}}{Z_{12}}. \tag{61}$$

If the networks under consideration are symmetrical, the distinction between the image impedance and the iterative impedance disappears. In this case the impedance is commonly spoken of as the characteristic impedance, since the analogy to a continuous transmission line now becomes complete. In symmetrical networks,

$$Z_{I_1} = Z_{I_2} = Z_0 = \sqrt{Z_{11}^2 - Z_{12}^2}, \tag{62}$$

and

$$\cosh \Gamma = \cosh \theta = \frac{Z_{11}}{Z_{12}}. \tag{63}$$

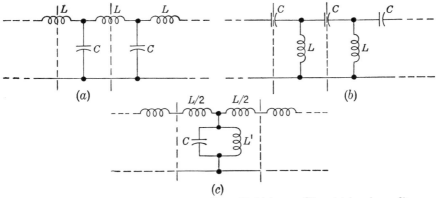

Fig. 4·29.—Simple filters: (a) Low-pass filter, (b) high-pass filter, (c) bandpass filter.

4·9. Filters.—A chain of two-terminal-pair networks connected in cascade constitutes a filter. Waves in certain definite bands of frequencies are propagated along the chain without attenuation but with a definite phase shift from section to section. Since no resistive loss is assumed to be present, Z_{11} and Z_{12} are pure imaginary. Equation (61) shows that $\cosh \theta$ is always real. If its value is between -1 and $+1$, θ must be pure imaginary. The range of frequencies for which this is true is called the pass band of the filter. When $|\cosh \theta|$ is greater than unity, θ is real and there is attenuated propagation without phase shift. By a suitable choice of the components of each network, it is possible for Z_{11} and Z_{12} to have a frequency dependence such that a given band of frequencies is passed without attenuation. It should be pointed out that frequencies which are rejected can be said to be reflected from the filter.

The presence of attenuation does not imply that the energy of the attenuated waves is dissipated in heat. It is, of course, possible to construct a device for which this is so, in which the attenuation of unwanted frequencies is accomplished by means of resistive elements. Such a device is usually called an equalizer. It is difficult, however, to have zero attenuation in the pass band when resistive elements are used. The microwave analogues of equalizers have, as yet, no important applications, and no further discussion of their properties will be presented here.

A simple example of a filter is shown in Fig. 4·29a. Each section of this filter may be taken to be a T-network with series inductance $L/2$ and shunt capacitance C. Hence

$$\left. \begin{aligned} Z_{11} &= \frac{j\omega L}{2} - \frac{j}{\omega C} \\ Z_{12} &= -\frac{j}{\omega C} \end{aligned} \right\} \quad (64)$$

The characteristic impedance is then

$$Z_0 = \sqrt{Z_{11}^2 - Z_{12}^2} = \sqrt{\frac{L}{C} - \frac{\omega^2 L^2}{4}}. \quad (65)$$

The characteristic impedance is real only for values of ω less than the cutoff value ω_c. The cutoff frequency is given by

$$\omega_c = \frac{2}{\sqrt{LC}}. \quad (66)$$

For angular frequencies below ω_c, Z_0 is real; θ is pure imaginary and is given by

$$\left. \begin{aligned} \cosh\theta &= 1 - \frac{\omega^2 LC}{2}, \\ \theta &= j\cos^{-1}\left(1 - \frac{\omega^2 LC}{2}\right). \end{aligned} \right\} \quad (67)$$

The filter is thus known as a low-pass filter. A simple high-pass filter is shown in Fig. 4·29b. Here

$$\left. \begin{aligned} Z_{11} &= -\frac{j}{2\omega C} + j\omega L, \\ Z_{12} &= j\omega L, \end{aligned} \right\} \quad (68)$$

$$Z_0 = \sqrt{\frac{L}{C} - \frac{1}{4\omega^2 C^2}}, \quad (69)$$

$$\omega_c = \frac{1}{2\sqrt{LC}}, \quad (70)$$

$$\theta = j\cos^{-1}\left(1 - \frac{1}{2\omega^2 LC}\right). \quad (71)$$

Figure 4.29c shows a bandpass filter. For this filter,

$$Z_{11} = j\omega \frac{L}{2} - j \frac{\frac{L'}{C}}{\omega L' - \frac{1}{\omega C}},$$

$$Z_{12} = -j \frac{\frac{L'}{C}}{\omega L' - \frac{1}{\omega C}}.$$
(72)

The pass band is given by

$$-1 \leq \frac{Z_{11}}{Z_{12}} \leq +1$$

or

$$-4 \leq \frac{L}{L'} - \omega^2 LC \leq 0.$$

The lower cutoff frequency is

$$\omega_c^{(1)} = \frac{1}{\sqrt{L'C}},$$
(73)

and the upper one is

$$\omega_c^{(2)} = \sqrt{\frac{4}{LC} + \frac{1}{L'C}}.$$
(74)

The characteristic impedance is given by

$$Z_0^2 = \frac{\frac{L}{C}}{1 - \frac{1}{\omega} 2L'C} - \frac{\omega^2 L^2}{4}.$$

For small values of ω, Z_0^2 will be negative and will be given approximately by

$$Z_0^2 \approx -\omega^2 L \left(L' + \frac{L}{4} \right),$$
(75)

which represents an inductive reactance.

By the use of more complicated structures, filters with several pass bands, or filters that eliminate a special band while passing all others, can be constructed. Only infinite chains of identical networks have been thus far considered. The problem of designing practical filters with only a few component networks depends very considerably on the manner in which the filter is terminated. Moreover, it is often important to have much larger attenuations near the pass band than can be obtained for the simple filters that have been used for illustration here. For a more complete discussion of filter design at microwave frequencies, the reader is referred to Chaps. 9 and 10 of Vol. 9.

If the series impedances and the shunt susceptances of the T-networks that compose the chain are decreased and made to approach zero, the chain of networks approaches the continuous transmission line that was discussed in Chap. 3. If the value of the series impedance of the T-network is $(Z/2)\,dz$ and the shunt admittance is $Y\,dz$, then

$$Z_{11} = \frac{Z}{2}\,dz + \frac{1}{Y\,dz}, \\ Z_{12} = \frac{1}{Y\,dz}, \quad \Bigg\} \tag{76}$$

and

$$\cosh \theta = \frac{Z}{2}\,Y(dz)^2 + 1$$

Since θ is now small, $\cosh \theta$ may be expanded and

$$\frac{\theta}{dz} = \pm \sqrt{ZY}. \tag{77}$$

By definition, however, θ/dz is identical with the propagation constant γ of a transmission line, and Eq. (77) is identical with Eq. (12) of Chap. 3. Similarly, the characteristic impedance becomes

$$Z_0 = \sqrt{\left(\frac{Z}{2}\,dz\right)^2 + \frac{Z}{Y}},$$

which approaches the value in Eq. (3·12) as dz approaches zero. The bandpass filter of Fig. 4·29c approaches a transmission line with characteristic impedance

$$Z_0 = \sqrt{\frac{Z}{Y}},$$

where

$$Z = j\omega L,$$
$$Y = j\omega C - \frac{j}{\omega L'},$$

or

$$Z_0 = \left(\frac{\dfrac{L}{C}}{1 - \dfrac{1}{\omega^2 L' C}}\right)^{1/2}. \tag{78}$$

As before, Z_0 is real for $\omega^2 > 1/L'C$ but remains real for all higher values of ω. The upper cutoff frequency has therefore moved off to infinity as the impedance of each section was decreased. The propagation constant is

$$\gamma = \frac{L}{L'} - \omega^2 LC. \tag{79}$$

This variety of high-pass transmission line is an exact analogue of a waveguide that is propagating an H-mode. If, as in Chap. 3, the voltage is taken to correspond to the transverse electric field and the current to the transverse magnetic field, then the impedance per unit length is a pure inductance, since there is no longitudinal electric field. Thus it may be assumed that

$$Z = j\omega\mu. \tag{80}$$

The shunt admittance per unit length will have two parts, the first contributed by the displacement current and therefore capacitive and the second arising from the longitudinal magnetic field. Let it be assumed that

$$Y = j\omega\epsilon - \frac{j}{\omega\mu} C_1, \tag{81}$$

where C_1 is a constant. To find the value of C_1, the propagation constant may be calculated as

$$\gamma^2 = ZY = -\omega^2\epsilon\mu + C_1. \tag{82}$$

It is evident that C_1 must be equal to the square of the cutoff wave number for the mode in question; for the dominant mode in rectangular

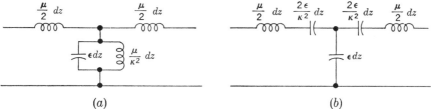

(a) (b)

FIG. 4·30.—(a) Equivalent circuit for dominant-mode transmission in rectangular wave guide; (b) equivalent transmission line for E-modes.

guide it is $(\pi/a)^2$. An equivalent circuit may thus be drawn as shown in Fig. 4·30a. By similar arguments, it can be shown that the equivalent transmission line for E-modes may be represented by Fig. 4·30b. It should be pointed out that if the waveguide is filled with a lossy dielectric, the circuits of Fig. 4·30 must be altered to have a resistance shunted across each condenser. Losses that arise from imperfectly conducting walls are not so simply represented.

4·10. Series and Parallel Connection of Networks.—Two-terminal-pair networks may be connected in other ways than in cascade. In Fig. 4·31b, two networks are shown connected in series. The network parameters of the combination may easily be found by a simple matrix calculation. If superscripts are used to distinguish the two networks, and symbols without superscripts to designate the parameters of the combination, then

$$\mathsf{Z} = \mathsf{Z}^{(1)} + \mathsf{Z}^{(2)}. \tag{83}$$

Figure 4·31a shows two networks connected in parallel. Here, the admittance matrices are most useful, and

$$\mathsf{Y} = \mathsf{Y}^{(1)} + \mathsf{Y}^{(2)}. \tag{84}$$

Combinations in which the input terminals are in series while the output terminals are in parallel, or vice versa, are also possible. They represent, however, obvious extensions of the simple cases just discussed.

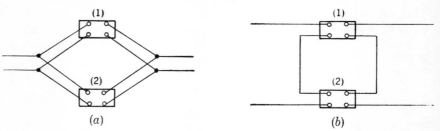

FIG. 4·31.—Two-terminal-pair networks connected (a) in parallel and (b) in series.

An important condition is always imposed in this type of interconnection. When the impedance matrix is set up, it is assumed that the same currents flow out of the lower terminals of the network as into the upper terminals. For the relations given by Eqs. (83) and (84) to be valid this must also be true of the combined network. Moreover, the potential between, for example, the upper terminals on the input and

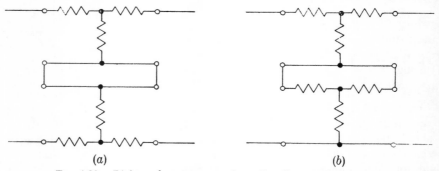

FIG. 4·32.—Right and wrong ways of coupling T-networks in series.

output sides must be undisturbed. Thus, in the series connection, if each network is represented by a T-network, the network must be arranged as in Fig. 4·32a and not as shown in Fig. 4·32b. In the arrangement b, the conditions are obviously violated by the short-circuiting of the lower network. For a further discussion of this question, the reader is referred to Guillemin[1] and the references there cited.

[1] E. A. Guillemin, *Communication Networks*, Vol. II, Wiley, New York 1935, Chap. 4, pp. 147*ff*.

4·11. Three-terminal-pair Networks.

Little has been done on the investigation of three-terminal-pair networks as low-frequency circuits. In the microwave field, many devices employ T-junctions whose equivalent circuits are of this type. The impedance matrix is of the third order and contains nine elements. From the reciprocity relations, the matrix must be symmetrical; therefore there are only six independent parameters,

$$Z = \begin{pmatrix} Z_{11} & Z_{12} & Z_{13} \\ Z_{12} & Z_{22} & Z_{23} \\ Z_{13} & Z_{23} & Z_{33} \end{pmatrix}.$$

Several convenient equivalent circuits exist by which the behavior of the network can be described. An obvious circuit is shown in Fig. 4·33. The

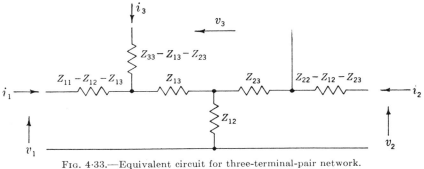

FIG. 4·33.—Equivalent circuit for three-terminal-pair network.

values of the circuit elements in terms of the matrix components are indicated in the figure. They may be verified by inspection. If this circuit represents a symmetrical T-junction with a plane of symmetry through the center line between terminals (3), the number of independent parameters is reduced to four and the impedance matrix takes the special form

$$Z = \begin{pmatrix} Z_{11} & Z_{12} & Z_{13} \\ Z_{12} & Z_{11} & Z_{13} \\ Z_{13} & Z_{13} & Z_{33} \end{pmatrix}.$$

Because many microwave junctions have a symmetry of this type, this represents an important case. Other equivalent circuits may be obtained from that shown in Fig. 4·33 by the transformation of portions of the circuit from T- to Π-networks or to some of the transmission-line forms shown in Figs. 4·16 to 4·23 inclusive (Sec. 4·6).

Since it is not required that the equivalent circuit represent the voltage between one terminal and another, the six input lines of Fig. 4·33 may be reduced to four by connecting three of the lines together. If this common point is designated as the ground point, then another circuit may be drawn, with four terminals and six independent parameters, as shown in Fig. 4·34. Here any pair of input terminals is to be taken as one of the numbered points and the ground point. The four terminal points may be thought of as the four corners of a tetrahedron, and the circuit elements then lie along the edges of the tetrahedron. This circuit also may be transformed to other forms by the usual T- to Π-transformation. It must be noted that although the circuits in Figs. 4·33 and 4·34 may be represented by the same impedance matrix, it is impossible to transform one of these circuits into the other. In this sense the two circuits are not equivalent. They differ in the relationship that exists between the voltage difference between, for example, the upper members of terminals (1) and (2) of Fig. 4·33 and the corresponding points, (1) and (2), of Fig. 4·34. The impedance matrix implies

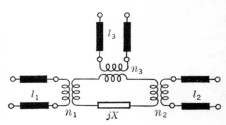

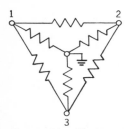

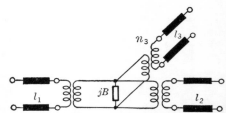

Fig. 4·34.—Circuit with four terminals and six independent parameters.

Fig. 4·35.—Transformer representation of series and shunt T-junctions.

nothing about this voltage difference but the circuits define it uniquely. This fact must be kept clearly in mind when equivalent circuits for microwave devices are employed.

The four-terminal circuit of Fig. 4·34, rather than the more usual T- or Π-network,[1] is the exact equivalent circuit for a low-frequency transducer. In the low-frequency region this problem is usually of importance only in connection with the exact equivalent-circuit representation of a practical transformer.

Other circuits whose components are transmission lines are often useful in microwave work. Figure 4·35 shows two extremely useful forms. A transformer and a length of transmission line have been included in each arm of each of the networks, although it is obvious that one transformer may be assigned an arbitrary turn ratio. These two circuits may be said to represent series and shunt T-junctions. It is to be

[1] See M. A. Starr, *Electric Circuits and Wave Filters*, 2d ed., Pitman, London, 1944, Chap. VI, and references there cited.

emphasized that either circuit is equally valid for any T-junction. The choice between the two circuits may depend, for example, upon the similarity between the physical configuration of the device and that of the circuit, and this might be a valid reason for choosing one in preference to the other.

Circuits of another type which may be useful in some circumstances are shown in Fig. 4·36. If three two-terminal-pair networks are connected in series or in shunt, equivalent circuits are obtained that have some

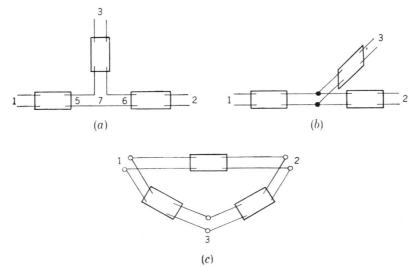

Fig. 4·36.—Combination of three two-terminal-pair networks to give a three-terminal-pair network.

extra parameters. By inserting the T- or Π-equivalents of the two-terminal-pair networks, it is easy to see which of the elements may be combined. By means of the Π- to T-transformation many other equivalent circuits may also be obtained. It should be pointed out that if Π-networks are used in Fig. 4·36c, the circuit in Fig. 4·34 is obtained. To show that the circuit of Fig. 4·36a is a valid one, note that the interconnections require that

$$v_5 = v_6 + v_7,$$
$$i_5 = -i_6 = i_7,$$

where the terminals (5), (6), and (7) are indicated in the figure. By means of these three equations i_5, i_6, and i_7 may be eliminated from the set of six equations that represent the three two-terminal-pair networks. The first member of this set is, for example,

$$v_1 = Z_{11}i_1 + Z_{15}i_5.$$

The result of eliminating these parameters will be three equations in the three currents i_1, i_2, i_3, and the impedance matrix is the matrix of these three equations. It might be thought that some general matrix method would be available to perform this elimination in a systematic fashion. Although such a method exists, it is usually too complicated to apply. A straightforward manipulation with the linear equations is much quicker and easier.

Circuits of the forms shown in Fig. 4·35 are particularly convenient for finding the effect on the power transfer from terminals (1) to (2), for example, of a load on terminal (3). It is immediately obvious that a reactive load of the proper value on terminals (3) will cause no voltage to appear across terminals (2) when voltage is applied to terminals (1). The same value of the load will also make the voltage across terminals (1) equal to zero when voltage is applied to terminals (2). Moreover, if the circuit is symmetrical about a plane through terminals (3), so that $n_1 = n_2$, then for some value of a reactive load on terminals (3), the input impedance seen from terminals (1) and (2) is the characteristic impedance.

The matrix manipulation that corresponds to the application of a load to one pair of terminals is again most easily seen from a consideration of the corresponding set of linear equations. If a load Z_3 is put on terminals (3), then

$$Z_3 = -\frac{v_3}{i_3},$$

where the negative sign arises from the convention that currents and voltages are always designated as positive when they represent power flow into the network. If i_3 is eliminated from the equation by means of this relation, there results the new second-order matrix whose elements are

$$Z'_{11} = Z_{11} - \frac{Z_{13}^2}{Z_{33} + Z_3},$$

$$Z'_{12} = Z'_{21} = Z_{12} - \frac{Z_{13}Z_{23}}{Z_{33} + Z_3},$$

$$Z'_{22} = Z_{22} - \frac{Z_{23}^2}{Z_{33} + Z_3}.$$

4·12. Circuits with N Terminal Pairs.—The equivalent circuits that have been briefly discussed for two- and three-terminal-pair networks can be generalized to circuits for the general case of N terminal pairs. The impedance or admittance matrix is of the Nth order and possesses $N + (N - 1) + (N - 2) + \cdots + 1 = N(N + 1)/2$ independent elements. The other $N^2 - N(N + 1)/2$ elements are equal, by the reciprocity relation, to corresponding independent elements so that $Z_{ji} = Z_{ij}$. There are three principal classes of equivalent circuits that can be constructed for this general case. A circuit analogous to that shown in Fig. 4·34 can be constructed. Let us choose $N + 1$ points. Then let us call one of these points the common terminal of all the N input lines. Now if an impedance element is connected between each pair of points, the number of such elements is given by the binomial coefficient.

$$\binom{N+1}{2} = \frac{N(N+1)}{2}.$$

This is just the number of independent elements needed. The well-known Π-network is the result when $N = 2$.

In Sec. 4·6, it was noted that a Π-network may be represented also by a line one-quarter wavelength long, of arbitrary impedance, shunted at each end by an arbitrary admittance. The circuit whose construction was described in the preceding paragraph may be thought of as made up of Π-networks connected between each pair of network terminals. If these Π-networks are replaced with lines shunted by admittances, the equivalent circuit shown in Fig. 4·37 is obtained for the four-terminal-

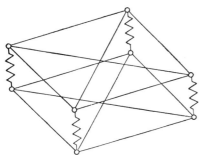

Fig. 4·37.—Equivalent circuit for a four-terminal-pair network constructed of quarter-wavelength lines.

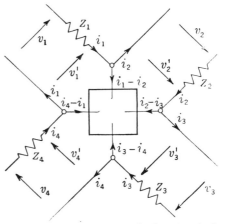

Fig. 4·38.—Reduction of a four-terminal-pair network to one with three terminal pairs and external impedances.

pair case. The parameters are the six characteristic impedances of the quarter-wavelength lines and the four shunt admittances, making, in all, the necessary 10 constants.

There is a third general approach to finding the equivalent circuit for N pairs of terminals. Suppose that it can be arranged that the sum of the applied potentials is zero. The lines can be arranged as shown schematically in Fig. 4·38, where the case for $N = 4$ is indicated. It should be noticed that the four-terminal network at the center of the figure has flowing into it currents whose sum is zero. This four-terminal device therefore satisfies the requirement that the voltages across its terminals are linearly related to the currents, and is equivalent to a three-terminal-pair network. The voltages v'_j must be derived from the applied voltages v_j by the relation

$$v'_j = v_j - Z_j i_j.$$

The value of the impedances Z_j must now be found. If $Z_j i_j$ is subtracted from each equation of the fundamental set, then

$$v_1 - Z_1 i_1 = (Z_{11} - Z_1)i_1 + Z_{12}i_2 + \cdots + Z_{1N}i_N$$
$$v_2 - Z_2 i_2 = Z_{21}i_1 + (Z_{22} - Z_2)i_2 + \cdots + Z_{2N}i_N$$
$$\cdots\cdots\cdots\cdots\cdots\cdots\cdots\cdots\cdots\cdots\cdots\cdots$$
$$v_N - Z_N i_N = Z_{N1}i_1 + Z_{N2}i_2 + \cdots + (Z_{NN} - Z_N)i_N.$$

To enforce the condition $\Sigma v'_j = 0$, it is required that the sum of these equations be zero, independently of the values of the currents. There-

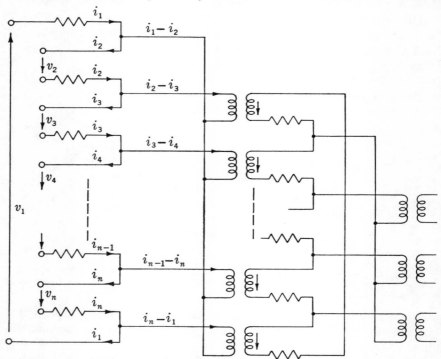

FIG. 4·39.—Reduction of equivalent circuit for N-terminal-pair network by means of ideal transformers.

fore let i_1 take a finite value while all the other i's are zero. Adding the equations results in

$$0 = (Z_{11} + Z_{21} + \cdots + Z_{N1})i_1 - Z_1 i_1$$

or, in general,

$$Z_j = \sum_k Z_{kj}.$$

Now if the equations are expressed in terms of v'_j whose sum is zero and $(i_j - i_{j+1})$ are used as independent variables, the set of N equations reduces to $N - 1$ equations, since one equation is redundant.[1]

To repeat this process, it should be remembered that the $N - 1$ equations correspond to $N - 1$ pairs of terminals, each pair having a member in common with all other pairs. This condition can be removed by the use of ideal transformers. Figure 4·39 shows schematically how this may be accomplished. The process of reduction can be continued until only a two-terminal-pair network remains. For this network any standard circuit form may be used. It should be noted that at each step an impedance in each line is removed and the total number of parameters is correct.

To conclude this section, the change in the elements of the impedance matrix when a load is placed on one terminal pair may be stated. If a load Z_k is placed across the kth pair of terminals, the new impedance elements are

$$Z'_{ij} = Z_{ij} - \frac{Z_{ik}Z_{kj}}{Z_{kk} + Z_k}.$$

The new impedance matrix has elements Z'_{ij}, and the kth row and column are struck out.

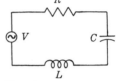

Fig. 4·40.—Series-resonant circuit.

4·13. Resonant Circuits.—In Sec. 4·3 it was shown that the input reactance of a network always increases with increasing frequency. At the frequencies at which the reactance is zero, the network is said to be resonant. The behavior of the network in the neighborhood of resonance is of considerable importance in many applications. The simplest case is that of a resistor, a capacitance, and an inductance in series with a generator, as shown in Fig. 4·40. The impedance is

$$Z = R + j\left(\omega L - \frac{1}{\omega C}\right).$$

At resonance,

$$\omega_0 L = \frac{1}{\omega_0 C},$$

$$\omega_0 = \frac{1}{\sqrt{LC}}.$$

For microwave applications other parameters are useful. The characteristic impedance Z_c of the circuit is defined as

$$Z_c = \sqrt{\frac{L}{C}},$$

and the parameter Q, sometimes called the "quality" or Q-factor of the

[1] *Cf.* Frank M. Starr, *Trans. AIEE*, **51**, 287 (June 1932).

circuit, is
$$Q = \frac{Z_c}{R} = \frac{\omega_0 L}{R} = \frac{1}{\omega_0 RC}.$$

In terms of these quantities,
$$Z = R + jZ_c\left(\frac{\omega}{\omega_0} - \frac{\omega_0}{\omega}\right)$$
$$= R\left[1 + jQ\left(\frac{\omega}{\omega_0} - \frac{\omega_0}{\omega}\right)\right].$$

Near the resonant frequency, the approximation
$$\omega = \omega_0 + \Delta\omega$$
may be made, where $\Delta\omega$ is small, and hence
$$Z = R\left(1 + 2jQ\frac{\Delta\omega}{\omega_0}\right).$$

If Q is large, then the resistance and the reactance are equal when
$$\Delta\omega = \pm\frac{\omega_0}{2Q}.$$

The power P absorbed by the circuit is $\frac{1}{2}R|I|^2$ or
$$P = \frac{1}{2}\frac{V^2}{R}\frac{1}{1 + \left(2Q\frac{\Delta\omega}{\omega_0}\right)^2},$$
where V is the generator voltage. Thus maximum power is absorbed at resonance. The absorbed power falls to half this maximum value at the frequencies where the reactance is equal to the resistance.

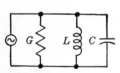
FIG. 4·41.—Shunt-resonant circuit.

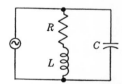
FIG. 4·42.—A second shunt-resonant circuit.

The duality principle may be invoked to obtain the corresponding relations for the shunt circuit shown in Fig. 4·41. These relations are
$$Y_c = \sqrt{\frac{C}{L}},$$
$$Q = \frac{Y_c}{G} = \frac{\omega_0 C}{G} = \frac{1}{\omega GL},$$
$$Y = G\left(1 + 2jQ\frac{\Delta\omega}{\omega_0}\right).$$

A second parallel circuit, shown in Fig. 4·42, may be reduced to the preceding circuit in the low-loss case. The admittance of the inductance and resistance in series is

$$Y = \frac{1}{R + j\omega L} = \frac{R - j\omega L}{R^2 + \omega^2 L^2}.$$

If $R^2 \ll \omega^2 L^2$,

$$Y = \frac{R}{\omega^2 L^2} - \frac{j}{\omega L},$$

and the circuits are equivalent if

$$G = \frac{R}{\omega_0^2 L^2}.$$

The resonance phenomenon just described is one of forced oscillations. If a network is short-circuited and excited by some transient process, free oscillations will occur whose frequency is given by the equation

$$Z(\omega) = R + j\left(\omega L - \frac{1}{\omega C}\right) = 0$$

in the simple series circuit. The roots of this equation are complex and are

$$j\omega = -\frac{R}{2L} \pm \sqrt{\frac{R^2}{4L^2} - \frac{1}{LC}}.$$

If

$$\frac{R}{2L} \geqq \frac{1}{\sqrt{LC}} \quad \text{or} \quad R \geqq 2Z_c,$$

$j\omega$ is real and no oscillations occur. There is only an exponential decay of the currents and the voltages. If the resistance is smaller, free oscillations occur whose frequency ω_f is

$$\omega_f = \sqrt{\frac{1}{LC} - \frac{R^2}{4L^2}}.$$

In terms of the resonant frequency ω_0

$$\omega_f = \omega_0 \sqrt{1 - \frac{1}{4Q^2}}.$$

Hence, only when Q is large is the frequency of free oscillation equal to the resonant frequency.

The shunt cases may be treated in a similar fashion, the case of free oscillation being given by the condition $Y(\omega) = 0$. Free oscillations in networks with more than two terminal pairs may be treated in an analogous fashion.

CHAPTER 5

GENERAL MICROWAVE CIRCUIT THEOREMS

By R. H. Dicke

5·1. Some General Properties of a Waveguide Junction.—A waveguide junction[1] in the generalized sense that is to be used here is defined as a region of space completely enclosed by a perfectly conducting metal surface except for one or more transmission lines that perforate the surface (Fig. 5·1). Nothing will be assumed about the interior of the junction except that the dielectric constant, permeability, and conductivity are everywhere within it independent of time and of the field quantities. It will be assumed for the present that there is only one propagating mode for each of the transmission lines. This definition will be extended later to include cases in which several modes are present.

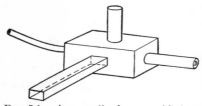

Fig. 5·1.—A generalized waveguide junction.

For reasons that seem to be largely historical, two different approaches to electrical problems have been developed. The first can be called the Maxwellian approach. It consists of the introduction of certain field quantities that are interrelated by a set of linear differential equations and the solution of these equations under particular boundary conditions. This approach is the natural outgrowth of the study of electrostatics which was the chief concern of early physicists interested in electrical phenomena.

The second approach, which may be called the electrical-engineering approach, may be said to have resulted from the discovery of the galvanic cell. The electric field strength produced by such a device is usually small; and as a result, the electric field loses much of its importance. Instead, the current flow in conductors becomes the primary physical quantity. In place of the electric field, its line integral, or potential

---

[1] For such a junction, the term "black box" has been used extensively, in recent years, by a certain small group of semimetamorphosed physicists at the Radiation Laboratory. This term is discarded here because it suggests a Hohlraum or something that is totally absorbing. The low-frequency term "network" is misleading in a discussion of waveguides, since it implies the presence of wires.

difference, becomes important. Kirchhoff's laws lead in a natural way to the solution of problems involving d-c networks. With the introduction of alternating current as a practical source of commercial power, it became necessary to make network calculations for a-c as well as d-c circuits. It was then found that Kirchhoff's laws could be extended to apply to a-c circuits through the artifice of representing alternating currents and voltages by means of complex numbers. Sometimes there is difficulty in reconciling these two methods of approach, and it is one of the purposes of this chapter to attempt to clarify their equivalence.

One of the difficulties with the Maxwellian approach lies in its ambition. It asks for a complete description of the field in a certain region subject to certain boundary conditions. This is usually much more information than one needs or wants. For example, one is usually not interested in knowing the distribution of the magnetic field in an inductor. Two numbers, the values of the inductance and the resistance, usually suffice. If this overabundant ambition of the Maxwellian approach is somewhat curtailed, the method gives perfectly reasonable, simple results. As will be seen presently, the curtailment consists in limiting the inquiry to certain energy integrals over the region under consideration.

Another lesson that the Maxwellians can learn from the engineers is the importance of a simply periodic solution. Historically, the engineers were interested in the periodic solution because it described the electromagnetic effect that was produced by their generators. In fact, this is still the most important reason for studying this solution. However, to pacify the Maxwellians, it should be pointed out that the solution of Maxwell's equations is simplified if the field quantities are replaced by Fourier integrals with respect to time. The resulting differential equations for the Fourier transforms are in general simpler than the original equations.

The engineering approach is built upon the concept of the lumped impedance element and, at first glance, would seem to break down at microwave frequencies where the lumped impedance element loses its significance. It is sometimes possible to regard the distributed system as composed of an infinite number of lumped parameters. An example of this is the coaxial line. Also it is always possible to introduce the displacement current as a fictitious current in an infinite number of fictitious networks to describe the electromagnetic field in the absence of conductors. This latter representation does not seem to be particularly novel. It leads to results of essentially the same type as does the field approach to the problem.

It has been pointed out that a complete solution of an electromagnetic problem is not always desired. Often a description of conditions at the terminals of a junction is sufficient. It is a purpose of the remainder of the chapter to make use of Maxwell's equations to obtain conditions that hold at the terminals of a general waveguide junction. In so doing, it will be found that the stored electric and magnetic energy and the dissipated power are the three most important field parameters for the description of conditions at the terminals of a junction.

THE TERMINATION OF A SINGLE TRANSMISSION LINE

If the generalized waveguide junction is connected to external regions by a single transmission line, it degenerates from a junction to a termination. The termination of a single transmission line is sometimes called, at low frequencies, an impedor or a single-terminal-pair network. This terminology may also be used at microwave frequencies. Referring to Fig. 5·2, the termination is imagined to be surrounded by a surface S. This surface is assumed to be completely exterior to the termination and to cut the transmission line (illustrated as a rectangular waveguide) perpendicular to its axis at some plane hereafter called the *terminals*. The portion of the transmission line inside the surface is now considered to be part of the termination.

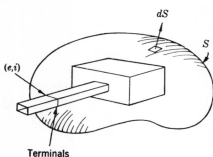

FIG. 5·2.—A waveguide termination in its enveloping surface.

The location of the terminals is completely arbitrary. However, it is desirable for purposes of clarity to consider them far enough from the actual end of the transmission line so that the amplitudes of the higher, nonpropagating modes that may exist within the termination are essentially zero at the terminals.

5·2. Poynting's Energy Theorem for a Periodic Field.—Maxwell's equations for a simply periodic field are

$$\begin{rcases} \operatorname{curl} \mathbf{H} - (j\omega\epsilon + \sigma)\mathbf{E} = 0, \\ \operatorname{curl} \mathbf{E} + j\omega\mu\mathbf{H} = 0, \\ \operatorname{div} \epsilon\mathbf{E} = \rho, \\ \operatorname{div} \mu\mathbf{H} = 0, \end{rcases} \quad (1)$$

where $\mathbf{E}$, $\mathbf{H}$, and ρ are complex numbers such that $\mathbf{E}'(t) = \operatorname{Re}(\mathbf{E}e^{j\omega t})$ and similarly for $\mathbf{H}$ and ρ.

A connecting link between field theory and electrical engineering can be found in Poynting's energy theorem limited to periodic fields. It was shown in Sec. 2·2 that if

$$\operatorname{div} \mathbf{E} \times \mathbf{H}^* = \mathbf{H}^* \cdot \operatorname{curl} \mathbf{E} - \mathbf{E} \cdot \operatorname{curl} \mathbf{H}^*$$
$$= -j\omega\mu\mathbf{H}^* \cdot \mathbf{H} - (\sigma - j\omega\epsilon)\mathbf{E}^* \cdot \mathbf{E} \quad (2)$$

Sec. 5·2] POYNTING'S ENERGY THEOREM FOR A PERIODIC FIELD

be integrated over the volume enclosed by a surface S and the volume integral converted to a surface integral in the usual way, then the surface integral may be written as the sum of three volume integrals, as shown in Eq. (3).

$$\int_S \mathbf{E} \times \mathbf{H}^* \cdot d\mathbf{S} = -j\omega \int \mu \mathbf{H}^* \cdot \mathbf{H}\, dv \\ + j\omega \int \epsilon \mathbf{E}^* \cdot \mathbf{E}\, dv - \int \sigma \mathbf{E}^* \cdot \mathbf{E}\, dv. \quad (3)$$

Assume for a surface S the one illustrated in Fig. 5·2. Since the field vanishes everywhere over the surface S except at the terminals, the surface integral reduces to an integral over the terminals. Since $d\mathbf{S}$ is perpendicular to the axis of the transmission line, only the transverse components of $\mathbf{E}$ and $\mathbf{H}$ enter into the integral. If the coordinates x and y in the terminal plane of the transmission line are introduced, then

$$\int_S \mathbf{E} \times \mathbf{H}^* \cdot d\mathbf{S} = \int (E_x H_y^* - E_y H_x^*)\, dx\, dy. \quad (4)$$

It is convenient to introduce a complex terminal "current" and "voltage" i and e as was done in Sec. 4·9.

Let

$$\left.\begin{array}{l} E_x = e f_x(x,y), \\ E_y = e f_y(x,y), \\ H_x = i g_x(x,y), \\ H_y = i g_y(x,y), \end{array}\right\} \quad (5)$$

where f_x, f_y, g_x and g_y are real functions of the coordinates which give the distribution of the field. It was shown in Chap. 2 that the transverse electric (or magnetic) components are in phase with each other, i.e., $\mathbf{E}$ and $\mathbf{H}$ at a given point are constant in direction. It is assumed that the functions are normalized in such a way that

$$\int_S (f_x g_y - f_y g_x)\, dx\, dy = -1. \quad (6)$$

Then

$$\int_S \mathbf{E} \times \mathbf{H}^* \cdot d\mathbf{S} = -ei^*, \quad (7)$$

and

$$ei^* = j\omega \left[\int \mu \mathbf{H}^* \cdot \mathbf{H}\, dv - \int \epsilon \mathbf{E}^* \cdot \mathbf{E}\, dv \right] + \int \sigma \mathbf{E}^* \cdot \mathbf{E}\, dv. \quad (8)$$

Equation (8) is rather fundamental. It may be written as

$$ei^* = 4j\omega(W_H - W_E) + 2P, \quad (9)$$

where W_H and W_E are the average stored magnetic and electric energies and P is the average dissipated power. It is to be remembered that it is the stored energies and dissipated power within the termination which enter into Eq. (9) and that the portion of the transmission line interior to the surface S must be included.

5·3. Uniqueness of Terminal Voltages and Currents.—It is evident that to any permissible field solution there corresponds a definite terminal voltage and current. It will now be shown that for any particular value of the voltage (or current) there corresponds a unique field distribution inside the termination.

Let us assume that there are two solutions of the wave equation which satisfy boundary conditions. It will be shown that identical terminal currents or voltages imply identical solutions. Let the two solutions be

$\mathbf{E}_1$, $\mathbf{H}_1$ with terminal parameters e_1, i_1,
$\mathbf{E}_2$, $\mathbf{H}_2$ with terminal parameters e_2, i_2.

From the linearity of Maxwell's equations, it follows that $(\mathbf{E}_1 - \mathbf{E}_2)$ and $(\mathbf{H}_1 - \mathbf{H}_2)$ with the terminal parameters $(e_1 - e_2)$ and $(i_1 - i_2)$ also form a solution. If this solution be substituted in Eq. (8), there follows

$$(e_1 - e_2)(i_1^* - i_2^*) = j\omega \left[\int \mu(\mathbf{H}_1^* - \mathbf{H}_2^*) \cdot (\mathbf{H}_1 - \mathbf{H}_2) \, dv \right.$$
$$\left. - \int \epsilon(\mathbf{E}_1^* - \mathbf{E}_2^*) \cdot (\mathbf{E}_1 - \mathbf{E}_2) \, dv \right] + \int \sigma(\mathbf{E}_1^* - \mathbf{E}_2^*) \cdot (\mathbf{E}_1 - \mathbf{E}_2) \, dv. \quad (10)$$

Now if either $e_1 = e_2$ or $i_1 = i_2$, the left side of Eq. (10) vanishes and the real and imaginary parts of the right side independently vanish. An inspection of the real part (the dissipated-power term) shows that the integrand is positive definite and the integral vanishes only if $\mathbf{E}_1 = \mathbf{E}_2$ when $\sigma \neq 0$. If the electromagnetic field is confined to the termination by real metal walls with *finite* conductivity, however, the nonvanishing electric fields, $\mathbf{E}_1$ and $\mathbf{E}_2$ must be identical for a finite distance inside these walls. Since these solutions are nonvanishing, identical, and have identical derivatives over the boundaries, the solutions are identical. Since $\mathbf{E}_1 = \mathbf{E}_2$ throughout the enclosure, the vanishing of the imaginary part of Eq. (10) requires that $\mathbf{H}_1 = \mathbf{H}_2$. These conditions, of course, require that $e_1 = e_2$ and $i_1 = i_2$.

The theorem breaks down in the case of completely lossless enclosures. Such enclosures are, of course, not physically realizable. However, a lossless termination is an artifice that is introduced for purposes of convenience in the mathematical treatment of many problems. For this reason it is interesting to examine the nature of the departures from the condition of uniqueness. When the termination is completely lossless, the dissipation term in Eq. (10) vanishes identically and the only condition imposed by Eq. (10) is the vanishing of the imaginary part of the right-hand side. This requires that the average stored magnetic and electric energies be equal and this condition may be defined as resonance. This resonance condition, however, can be satisfied only at certain discrete frequencies, the natural resonances of the termination. For all other frequencies $\mathbf{E}_1 = \mathbf{E}_2, \mathbf{H}_1 = \mathbf{H}_2$, and the uniqueness theorem is satisfied.

In completely lossless terminations resonances of two different types may be considered. In the first type there is no coupling between the terminals and the resonance fields, as though there were an isolated cavity somewhere inside the termination. Clearly a resonance of this type is of no particular importance. The second type of resonance couples with the input terminals, and in this case the terminal current is not given uniquely by the terminal voltage. Of course, it must be emphasized that these conditions are never encountered in practice; therefore whenever lossless terminations are discussed, it will be assumed that the terminal voltage is uniquely related to the terminal current and the field quantities.

It is evident from the uniqueness theorem and from the linearity of Maxwell's equations that a change in the argument or modulus of e_1 will result in a corresponding change in the argument or modulus of the field quantities inside the termination. Thus the field quantities inside the termination are proportional to the voltage or current at the terminals. It follows from this that the terminal current i is proportional to the terminal voltage e. Thus, as before, values $Z(\omega)$ and $Y(\omega)$ may be defined such that

$$\frac{e}{i} = Z(\omega) = \frac{1}{Y(\omega)}. \tag{11}$$

They are called respectively the impedance and admittance of the termination. Z and Y are complex numbers that depend only on the frequency of the periodic field and the nature of the termination.

5·4. Connections between Impedance and Stored and Dissipated Energy.—If iZ is substituted for e in Eq. (8), there results

$$i^*iZ = j\omega \left(\int \mu \mathbf{H}^* \cdot \mathbf{H}\, dv - \int \epsilon \mathbf{E}^* \cdot \mathbf{E}\, dv \right) + \int \sigma \mathbf{E}^* \cdot \mathbf{E}\, dv \tag{12}$$

or

$$Z = \frac{2j\omega(W_H - W_E) + P}{\frac{1}{2}ii^*}. \tag{13}$$

W_H and W_E are the mean stored magnetic and electric energies; P is the average power dissipated in the termination. In the same way,

$$Y = \frac{2j\omega(W_E - W_H) + P}{\frac{1}{2}ee^*}. \tag{14}$$

From Eqs. (13) and (14) it is seen that

$$Z^*(-\omega) = Z(\omega), \tag{15}$$

and
$$Y^*(-\omega) = Y(\omega). \tag{16}$$

Several things are to be noticed about Eqs. (13) and (14). Let
$$Z = R + jX$$
(R and X real).
1. Since $P \geqq 0$, $R \geqq 0$.
2. If $P = 0$, then $R = 0$, or Z is purely imaginary.
3. If $X = 0$, $W_E = W_H$; this is the resonance case.
4. From Eqs. (15) and (16) it can be seen that for a lossless termination the reactance and the susceptance are both odd functions of frequency.

Equations (13) and (14) indicate the steps that may be taken to produce desired impedance effects at microwave frequencies. For instance, any change in configuration that increases the amount of stored magnetic energy in a termination automatically increases the reactance at the terminals.

5·5. Field Quantities in a Lossless Termination.—One of the results obtainable from the uniqueness theorem concerns a lossless termination. It will be shown that in such a termination the electric field is everywhere in phase and the magnetic field is everywhere 90° out of phase with the electric field. Let us assume that **E** and **H**, with corresponding terminal quantities e and i, are a permissible solution of the field equations for a particular lossless termination. Substitute for **E** and **H** in Maxwell's equations [Eqs. (1)] the following quantities:

$$\mathbf{E} = \mathbf{E}_r + \mathbf{E}_i,$$
$$\mathbf{H} = \mathbf{H}_r + \mathbf{H}_i.$$

The subscripts r and i denote, respectively, that the quantities are the real and imaginary portions of the field vectors. Since the termination is assumed to be lossless,
$$\sigma = 0$$
for
$$\mathbf{E} \neq 0.$$

The imaginary portion of the first equation, the real portion of the second, the real portion of the third, and the imaginary portion of the fourth are

$$\left.\begin{array}{l} \text{curl } \mathbf{H}_i - j\omega\epsilon\mathbf{E}_r = 0, \\ \text{curl } \mathbf{E}_r + j\omega\mu\mathbf{H}_i = 0, \\ \text{div } \epsilon\mathbf{E}_r = \rho_r, \\ \text{div } \mu\mathbf{H}_i = 0. \end{array}\right\} \tag{17}$$

It is evident that $\mathbf{E}_r$ and $\mathbf{H}_i$ represent a particular solution of Maxwell's equations satisfying the boundary conditions. To the solution

E_r, H_i there correspond terminal voltage and current e_r and i_i. By the uniqueness theorem, the above particular solution is unique. Any other solution can be obtained from this solution by the multiplication of the field quantities by some complex number. E_r is a solution with the electric field everywhere in phase or 180° out of phase, and the phase of the magnetic field is 90° or 270° with respect to the electric field.

5·6. Wave Formalism.—It has been shown that the field quantities inside a termination, or single-terminal-pair network, are uniquely determined by either the current or voltage at the terminals. Voltages and currents are not the only useful parameters that can be used as representations for the fields inside a termination. Another very useful representation can be obtained from the amplitudes of the incident and scattered waves.

The amplitude and phase of the transverse component of the electric field in the incident wave, measured at the terminals, will be designated by a, which will be so normalized that $\frac{1}{2}aa^*$ represents the incident power. In a similar way b will be used to designate the amplitude and phase of the reflected wave. It is easily seen that the uniqueness theorem also applies to the representation in terms of incident and reflected waves. For any incident or reflected wave the fields inside the termination are uniquely defined. As was shown above, the impedance $Z = e/i$ is a quantity that is a function only of the frequency and the shape of the termination. In an analogous way one can define the *reflection coefficient*,

$$\Gamma = \frac{b}{a}. \qquad (18)$$

It is evident that the reflection coefficient is defined for a particular reference plane or terminal pair.

A connection between the representation in terms of currents and voltages and the representation by incident and reflected waves is easily shown. Since e is a measure of the total transverse electric field,

$$e = g(a + b) = ga(1 + \Gamma), \qquad (19)$$

where g is some proportionality factor. In the same way,

$$i = \frac{1}{g}(a - b) = \frac{1}{g}a(1 - \Gamma), \qquad (20)$$

where b has a negative sign because the magnetic field is reversed in the reflected wave. The proportionality factor is $1/g$ in Eq. (20) because of the way in which the quantities are normalized with respect to power [see Eq. (7)]. The introduction of wave amplitudes a and b defined in Eqs. (19) and (20) is a convenient artifice even in low-frequency circuits where the wave nature of the solutions is not obvious.

5·7. Connection between the Reflection Coefficient and Stored Energy.

—If Eqs. (19) and (20) are substituted in Eq. (8),

$$aa^*(1 + \Gamma)(1 - \Gamma^*)$$
$$= j\omega \left(\int \mu \mathbf{H}^* \cdot \mathbf{H}\, dv - \int \epsilon \mathbf{E}^* \cdot \mathbf{E}\, dv \right) + \int \sigma \mathbf{E}^* \cdot \mathbf{E}\, dv, \quad (21)$$

and

$$(1 + \Gamma)(1 - \Gamma^*) = \frac{2j\omega(W_H - W_E) + P}{\tfrac{1}{2}a^*a}. \quad (22)$$

If Eq. (22) is broken into its real and imaginary parts

$$1 - \Gamma^*\Gamma = \frac{P}{\tfrac{1}{2}a^*a}, \quad (23)$$

and

$$\tfrac{1}{2}(\Gamma - \Gamma^*) = j\,\mathrm{Im}(\Gamma) = \frac{j\omega(W_H - W_E)}{\tfrac{1}{2}a^*a}. \quad (24)$$

In Eq. (23), since P is positive,

$$\Gamma^*\Gamma \leq 1. \quad (25)$$

Condition (25) is equivalent to the statement that the reflected power is always equal to or less than the incident power. If the termination is lossless, $P = 0$ and

$$\Gamma^*\Gamma = 1. \quad (26)$$

Equation (24) relates the imaginary part of the reflection coefficient to the stored electric and magnetic energy. Since $|\Gamma| \leq 1$,

$$|W_H - W_E| \leq \tfrac{1}{2}\tfrac{1}{\omega} a^*a. \quad (27)$$

Stated in words Eq. (27) says that for 1 watt of power incident on a termination, the difference between the magnetic and electric stored energy is always less than or equal to $1/\omega$ joules.

If Eqs. (23) and (24), are solved for Γ,

$$\left.\begin{array}{l} \Gamma = \sqrt{1 - P'}\, e^{j\phi}, \\ \phi = \sin^{-1}\left[\dfrac{\omega(W'_H - W'_E)}{\sqrt{1 - P'}}\right], \end{array}\right\} \quad (28)$$

where

$$W'_E = \frac{W_E}{\tfrac{1}{2}a^*a},$$
$$W'_H = \frac{W_H}{\tfrac{1}{2}a^*a},$$
$$P' = \frac{P}{\tfrac{1}{2}a^*a}.$$

THE JUNCTION OF SEVERAL TRANSMISSION LINES

In the previous section were considered some of the general properties of the termination of a single transmission line. The extension of these results to cases involving more than one transmission line is straightforward. Let us consider a junction having N pairs of terminals. As in the previous case, the junction may be enclosed in a surface S which cuts the various transmission lines perpendicular to their axes (see Fig. 5·3). As before use will be made of Poynting's energy theorem for periodic fields [Eq. (3)].

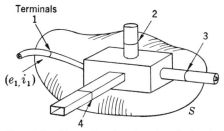

FIG. 5·3.—Example of a four-terminal-pair junction.

The integral of the Poynting vector over the surface S can be evaluated in terms of terminal voltages and currents.

$$\int_S \mathbf{E} \times \mathbf{H}^* \cdot d\mathbf{S} = -\sum_n e_n i_n, \qquad (29)$$

where e_n and i_n are the *voltage* and *current* of the nth terminal pair and are defined in the same way as in the previous section. Substituting Eq. (9) in Eq. (25),

$$\sum_n e_n i_n^* = 4j\omega(W_H - W_E) + 2P, \qquad (30)$$

where, as before, W_H and W_E are the average electric and magnetic energies in the junction and P is the average dissipated power. Equation (30) provides a connection between the terminal quantities and the field quantities.

5·8. Extension of the Uniqueness Theorem to N-terminal-pair Junctions.—The uniqueness theorem of Sec. 5·3 is easily extended to the case of N-terminal-pair junctions. It leads to the result that to a particular value of the current or voltage at each of the N pairs of terminals there corresponds a unique field distribution inside the junction. This result is valid provided the junction is not completely lossless. As before, the theorem will be extended to include the idealized lossless junction. Another simple extension of the previous results which follows from the linearity of Maxwell's equations and from the uniqueness theorem is the following: The electric and magnetic fields at any point inside the junction are linear functions of the currents or voltages applied to the N terminals. As a corollary to the above, it can be stated that the N terminal currents are linearly related to the N terminal voltages.

5·9. Impedance and Admittance Matrix.—Since the N terminal currents and voltages of the junction are connected by linear equations, N^2 quantities Z_{pq} can be defined such that

$$e_p = \sum_q Z_{pq} i_q. \tag{31}$$

The N^2 components Z_{pq} can be regarded as forming an Nth-order square matrix,

$$\mathsf{Z} = \begin{pmatrix} Z_{11} & Z_{12} & \cdots \\ Z_{21} & Z_{22} & \cdots \\ \cdot & \cdot & \\ \cdot & \cdot & \\ \cdot & \cdot & \end{pmatrix} \tag{32}$$

Matrix (32) will be called the impedance matrix. The N components i_q and e_p can be arranged as column vectors,

$$\mathsf{i} = \begin{pmatrix} i_1 \\ \cdot \\ \cdot \\ \cdot \\ i_N \end{pmatrix}, \quad \mathsf{e} = \begin{pmatrix} e_1 \\ \cdot \\ \cdot \\ \cdot \\ e_N \end{pmatrix}. \tag{33}$$

The vectors i and e will in the future be called the current and voltage vectors of the junction. The matrix formulation of Eq. (31) is

$$\mathsf{e} = \mathsf{Z}\mathsf{i}. \tag{34}$$

The linear relation between the terminal currents and voltages can be expressed in another way if Eq. (31) is solved for the N currents. Then

$$i_p = \sum_q Y_{pq} e_q. \tag{35}$$

Again a matrix Y can be defined such that

$$\mathsf{Y} = \begin{pmatrix} Y_{11} & Y_{12} & \cdots \\ Y_{21} & Y_{22} & \cdots \\ \cdot & \cdot & \\ \cdot & \cdot & \\ \cdot & \cdot & \end{pmatrix}. \tag{36}$$

This matrix will be called the admittance matrix;

$$\mathsf{i} = \mathsf{Y}\mathsf{e}. \tag{37}$$

SEC. 5·10] SYMMETRY OF MATRICES 141

If it is assumed that Z is nonsingular,[1] Eq. (34) can be multiplied by Z^{-1},

$$Z^{-1}e = Z^{-1}Zi = i,$$

and

$$Z^{-1} = Y. \tag{38}$$

It is worth while to notice the analogy between Eqs. (34) and (11). They are formally identical. The impedance Z of Eq. (11) has been generalized to an impedance matrix. The current and voltage have been generalized to current and voltage column vectors. For a termination, that is, a single-terminal-pair network, Eq. (34) reduces to Eq. (11).

5·10. Symmetry of Impedance and Admittance Matrices.—It will now be shown that the impedance and admittance matrices, matrices (32) and (36), are symmetrical. By a symmetrical matrix is meant one for which

$$\left. \begin{array}{l} Z_{mn} = Z_{nm}, \\ Y_{mn} = Y_{nm}. \end{array} \right\} \tag{39}$$

Let there be two solutions of Maxwell's equations that satisfy the boundary conditions imposed by the junction. The field quantities and terminal quantities of the two solutions will be distinguished by superscripts 1 and 2. From Eqs. (1),

$$\left. \begin{array}{l} \text{curl } \mathbf{H}^{(1)} - (j\omega\epsilon + \sigma)\mathbf{E}^{(1)} = 0, \\ \text{curl } \mathbf{E}^{(1)} + j\omega\mu\mathbf{H}^{(1)} = 0; \end{array} \right\} \tag{40}$$

$$\left. \begin{array}{l} \text{curl } \mathbf{H}^{(2)} - (j\omega\epsilon + \sigma)\mathbf{E}^{(2)} = 0, \\ \text{curl } \mathbf{E}^{(2)} + j\omega\mu\mathbf{H}^{(2)} = 0. \end{array} \right\} \tag{41}$$

Likewise

$$\text{div } [\mathbf{E}^{(1)} \times \mathbf{H}^{(2)} - \mathbf{E}^{(2)} \times \mathbf{H}^{(1)}]$$
$$= \mathbf{H}^{(2)} \text{ curl } \mathbf{E}^{(1)} - \mathbf{E}^{(1)} \text{ curl } \mathbf{H}^{(2)} - \mathbf{H}^{(1)} \text{ curl } \mathbf{E}^{(2)} + \mathbf{E}^{(2)} \text{ curl } \mathbf{H}^{(1)}. \tag{42}$$

From Eqs. (40) and (41),

$$\text{div } [\mathbf{E}^{(1)} \times \mathbf{H}^{(2)} - \mathbf{E}^{(2)} \times \mathbf{H}^{(1)}] = 0. \tag{43}$$

If Eq. (43) is integrated over the volume enclosed by the surface,

$$0 = \int \text{div } [\mathbf{E}^{(1)} \times \mathbf{H}^{(2)} - \mathbf{E}^{(2)} \times \mathbf{H}^{(1)}] \, dv$$
$$= \int [\mathbf{E}^{(1)} \times \mathbf{H}^{(2)} - \mathbf{E}^{(2)} \times \mathbf{H}^{(1)}] \cdot d\mathbf{S}. \tag{44}$$

The right-hand side of Eq. (44) can be expressed in terms of terminal voltages and currents [see Eq. (29)],

$$0 = \int [\mathbf{E}^{(1)} \times \mathbf{H}^{(2)} - \mathbf{E}^{(2)} \times \mathbf{H}^{(1)}] \cdot d\mathbf{S} = \sum_j [e_j^{(1)} i_j^{(2)} - e_j^{(2)} i_j^{(1)}]. \tag{45}$$

---

[1] If Z were singular, the connection between the currents and voltages would not be unique.

Equation (45) holds for any two sets of applied voltages at the terminals. In particular let

$$e_j^{(1)} = 0, \quad j \neq 1, \\ e_j^{(2)} = 0, \quad j \neq 2. \quad (46)$$

The sum given in Eq. (45) reduces, for this special case, to

$$e_1^{(1)} i_1^{(2)} - e_2^{(2)} i_2^{(1)} = 0. \quad (47)$$

However, for the special case considered

$$i_1^{(2)} = Y_{12} e_2^{(2)}, \\ i_1^{(1)} = Y_{21} e_1^{(1)}. \quad (48)$$

Substituting (48) in (47), the result is

$$Y_{12} = Y_{21}. \quad (49)$$

It is obvious that there is nothing special about the indices 1 and 2 and that

$$Y_{kl} = Y_{lk}, \quad \text{for any } k \text{ and } l. \quad (50)$$

It can be shown in a similar way that

$$Z_{kl} = Z_{lk}. \quad (51)$$

Therefore, the impedance and admittance matrices are always symmetrical.

5·11. Physical Realizability.—Certain conditions are imposed on the impedance and admittance matrices by the energy theorem [Eq. (30)]. If $e_n = \Sigma Z_{nm} i_m$ is substituted in Eq. (30),

$$\sum_{n,m} i_n^* Z_{nm} i_m = 4j\omega(W_H - W_E) + 2P. \quad (52)$$

In order to simplify the discussion, a two-terminal-pair junction will be considered first. The real part of Eq. (52) can be written as

$$i_1^* i_1 R_{11} + (i_2^* i_1 + i_2 i_1^*) R_{12} + i_2^* i_2 R_{22} = 2P, \quad (53)$$

where R_{ij} is the real part of Z_{ij}. If either i_1 or $i_2 = 0$, since $P \geq 0$, it is seen that

$$R_{11} \geq 0, \\ R_{22} \geq 0.$$

It is evident that this condition is the same as that for the termination, as indeed it should be.

Since $P \geq 0$ for any values of i_1 and i_2,

$$i_1^* i_1 R_{11} + (i_2^* i_1 + i_2 i_1^*) R_{12} + i_2^* i_2 R_{22} \geq 0. \quad (54)$$

It is evident from inspection of the coefficient of R_{12} in Eq. (54) that the minimum value of the left side of Eq. (54) occurs when i_1 and i_2 have arguments that are the same or that differ by π. Then i_1/i_2 is real, and

$$\left(\frac{i_1}{i_2}\right)^2 R_{11} + 2\left(\frac{i_1}{i_2}\right) R_{12} + R_{22} \geqq 0. \tag{55}$$

This condition can be met by requiring that the equation

$$X^2 R_{11} + 2X R_{12} + R_{22} = 0 \tag{56}$$

has no single real roots. The condition for this is

$$R_{12}^2 - R_{11} R_{22} \leqq 0. \tag{57}$$

This can be expressed as

$$\begin{vmatrix} R_{11} & R_{12} \\ R_{21} & R_{22} \end{vmatrix} \geqq 0. \tag{58}$$

The above arguments apply also to the admittance matrix and can be extended to junctions having more than two pairs of terminals.

An extension of the theorem to junctions with more than two terminal pairs yields the result that conditions imposed by Eq. (52) require the determinant of the real part of the impedance or admittance matrix and the determinant of each of its minors obtained by successively removing diagonal elements in any order to be greater than or equal to zero.

5·12. The Polyterminal-pair Lossless Junction.—Usually in practical microwave applications, a junction having more than one pair of terminals is essentially lossless. This is not always true; but usually, such things as tuners, T-junctions, and directional couplers have low loss. For this reason the lossless case is of considerable importance. In Eq. (52), if $P = 0$, the equation is purely imaginary for all applied currents i_n. Consider the special case

$$i_n = 0, \quad \text{for } n \neq k.$$

It follows from Eq. (52) that

$$\text{Re } (Z_{kk}) = 0. \tag{59}$$

That is to say, all the diagonal terms of the impedance matrix are pure imaginary. Consider now a special case in which all the applied currents vanish except two, the kth and mth, for example,

$$i_n = 0, \quad \text{for } n \neq k, m.$$

From Eq. (52),

$$\text{Re } [i_k i_k^* Z_{kk} + (i_m i_k^* + i_k i_m^*) Z_{km} + i_m i_m^* Z_{mm}] = 0, \tag{60}$$

and hence

$$\text{Re } (Z_{km}) = 0, \quad \text{for any } k \text{ and } m. \tag{61}$$

Thus, for a lossless junction, *all the terms in the impedance matrix are pure imaginary*. The above conditions apply also, in an analogous way, to the admittance matrix, and *all the terms of the admittance matrix for a lossless junction are pure imaginary*. It is seen that the statements in Sec. 5·4 are special cases of the above.

5·13. Definition of Terminal Voltages and Currents for Waveguides with More than One Propagating Mode.—The original definition of a waveguide junction was limited to one excited by transmission lines supporting a single propagating mode. In order to extend this definition to transmission lines with more than one propagating mode without invalidating the previous results, it is natural to impose the condition that Eq. (30) be valid in the new system and that the resulting impedance and admittance matrices be symmetrical. Whereas previously it was necessary to introduce a single voltage and current to describe completely the conditions in a given transmission line, now it will be necessary to introduce a voltage and current for each mode in the guide. There is no unique way of introducing these voltages and currents, but there is one way that is a little more natural than the others. It is to let each voltage and current be a description of one particular mode in the transmission line. To make this more definite, in the derivation of Eq. (30) a surface integral Eq. (29) is encountered. This is the same surface integral that occurs in Eq. (4).

To simplify the discussion, let us consider the junction to be excited by a single transmission line along which N modes may propagate. Equation (4) is applicable; but because of the N modes, Eq. (5) must be generalized to

$$\left. \begin{aligned} E_x &= \sum e_n f_x^{(n)}(x,y) \text{s} \\ E_y &= \sum e_n f_y^{(n)}(x,y), \\ H_x &= \sum i_n g_x^{(n)}(x,y), \\ H_y &= \sum i_n g_y^{(n)}(x,y). \end{aligned} \right\} \quad (62)$$

In Eq. (62), $f_x^{(n)}(x,y)$ is a real function of the coordinates which describes the distribution of the x-component of electric field of the nth mode over the terminals; similar statements apply to the other functions. The normalizing parameters e_n and i_n are so chosen that Eq. (6) is satisfied:

$$\int [f_x^{(n)} g_y^{(n)} - f_y^{(n)} g_x^{(n)}] \, dx \, dy = -1, \quad \text{for all } n. \quad (63)$$

It was shown in Sec. 2·18 that the transverse electric and magnetic fields

SEC. 5·13]   DEFINITIONS   145

for one mode are orthogonal to the transverse magnetic and electric fields, respectively, of any other mode. By the use of this fact together with Maxwell's equations it can be shown that

$$\int [f_x^{(n)} g_y^{(m)} - f_y^{(m)} g_x^{(n)}] \, dx \, dy = 0, \quad \text{for } n \neq m. \tag{64}$$

If Eqs. (62) are substituted in Eq. (4) and then Eqs. (63) and (64) are used, Eq. (30) results. In other words, the particular choice of parameters e_n and i_n introduced in Eq. (62) results in the same connection [Eq. (30)] between terminal energy quantities that was obtained for single-mode guides. Thus the terminal-parameter description of a transmission line with N propagating modes is, at least to this extent, equivalent to the description of N single-mode guides. It can be seen in an analogous way that the reciprocity condition is also satisfied and that the impedance and admittance matrices are symmetrical. Thus all the preceding results are valid for this case also.

As was pointed out earlier, the currents and voltages introduced in Eq. (62) are not the only permissible ones. To show this, new currents and voltages that are linearly related to the currents and voltages of Eq. (62) may be defined. If i' and e' are column vectors representing the new currents and voltages, then the linear relation may be expressed as

$$\left. \begin{array}{l} \mathsf{i} = \mathsf{T}\mathsf{i}', \\ \mathsf{e} = \mathsf{T}\mathsf{e}', \end{array} \right\} \tag{65}$$

where T is a matrix expressing the linear relation.

The left-hand side of Eq. (30) may be expressed in matrix form as

$$\sum_n i_n^* e_n = \tilde{\mathsf{i}}^* \mathsf{e}. \tag{66}$$

Here $\tilde{\mathsf{i}}$ represents the transpose of i; in other words, $\tilde{\mathsf{i}}$ is a row vector:

$$\tilde{\mathsf{i}} = (i_1, \cdots, i_n). \tag{67}$$

From Eq. (65),

$$\tilde{\mathsf{i}} = \tilde{\mathsf{i}}' \tilde{\mathsf{T}}. \tag{68}$$

If Eqs. (68) and (65) are substituted in Eq. (66),

$$\tilde{\mathsf{i}}^* \mathsf{e} = \tilde{\mathsf{i}}'^* \tilde{\mathsf{T}}^* \mathsf{T} \mathsf{e}'. \tag{69}$$

If e' and i' are to be a permissible representation of the left-hand side of Eq. (30), then Eq. (66) must be invariant under the transformation. The necessary and sufficient condition for this is that

$$\tilde{\mathsf{T}}^* \mathsf{T} = \mathsf{I}, \tag{70}$$

where I is the unit matrix. Equation (70) is equivalent to
$$\tilde{\mathsf{T}}^* = \mathsf{T}^{-1}. \tag{71}$$
A matrix that satisfies Eq. (71) is said to be unitary.

Condition (71) guarantees that the transformed voltages and currents will satisfy Eq. (30). However, condition (71) is not sufficient to guarantee the symmetry of the impedance and admittance matrices. The impedance matrix is defined by
$$\mathsf{e} = \mathsf{Z}\mathsf{i}. \tag{5.34}$$
The substitution of Eq. (65) in this equation gives
$$\left.\begin{aligned}\mathsf{T}\mathsf{e}' &= \mathsf{Z}\mathsf{T}\mathsf{i}',\\ \mathsf{e}' &= \mathsf{T}^{-1}\mathsf{Z}\mathsf{T}\mathsf{i}'.\end{aligned}\right\} \tag{72}$$
In order for T to be a permissible transformation, the matrix $\mathsf{T}^{-1}\mathsf{Z}\mathsf{T}$ must be symmetrical,
$$\mathsf{T}^{-1}\mathsf{Z}\mathsf{T} = \tilde{\mathsf{T}}\tilde{\mathsf{Z}}\tilde{\mathsf{T}}^{-1} = \tilde{\mathsf{T}}\mathsf{Z}\tilde{\mathsf{T}}^{-1}. \tag{73}$$
In order for this equation to hold for all Z,
$$\mathsf{T}^{-1} = \tilde{\mathsf{T}}. \tag{74}$$
A matrix that satisfies Eq. (74) is said to be orthogonal. Conditions (74) and (71) lead to the result
$$\mathsf{T} = \mathsf{T}^*. \tag{75}$$
To state the results in words: If the terminal currents and voltages are transformed by a real orthogonal transformation, the resulting impedance matrix is symmetrical and Eq. (30) is left invariant. Because all the preceding results stem from the symmetry of the impedance and admittance matrices and from Eq. (30), it follows that the new set of currents and voltages are a permissible set.

5·14. Scattering Matrix.—The wave formalism introduced in Sec. 5·6 can be extended to a junction with many terminal pairs. As in the previous restricted case the incident wave will be represented by **a** and the reflected wave by **b**.

Let a_n be a complex number representing the amplitude and phase of the transverse electric field of the incident wave at the nth terminal pair. Let b_n be the corresponding measure of the emergent wave. It is assumed that a_n and b_n are normalized in such a way that $\frac{1}{2}a_n^* a_n$ is the average incident power, and correspondingly for b_n.

As in Eqs. (19) and (20)
$$\left.\begin{aligned}e_n &= g_n(a_n + b_n),\\ i_n &= \frac{1}{g_n}(a_n - b_n),\end{aligned}\right\} \tag{76}$$
where g_n is a constant for that terminal pair.

The characteristic impedance of the nth guide is connected in a simple way with g_n. To show this connection, let $b_n = 0$; then

$$\left.\begin{aligned} e_n &= g_n a_n, \\ i_n &= \frac{1}{g_n} a_n, \\ Z_0^{(n)} &= \frac{e_n}{i_n} = g_n^2, \\ g_n &= \sqrt{Z_0^{(n)}}. \end{aligned}\right\} \quad (77)$$

Thus g_n is the square root of the characteristic impedance of the nth guide.

As can be seen from Eqs. (62) and (63), it is always possible to choose e_n and i_n in such a way as to make $Z_0^{(n)} = 1$. It will be assumed that this has been done. Then $g_n = 1$, and

$$\left.\begin{aligned} e_n &= a_n + b_n, \\ i_n &= a_n - b_n; \end{aligned}\right\} \quad (78)$$

$$\left.\begin{aligned} a_n &= \tfrac{1}{2}(e_n + i_n), \\ b_n &= \tfrac{1}{2}(e_n - i_n). \end{aligned}\right\} \quad (79)$$

If e_n is substituted from Eq. (31),

$$\left.\begin{aligned} a_n &= \tfrac{1}{2}\sum_m (Z_{nm} + \delta_{nm}) i_m, \\ b_n &= \tfrac{1}{2}\sum_m (Z_{nm} - \delta_{nm}) i_m, \end{aligned}\right\} \quad (80)$$

where

$$\begin{aligned} \delta_{nm} &= 0, \quad n \neq m, \\ \delta_{nm} &= 1, \quad n = m. \end{aligned}$$

Using the matrix notation, Eq. (80) becomes

$$\mathbf{a} = \tfrac{1}{2}(\mathbf{Z} + \mathbf{1})\mathbf{i}, \quad (81)$$
$$\mathbf{b} = \tfrac{1}{2}(\mathbf{Z} - \mathbf{1})\mathbf{i}, \quad (82)$$

where

$$\begin{pmatrix} a_1 \\ \cdot \\ \cdot \\ \cdot \\ a_n \end{pmatrix} = \mathbf{a},$$

and so forth. Equation (81) can be solved for $\mathbf{i}$:

$$\mathbf{i} = 2(\mathbf{Z} + \mathbf{1})^{-1}\mathbf{a}. \quad (83)$$

If this is substituted in Eq. (82),
$$b = (Z - I)(Z + I)^{-1}a. \tag{84}$$
The matrix connecting a and b in Eq. (84) will be called the scattering matrix S;
$$b = Sa, \tag{85}$$
and
$$S = (Z - I)(Z + I)^{-1}. \tag{86}$$
It can be shown in an analogous way that
$$S = (I - Y)(I + Y)^{-1}. \tag{87}$$
The importance of the scattering matrix will become evident when examples are discussed.

5·15. Symmetry.—The matrix S has several general properties of importance. One of these is its symmetry. It will be shown that the transpose of S is equal to S.

To show this, let
$$\left. \begin{array}{l} Z - I = G, \\ Z + I = H, \\ S = GH^{-1}. \end{array} \right\} \tag{88}$$
It is evident that
$$GH = HG. \tag{89}$$
If Eq. (89) is multiplied on the left and right by H^{-1},
$$H^{-1}GHH^{-1} = H^{-1}HGH^{-1}, \tag{90}$$
$$H^{-1}G = GH^{-1}.$$
If this result is substituted in Eq. (88),
$$S = H^{-1}G,$$
and
$$\tilde{S} = \widetilde{H^{-1}G} = \tilde{G}\tilde{H}^{-1} = GH^{-1} = H^{-1}G, \tag{91}$$
since G and H are symmetrical.
Therefore
$$\tilde{S} = S, \tag{92}$$
and S is symmetrical.

5·16. Energy Condition.—Some additional conditions are imposed by Eq. (30). If Eq. (78) is substituted in Eq. (30), there follows
$$\sum_n (a_n^* - b_n^*)(a_n + b_n) = 4j\omega(W_H - W_E) + 2P. \tag{93}$$
The real and imaginary parts of Eq. (93) are

$$\sum (a_n^* a_n - b_n^* b_n) = 2P, \tag{94}$$

$$\sum (a_n^* b_n - a_n b_n^*) = 4j\omega(W_H - W_E). \tag{95}$$

In matrix notation, Eqs. (94) and (95) become

$$\tilde{\mathsf{a}}^*(\mathsf{I} - \mathsf{S}^*\mathsf{S})\mathsf{a} = 2P, \tag{96}$$
$$\tilde{\mathsf{a}}^*(\mathsf{S} - \mathsf{S}^*)\mathsf{a} = 4j\omega(W_H - W_E). \tag{97}$$

Since $P \geqq 0$, the same conditions are imposed on $(\mathsf{I} - \mathsf{S}^*\mathsf{S})$ as were imposed on Re (Z) in Sec. 5·11. These conditions were that

$$\det(\mathsf{I} - \mathsf{S}^*\mathsf{S}) \geqq 0, \tag{98}$$

and the same for each of the principal minors.

The case of a lossless junction deserves special notice. In this case $P = 0$ for all a in Eq. (96). This leads to the result

$$\mathsf{I} = \mathsf{S}^*\mathsf{S} \tag{99}$$

or

$$\mathsf{S}^{-1} = \mathsf{S}^*. \tag{100}$$

Since S is symmetrical, $\tilde{\mathsf{S}}^* = \mathsf{S}^*$ and

$$\mathsf{S}^{-1} = \tilde{\mathsf{S}}^*. \tag{101}$$

Equation (101) is the definition of a unitary matrix. *Thus the scattering matrix is symmetrical and unitary for a lossless junction.*

5·17. Transformation of the Scattering Matrix under a Shift in Position of the Terminal Reference Planes.—The transformation introduced in an impedance or admittance matrix by a shift in the reference planes was discussed in Chap. 4. In the case of the scattering matrix this transformation is almost trivial. This can be seen if it is remembered that the time required for a wave to enter a junction by way of a transmission line is increased if the reference plane of that line is moved away from the junction. Also the time required for a wave to leave the junction is increased if the phase plane is moved away from the junction.

To put this in quantitative terms, if the terminal (reference) plane of the kth line is moved out from the junction by a distance l_k, then the transformed scattering matrix is

$$\mathsf{S}' = \mathsf{PSP}, \tag{102}$$

where S is the original scattering matrix and P is a matrix with nonzero elements on the principal diagonal only, a *diagonal* matrix. The kth diagonal element of P is

$$P_{kk} = e^{-2\pi j \frac{l_k}{\lambda_k}} \tag{103}$$

where λ_k is the wavelength in the kth transmission line.

5·18. The T-matrix of a Series of Junctions Connected in Cascade.—
Let a series of two-terminal-pair junctions be connected end to end so as to form a chain. The problem is to find the scattering matrix for the chain. The scattering matrix is actually not the most convenient representation of the properties of such a cascade of junctions. It is more convenient to introduce a matrix that relates the conditions at the output terminals to those at the input terminals. This matrix will be called the T-matrix.

If
$$b_1 = S_{11}a_1 + S_{12}a_2 \tag{104}$$
and
$$b_2 = S_{21}a_1 + S_{22}a_2,$$
then
$$b_2 = \frac{S_{22}}{S_{21}} b_1 + \left(S_{12} - \frac{S_{22}S_{11}}{S_{21}} \right) a_1 \tag{105}$$
and
$$a_2 = \frac{1}{S_{12}} b_1 - \frac{S_{11}}{S_{21}} a_1.$$

This may be written as
$$\mathsf{g} = \mathsf{Th}, \tag{106}$$
where
$$\mathsf{g} = \begin{pmatrix} b_2 \\ a_2 \end{pmatrix}, \qquad \mathsf{h} = \begin{pmatrix} a_1 \\ b_1 \end{pmatrix} \tag{107}$$
and
$$\mathsf{T} = \begin{pmatrix} S_{12} - \dfrac{S_{22}S_{11}}{S_{21}} & \dfrac{S_{22}}{S_{21}} \\ -\dfrac{S_{11}}{S_{21}} & \dfrac{1}{S_{12}} \end{pmatrix}. \tag{108}$$

If the junctions are numbered in the same order in which they occur in the chain, then
$$\mathsf{g}_k = \mathsf{T}_k \mathsf{h}_k, \tag{109}$$
and
$$\mathsf{g}_k = \mathsf{h}_{k+1}, \tag{110}$$
where the subscript refers to the number of the junction. If the equations are combined, it is found that
$$\mathsf{T}_n \cdots \mathsf{T}_1 \mathsf{h}_1 = \mathsf{g}_n. \tag{111}$$
However, this implies that the resultant T-matrix for the chain is
$$\mathsf{T} = \mathsf{T}_n \mathsf{T}_{n-1} \cdots \mathsf{T}_1. \tag{112}$$
If desired, the components of T may be used to obtain the scattering matrix assuming a lossless chain by the relation

$$S = \begin{pmatrix} -\dfrac{T_{21}}{T_{22}} & \dfrac{1}{T_{22}} \\ \dfrac{1}{T_{22}} & \dfrac{T_{12}}{T_{22}} \end{pmatrix}. \tag{113}$$

5·19. The Scattering Matrix of a Junction with a Load Connected to One of the Transmission Lines.—Let one of the transmission lines (the kth) of the junction be terminated by a load of reflection coefficient Γ referred to the reference terminal of the transmission line. This load imposes the condition

$$\Gamma b_k = a_k. \tag{114}$$

But

$$b_k = \sum_j S_{kj} a_j. \tag{115}$$

Hence, if these expressions are combined,

$$a_k = \frac{\Gamma}{1 - \Gamma S_{kk}} \sum_j{'} S_{kj} a_j, \tag{116}$$

where the prime denotes that the kth term is eliminated in the sum. Equation (116) can then be substituted in the remainder of the scattering matrix to eliminate a_k. The kth terminal is thus completely eliminated from the scattering matrix.

FREQUENCY DEPENDENCE OF A LOSSLESS JUNCTION

The energy integrals of the previous sections gave information about terminal quantities at one frequency. A new energy integral will now be formulated which relates the rate of change of terminal parameters with respect to frequency to the stored energies in a lossless junction.

5·20. Variational Energy Integral.—Let us consider a junction to which is connected several transmission lines. Maxwell's equations for a lossless junction are

$$\left. \begin{array}{l} \operatorname{curl} \mathbf{H} - j\omega\epsilon\mathbf{E} = 0, \\ \operatorname{curl} \mathbf{E} + j\omega\mu\mathbf{H} = 0. \end{array} \right\} \tag{117}$$

Let us now consider a solution of Eq. (117) which satisfies the boundary conditions of the junction, and let us introduce a variation of the frequency and field quantities consistent with the boundary conditions. The variations satisfy the equations

$$\begin{array}{l} \operatorname{curl} \delta\mathbf{H} - j\epsilon(\omega\delta\mathbf{E} + \mathbf{E}\delta\omega) = 0, \\ \operatorname{curl} \delta\mathbf{E} + j\mu(\omega\delta\mathbf{H} + \mathbf{H}\delta\omega) = 0. \end{array} \tag{118}$$

If the quantity

$$\text{div } [\mathbf{E} \times \delta\mathbf{H} - \delta\mathbf{E} \times \mathbf{H}] = [\delta\mathbf{H} \cdot \text{curl } \delta\mathbf{E} - \mathbf{E} \cdot \text{curl } \delta\mathbf{H}]$$
$$- [\mathbf{H} \cdot \text{curl } \delta\mathbf{E} - \delta\mathbf{E} \cdot \text{curl } \mathbf{H}] \quad (119)$$

is introduced, and if, in the right side of this expression, quantities from Eqs. (117) and (118) are substituted, there results

$$\text{div } [\mathbf{E} \times \delta\mathbf{H} - \delta\mathbf{E} \times \mathbf{H}] = j(\mu H^2 - \epsilon E^2)\delta\omega. \quad (120)$$

If Eq. (120) is integrated over the volume of the junction and the left side of the equation converted to a surface integral, there results

$$\int_S (\mathbf{E} \times \delta\mathbf{H} - \delta\mathbf{E} \times \mathbf{H}) \cdot d\mathbf{S} = j\delta\omega \int_v (\mu H^2 - \epsilon E^2) \, dv. \quad (121)$$

5·21. Application to Impedance and Admittance Matrix.—The left-hand side of Eq. (121) is an integral over the various terminals of the waveguide junction and leads to the result

$$\sum_n (e_n \delta i_n - i_n \delta e_n) = j\delta\omega \int_v (\mu H^2 - \epsilon E^2) \, dv, \quad (122)$$

where, as before, i_n and e_n are the current and voltage at the nth terminals. It is to be noted that Eq. (122) relates a variation of the terminal voltages and currents to an energy integral times a variation in frequency. It is evident from Sec. 5·10 that if e_n is real for all n, then $\mathbf{E}$ is real, $\mathbf{H}$ is imaginary, and i_n is imaginary. If Eq. (122) be limited to real terminal voltages, it becomes

$$\sum_n (e_n \delta i_n - i_n \delta e_n) = -j\delta\omega \int_v (\mu \mathbf{H}^* \cdot \mathbf{H} + \epsilon \mathbf{E}^* \cdot \mathbf{E}) \, dv$$
$$= -4j\delta\omega(W_E + W_H), \quad (123)$$

when e_n is real, independently of n. If the variation of Eq. (31) is taken

$$\delta e_n = \sum_m (Z_{nm} \delta i_m + \delta Z_{nm} i_m). \quad (124)$$

If this and Eq. (31) are substituted in Eq. (123), since $i_n^* = -i_n$,

$$\sum_{n,m} i_n^* \delta Z_{nm} i_m = 4j\delta\omega(W_E + W_H), \quad (125)$$

where, as before, W_E and W_H are the average electric and magnetic energies. The impedance elements Z_{nm} are functions of frequency only, and if

$$\frac{\delta Z_{nm}}{\delta \omega} = \frac{dZ_{nm}}{d\omega} = Z'_{nm},$$

$$\sum_{n,m} i_n^* Z'_{nm} i_m = 4j(W_E + W_H). \tag{126}$$

Similarly, from Eq. (35),

$$\sum_{n,m} e_n^* Y'_{nm} e_m = 4j(W_E + W_H). \tag{127}$$

Equations (126) and (127) are the starting point of the discussion of the frequency dependence of impedance and admittance matrices.

In matrix notation, Eqs. (126) and (127) become

$$\tilde{\text{i}}^* \text{Z}' \text{i} = 4jW, \tag{128}$$
$$\tilde{\text{e}}^* \text{Y}' \text{e} = 4jW, \tag{129}$$

where W is the total stored energy corresponding to the particular terminal conditions. In Eqs. (128) and (129) it is necessary that all the terminal voltages have the same phase angle.

Since $W > 0$, the conditions that are imposed on the reactance matrix and the susceptance matrix by Eqs. (128) and (129) are identical with the conditions imposed on the real part of the impedance matrix by Eq. (52).

In order for Eq. (128) or (126) to be satisfied for any pure real currents i_m,

$$\det(X'_{nm}) \geqq 0.$$

Also, all the principal minors must be greater than zero.

5·22. Application to Scattering Matrix.—If Eq. (78) is substituted in Eq. (122), there results

$$2 \sum_n (b_n \, \delta a_n - a_n \, \delta b_n) = j(\epsilon E^2 - \mu H^2) \, \delta \omega. \tag{130}$$

If E^2 has the phase angle β throughout the junction, then, as above, H^2 will have the phase angle $\beta + \pi$, and

$$2 \sum_n (b_n \, \delta a_n - a_n \, \delta b_n) = je^{j\beta}(\epsilon \mathbf{E}^* \cdot \mathbf{E} + \mu \mathbf{H}^* \cdot \mathbf{H}) \, \delta \omega. \tag{131}$$

In matrix notation

$$2(\tilde{\text{b}} \delta \text{a} - \tilde{\text{a}} \delta \text{b}) = 4je^{j\beta} W \, \delta \omega, \tag{132}$$

where, as above, W is the total stored energy. Since E^2 has the phase angle β, e_n has, except for a possible change in sign, a phase angle $\beta/2$ for all n. Also i_n has a phase angle $(\beta/2) + (\pi/2)$.

Therefore,

$$e^{-\frac{j\beta}{2}}\mathsf{e} \tag{133}$$

is pure real and

$$e^{-\frac{j\beta}{2}}\mathsf{i}$$

is pure imaginary.

From Eqs. (78) and (85),

$$\begin{aligned}\mathsf{e} &= (\mathsf{I} + \mathsf{S})\mathsf{a},\\ \mathsf{i} &= (\mathsf{I} - \mathsf{S})\mathsf{a},\end{aligned} \tag{134}$$

where S is the scattering matrix.
From Eqs. (133) and (134),

$$\begin{aligned} e^{-\frac{j\beta}{2}}(\mathsf{I} + \mathsf{S})\mathsf{a} &= e^{\frac{j\beta}{2}}(\mathsf{I} + \mathsf{S}^*)\mathsf{a}^*, \\ e^{-\frac{j\beta}{2}}(\mathsf{I} - \mathsf{S})\mathsf{a} &= -e^{\frac{j\beta}{2}}(\mathsf{I} - \mathsf{S}^*)\mathsf{a}^*. \end{aligned} \tag{135}$$

From Eq. (135),

$$\mathsf{a} = e^{j\beta}\mathsf{S}^*\mathsf{a}^*. \tag{136}$$

Equation (136) is the condition that a must satisfy in order that

$$\arg E^2 = \beta.$$

If Eq. (85) is substituted in Eq. (132),

$$-\tfrac{1}{2}\tilde{\mathsf{a}}(\delta \mathsf{S})\mathsf{a} = je^{j\beta}(W_E + W_H)\,\delta\omega. \tag{137}$$

Let

$$\mathsf{S}' = \frac{d\mathsf{S}}{d\omega}. \tag{138}$$

Then if Eqs. (138) and (136) are substituted in Eq. (137),

$$\tfrac{1}{2}j\tilde{\mathsf{a}}^*\mathsf{S}^*\mathsf{S}'\mathsf{a} = W. \tag{139}$$

Equations (139) and (136) are the starting point for the investigation of the frequency dependence of the scattering matrix of a general junction.

5·23. Transmission-line Termination.—Consider a lossless termination of a single transmission line. The matrix equations [Eqs. (128) and (129)] reduce to the scalar equations

$$Z' = 2j\frac{W}{\tfrac{1}{2}i^*i} \tag{140}$$

and

$$Y' = 2j\frac{W}{\tfrac{1}{2}e^*e}. \tag{141}$$

Stated in words, the rate of change of reactance with frequency is always positive and is equal to four times the stored energy divided by the square of the magnitude of the current. A similar statement holds for the susceptance.

The matrix equation [Eq. (139)] reduces to the scalar equation

$$jS^*S' = \frac{W}{\frac{1}{2}a^*a}. \qquad (142)$$

Equation (136) is satisfied for any a. Since the termination is lossless,

$$S = e^{j\phi}. \qquad (143)$$

Also $\frac{1}{2}a^*a$ is the incident power P. If Eq. (143) is substituted in Eq. (142),

$$-\frac{d\phi}{d\omega} = +\frac{W}{P}, \qquad (144)$$

where $-\phi$ is the phase delay in the wave after reflection. Equation (144) states that the electrical line length into the termination and out again always increases with frequency, and the rate of increase is equal to the stored energy per unit incident power.

The physical significance of Eq. (144) is rather interesting. If a pulse, represented by

$$g(t) = \int a(\omega)e^{j\omega t}\, d\omega, \qquad (145)$$

is introduced at the terminals, then some time later a pulse will be reflected out of the termination. This pulse will have the form

$$h(t) = \int b(\omega)e^{j\omega t}\, d\omega, \qquad (146)$$

where

$$b(\omega) = S(\omega)a(\omega). \qquad (147)$$

If the pulse contains only a small band of frequencies, it is not distorted by the termination and there is the relation

$$h(t) = g(t - \tau), \qquad (148)$$

where τ is the delay introduced by the termination. From the combination of Eqs. (145), (146), and (148), there results

$$a(\omega)e^{-j\omega\tau} = b(\omega). \qquad (149)$$

From this and Eq. (147), it is seen that

$$S(\omega) = e^{-j\omega\tau}. \qquad (150)$$

Equation (150) is, of course, valid only over the small range of frequencies included in the pulse.

If Eq. (150) is compared with Eq. (144),

$$\tau = \frac{W}{P}. \qquad (151)$$

Equation (151) states that the time required for a pulse of energy to enter the termination and leave again is just the average stored energy per unit incident c-w power.

5·24. Foster's Reactance Theorem.—From Eqs. (140) and (141),

$$X' = 2 \frac{W_E + W_H}{\frac{1}{2} i^* i}, \qquad (152)$$

$$B' = 2 \frac{W_E + W_H}{\frac{1}{2} e^* e}. \qquad (153)$$

If $W = W_E + W_H$ vanishes, **E** and **H** vanish throughout the termination including the terminals. Hence X' and B' must always be greater than zero for a positive W.

If it is assumed that X has a zero at $\omega = \omega_0$, then X can be expanded in a power series about ω_0

$$X = a_1(\omega - \omega_0) + a_2(\omega - \omega_0)^2 + \cdots . \qquad (154)$$

This expansion is valid in the neighborhood of ω_0, and hence

$$X' = a_1 + 2a_2(\omega - \omega_0) + \cdots . \qquad (155)$$

From Eq. (152), $X' > 0$; therefore, $a_1 > 0$.

Since $a_1 > 0$, for ω nearly equal to ω_0, the first term in Eq. (154) is the dominant term, and in this region

$$B(\omega) = \frac{-1}{X} = \frac{-\frac{1}{a_1}}{(\omega - \omega_0)}. \qquad (156)$$

Thus if $X(\omega)$ has a zero at a certain frequency, $B(\omega)$ has a simple pole with a negative residue at the same point. Conversely a zero in $B(\omega)$ leads to a simple pole in $X(\omega)$.

From Eqs. (13) and (14),

$$X = 2\omega \frac{W_H - W_E}{\frac{1}{2} i^* i}, \qquad (157)$$

$$B = 2\omega \frac{W_E - W_H}{\frac{1}{2} e^* e}. \qquad (158)$$

It is evident from Eq. (157) that for $\omega = 0$, $X = 0$ unless the stored electric or magnetic energy becomes divergent for a given finite current. If the stored energy becomes infinite, from Eq. (152) $\omega = 0$ is a singular point. If this singular point is a pole, it is clear from the foregoing discussion that the pole is simple and has a negative residue. In this case $B(0) = 0$. Thus, at $\omega = 0$, either $X(\omega)$ or $B(\omega)$ has a zero.

Let it be assumed for the present that $X(\omega)$ has a zero at $\omega = 0$. Remembering that $X(\omega) = -X(-\omega)$,

$$X = b_1\omega + b_3\omega^3 + b_5\omega^5 + \cdots. \quad (159)$$

This expansion is valid for $|\omega| < |\omega_1|$, where ω_1 is the location of the first pole.[1] Since this pole must be simple and must have a negative residue, its principal part is

$$-\frac{V_1}{\omega - \omega_1}, \quad (160)$$

where V_1 is positive.

In order for X to be odd, however, there must be a singularity at $\omega = -\omega_1$. If the principal parts of both these singularities are subtracted from X, the remainder of the function is regular at the points $\omega - \omega_1$ and $\omega = -\omega_1$ and can be expanded in a power series that is valid for $|\omega| < |\omega_2|$ where ω_2 is the next singularity.

If there are only a finite number of poles, this process can be continued until the power-series expansion is valid for all finite frequencies. Then

$$X(\omega) = -\sum_n r_n \left(\frac{1}{\omega - \omega_n} + \frac{1}{\omega + \omega_n}\right) + a_1\omega + a_3\omega^3 + \cdots. \quad (161)$$

If the termination is a network composed of a finite number of impedance elements, then the power series in Eq. (161) can have only a finite number of terms, and $X(\omega)$ may have at most a pole at infinity. All poles, however, must be simple with a positive residue. Therefore, all terms in the power series after the first are zero. If these terms are set equal to zero, there finally results

$$X(\omega) = -2\sum r_n \frac{\omega}{\omega^2 - \omega_n^2} + a_1\omega. \quad (162)$$

In the above development it was assumed that there were a finite number of poles. This is true only for networks composed of lumped impedance elements. A distributed system has an infinite number of poles which form a condensation at the point at infinity. In this case the sum in Eq. (162) is taken over all poles up to m. Equation (162) is then a good approximation in the range $|\omega| \ll |\omega_m|$.

It is to be noted that the last term in Eq. (162) is the reactance of an

---

[1] An essential singularity causes a function to behave very wildly in its vicinity. It is unreasonable to expect a function representing a physical quantity to have an essential singularity in the finite plane. It will be assumed that all singularities in the finite plane of a reactance or susceptance function are poles. It will be shown later that all poles must lie on the imaginary impedance axis.

inductance. If there is a nonvanishing r_0 for $\omega_0 = 0$, then this term represents a capacitor. Also the nth term in the sum is the reactance of a shunt combination of an inductor and a capacitor whose resonant frequency is ω_n and whose capacitance is $1/2r_n$. It is evident that the reactance function given in Eq. (162) can be synthesized by the circuit in Fig. 5·4.

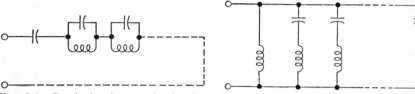

FIG. 5·4.—Synthesis of a termination by shunt-tuned elements.

FIG. 5·5.—Synthesis of a termination by series-tuned elements.

As was pointed out above, an expression of the same form as Eq. (162) can be obtained for the susceptance function. It can be synthesized by the circuit of Fig. 5·5. It is clear that a circuit of either of these two types can be used to synthesize any lossless termination, provided the frequency is not too high.

It has been assumed that poles of the reactance or susceptance function lie only on the real axis in the complex ω-plane. This will now be proved. Assume that there are poles lying on the real axis as well as poles that are not on the real axis. The poles on the real axis can be removed by a circuit of either of the types shown in Figs. 5·4 and 5·5. That part of the circuit which remains within the termination is assumed to consist at most of poles not lying on the real frequency axis. Such poles have zero susceptance at infinity. Also the susceptance is a continuous function along the real axis, since there are no singularities on the real axis. Since the susceptance of the termination in Fig. 5·5 is a continuous function of frequency vanishing at $\pm \infty$ and must have a zero or positive slope, it must vanish everywhere. Thus all poles of a reactance or susceptance function must lie on the real frequency axis.

5·25. Frequency Variation of a Lossless Junction with Two Transmission Lines.—If a two-transmission-line junction is matched, the scattering matrix is

$$S = \begin{pmatrix} 0 & 1 \\ 1 & 0 \end{pmatrix}. \tag{163}$$

In order to simplify the discussion, the terminals have been located in the transmission line at such a position that $S_{12} = 1$. This represents no important restriction.

In general none of the elements of S' vanish, but S' must satisfy Eq. (139), where a satisfies Eq. (136). Let

$$\left.\begin{aligned} \mathbf{a}_1 &= \sqrt{2}\begin{pmatrix}1\\1\end{pmatrix}, \\ \mathbf{a}_2 &= \sqrt{2}\begin{pmatrix}1\\j\end{pmatrix}, \\ \mathbf{a}_3 &= \sqrt{2}\begin{pmatrix}1\\-1\end{pmatrix}, \\ \mathbf{a}_4 &= \sqrt{2}\begin{pmatrix}1\\-j\end{pmatrix}. \end{aligned}\right\} \quad (164)$$

These four column vectors satisfy Eq. (136), as can be seen by inspection. Let W_k be the average stored energy in the junction corresponding to $\mathbf{a}_k$.

If $\mathbf{a}_1$ is substituted in Eq. (139),

$$j(1,1)\begin{pmatrix}0&1\\1&0\end{pmatrix}\mathbf{S}'\begin{pmatrix}1\\1\end{pmatrix} = W_1. \quad (165)$$

This reduces to
$$j(S'_{11} + 2S'_{12} + S'_{22}) = W_1, \quad (166)$$
where
$$\mathbf{S}' = \begin{pmatrix}S'_{11} & S'_{12}\\S'_{12} & S'_{22}\end{pmatrix}. \quad (167)$$

In a similar way the remainder of the $\mathbf{a}$'s may be substituted in Eq. (139) to yield

$$\left.\begin{aligned} -(S'_{11} + 2jS'_{12} - S'_{22}) &= W_2, \\ -j(S'_{11} - 2S'_{12} + S'_{22}) &= W_3, \\ (S'_{11} - 2jS'_{12} - S'_{22}) &= W_4. \end{aligned}\right\} \quad (168)$$

If the column vectors,

$$\begin{aligned} \mathbf{a}'_1 &= \tfrac{1}{2}(\mathbf{a}_1 + \mathbf{a}_3) = \tfrac{1}{2}(\mathbf{a}_2 + \mathbf{a}_4), \\ \mathbf{a}'_2 &= \tfrac{1}{2}(\mathbf{a}_1 - \mathbf{a}_3) = -\tfrac{1}{2}j(\mathbf{a}_2 - \mathbf{a}_4), \end{aligned} \quad (169)$$

are formed for which

$$\left.\begin{aligned} \mathbf{a}'_1 &= \sqrt{2}\begin{pmatrix}1\\0\end{pmatrix}, \\ \mathbf{a}'_2 &= \sqrt{2}\begin{pmatrix}0\\1\end{pmatrix}, \end{aligned}\right\} \quad (170)$$

then $\mathbf{a}'_1$ and $\mathbf{a}'_2$ represent waves incident in one line only. They differ only in the direction of transit through the junction. Hence the stored energies are equal for $\mathbf{a}'_1$ and $\mathbf{a}'_2$. Let this stored energy be W. It should be noted that

$$\mathbf{a}_k = \mathbf{a}'_1 + (j)^{k-1}\mathbf{a}'_2. \quad (171)$$

Thus each of the column vectors of Eq. (164) can be represented as a linear combination of the vectors of Eq. (170). If the electric and magnetic fields in the junction corresponding to the incident waves a_1' and a_2' are denoted by the subscripts 1 and 2 respectively, then

$$W = \tfrac{1}{2} \int (\mu H_1^* H_1 + \epsilon E_1^* E_1)\, dv = \tfrac{1}{2} \int (\mu H_2^* H_2 + \epsilon E_2^* E_2)\, dv. \quad (172)$$

Also

$$W_k = \tfrac{1}{2} \int [\mu(H_1^* + j^{1-k} H_2^*)(H_1 + j^{k-1} H_2)$$
$$+ \epsilon(E_1^* + j^{1-k} E_2^*)(E_1 + j^{k-1} E_2)]\, dv. \quad (173)$$

Thus
$$W_k = 2W + \operatorname{Re}(j^{k-1} A), \quad (174)$$
where

$$A = \int (\mu H_1^* H_2 + \epsilon E_1^* E_2)\, dv. \quad (175)$$

It is important to note that
$$|A| \leq 2W. \quad (176)$$

From Eqs. (166) and (168),
$$W_1 + W_3 = W_2 + W_4 = 4j S_{12}'. \quad (177)$$

From Eqs. (174) and (177),
$$j S_{12}' = W. \quad (178)$$

Since a_1' and a_2' represent 1 watt of power incident on the junction, Eq. (178) may be generalized to

$$j S_{12}' = \frac{W}{P}, \quad (179)$$

where P is the power flowing through the junction.
Let
$$S_{12} = e^{j\phi}.$$

For $\phi = 0$, Eq. (179) may be written as

$$\phi' = -\frac{W}{P}. \quad (180)$$

It will be noted that this expression is completely equivalent to Eq. (144). The discussion following Eq. (144) is also applicable to Eq. (180). In particular, the delay T introduced by a matched delay line is, in seconds,

$$T = \frac{W}{P}. \quad (181)$$

From Eqs. (166) and (168),

$$S_{11}' + S_{22}' = \tfrac{1}{2} j (W_3 - W_1), \quad (182)$$
$$S_{11}' - S_{22}' = \tfrac{1}{2}(W_4 - W_2). \quad (183)$$

From Eqs. (182) and (183),
$$S'_{11} = -S'^{*}_{22}, \tag{184}$$
$$S'_{11} = \tfrac{1}{4}[(W_4 - W_2) + j(W_3 - W_1)]. \tag{185}$$

From Eq. (174),
$$\begin{aligned}S'_{11} &= -\tfrac{1}{2}j[\operatorname{Re}(A) - j\operatorname{Re}(jA)] \\ &= -\tfrac{1}{2}jA.\end{aligned} \tag{186}$$

From Eqs. (176) and (184),
$$|S'_{11}| = |S'_{22}| \leq \frac{W}{P} = |S'_{12}|. \tag{187}$$

Equation (187) expresses important restrictions on the frequency sensitivity of the phase shift through the junction and of the match at each terminal.

To illustrate the significance of Eq. (187), a waveguide junction may be considered that may be represented by the simple equivalent circuit of a shunt inductance and a shunt capacitance at resonance, in a transmission line of unit characteristic impedance. Then

$$S_{11} = j\frac{\omega C - \dfrac{1}{\omega L}}{2 + j\left(\omega C - \dfrac{1}{\omega L}\right)} = S_{22}. \tag{188}$$

At the resonant frequency,
$$\left.\begin{aligned}S_{11} &= 0, \\ S'_{11} &= jC.\end{aligned}\right\} \tag{189}$$

The value of W/P is C, and therefore the equal sign in Eq. (187) holds. A simple shunt-tuned circuit thus has the maximum value of frequency sensitivity.

CHAPTER 6

WAVEGUIDE CIRCUIT ELEMENTS

By C. G. Montgomery

In preceding chapters the normal modes of propagation along a continuous waveguide have been described and the production of reflected waves by a discontinuity has been discussed. It was shown that the low-frequency transmission-line formulas were valid to describe the propagation of the effects of the discontinuity along the waveguide. In Chap. 5 general theorems were developed that formed extensions of the low-frequency network theorems to waveguide transmission lines. In the present chapter particular examples of discontinuities will be discussed and these general theorems applied.

6·1. Obstacles in a Waveguide.—One of the most common forms of discontinuity used in waveguide circuits is a metallic partition extending partially across the guide in a plane perpendicular to the axis. The thickness of the partition is usually small compared with a wavelength, but the effects of the thickness cannot always be neglected. The opening in the partition may be of any shape. Such a partition is called a diaphragm or an iris. In the neighborhood of the iris higher-mode fields are set up when a wave is incident, so that the total field satisfies the proper boundary conditions. A dominant-mode wave is reflected from the iris, and some of the incident power is transmitted through the opening. Consider first that the waveguide to which the power is transmitted is infinite in length or is terminated in a reflectionless absorber. The iris and the reflectionless termination may now be considered together to be some impedance or reactance terminating the transmission line. Equations (5·13) and (5·14) give the values of this impedance or admittance in terms of the stored electric and magnetic energies W_E and W_H and the dissipated power P. If the losses in the metal diaphragm are negligible, as is usually the case, power is dissipated only in the absorbing load. Thus the metal diaphragm is responsible for the imaginary part of the impedance or the reactance, and the magnitude of the imaginary part is proportional to the difference between the stored electric and magnetic energies in the neighborhood of the iris. In the waveguide far from the iris, the stored electric and magnetic energies are equal as shown in Sec. 2·18.

Irises are classified according to the sign of the imaginary part of the impedance to conform to the terminology of low-frequency networks. Thus an iris that contributes a negative imaginary term to the admittance is called inductive; one that contributes a negative imaginary part to the impedance is called capacitive. It should be noted that this classification depends only upon the sign and not upon the frequency variation of the reactance or susceptance.

THIN DIAPHRAGMS AS SHUNT REACTANCES

6·2. Shunt Reactances.—Since a metal diaphragm in a waveguide may be considered as a junction with two emergent transmission lines, a proper equivalent-circuit representation would be that of a two-terminal-pair network at low frequencies, for example, a T- or Π-network. If the metal diaphragm is sufficiently thin, the series elements vanish and the circuit representation reduces to that of a simple shunt element. This is most easily demonstrated by the argument already used in connection with the bisection theorem (Sec. 4·7). If equal and opposite electric fields are applied to either side of the diaphragm, then the electric field in the plane of the diaphragm must be zero. The short-circuit impedance of half of the network Z_{sch} is therefore zero; $Z_{11} = Z_{12}$; and the series elements of the T-representation vanish. The open-circuit impedance Z_{och} of half the network is just twice the impedance of the shunt element or twice the shunt impedance of the diaphragm. In most cases, the susceptance of the diaphragm is a more useful quantity and is usually the parameter specified. From Eq. (5·14) the susceptance B may be written as

$$B = \frac{2\omega(W_E - W_H)}{\frac{1}{2}ee^*}. \tag{1}$$

To find the value of the susceptance B it is necessary to solve an electromagnetic problem for the geometrical configuration under consideration. A discussion of the solution of problems of this nature will not be given here. It may be noted, however, that a complete solution giving the electric and magnetic fields at every point is not necessary, since the susceptance depends only on the *total* stored electric and magnetic energy. Variational methods[1] are found to be very powerful tools for the solutions of such problems.

Since the characteristic impedance of a waveguide can be defined only as a quantity proportional to the ratio of the transverse electric and magnetic fields, the absolute value must remain arbitrary. In a similar manner the absolute value of the susceptance of an iris is undefined by an unknown factor of proportionality. It is customary therefore always to express the susceptance of an iris relative to the characteristic imped-

[1] David S. Saxon, "Notes on Lectures by Julian Schwinger: Discontinuities in Waveguides," February 1945.

ance of the waveguide. The relative susceptance is then a definite quantity. It has been shown in Chap. 3 that relative admittances or impedances are all that are necessary for most problems in waveguide circuits; the absolute values are not important.

From a consideration of the special cases about to be discussed, it will be evident that the frequency variation of the susceptance of a diaphragm in waveguide is different from the variation with frequency of the susceptance of a coil or a condenser. Even the simplest diaphragms have a complicated dependence of susceptance on wavelength or frequency. This fact arises from two circumstances. (1) The absolute value of the susceptance must depend on the frequency variations of contributions

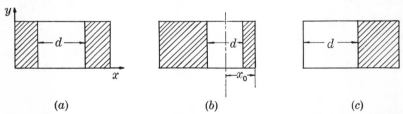

Fig. 6·1.—Inductive slits in rectangular waveguide. The metal partitions are shaded (a) A symmetrical opening, (b) an unsymmetrical slit, and (c) the partition on one side only, $x_0 = d/2$.

from many higher modes, each of which differs from the others. (2) The relative susceptance contains the frequency variation of the characteristic admittance of the waveguide. The characteristic admittance is proportional to the wave admittance of the guide and therefore contains the factor λ/λ_g. Thus if an absolute inductive susceptance contained the factor $1/\omega$ or λ, as the susceptance of a coil of wire at low frequencies, the relative susceptance would be proportional to λ_g. A capacitance independent of λ in absolute value yields a relative susceptance proportional to λ_g/λ^2. However, since Foster's theorem remains valid for waveguide terminations, no radical departures from the accustomed frequency variation are to be expected.

6·3. The Inductive Slit.—If, in rectangular waveguide capable of propagating the dominant mode only, a thin metal partition is inserted in such a way that the edge of the partition is parallel to the electric field, the iris formed is equivalent to a shunt inductance. The iris may be formed symmetrically as in Fig. 6·1a, or the slit may be asymmetrically placed as in Fig. 6·1b and c. Since the electric field is in the y-direction, the higher modes excited by the diaphragm are all H-modes and the stored energy is therefore predominantly magnetic. According to Eq. (5·14) the shunt susceptance is negative, and the diaphragm is a shunt inductance. The value of the susceptance has been accurately calculated and the exact formula may be found in *Waveguide Handbook*, Vol. 10 of this series. An approximate expression for the symmetrical case (Fig. 6·1a) is

$$B = -\frac{\lambda_g}{a}\cot^2\frac{\pi d}{2a} + f\left(\frac{a}{\lambda}\right), \qquad (2)$$

where $f(a/\lambda)$ is a small term. Equation (2) gives the susceptance relative to the characteristic admittance of the waveguide. The principal term is proportional to λ_g, and hence the susceptance has very nearly the frequency dependence of the susceptance of a coil of wire at low frequencies. The correction term f is, however, not proportional to λ_g but has a different frequency dependence. The magnitude of f and its variation with frequency are illustrated by Fig. 6·2. At large values of λ_g, f can be neglected; when λ_g is small, the correction term contributes appreciably. The short vertical lines indicate the values of λ_g/a for which the H_{20}- and the H_{30}-modes may first propagate. Since the slit is symmetrical, the H_{20}-mode is not excited by it. For wavelengths short enough for the H_{30}-mode to

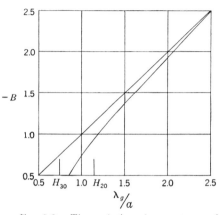

FIG. 6·2.—The variation of susceptance of an inductive slit of width $d = a/2$. The straight line with a slope of unity is the cotangent term in Eq. (2); the curve gives the exact value of B.

propagate, the slit no longer behaves as a simple shunt element but excites some of the H_{30}-mode.

The susceptances of the asymmetrical cases of Fig. 6·1b and c may also be expressed to a good approximation by simple formulas. The susceptance of the diaphragm of Fig. 6·1b is given by

$$B = -\frac{\lambda_g}{a}\cot^2\frac{\pi d}{2a}\left(1 + \sec^2\frac{\pi d}{2a}\cot^2\frac{\pi x_0}{a}\right). \qquad (3)$$

When one-half of the partition is absent, as in Fig. 6·1c, $x_0 = d/2$, and Eq. (3) reduces to

$$B = -\frac{\lambda_g}{a}\cot^2\frac{\pi d}{2a}\left(1 + \csc^2\frac{\pi d}{2a}\right). \qquad (4)$$

The expressions given in Eqs. (3) and (4) are not so exact as the corresponding approximation for the symmetrical slit. Asymmetrical diaphragms excite the H_{20}-mode and other even modes as well as the H_{30}-mode and the other odd modes. The correction terms to be added are therefore larger and the frequency dependence correspondingly greater than for the symmetrical case. For practical applications the diaphragm of Fig. 6·1c is often used, since it is the simplest possible construction.

The approximate formulas just stated and the exact curve shown in Fig. 6·2 are all valid only for a metal partition that is infinitely thin. It is usually necessary to use metal thick enough so that some correction is needed. Although the theoretical correction has not been worked out, an empirical correction has been found that is fairly exact. If the value of d of the thick slit is reduced by the thickness t, the value of B is increased to compensate for the thickness effect; thus

$$B_{\text{thick}}(d) = B_{\text{thin}}(d - t). \tag{5}$$

6·4. The Capacitive Diaphragm.—If metal partitions are introduced from the broad faces of a rectangular waveguide so that the edges of the

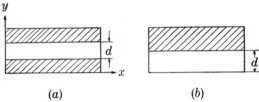

Fig. 6·3.—Capacitive diaphragms in rectangular waveguide. The metal partitions are shown shaded.

partitions are perpendicular to the electric field, then a capacitive susceptance is produced. Such diaphragms are shown in Fig. 6·3. These discontinuities excite only higher E-modes; the stored electric energy exceeds the magnetic energy, and B of Eq. (1) is positive. The susceptance for these diaphragms has also been calculated and is given by

$$B = \frac{4b}{\lambda_g} \ln \csc \frac{\pi d}{2b}, \tag{6}$$

for the symmetrical opening of Fig. 6·3a. Correction terms that are important at high frequencies are omitted from Eq. (6). The frequency variation of B is similar to that of the susceptance of a condenser at low frequencies except that λ_g is substituted for λ. It does not have the frequency variation of the relative susceptance of a lumped capacitance, which would be proportional to λ_g/λ^2 as mentioned in Sec. 6·2. The importance of the high-frequency correction terms to Eq. (6) may be judged from Fig. 6·4 in which the susceptance of a diaphragm is plotted as a function of b/λ_g for an opening $d = b/2$. The straight line represents Eq. (6); the accurate value of the susceptance is given by the curve. It should be remembered that the dimensions of the waveguide are usually chosen so that b/λ_g is about $\frac{1}{4}$.

The E-modes excited by the symmetrical slit will have longitudinal fields E_z which are odd about the center of the slit. The next mode will therefore propagate in the waveguide when $\lambda_g = b$. Since E_z is zero along the plane passing through the center of the slit, a sheet of metal of zero thickness may be placed along the center of the waveguide and the fields will not be disturbed.[1] The waveguide is then divided into halves, each half containing an asymmetrical slit like the one shown in Fig. 6·3b. The height of each half has become $b/2$, and the aperture of the slit has become $d/2$. The relative susceptance must, however, remain unaltered. Since the susceptance must be a function of the parameters d/b and b/λ_g, the value of B for the asymmetrical slit can be obtained from Eq. (6) by replacing λ_g by $\lambda_g/2$. Hence

$$B = \frac{8b}{\lambda_g} \ln \csc \frac{\pi d}{2b}. \quad (7)$$

The capacitive slit is not often used in high-power microwave applications, since the breakdown strength of the waveguide is greatly reduced by it. The effect of a finite thickness of the partition is much larger than for the inductive slit and will be discussed in a following section.

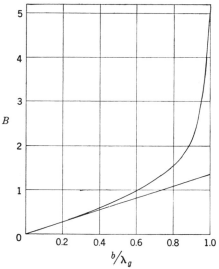

Fig. 6·4.—Relative susceptance of a thin symmetrical capacitive slit with an opening equal to one-half the height of the waveguide. The straight line represents Eq. (6); the curve shows the values calculated from the accurate expression.

6·5. The Thin Inductive Wire.—A thin wire extending across a rectangular waveguide between the center lines of the two broad faces of the guide forms an obstacle that acts as a shunt inductance. If the radius of the wire is small and the resistivity large, the skin depth in the wire may be made comparable with the radius of the wire. The relative impedance of the wire then contains a resistive component, and power is absorbed in heating the wire. Such a device forms a bolometer element and is commonly used to measure microwave power. The relative impedance of the wire is given by

$$Z = R + jX = \frac{\lambda}{\lambda_g} \frac{a}{2\pi\sigma \sqrt{\frac{\mu}{\epsilon}} r^2} + j \frac{a}{2\lambda_g} \ln \frac{2a}{\pi e^2 r}, \quad (8)$$

[1] See Sec. 2·5.

where r is the radius of the wire, σ is the conductivity, and e is the base of natural logarithms. The d-c resistance R_0 of the wire is given by

$$R_0 = \frac{b}{\pi \sigma r^2}. \tag{9}$$

The relative impedance may therefore be written as

$$Z = \frac{1}{Z_0}\left(R_0 + j\frac{b}{\lambda}\sqrt{\frac{\mu}{\epsilon}}\ln\frac{2a}{\pi e^2 r}\right), \tag{10}$$

where

$$Z_0 = \frac{2b}{a}\frac{\lambda_g}{\lambda}\sqrt{\frac{\mu}{\epsilon}}. \tag{11}$$

Thus if the characteristic impedance of the waveguide is chosen to be that given by Eq. (11), the relative resistance of the wire is just R_0/Z_0. The reactance, moreover, corresponds to an absolute inductance L per unit length of

$$L = \frac{1}{2\pi}\mu\ln\frac{2a}{\pi e^2 r}, \tag{12}$$

which is similar to the inductance L_0 per unit length of a straight wire in free space:

$$L_0 = \frac{1}{2\pi}\mu\ln\frac{2l}{er}. \tag{13}$$

This circumstance strengthens the belief by some that the most reasonable choice for the characteristic impedance of a waveguide is that given by Eq. (11). It will be recognized that this impedance is the proper value to choose in order to obtain the correct value of the power flow W from the expression

$$W = \frac{1}{2}\frac{V^2}{Z_0},$$

where the voltage V is defined in the natural manner as

$$V = bE_y.$$

It has been repeatedly emphasized, however, that any choice of the absolute value of Z_0 for a waveguide must be an arbitrary one. Choices other than that of Eq. (11) are more suitable for other situations.

6·6. Capacitive Tuning Screw.—A metallic post of small diameter introduced from the broad side of a rectangular waveguide but not extending completely across the guide forms a shunt capacitance. A variable susceptance of this type can be made by simply inserting a screw into the waveguide, and such a tuning screw is often employed in low-power microwave equipment. Currents flow from the broad face of the waveguide down the screw, and consequently it is necessary that good electrical contact be made between the screw and the guide. As the screw is inserted, the capacitive susceptance increases in very much the same manner as does the susceptance of a condenser whose plates are the top and bottom walls of the waveguide.

For small distances of insertion, the screw should behave much as a lumped capacitance, and the relative susceptance should vary as λ_g/λ^2. As the distance of insertion increases, however, and becomes an appreciable fraction of a wavelength, the currents flowing along the length of the post are no longer constant and the screw acts as an inductance and a capacitance in series shunted across the waveguide transmission line. When the length of the screw is approximately one-quarter of a freespace wavelength, resonance occurs and the susceptance of the screw becomes infinite. With still greater distance of insertion, the susceptance becomes negative; and when contact with the opposite wall is made, the susceptance becomes that given approximately by the inverse of Eq. (8). No adequate theoretical treatment has been given of the susceptance of tuning screws, but the behavior just described is illustrated by the experimental data of Fig. 6·5. The dimensions of the screw and the equivalent circuit are indicated in the figure. The value of B for $l/b = 1.0$ calculated from Eq. (8) is -3.45.

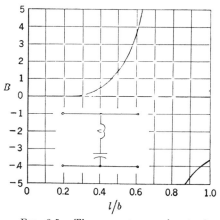

FIG. 6·5.—The susceptance of a tuning screw 0.050 in. in diameter as a function of the depth of insertion in waveguide 0.9 by 0.4 in. ID at a wavelength of 3.2 cm.

No data exist on the frequency sensitivity of the susceptance. In the neighborhood of resonance, the susceptance seems to depend critically on the dimensions and on the frequency. The resonant length of the screw of Fig. 6·5 appears to be about $0.75b$. One-quarter of a free-space wavelength is $0.79b$.

Other resonant structures that totally reflect the incident power exist in many forms. A common one is a rectangular ring, the perimeter of which must be about 1.1λ for resonance to occur; a dumbbell-shaped antenna is another. No extensive investigations have been made of the properties of such structures.

6·7. Resonant Irises.—Since both capacitive and inductive diaphragms exist, it should be possible to combine them and obtain a shunt-resonant circuit such as that of Fig. 6·6. No theoretical treatment of the properties of such a diaphragm has been made. It has been found empirically that for resonance the dimensions a', b' of the rectangular opening are given by

$$\frac{a}{b}\sqrt{1 - \left(\frac{\lambda}{2a}\right)^2} = \frac{a'}{b'}\sqrt{1 - \left(\frac{\lambda}{2a'}\right)^2} \qquad (14)$$

If the origin of rectangular coordinates is taken at the center of the waveguide, then Eq. (14) is the equation of an hyperbola in the variables a' and b'. The hyperbola passes through the corners of the waveguide, and the branches are separated by $\lambda/2$ at the closest point. Equation

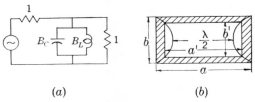

(a) (b)

Fig. 6·6.—(a) Equivalent circuit of a shunt-resonant thin diaphragm; (b) resonant diaphragm.

(14) is obtained if the characteristic impedance of a waveguide of dimensions a and b given by Eq. (11) is equated to that of a guide of dimensions a' and b'. This condition cannot be derived rigorously, but qualitative arguments that make it appear reasonable have been given by Slater.[1] Equation (14) fits the experimental data very well for apertures in metal walls that are thin compared with the dimension b'. An increase in the thickness of the partition decreases the resonant wavelength.

A resonant circuit such as that of Fig. 6·6 has a frequency sensitivity characterized by the parameter Q. As shown in Sec. 4·13, near resonance the admittance of the combination of the resonant aperture and the matched load is given by

$$Y = 1 + 2jQ\frac{\Delta\omega}{\omega_0},$$

where ω_0 is the resonant angular frequency and $\Delta\omega$ is the deviation from the resonant value. The Q-values for resonant apertures are low, of the order of magnitude of 10, and increase as b' decreases. This is to be expected, since a decrease in b' increases the capacitive susceptance B_C.

Another commonly used resonant aperture is obtained by combining a symmetrical inductive diaphragm and a capacitive tuning screw. The resonant frequency may be conveniently changed by means of the screw, and the Q altered by changing the aperture of the diaphragm. Such

[1] J. C. Slater, *Microwave Transmission*, McGraw-Hill, New York, 1942, pp. 184*ff*.

resonant devices have been employed in the construction of filters.[1] A variable susceptance of this type is used in waveguide crystal mixers.[2] Considerable experimental information about resonant irises of this and other kinds is to be found in Chap. 3 of Vol. 14 of this series, where the use of several such circuits in cascade to make a TR switch with a band-pass characteristic is described.

6·8. Diaphragms in Waveguides of Other Cross Sections.—Although only rectangular waveguide has been discussed, diaphragms in waveguides of other cross sections also act as shunt susceptances, provided that the metal partitions are thin and that the waveguide is capable of supporting only one mode. In round waveguide carrying the TE_{11}-mode, a centered circular aperture is a shunt inductive suscept-

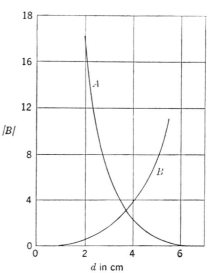

FIG. 6·7.—Susceptance of inductive apertures (Curve A) and capacitive disks (Curve B) in round waveguide of 2.5-in. inside diameter at a wavelength of 9.1 cm, TE_{11}-mode.

ance. A capacitive susceptance is formed if a circular disk is centered on the waveguide axis. In Fig. 6·7 some rather old experimental data are pre-

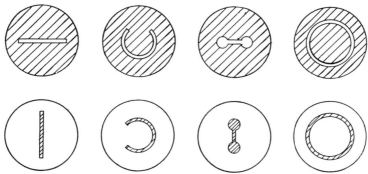

FIG. 6·8.—Resonant obstacles and apertures in waveguide of circular cross section. The obstacles are totally reflecting at resonance; the apertures totally transmitting. The metallic portions are shaded.

sented to illustrate the variation of the susceptance with the diameter of the aperture or disk. No theoretical estimates of the susceptance are available.

[1] *Microwave Transmission Circuits*, Vol. 9, Chap. 10, Radiation Laboratory Series.
[2] *Microwave Mixers*, Vol. 16, Radiation Laboratory Series.

There exist also both series-resonant obstacles that are totally reflecting and shunt-resonant apertures. A group of these is shown in Fig. 6·8. In the use of diaphragms in waveguide of circular cross section it must be remembered that modes of two polarizations can exist (Sec. 2·13) and the diaphragm must be symmetrical with respect to the electric field so that the second polarization will not be excited.

A close correspondence exists between a capacitive slit in rectangular waveguide and a slit in a parallel-plate transmission line. In rectangular waveguide the x-dependence of the fields, both near the obstacle and far from it, is determined by the x-dependence of the incident field; in particular E_x is zero, and E_y, E_z and H_x vary as $\sin \pi x/a$. Each component of the electric field satisfies the wave equation; for example,

$$\frac{\partial^2 E_y}{\partial x^2} + \frac{\partial^2 E_y}{\partial y^2} + \frac{\partial^2 E_y}{\partial z^2} + k^2 E_y = 0,$$

where $k = 2\pi/\lambda$ is the wave number in free space. The x-derivative can, however, be evaluated, and the equation becomes

$$\frac{\partial^2 E_y}{\partial y^2} + \frac{\partial^2 E_y}{\partial z^2} + \left[k^2 - \left(\frac{\pi}{a}\right)^2\right]E_y = 0, \qquad (15)$$

and similar equations hold for the other components. In a parallel-plate transmission line also, E_x is zero and the other components satisfy the wave equation; thus,

$$\frac{\partial^2 E_y}{\partial y^2} + \frac{\partial^2 E_y}{\partial z^2} + k^2 E_y = 0. \qquad (16)$$

Since the boundary conditions are independent of x for a capacitive obstacle, the solutions of Eqs. (15) and (16) differ only in that where k occurs in a parallel-plate solution $k^2 - (\pi/a)^2$ occurs in the waveguide solution Consequently, one result may be derived from the other by replacing $k^2 - (\pi/a)^2$ with k^2 or equivalently by replacing λ_g with λ. Thus the susceptance of a symmetrical capacitive slit in a parallel-plate transmission line follows immediately from Eq. (6) and is

$$B = \frac{4b}{\lambda} \ln \csc \frac{\pi d}{2b}. \qquad (17)$$

A coaxial transmission line with a thin disk on either the inner or the outer conductor behaves very similarly to a parallel-plate transmission line with a capacitive slit. Accurate values of the susceptance for capacitive disks are to be found in *Waveguide Handbook*. Some values calculated in a different manner have been given by Whinnery and others.[1] A wire extending from the inner to the outer conductor of a

[1] J. R. Whinnery and H. W. Jamieson, *Proc. IRE*, **32**, 98 (1944); J. R. Whinnery, H. W. Jamieson, and T. E. Robbins, *Proc. IRE*, **32**, 695 (1944).

coaxial line forms an inductive susceptance. Although no measurements or calculations are available for this case, an approximate value of the shunt admittance can be found from Eq. (8) if λ_g is replaced by λ and a replaced by the mean circumference of the coaxial line.

6·9. The Interaction between Two Diaphragms.—When a wave is incident upon a diaphragm, the field near the diaphragm consists of the dominant-mode wave together with enough higher-mode waves to satisfy the boundary conditions at the aperture. The intensities of the higher-mode waves fall off in both directions away from the aperture, and at a sufficient distance only the dominant wave is left. The simple equivalent-circuit picture gives a valid description of the fields only at distances large enough for the higher-mode waves to be neglected. The effective range of the higher-mode fields thus depends on the attenuation constants of these higher modes and on the amplitude of their excitation. The attenuation constants are given by

$$\alpha = 2\pi \sqrt{\left(\frac{1}{\lambda_c}\right)^2 - \left(\frac{1}{\lambda}\right)^2}, \qquad (18)$$

which is independent of λ if $\lambda \gg \lambda_c$. The range of interaction is thus about $1/\alpha$ or $\lambda_c/2\pi$. For an asymmetrical inductive slit the range is therefore $a/2\pi$; for a symmetrical inductive slit the range is $a/3\pi$; for a symmetrical capacitive slit, $b/2\pi$.

If two thin apertures are separated along the waveguide by a distance comparable to the range, there will be an interaction between them that is not given by the simple equivalent-circuit picture. The magnitude of this interaction can be judged from some calculations of Frank.[1] Frank considered the case of two symmetrical inductive slits. The effect of interaction can be expressed in terms of the "effective" admittance of the diaphragm nearest the load. The effective admittance is the admittance that the diaphragm would have to have in order that the admittance of the combination of the two diaphragms and the load could be correctly calculated if no interaction were assumed. If s is the distance between the windows, the effective admittance is given by

$$Y_2' = jB_2 - 6\frac{a}{\lambda_g}\left(\frac{B_1}{1 + \frac{a}{\lambda_g}B_1}\right)\left(\frac{B_2}{1 + \frac{a}{\lambda_g}B_2}\right)e^{-\frac{3\pi s}{a}}Y_L \sin\frac{2\pi s}{\lambda_g}$$
$$-j\left(B_2 \sin\frac{2\pi s}{\lambda_g} + \cos\frac{2\pi s}{\lambda_g}\right), \quad (19)$$

where $-B_1$ and $-B_2$ are the susceptances of the two windows and Y_L is the admittance of the load at the second window. The magnitude and

[1] N. H. Frank, RL Report No. 197, February 1943.

phase of Y_2'/Y_2 are plotted in Fig. 6·9 as functions of s/λ_g for the case where $Y_L = 1$, $\lambda_g/a = 1.96$, and $B_1 = B_2 = 0.5$. The effect of the

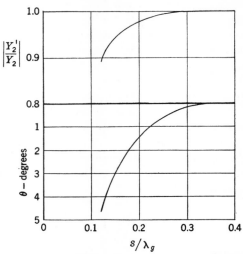

Fig. 6·9.—The magnitude and phase angle of the relative effective admittance Y_2'/Y_2 as a function of the separation s of two symmetrical inductive apertures of susceptance $B = -0.5$.

interaction is important at small separations and is principally a reduction in the magnitude of the susceptance.

To describe these effects by means of an equivalent circuit, it is necessary to consider each aperture as a three-terminal-pair network and the two networks connected together as shown in Fig. 6·10. Terminals 2 of the two networks are connected by the transmission line A for the dominant mode, and terminals 3 are joined by the transmission line B for the next higher mode excited. Line B is attenuating, since the waveguide is beyond cutoff for the higher mode. Three-terminal-pair structures are discussed more completely in Chap. 9. Unfortunately, the necessary data are not available for most situations, and interaction effects must be determined by experiment.

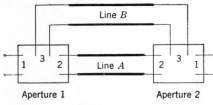

Fig. 6·10.—Equivalent circuit of two apertures so close that interaction effects must be taken into account.

6·10. Babinet's Principle.—Useful results for waveguide structures with a high degree of symmetry can be obtained by the application of Babinet's principle (Sec. 2·10). An example of this has been given by Schwinger.[1] The actual electromagnetic problem that is solved to find

[1] David S. Saxon, "Notes on Lectures by Julian Schwinger, Discontinuities in Waveguides," February 1945.

the impedance of a symmetrical inductive diaphram is that of finding the terminating impedance Z_{och} of the structure of Fig. 6·11a in which a solid line represents an electric wall and the dashed line a magnetic wall. The electric field is outward from the paper. The terminating impedance of this configuration is $2jX$, where X is the shunt reactance of the diaphragm. Since the fields are independent of the y-coordinate,

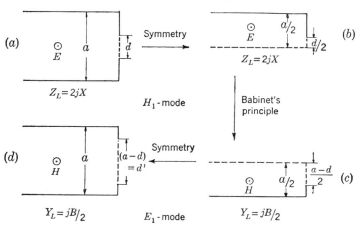

FIG. 6·11.—Application of Babinet's principle to obtain the susceptance of one aperture from that of another.

the guide can be considered of infinite height. The value of X was given in Sec. 6·3 and is approximately

$$X = \frac{a}{\lambda_g} \tan^2\left(\frac{\pi d}{2a}\right). \tag{20}$$

Since the magnetic field is zero along the plane through the center of the aperture, a magnetic wall may be inserted there and the configuration of Fig. 6·11b is obtained. This configuration has the same terminating impedance $2jX$. Babinet's principle may now be employed to obtain Fig. 6·11c. The electric and magnetic walls are interchanged, and E is replaced by H. The terminating admittance $j(B/2)$ of this structure must be equal to the terminating impedance of Fig. 6·11b, and

$$j\frac{B}{2} = 2jX. \tag{21}$$

The magnetic wall of the waveguide may now be removed by symmetry considerations, and the terminating admittance is unchanged. The susceptance B is now the shunt susceptance of a capacitive aperture for the first E-mode between parallel plates of separation a. From Eq. (21)

$$B = \frac{4a}{\lambda_g} \tan^2 \frac{\pi d}{2a},$$

or in terms of the opening d' of the diaphragm,

$$B = \frac{4a}{\lambda_g} \cot^2 \frac{\pi d'}{2a}. \tag{22}$$

The guide wavelength λ_g in Eq. (22) is given by

$$\left(\frac{1}{\lambda_g}\right)^2 = \left(\frac{1}{\lambda}\right)^2 - \left(\frac{1}{2a}\right)^2.$$

The susceptance of a slit in rectangular waveguide operating in the E_{11}-mode is obtained by the arguments of Sec. 6·8 by using in Eq. (22) the appropriate value of λ_g given by

$$\left(\frac{1}{\lambda_g}\right)^2 = \left(\frac{1}{\lambda}\right)^2 - \left(\frac{1}{2a}\right)^2 - \left(\frac{1}{2b}\right)^2.$$

Babinet's principle in the form of Eq. (21) has been applied also to corresponding resonant structures such as those shown in Fig. 6·8. This application is not a rigorous one, however, since the electric walls that form the waveguide are not transformed. Nevertheless, if the size of aperture is not too large, Eq. (21) applies approximately. If the apertures are large or close to the walls of the waveguide, as those in the capacitive and inductive obstacles of Fig. 6·7, the deviations from Eq. (21) are large. The product of the inductive and capacitive susceptances is not -4 as required by Eq. (21) but varies from -2 to -10 over the range of disk diameters from 2.0 to 5.5 cm.

6·11. The Susceptance of Small Apertures.—The transmission of radiation through an aperture may be expressed in very general terms[1] if the size of the aperture is small enough. On a metallic wall, the normal magnetic field and the tangential electric field vanish, but a tangential magnetic field $\mathbf{H}_0$ and a normal electric field $\mathbf{E}_0$ may be present. If there is a small hole in the wall, within the hole there will be a tangential electric field and a component of the magnetic field perpendicular to the wall. If a linear dimension x of the hole satisfies the relation that $x \ll \lambda/2\pi$, then the fields in the neighborhood of the hole are closely approximated by the unperturbed fields $\mathbf{H}_0$ and $\mathbf{E}_0$ plus the fields from an electric dipole and a magnetic dipole within the hole. The strength of the electric dipole is proportional to E_0, and the dipole is directed normally to the wall. Similarly, the magnetic dipole is in the plane of the wall and of a strength proportional to H_0. The constants of proportionality are the polarizabilities of the hole. The electric polarizability P is simply a constant, since the dipole and the field are parallel. The directions of the magnetic dipole and the exciting magnetic field are, however, not necessarily the same. The magnetic polarizability is

[1] H. A. Bethe, *Phys. Rev.*, **66**, 163, (1944).

Sec. 6·11]   THE SUSCEPTANCE OF SMALL APERTURES   177

therefore a dyadic quantity. In the usual manner, principal axes of the hole may be chosen that correspond to axes of symmetry of the hole if these exist, and the magnetic polarizability is then determined by two scalar numbers M_1 and M_2 that relate the components of $\mathbf{H}_0$ to the components of the magnetic dipole. The values of the constants P, M_1 and M_2 can be calculated if it is assumed that the hole is small and far away from any metallic objects that have radii of curvature comparable with λ. When the strengths of the dipoles are known, then the transmission through the hole may be calculated by finding the radiation from the dipoles into the waveguide on the other side of the wall. The calculation of the transmission through a hole is thus performed in two distinct steps. The strengths of the equivalent dipoles within the hole are first calculated, and then the radiation from the dipoles is found. The theory therefore applies to holes in the side walls of the waveguide as well as to holes in a metallic wall perpendicular to the waveguide axis. The second case only will be considered here. The properties of holes in the side walls are treated in Chap. 9. On the other hand, the theory neglects entirely all reaction of the load upon the generator. For small holes or large susceptances this neglect is justified.

The radiation from the hole may be expressed in terms of the normal-mode functions of the waveguide into which it radiates. If the amplitude is measured by the transverse electric field, the amplitude transmission coefficient A is given by

$$AS_0 = \int \mathbf{E}_1 \times \mathbf{H}_t \cdot \mathbf{n}\, d\sigma \tag{23}$$

for the magnetic-dipole radiation, where $\mathbf{E}_1$ is the field in the hole, $\mathbf{H}_t$ the transverse normal-mode magnetic field, $\mathbf{n}$ a unit vector normal to the plane of the hole, and S_0 a normalizing factor. The integral is taken over the area of the hole. The quantity S_0 is

$$S_0 = \int \mathbf{n} \cdot \mathbf{E} \times \mathbf{H}_t\, ds, \tag{24}$$

where $\mathbf{E}$ is the normal-mode electric field. Equation (23) may be written as

$$AS_0 = \int \mathbf{n} \times \mathbf{E}_1 \cdot \mathbf{H}_t\, d\sigma = \mathbf{H}_t \cdot \int \mathbf{n} \times \mathbf{E}_1\, d\sigma,$$

where $\mathbf{H}_t$ is evaluated at the hole. The integral of $\mathbf{n} \times \mathbf{E}_1$ is proportional to the magnetic dipole moment and in turn proportional to the incident field. A similar relation is true for the electric moment, and the total value of A is

$$AS_0 = \frac{2\pi j}{\lambda}(M_1 H_{0l} H_l + M_2 H_{0m} H_m + P E_{0n} E_n), \tag{25}$$

where the l- and m-components are taken in the directions of the principal axes of the hole and the n-component is normal to the hole.

The amplitude transmission coefficient may be expressed in terms of the susceptance of the diaphragm, since

$$A = 1 + \Gamma = \frac{2}{2 + jB}, \tag{26}$$

if the waveguide is the same on both sides of the diaphragm. Since the hole is small, B is very large and

$$B = -\frac{2j}{A}. \tag{27}$$

Finally, since the normal-mode fields are the same on both sides of the diaphragm, the zero subscripts may be dropped provided that the amplitude of the fields is doubled. Hence

$$\frac{1}{B} = -\frac{2\pi}{\lambda S_0}(M_1 H_l^2 + M_2 H_m^2 + P E_n^2). \tag{28}$$

For a waveguide operating in the dominant mode, the longitudinal electric field E_n is always zero, and the susceptance of a small hole is therefore inductive regardless of the shape of the hole or its location. The theory does not apply to a capacitive slit, since this cannot be considered as a small hole.

TABLE 6·1.—VALUES OF THE POLARIZABILITIES OF SMALL HOLES

	M_1	M_2	P
Circle of radius r...............	$\frac{4}{3}r^3$	$\frac{4}{3}r^3$	$\frac{2}{3}r^3$
Ellipse* of eccentricity $\epsilon = \sqrt{1 - \left(\frac{b}{a}\right)^2}$	$\frac{\pi}{3}\frac{ab^2\epsilon^2}{(1-\epsilon^2)(F-E)}$	$\frac{\pi}{3}\frac{ab^2\epsilon^2}{E - (1-\epsilon^2)F}$	$\frac{\pi}{3}\frac{ab^2}{E}$
Long narrow ellipse $(a \gg b)$........	$\dfrac{\pi}{3}\dfrac{a^3}{\ln\left(\dfrac{4a}{b}\right) - 1}$	$\frac{\pi}{3}ab^2$	$\frac{\pi}{3}ab^2$
Slit† of width d and length l........	....	$\frac{\pi}{16}ld^2$	$\frac{\pi}{16}ld^2$

* F and E are the complete elliptic integrals of the first and second kind, respectively:

$$F(\epsilon) = \int_0^{\pi/2} \frac{d\varphi}{\sqrt{1 - \epsilon^2 \sin^2 \varphi}},$$

$$E(\epsilon) = \int_0^{\pi/2} d\varphi \sqrt{1 - \epsilon^2 \sin^2 \varphi}.$$

The polarizability M_1 is for the magnetic field parallel to the major semiaxis a; M_2 is for the field parallel to the minor semiaxis b.

† The magnetic field is transverse to the slit and constant along the length.

Values of the polarizabilities have been calculated for several cases, and the results are given in Table 6·1. The values of the fields and the normalizing factor S_0 are easily found. For example, in rectangular waveguide in the dominant mode, the susceptance is given by

$$\frac{1}{B} = -\frac{4\pi}{ab\lambda_g} \sin^2 \frac{\pi x_0}{a} [M_1 \cos^2 (l,x) + M_2 \cos^2 (m,x)], \quad (29)$$

where x_0 is the x-coordinate of the center of the iris and (l,x) is the angle between one principal axis of the aperture and the x-axis. The susceptance of a narrow inductive slit that is centered may be calculated by inserting the proper value of M_1, and it is found to be

$$B = -\frac{\lambda_g}{a}\left(\frac{2a}{\pi d}\right)^2,$$

which is the asymptotic form of Eq. (2) for $d \ll a$.

For circular waveguide of radius R in the dominant mode, the susceptance of a centered iris is

$$\frac{1}{B} = -\frac{4\pi}{0.955(\pi R^2)} M, \quad (30)$$

which has been written to exhibit the similarity to Eq. (29). For circular waveguide operating in the E_0-mode, where the iris is at the center of the guide and has sufficient symmetry so that no other propagating modes are excited, the only field at the iris is the normal electric field. The quantity E_n^2 in Eq. (28) is negative, since the longitudinal field in a waveguide is 90° out of phase with the transverse field; B is positive, and the iris is capacitive. The susceptance is

$$B = \frac{0.92R^4}{P\lambda_g}. \quad (31)$$

It should be noted that the variation with frequency of both the inductive and capacitive small holes is similar to that of the larger irises.

IMPEDANCE MATCHING WITH SHUNT SUSCEPTANCES

6·12. Calculation of the Necessary Susceptance.—Susceptive irises are widely used in transmission lines to match a load impedance to the characteristic impedance of the line. In Sec. 4·3 it was shown that maximum power is delivered to the load if the load resistance is equal to the generator resistance and the load admittance is equal to the negative of the generator admittance. If the load and generator are connected together by a transmission line that is many wavelengths long, as is always the case in microwave transmission, then the best practice is to match both the load impedance and the generator impedance separately to the characteristic impedance of the transmission line. If the load were matched to the generator directly through a long line, then a small change in frequency would produce a large change in the impedance of the load as seen through the line, and a matched condition would no longer exist. Moreover, standing waves would exist along the transmission line even at the correct frequency, and the resistive losses in the line would be correspondingly great. By the insertion of a shunt susceptance at the proper place a short distance from the load, it is possible to make the relative admittance of the combination equal to unity, and the frequency sensitivity of the match is small.

The load admittance usually must be determined by a standing-wave measurement, and it is therefore convenient to express the matching conditions in terms of the observed standing-wave ratio r and the position of a minimum in the standing-wave pattern. If the losses in the matching elements are neglected, the magnitude of the susceptance necessary to match a load that produces a standing-wave ratio of r is the susceptance that produces a voltage standing-wave ratio of r when inserted in a matched line (Sec. 4·3). The relation between B and r is given by Eq. (3·28),

$$r = \frac{\sqrt{4 + B^2} + |B|}{\sqrt{4 + B^2} - |B|}, \tag{32}$$

or

$$|B| = \frac{r - 1}{\sqrt{r}}. \tag{33}$$

The correct placement of the susceptance is easily determined. A convenient reference plane is the minimum of the standing-wave pattern nearest the load. Here the load admittance is $G = r$. This admittance is transformed along the line according to the equation

$$Y = \frac{G + j \tan \beta l}{1 + jG \tan \beta l}. \tag{34}$$

If the length of line is chosen so that $Y = 1 + jB$, a susceptance can then be inserted at that point to cancel the susceptance of the load. If the real part of Eq. (34) is set equal to unity, then it is found that

$$\tan \beta l = -\frac{1}{\sqrt{G}} = -\frac{1}{\sqrt{r}}, \tag{35}$$

or

$$\frac{l}{\lambda_g} = -\frac{\tan^{-1} \frac{1}{\sqrt{r}}}{2\pi}. \tag{36}$$

The negative sign of the square root in Eq. (35) was chosen, and therefore l/λ_g is less than $\frac{1}{8}$, and a positive, capacitive susceptance must be added to produce a match. If it is desired to match with a negative, inductive susceptance, then l/λ_g is between $\frac{3}{8}$ and $\frac{1}{2}$ and is given by

$$\frac{l}{\lambda_g} = \frac{\pi - \tan^{-1}\frac{1}{\sqrt{r}}}{2\pi}. \tag{37}$$

The impedance charts described in Chap. 3 are especially useful in making matching calculations. The calculations just described are illustrated in Fig. 6·12. The standing-wave minimum occurs at the point $Y = G$; the angle βl is determined by moving toward the generator along the circle with the center at O to the point A on the circle $Y = 1 + jB$. A positive susceptance added here moves the admittance to O along the circle $Y = 1 + jB$. If an inductive susceptance is used, the angle is determined by moving from $Y = G$ through A to the point C on the circle $Y = 1 + jB$. The addition of a negative susceptance transforms the admittance to the point O.

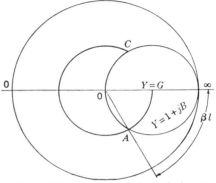

FIG. 6·12.—The calculation of the susceptance and position of a matching window by means of the admittance circle diagram.

It is obvious from the diagram that the point C can also be reached by moving from the minimum toward the *load* a distance given by Eq. (36).

To obtain a match that has a small frequency sensitivity, the matching window should usually be placed as close as possible to the load. It should be remembered, however, that interaction effects are often important. It may be desirable to resort to experiment in order to determine the optimum size and position of a diaphragm after the first approximation has been obtained by the procedure just described.

6·13. Screw Tuners.—By a technique similar to that described in the preceeding section, it is possible to design a variable tuner that can be inserted in a transmission line. To provide a variable susceptance that can be adjusted in position along the line, a capacitive tuning screw may be used. The screw is mounted in a closely fitting sleeve that may be slid along the line. Since a longitudinal slot along the center of the broad face of a rectangular waveguide or along a coaxial line does not disturb the fields within the pipe, the screw can be set easily in any position. When the screw penetrates the waveguide, currents flow along the screw. Consequently, the sliding sleeve and the screw threads must make good electrical contact. Such a device is extremely easy to use. Both the position and the depth of insertion are always varied in the direction to decrease the standing-wave ratio in the line. Eventually the proper position for match will be reached. The breakdown power of the line is lowered considerably by such a tuner, and its use is practical only for small standing-wave ratios. The necessity for a good electrical contact makes the mechanical design difficult.

To eliminate the sliding sleeve and the contact difficulties that it causes, several fixed screws may be used as a variable tuner. For example, three tuning screws separated from each other by one-quarter of a guide wavelength is a commonly used combination. To match a load to the line, the center screw and only one of the outer ones are employed. The range of admittances that such a tuner can produce is best illustrated on an admittance diagram. If a susceptance of unity is the maximum value that a screw can introduce, then the shaded area in Fig. 6·13 is the range of admittances of the tuner using one pair of screws. Since the other pair of screws is one-quarter wavelength away, the region enclosed in the dashed lines is accessible by the second pair. A standing-wave ratio of 2 in all phases can therefore be matched by this tuner, and slightly larger values of r can be matched if the phase is correct.

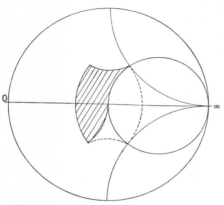

Fig. 6·13.—The range of admittance produced by a triple-screw tuner with the screws separated by one-quarter wavelength, if the maximum susceptance of a screw is unity.

If easily adjustable inductive susceptances were available, only two screws would be necessary. A tuner[1] employing an "inductive screw" has been used in waveguide operating at a 10-cm wavelength. The inductance is a short section of circular waveguide attached to the center of the broad face of the guide. The inductance is adjusted by a screw that fills the circular guide.

6·14. Cavity Formed by Shunt Reactances.—The process of matching a very large standing-wave ratio leads to the formation of a resonant cavity. In fact it is extremely useful to regard all cavities as made up of irises of large susceptance in a transmission line. Often the transmission line is a radial line; the properties of such lines are treated in Chap. 8. At present only lines that are cylindrical (boundary conditions independent of z) will be discussed. If, for example, it is desired to match a short section of line terminated in a metal plate or short circuit, then a

[1] See Vol. 9, Chap. 6, of this series.

diaphragm can be found of such susceptance that the admittance of the cavity is real. In rectangular waveguide such a process results in a rectangular box that is resonant.

The admittance of a short-circuited line is $Y = \coth \gamma l$. The proper value of a susceptance B to make the total admittance real is given by

$$\text{Im}(Y + jB) = 0. \tag{38}$$

If the line is lossless and B is finite, then the conductance of the cavity is zero. For resonance, the line is nearly one-half wavelength long if B is large and negative, and the length approaches one-quarter wavelength as the magnitude of B decreases. In practice, however, the losses must be taken into account; but since the length l is small, it will be assumed that αl is small compared with unity, where α is the attenuation coefficient. The magnitude of the susceptance B will be assumed to be large compared with unity. The length of the cavity is near n half wavelengths, and the dimensionless quantity ϵ is defined by

$$l = n \frac{\lambda_g}{2} - \epsilon \frac{\lambda_g}{2\pi}, \tag{39}$$

where n is a small integer. The admittance of the line then becomes

$$Y = \frac{1}{\alpha l - j\epsilon}, \tag{40}$$

and Eq. (38) becomes

$$B = -\frac{\epsilon}{(\alpha l)^2 + \epsilon^2} \approx -\frac{1}{\epsilon}. \tag{41}$$

For an inductive susceptance, ϵ is positive and the cavity is a trifle less than n half wavelengths long. The conductance of the cavity is

$$G = \frac{\alpha l}{(\alpha l)^2 + \epsilon^2} \approx B^2 \alpha l. \tag{42}$$

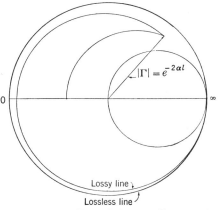

Fig. 6·14.—The admittance diagram of a length of short-circuited line plus an inductive susceptance of such a magnitude as to produce resonance.

It is instructive to follow the various transformations on an admittance chart as in Fig. 6·14. In a lossless line, as the point of observation is moved away from the short circuit toward the generator, the admittance point travels along the outer circle in a clockwise direction. If a large negative susceptance is added after the point has traveled almost half a wavelength, the total admittance can be made zero and a resonant condition exists. If the line is lossy, the admittance point spirals inward instead of traveling on the outside circle. It will be remembered that the magnitude of the reflection coefficient varies as $e^{-2\alpha l}$. After nearly a half wavelength of travel the addition of a negative susceptance brings the admittance to the positive real value given by Eq. (42).

Of primary interest in resonance phenomena is the frequency sensitivity of the admittance. The conductance varies as $1/\lambda_g^2$, since B varies as $1/\lambda_g$ if α is assumed to be constant, and the derivative of G is

$$\frac{dG}{d\lambda_g} = B^2 \alpha n, \tag{43}$$

since

$$\frac{dB}{d\lambda_g} = \frac{B}{\lambda_g}, \tag{44}$$

for an inductive susceptance. The susceptance varies much more rapidly with wavelength. It is easy to show that if B is large, the variation of G with wavelength is negligible compared with the variation of the susceptance of the line. If the losses are neglected for this calculation, the susceptance B_L of the line is $-\cot \beta l$ and

$$\frac{dB_L}{d\lambda_g} = -\frac{\beta l}{\lambda_g} \csc^2 \beta l = -\frac{\beta l}{\lambda_g}(1 + B^2), \tag{45}$$

since $B = \cot \beta l$ at resonance. Therefore

$$\frac{dB_L}{d\lambda_g} \approx -\frac{n\pi}{\lambda_g} B^2. \tag{46}$$

That it is correct to neglect the losses can be verified by differentiating Eq. (40) directly. The coefficient of B^2 in Eq. (46) is much larger than the coefficient of B^2 in Eq. (43), and the conductance can be assumed to be constant. The frequency sensitivity of the cavity can therefore be described by the Q of the equivalent shunt-resonant circuit (Sec. 4·13),

$$Q = \frac{\omega_0}{2G}\frac{dB}{d\omega} = -\frac{\lambda_g}{2G}\left(\frac{\lambda_g}{\lambda}\right)^2 \frac{dB}{d\lambda_g}. \tag{47}$$

The various values of Q correspond to various choices for G. If the value given by Eq. (43) is inserted, Q_0, the unloaded Q, is obtained:

$$Q_0 = \frac{\pi}{\alpha \lambda_g}\left(\frac{\lambda_g}{\lambda}\right)^2, \tag{48}$$

which is the value of Q_0 given in Eq. (2·83) with the losses in the end walls omitted. The unloaded Q is independent of n. If the cavity is matched, $G = 1$ and the loaded Q is

$$Q = \frac{n\pi}{2}\left(\frac{\lambda_g}{\lambda}\right)^2 B^2. \tag{49}$$

SEC. 6·14]   CAVITY FORMED BY SHUNT REACTANCES

The action of the susceptance B is very similar to that of a transformer of turn ratio $|B|$. This is most obvious from the equivalent circuit of Fig. 4·21. For a pure shunt element, the elements of the impedance matrix are all equal and

$$Z_{11} = Z_{12} = Z_{22} = -\frac{j}{B}.$$

The circuit parameters are the two line lengths

$$\left. \begin{array}{l} \tan \beta l_1 = \dfrac{B}{2} + \sqrt{\dfrac{B^2}{4} + 1} \approx B, \\ \tan \beta l_2 = -\dfrac{B}{2} - \sqrt{\dfrac{B^2}{4} + 1} \approx -\dfrac{1}{B}, \end{array} \right\} \quad (50)$$

where the approximate values are for $B \gg 1$. The transformer ratio N is given by

$$N^2 = \frac{B^2}{2} + B\sqrt{\frac{B^2}{4} + 1} + 1 \approx B^2. \quad (51)$$

For resonance the total length of line should be an integral number of half wavelengths or

$$l_2 + l = \frac{n\lambda_g}{2},$$

and this is the same condition as that given by Eq. (41).

The transformer ratio N is the stepup in voltage in the cavity over the voltage in the line if the cavity is matched. The value of N may be expressed in terms of the Q as

$$N = \frac{\lambda}{\lambda_g} \sqrt{\frac{2}{n\pi}} \sqrt{Q}. \quad (52)$$

The relation of Eq. (52) can be obtained easily by direct calculation. The quantity Q may be defined by

$$Q = \omega \frac{\text{energy stored}}{\text{energy lost per sec}}.$$

If the cavity is matched, the energy lost per second is equal to the incident power. Hence

$$Q = \omega \frac{\int \frac{1}{2}(\epsilon E^2 + \mu H^2)\, dV}{\frac{1}{2} \int E_t H_t\, dS}, \quad (53)$$

where the volume integral is taken over the cavity and the surface integral over the cross section of the waveguide. For a TE-wave

$$\frac{1}{2} E_t H_t = \sqrt{\frac{\epsilon}{\mu}} \frac{\lambda}{\lambda_g} E_o^2,$$

where E_o denotes the field outside the cavity. At resonance $\epsilon E_i^2 = \mu H_i^2$, where E_i is the field inside the cavity. Equation (53) can therefore be written as

$$Q = \omega \frac{\lambda_g}{\lambda} \sqrt{\frac{\mu}{\epsilon}} \int_0^{n\lambda_g/2} \sin^2 \beta z \, dz \, \frac{\int E_i^2 \, dS}{\int E_o^2 \, dS}.$$

or

$$Q = \frac{\pi n}{2} \left(\frac{\lambda_g}{\lambda}\right)^2 \frac{\int E_i^2 \, dS}{\int E_o^2 \, dS}.$$

Consequently if the voltage is taken proportional to the electric field, the voltage stepup is given by Eq. (52).

More complicated cases may be treated in a similar fashion. Thus a transmission cavity may be formed by placing two inductive susceptances slightly less than a half wavelength apart. In fact, it is immediately evident from the equivalent circuit (Fig. 4·21) of the iris and from Eq. (50) that the length is given by

$$B \tan \beta l = 2. \tag{54}$$

The loaded Q has just one-half the value for a cavity with a single window; G is, of course, now equal to 2. It is easy to see from the admittance diagram of Fig. 6·15 that for a lossless line, the conductance is unity if the two windows are equal. If the line is lossy, the conductance is larger. To obtain a matched transmission cavity in a lossy line, the input window must be slightly larger (smaller $|B|$) than the output window.

It should perhaps be mentioned explicitly that calculations of the kind just described neglect the losses in the iris itself. For accurate calculations of high-Q cavities, some estimate of the loss should be included. It has been suggested that a suitable estimate for the losses produced by the presence of a hole in a metal wall is made by assuming that the losses over the area of the hole are twice the losses in the wall before the hole was cut. It is evident that this is an extremely uncertain approximation.

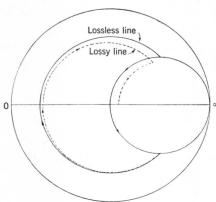

FIG. 6·15.—The admittance diagram of a cavity formed by two inductive windows one-half wavelength apart.

CHANGES IN THE CHARACTERISTIC IMPEDANCE OF A TRANSMISSION LINE

6·15. Diameter Changes in Coaxial Lines.—If the usual definition of the characteristic impedance of a coaxial line is adopted, a change in the diameter of the inner or outer conductor results in a change in the line impedance. Such a change in impedance produces a discontinuity in the voltage and current, and a reflected wave is produced. At long wavelengths, no other effects are important. An equivalent circuit of the junction is therefore simply an ideal transformer that has a turn ratio n given by

$$n^2 = \frac{Z_2}{Z_1}. \tag{55}$$

At short wavelengths or high frequencies other effects become important and the equivalent circuit becomes more complicated. Although any one of the general forms of a two-terminal-pair network is suitable, for example, a T- or Π-network, the circuit of Fig. 4·23, which contains an ideal transformer and two reactive elements, is particularly suitable for the representation of junctions. For all of the junctions in common use both in coaxial line and in waveguide, the series element in this circuit has negligible impedance and the circuit reduces to a shunt element and an ideal transformer. If the proper ratio of characteristic impedances is chosen for the two lines, the transformer can be made to have a turn ratio of unity. The equivalent circuit is therefore simply two transmission lines connected together with a shunt impedance as shown in Fig. 6·16. The shunt element in the circuit is called the "junction effect."

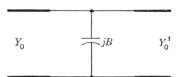

FIG. 6·16.—The equivalent circuit for a junction between two transmission lines. The shunt susceptance is called the "junction effect."

In coaxial lines, a change in diameter is associated with a distortion of the fields of the normal principal mode. Higher modes are excited in the neighborhood of the junction. It is evident that the higher modes are E-modes, and consequently the stored energy in these modes corresponds to a capacitive junction effect as indicated in Fig. 6·16. The capacitive susceptance should be inversely proportional to λ and hence negligible for low frequencies. At high frequencies, B becomes appreciable, although with the coaxial line in use at 10 cm it can often still be neglected. In accord with the general principles stated in Sec. 6·8, the magnitude of the junction effect should be nearly the same as the effect of the corresponding change in height of a parallel-plate transmission line or of a rectangular waveguide. The waveguide effect is discussed in the following section.

6·16. Change in the Dimensions of a Rectangular Waveguide. *Change in Waveguide Height.*—The junction of two rectangular waveguides of heights b and b' is rigorously represented by the equivalent circuit of Fig. 6·16. The characteristic admittance of the waveguide should be chosen inversely proportional to the height. In terms of the parameter

$$\alpha = \frac{Y_0}{Y_0'} = \frac{b'}{b}, \qquad (56)$$

where α is chosen smaller than unity, the junction susceptance is given by the approximate formula

$$\frac{B}{Y_0} = \frac{2b}{\lambda_g}\left[\ln\frac{1-\alpha^2}{4\alpha}\left(\frac{1+\alpha}{1-\alpha}\right)^{\frac{1}{2}\left(\alpha+\frac{1}{\alpha}\right)} - f\left(\frac{b}{\lambda_g}\right)\right], \qquad (57)$$

for the symmetrical change in height. The function $f(b/\lambda_g)$ is the high-frequency correction term. Since the susceptance is positive, the junction is sometimes termed a capacitive change in cross section. In the larger of the two waveguides the field configuration is very similar to that near a symmetrical capacitive slit, and in the smaller guide the field is not greatly different from that of the dominant mode. It might be expected that the stored energy and therefore the susceptance given by Eq. (57) should be approximately half that given by Eq. (6). Although it is not at all evident from the form of the expressions, insertion of numerical values shows that the difference is indeed small, of the order of 10 percent.

By an argument identical with that given in Sec. 6·4, the junction susceptance for the completely asymmetrical change in height can be obtained from the symmetrical case if λ_g is replaced by $\lambda_g/2$. Similarly, the junction effect for a change of height in a parallel-plate transmission line is to be found by replacing λ_g by λ.

Change in Waveguide Width.—The inductive change in cross section of a waveguide leads to a junction effect that is approximately one-half the susceptance of the corresponding thin inductive aperture. The proper choice of the characteristic admittance allows the equivalent circuit to be that of Fig. 6·16. Junctions of two types are possible. If the change in width is symmetrical and from a to a', the waveguide of smaller width is beyond cutoff if $\lambda > 2a'$ and the characteristic admittance of the smaller guide becomes imaginary. The proper value of admittance is

$$\frac{Y_0'}{n^2 Y_0} = -j\frac{\lambda_g a}{|\lambda_g'|a'}, \qquad (58)$$

where

$$\lambda'_g = j\frac{2a'}{\sqrt{1-\left(\frac{2a'}{\lambda}\right)^2}} \qquad (59)$$

and

$$n = \frac{4}{\pi}\frac{\cos\frac{\pi}{2}\frac{a'}{a}}{1-\left(\frac{a'}{a}\right)^2}. \qquad (60)$$

For $a'/a \ll 1$, Eq. (58) reduces to

$$\frac{Y'_0}{n^2 Y_0} = -j\frac{\pi^2}{32}\frac{\lambda_g}{a}\left(\frac{a}{a'}\right)^2. \qquad (61)$$

For an unsymmetrical change in cross section, n is replaced by

$$n = \frac{2}{\pi}\frac{\sin \pi \frac{a'}{a}}{1-\left(\frac{a'}{a}\right)^2}. \qquad (62)$$

If $\lambda < 2a'$, the small waveguide is not beyond cutoff, and the proper characteristic admittance is

$$\frac{Y'_0}{Y_0} = \frac{\lambda_g a}{\lambda'_g a'}, \qquad \left(1 - \frac{a'}{a}\right) \ll 1. \qquad (63)$$

The rough approximation that the junction effect is half the susceptance of the thin iris can be replaced by a more accurate expression given in *Waveguide Handbook*.

Resonant Change in Cross Section.—A combination of an inductive and a capacitive change in cross section results in a junction that is matched. The condition for this seems to be very nearly the same as the condition for a resonant rectangular iris, which was given as Eq. (14). Little exact information is available for junctions of this type.

6·17. Quarter-wavelength Transformers.—The reflections produced at a change in cross section of a transmission line can be used for matching a load to the line. If a section of line of impedance Z_1 a quarter wavelength long is inserted in a line of impedance Z_0, the impedance seen at the input end of the quarter-wavelength section is

$$Z_\text{in} = \frac{Z_1^2}{Z_0} \qquad (64)$$

if the junction effects are negligible (see Sec. 3·3). Such a section of line is commonly called a transformer. If the load end of such a transformer is placed at a voltage minimum in the standing-wave pattern produced by the load, the input impedance is Z_0 provided that

$$Z_0^2 = Z_1^2 r,$$

or

$$Z_1 = Z_0 \sqrt{\frac{1}{r}}, \tag{65}$$

where r is the voltage standing-wave ratio. Quarter-wavelength transformers are frequently used in coaxial line for matching in this way. The impedance change is produced either by increasing the diameter of the inner conductor by placing a metal sleeve over it or by decreasing the outer-conductor diameter with a sleeve, since $Z_1 < Z_0$. If the inner and outer radii of the coaxial line are a and b respectively and the radius of the sleeve on the inner conductor is r_1, then, since the characteristic impedance is proportional to the logarithm of the ratio of the radii of the inner and outer conductors, the proper value of r_1 is

$$r_1 = b \left(\frac{a}{b}\right)^{r - \frac{1}{2}}. \tag{66}$$

If the sleeve is placed inside the outer conductor, match is obtained with a sleeve of inner radius r_2, where

$$r_2 = a \left(\frac{b}{a}\right)^{r - \frac{1}{2}}. \tag{67}$$

The frequency sensitivity of the match is usually not large, but this sensitivity can be reduced by using several quarter-wavelength transformers.[1]

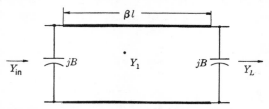

FIG. 6·17.—The equivalent circuit of a quarter-wavelength transformer when the junction effects are included.

If the frequency is high and the junction effect can no longer be neglected, it is still possible to use quarter-wavelength transformers effectively. If it is desired to transform a load admittance Y_L in this manner, the equivalent circuit shown in Fig. 6·17 must be used. The input admittance relative to the main line is

$$Y_\text{in} = jB + Y_1 \frac{Y_L + j(B + Y_1 \tan \beta l)}{Y_1 - B \tan \beta l - jY_L \tan \beta l}. \tag{68}$$

[1] J. C. Slater, *Microwave Transmission*, McGraw-Hill, New York, 1942, pp. 57*ff*.

If the length of the transformer is chosen so that

$$\tan \beta l = \frac{Y_1}{B},\tag{69a}$$

then

$$Y_{\text{in}} = \frac{B^2 + Y_1^2}{Y_L}.\tag{69b}$$

Equation (69b) thus determines the admittance of the transformer section, and Eq. (69a) determines the length of the transformer. If B is positive, then both Y_1 and l are smaller than the corresponding values when the junction effect is neglected.

Quarter-wavelength transformers have also been used in waveguide matching. For this purpose the capacitive change in cross section is most suitable.

6·18. Tapered Sections of Line.—The quarter-wavelength transformers for matching two transmission lines are of a characteristic impedance intermediate between those of the two lines. This circumstance suggests that two lines might be matched by a gradual taper of the dimensions of one line to those of the other. Such is indeed the case, but it is found that for many applications such a tapered transformer is too long if the taper is gradual enough to be reflectionless. It is often desirable to use short tapers and to arrange that the reflection from one end of the tapered section cancels that from the other. The investigation of tapered lines proceeds from the transmission-line equations

$$\frac{dV}{dz} = -ZI, \qquad \frac{dI}{dz} = -YV,$$

where Z and Y are no longer assumed to be constant. If I is eliminated from these two equations, V is found to satisfy

$$\frac{d^2V}{dz^2} - \frac{d \ln Z}{dz}\frac{dV}{dz} + ZYV = 0.\tag{70}$$

Hence the properties of the derivative of $\ln Z$ determine the behavior of the taper. For a gradual taper, it can be shown[1] that the voltage reflection coefficient is approximately

$$\Gamma = \frac{1}{4\gamma_0}\left(\frac{d \ln Z}{dz}\right)_0 - \frac{1}{4\gamma_1}\left(\frac{d \ln Z}{dz}\right)_1 e^{-2\int_{z_0}^{z_1}\gamma\,dz}.\tag{71}$$

The subscripts 0 and 1 refer to the values of the quantities at the beginning and end of the taper, respectively. If the derivatives are not very different in value, a length l of the taper can be chosen to make Γ a minimum. For example, if $\ln Z$ varies linearly along the line, the reflection

[1] Slater, *op. cit.*, pp. 71*ff.*

is given by

$$\Gamma = \frac{1}{4j\beta l} \ln \frac{Z_1}{Z_0} (1 - e^{-2j\beta l}) \tag{72}$$

if $\gamma = j\beta$ does not change over the length of the taper. The quantity Γ is thus zero when $\beta l = n\pi$ or $l = \lambda_g/2$.

6·19. The Cutoff Wavelength of Capacitively Loaded Guides.—Values of impedance changes and shunt susceptances can be used to compute the cutoff wavelengths of waveguides with complicated cross sections. If thin metal fins are inserted from the top and bottom faces of a rectangular waveguide, the cross section becomes that shown in Fig. 6·18a.

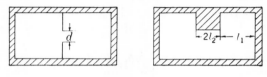

(a)   (b)
FIG. 6·18.—Capacitively loaded waveguides.

The boundary conditions in the waveguide are independent of the z-coordinate except for the discontinuity introduced by the metal fins. The variation of the fields in the z-direction is known and must be proportional to $e^{-j\kappa z}$, where $\kappa = 2\pi/\lambda_g$. Consequently, by arguments similar to those used in Sec. 6·8, the problem can be considered as that of a waveguide in which the x-direction is the direction of propagation, and this waveguide is terminated at each end by a short circuit and contains a capacitive slit at the center. A pure standing wave exists in this guide. The condition for this standing wave to exist is that the admittance become infinite at the two metal walls. The admittance Y' seen to the right just at the left side of the metal fin is

$$Y' = -j \cot k_x \frac{a}{2} + jB_1 \tag{73}$$

where k_x is the wave number in the x-direction and B is that susceptance given by Eq. (6), neglecting the high-frequency corrections, except that k_x has been substituted for $\kappa = 2\pi/\lambda_g$; thus

$$B = \frac{2k_x b}{\pi} \ln \csc \frac{\pi d}{2b}. \tag{74}$$

The admittance Y' transformed through a length $a/2$ of line must be infinite; therefore

$$1 + jY' \tan k_x \frac{a}{2} = 0,$$

or

$$\tan k_x \frac{a}{2} = \frac{2}{B}. \qquad (75)$$

The value of $k_x = 2\pi/\lambda_c$ is obtained by eliminating B from Eqs. (74) and (75). For $d/b \ll 1$,

$$\lambda_c^2 = 2\pi ab \ln \frac{2b}{\pi d}. \qquad (76)$$

The cutoff wavelength therefore increases without limit as d/b approaches zero, and the guide wavelength approaches the wavelength in free space. For d/b very small, the presence of the side walls of the guide has little effect and the field is concentrated between the two metal fins.

The waveguide cross section of Fig. 6·18b can be treated similarly. The discontinuity is now a change in the height of the waveguide. The admittance seen at the center of the guide must be zero, and hence

$$-j \cot k_x l_1 + jB - j\frac{b}{d} \tan k_x l_2 = 0, \qquad (77)$$

where B is the junction susceptance given by Eq. (57) with $k_x/2$ substituted for κ.

In all calculations of this type the possible effects of interaction must not be forgotten. Thus if the distance l_1 is too small, the wall of the waveguide will interact with the junction and the results of the calculation will be inaccurate.

BRANCHED TRANSMISSION LINES

6·20. Shunt Branches in Coaxial Lines.—If the wavelength of the radiation in a coaxial line is long compared with the diameter, the junction effects can be neglected in all cases. Even where the junction effects are not small, the behavior of a configuration is not greatly altered by their presence. It has been the practice in the design of components in coaxial line at 10-cm wavelength to ignore the junction effects in the first approximation and then to make small alterations in the critical dimensions to take account of the more complicated behavior at high frequencies. Thus a shunt branch in a coaxial line behaves very much as a shunt circuit at low frequencies. The currents at the junction divide in the ratio of the two admittances, and a reflection occurs at the junction that is equal to the reflection produced by the sum of the admittances of the branches.

A common application of a shunt branch is to form a stub to support the inner conductor of a coaxial line. The stub is a short-circuited shunt line of a length approximately one-quarter of a wavelength. The admittance of the stub is therefore zero at the junction, and no reflection is produced. If the frequency of the radiation is high, the stub should not be exactly a quarter wavelength long, but the correct length must be found experimentally. Such a simple stub is rather frequency-sensitive. If a standing-wave ratio of 1.05 is allowed, which is a rather large value, the admittance Y of the stub must be less than $\pm 0.05j$. Since

$$\lambda \frac{dY}{d\lambda} = -\lambda \left(\frac{d}{d\lambda} j \cot \beta l\right)_{\beta l = \pi/2} = j\frac{\pi}{2},$$

the usable wavelength band $d\lambda/\lambda$ is only a little over 3 per cent.

A stub support that is usable over a much broader band can be made by adding two quarter-wavelength transformers, one on each side of the stub, as shown in Fig. 6·19. At the center frequency, where the stub admittance is zero, the quarter-wavelength transformers are, together, one-half wavelength long and the whole device is reflectionless. At a lower frequency, the stub presents an inductive susceptance, but the transformers are less than a quarter wavelength long and the net reflection is again zero. This is indicated in the rectangular admittance diagram shown in Fig. 6·20a. The first transformer moves the admittance point clockwise along the circle from Y_0 to A; the inductive susceptance of the stub takes it from A to B; and the second transformer moves the admittance from B back to Y_0. At a shorter wavelength, the admittance

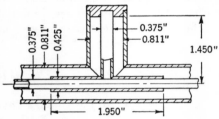

Fig. 6·19.—Broadband stub support.

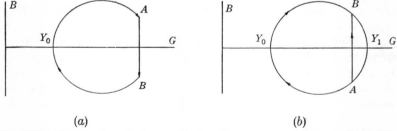

Fig. 6·20.—Rectangular admittance diagram illustrating the action of the broadband stub: (a) The path of the admittance point for a long wavelength, (b) the path for a wavelength shorter than the center wavelength.

diagram is shown in Fig. 6·20b. The first transformer moves the admittance point from Y_0 to A through more than a quarter wavelength; the stub is capacitive and moves the admittance from A to B; and the second transformer moves it from B back to Y_0. Here again, to obtain compensation for the junction effects, the stub length must be adjusted experimentally. The dimensions for a broadband stub for a wavelength of 10 cm are shown in the figure. Such a stub will give a standing-wave ratio of less than 1.02 from about 8- to 12-cm wavelength.

Stubs may also be used for matching in exactly the same manner as susceptive diaphragms. The susceptance may be adjusted by changing the stub length, and the stub should be placed a distance from a voltage minimum given by Eq. (36) or (37).

Stub tuners that are similar to the screw tuner described in Sec. 6·13 are also in common use. Since the susceptance of a stub may be either positive or negative, only two stubs are necessary. If the stub susceptance is limited to ± 1, the area of the admittance chart covered by two stubs one eighth of a wavelength apart is shown in Fig. 6·21. This diagram should be compared with Fig. 6·13 for a triple-screw tuner.

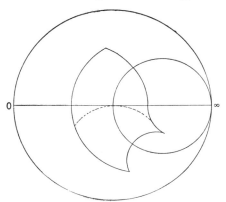

FIG. 6·21.—The range of admittance covered by a double-stub tuner with the stubs separated one-eighth of a wavelength and with a maximum susceptance of ± 1.

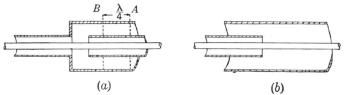

FIG. 6·22.—Series branches in coaxial lines.

6·21. Series Branches in Coaxial Lines.—A series branch may be made in coaxial transmission lines as illustrated in Fig. 6·22. In (a) the outer conductor is broken, the branch line is formed with a third cylinder as the outer conductor, and the original outer conductor is also the inner conductor of the branch line. The voltage across one line is equal to the sum of the voltages across the other two, if the positive directions are chosen properly, and the same currents flow in each line. At low frequencies, the lines are therefore in series, with negligible junction effect. There is a change of impedance and consequently a reflection produced at such a junction. If the impedances are adjusted, a series branch can be made with no reflection as indicated in Fig. 6·22b.

If a short circuit is placed in the branch line a half wavelength from the junction, as at A in Fig. 6·22a, there is a short circuit at the junction and power is transmitted without reflection. Current still flows in the branch line, however, and a very large standing-wave ratio is present. The current flowing along the branch is a maximum and the voltage across the branch is zero at the short circuit A and at the junction; at B, a quarter wavelength from A, the voltage is a maximum and the current is zero. Since the current is zero at B, the circuit may be broken there. No radiation will occur from a small gap, since the transverse magnetic field is zero. Such a joint is known as a half-wavelength choke, or choke joint. Choke joints are commonly used when it is desired to rotate one section of the transmission line with respect to another section. A sche-

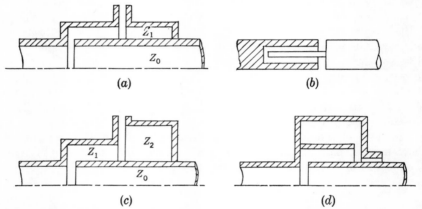

FIG. 6·23.—Choke joints in coaxial line: (a) a joint in the outer conductor shown schematically, (b) a joint in the inner conductor, (c) and (d) choke joints in the outer conductor which have reduced frequency sensitivity.

matic representation of the joint in the outer conductor of a coaxial line is shown in Fig. 6·23a. A similar joint can be made in the inner conductor, as in Fig. 6·23b. In all cases the short-circuited line is only approximately a half wavelength long, and the exact length must be determined experimentally.

Since a choke joint is a resonant configuration, the frequency sensitivity of the impedance at the junction must often be taken into consideration. If the main line has a characteristic impedance Z_0 and the branch line an impedance Z_1 as in Fig. 6·23a, the percentage rate of change of impedance at the junction is

$$\beta \frac{dZ}{d\beta} = \beta \frac{d}{d\beta}\left(j \frac{Z_1}{Z_0} \tan \beta l\right)_{\beta l = \pi} = j \frac{Z_1}{Z_0} \pi. \tag{78}$$

Thus it is desirable to keep Z_1 small compared with Z_0. Further improvement is possible by making the parts of the choke section of lines having different characteristic impedances, as shown in Fig. 6·23c. The relative impedance Z at the junction is

$$Z = j\frac{Z_1}{Z_0}\frac{Z_1 \tan \beta l_1 + Z_2 \tan \beta l_2}{Z_1 - Z_2 \tan \beta l_1 \tan \beta l_2}. \qquad (79)$$

The impedance derivative when $\beta l_1 = \beta l_2 = \pi/2$ is

$$\beta \frac{dZ}{d\beta} = j\frac{Z_1}{Z_0}\frac{\pi}{2}\left(1 + \frac{Z_1}{Z_2}\right). \qquad (80)$$

Therefore the optimum condition is to have $Z_1 \ll Z_0$ and $Z_2 \gg Z_1$. This desirable low-impedance—high-impedance condition reduces the frequency sensitivity by almost a factor of 2, and it is employed in the design of nearly all half-wavelength chokes. A more compact choke design is shown in Fig. 6·23d. For low frequencies, the "folding" of the choke has little effect on the length or frequency sensitivity.

6·22. Series Branches and Choke Joints in Waveguide.—A junction that behaves much as a series branch may be made in rectangular wave-

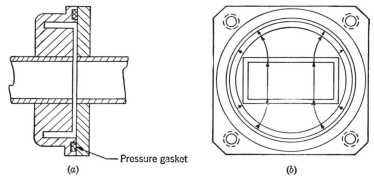

Fig. 6·24.—Choke-joint connector for rectangular waveguide.

guide with a secondary guide branching from the broad face of the main transmission line. Since waveguide dimensions are much longer compared with the wavelength than are the dimensions of coaxial line, the junction effects are large and cannot be neglected. If the height b' of the branch guide and the height b of the main line are both small compared with λ_g, a pure series junction is closely approximated. A choke joint may be made in much the same manner as in coaxial line. A gap in the form of a vertical slit in the narrow face of the waveguide has only a small effect, since the currents in the wall are in the direction of the length of the slit. A gap in the broad face of the waveguide would produce, however, a large disturbance. A short-circuited stub line one-half wavelength long, which is broken at the quarter-wavelength point, would form a good choke joint. A practical design of a choke connector for rectangular waveguide is shown in Fig. 6·24. The circular choke groove is easy to manufacture, and the high- and low-impedance stub sections are incorporated in the design. The proper diameter of the groove must, of course, be determined experimentally; the proper depth of the groove is very close to one-quarter of a free-space wavelength.

Stub tuners can also be constructed in waveguide through the use of series junctions. The stubs can be waveguides of either round or rectangular cross section. Such a stub introduces a series reactance in the line rather than a shunt susceptance. The application of the duality

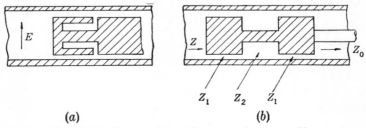

(a)   (b)

Fig. 6·25.—Plungers for use in rectangular waveguide.

principle, however, allows the matching formulas and the tuning range that apply for the shunt circuit to be readily converted to the series case. To vary the length of a stub line, some form of adjustable short circuit or plunger is necessary. Plungers are usually designed with choke joints as indicated in Fig. 6·25a. Another design that is very similar employs three quarter-wavelength sections as shown in Fig. 6·25b. If the characteristic impedances of the sections, which are proportional to the waveguide heights, are those designated in the figure, the input impedance Z of the plunger is

$$Z = \frac{\left(\dfrac{Z_1}{Z_0}\right)^4}{\left(\dfrac{Z_2}{Z_0}\right)^2}. \tag{81}$$

Since Z_1/Z_0 can be made small, perhaps 0.1, and Z_2/Z_0 can be 0.5, Z can easily be as small as 4×10^{-4}, and the power loss is therefore about 0.01 db. These values are adequately small for nearly every application.

DISCONTINUITIES WITH SHUNT AND SERIES ELEMENTS

6·23. Obstacles of Finite Thickness.—If the thickness of the metal partition forming an iris in a waveguide is not small compared with a wavelength, the equivalent circuit is not a pure shunt element but must contain three independent elements or, if the device is symmetrical, two elements. Thus a thick wire or post extending across the waveguide between the broad faces can no longer be represented by an inductive susceptance alone, but the T-network that describes its behavior contains series elements also. The electromagnetic problem that must be solved to find the elements of the equivalent circuit may be reduced as before to two problems by symmetry arguments. Thus if equal voltages are applied to the two terminal pairs of the device, a magnetic wall is

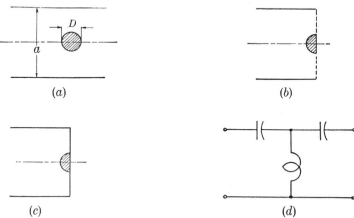

Fig. 6·26.—Decomposition of the problem of the thick post across a waveguide (a) into the even (b) and the odd (c) problems. The cross sections are taken in the magnetic plane.

effectively inserted in the plane of symmetry as shown in Fig. 6·26b. The terminating impedance of the waveguide is $Z_{oc}^{(1/2)}$ (Sec. 4·6), where

$$Z_{oc}^{(1/2)} = Z_{11} + Z_{12}.$$

For the odd case, an electric wall is in the plane of symmetry as in Fig. 6·26c and the terminating impedance is $Z_{sc}^{(1/2)}$, where

$$Z_{sc}^{(1/2)} = Z_{11} - Z_{12},$$

or just the value of the series element in the T-network representation. If the reference planes for the obstacle are chosen as the plane of symmetry, then the position of the effective short circuit for the odd case will be slightly in front of the reference plane. The series elements of the T-network are therefore capacitive. An accurate calculation gives

$$X_{11} - X_{12} = -\frac{a}{\lambda_g}\frac{\left(\dfrac{\pi D}{a}\right)^2}{1 + \dfrac{11}{24}\left(\dfrac{\pi D}{a}\right)^2}. \tag{82}$$

The shunt element of the T-network is inductive, as for the thin wire, and the equivalent circuit is that shown in Fig. 6·26d. If the high-frequency correction terms are omitted, $X_{11} + X_{12}$ is given by the imaginary part of Eq. (8). For accurate results these correction terms should be included and the data given in *Waveguide Handbook* should be used.

The thick tuning screw must be described similarly by a T-network with both shunt and series elements. For small insertions of the screw, the circuit elements are all capacitive. With increasing insertion, the absolute value of the shunt reactance decreases and the magnitude of the series reactance increases. No theoretical treatment of the behavior is available, but experiments indicate that when the reactance of the shunt element is zero, the series element is approximately $-0.2j$. The effective short circuit is therefore slightly in front of the reference plane.

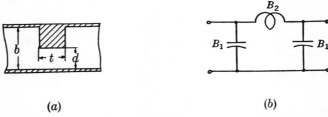

(a) (b)

Fig. 6·27.—A thick capacitive slit and the equivalent circuit.

A thick capacitive slit must also be described by a general two-terminal-pair network, but a somewhat different treatment is useful in this case. If the slit is very thick, the configuration can be considered as a change in the height of the waveguide plus a transmission line of length equal to the thickness of the slit and then another change in height. The natural reference planes to choose from this point of view are the entrance planes of the slit, and a Π-network is the most convenient. Thus if the dimensions are denoted by the symbols of Fig. 6·27a, the elements of the equivalent circuit in Fig. 6·27b are given by

$$\left. \begin{array}{l} B_1 = B - \dfrac{b}{d} \tan \dfrac{\pi t}{\lambda_g}, \\ B_2 = -\dfrac{b}{d} \csc \dfrac{2\pi t}{\lambda_g}, \end{array} \right\} \quad (83)$$

where B is the junction-effect susceptance for the change in height of the waveguide (Sec. 6·16). Equations (83) are based on the assumption that the interaction of the two changes in height is completely negligible. This assumption is certainly justified when $t \gg d$. The experimental data indicate, however, that these expressions are approximately true even for irises of very small thickness.

For all thick obstacles, the choice of the reference planes is arbitrary to some extent, as well as the form of the equivalent circuit. Two circumstances should be considered in the choice. (1) A certain set of reference planes might lead to an equivalent circuit that is particularly convenient to use for the application at hand. (2) It might be expected that a particular set of reference planes or a particular form of circuit would yield elements whose sensitivity to frequency would be most reasonable. Thus a T-network seems a suitable representation for a tuning screw, since the series reactances are small and have a simple dependence on λ_g, and since the single shunt element is resonant. On the other hand, the entrance reference planes and the Π-network seem a most reasonable representation of the thick capacitive slit.

6·24. Radiation from Thick Holes.—The radiation from small holes in thick metal walls should be treated similarly to the thick capacitive slit. The hole should be regarded as a transition from the normal waveguide to waveguide whose cross section is that of the hole, and back again. Within the hole the waveguide has a certain characteristic admittance, and there is a shunt susceptance at the junction. Since the hole is small, the small waveguide will be beyond cutoff, the characteristic admittance will be imaginary, and the propagation constant real. Thus, power is attenuated through the hole without a change in phase. If the hole is circular, the attenuation constant is

$$\alpha = \frac{2\pi}{1.706d} \sqrt{1 - \left(\frac{1.706d}{\lambda}\right)^2} \approx \frac{16.0}{d} \quad \text{db/meter}, \tag{84}$$

where d is the diameter of the hole. The junction susceptance should be approximately one-half the susceptance of a thick hole. Such accurate calculations have been made for only one case: a circular hole centered in rectangular waveguide. *Waveguide Handbook* should be consulted for the results.

A useful empirical rule is in common use for the calculation of the power transmission through a small circular hole. The transmission T in decibels is given by

$$T = T' + 2\alpha t, \tag{85}$$

where t is the thickness of the hole, α is given by Eq. (84), and T' is the transmission through a thin hole,

$$T' = 10 \log_{10} \frac{B^2}{4}.$$

The accurate calculations for the centered circular hole verify this empirical relation rather closely, and Eq. (85) is extremely valuable for design calculations. It may be relied upon to be accurate to about 1 db when used with the formulas of Sec. 6·11 for T'.

6·25. Bends and Corners in Rectangular Waveguide.—The transition from a straight waveguide to a smooth circular bend, either in the E- or in the H-plane, is another junction problem. The bend has a slightly larger impedance[1] than the straight portion, and there is a characteristic junction susceptance. If the radius of curvature is large, these effects are small, and negligible reflections occur at the bend. Thus a smooth bend is a practical method of changing the direction of a transmission line. If the bend has a small radius of curvature, it is still possible to have a reflectionless device by arranging the reflection from one transition to cancel that from the other. Waveguide bends of small radius are

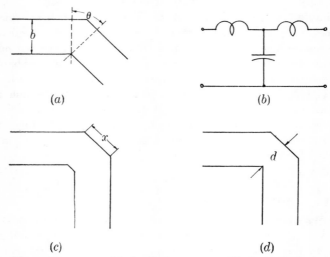

FIG. 6·28.—Waveguide corners: (a) An E-plane corner with the equivalent circuit shown in (b). Reflectionless corners are shown in (c) and (d).

difficult to manufacture, however, and special techniques must be employed.[2]

A sharp corner in waveguide is a more interesting device. If the reference planes are chosen as indicated by the dashed lines in Fig. 6·28a, the equivalent circuit of the corner is that shown in b of the same figure for both E- and H-plane bends[3]. The magnitudes of the series reactances and of the shunt susceptances are large, and large reflections occur from a sharp corner in a matched waveguide. At a certain angle a resonance occurs, the series reactances become infinite, and the shunt susceptance zero. For an E-plane corner at 3.2 cm in waveguide 0.4 by 0.9 in. ID, the angle θ for resonance is about 112°; for an H-plane corner, the angle is 106°.

[1] S. A. Schelkunoff, *Electromagnetic Waves*, Van Nostrand, New York, 1943, Chap. 8, p. 324.
[2] R. M. Walker, RL Report No. 585, July 1944.
[3] Pickering, NDRC Report 14–460, 14 July 1945.

A reflectionless change in direction can be made by combining two sharp corners[1] so that the reflections cancel as indicated in Fig. 6·28c. The proper distance x cannot be determined, however, from the equivalent circuit of a single corner, since x is small enough for the interaction effects to be important. Such mitered corners are in common use, since they are easy to make and have low reflections over a fairly broad band.

A reflectionless device can also be made by cutting off the edge of a sharp corner as indicated in Fig. 6·28d. If the distance d has the proper value, the reflection can be made zero for any angle of bend. This value of d is, however, very critical, and a bend of this type is difficult to manufacture.

6·26. Broadbanding.—The techniques that have been described in this chapter for matching a load to a transmission line are straightforward and can be definitely formulated. With lumped susceptances or quarter-wavelength transformers a matched condition is obtained at one frequency only. It is usually desired, however, that the load be matched to the line over a band of frequencies. Specified bandwidths vary from narrow, perhaps 1 per cent, to perhaps 20 per cent. Over a 20 per cent band the value of a matching susceptance also varies 20 per cent; and if a high standing-wave ratio exists, it can be canceled out only over a small band. The techniques for matching a microwave device over a broad band are not well defined, and no practical general procedure has been developed for "broadbanding" a piece of microwave equipment. Certain methods that have been successful for particular applications will now be described. It should be pointed out that although when expressed as percentages the microwave bandwidths under consideration are small compared with easily attainable bandwidths at audio or video frequencies, in terms of megacycles per second they are much larger.

In the previous discussion it has always been assumed that the problem was one of matching a dissipative load to the transmission line. A two-terminal-pair device is also said to be matched if there is no reflected wave at one terminal pair when the transmission line at the other terminal pair is terminated in its characteristic admittance. Lossless two-terminal-pair junctions can be matched by placing irises or transformers in either transmission line or in both. It is necessary to match a two-terminal-pair junction for power flowing in only one direction, since no standing waves exist in either the input or the output lines.

It has already been pointed out that the matching diaphragm should be placed as close as possible to the load. The extent to which an added length of transmission line contributes to the frequency sensitivity can be easily calculated by differentiating the equation of transformation of an admittance by a line,

[1] R. M. Walker, *loc. cit.*

$$Y_{in} = \frac{Y_L + j \tan \beta l}{1 + jY_L \tan \beta l}.$$

The result can be expressed as

$$\frac{\omega}{1 + (jY_{in})^2} \frac{dY_{in}}{d\omega} = j\beta l \left(\frac{\lambda_g}{\lambda}\right)^2 + \frac{\omega}{1 + (jY_L)^2} \frac{dY_L}{d\omega}. \quad (86)$$

This equation is easily interpreted if it is realized that

$$\frac{dY}{1 - (jY)^2} = d\left(-\frac{1}{2} \ln \Gamma\right),$$

where

$$\Gamma = \frac{1 - Y}{1 + Y}.$$

Equation (86) thus expresses the increase in the frequency sensitivity of the phase of the reflection coefficient from the added length of line. It appears reasonable, therefore, that a matching susceptance should be placed as close as possible to the load.

Resonant irises can sometimes be used to extend the bandwidth of a microwave device. If the device is matched at one frequency ω_1, then at a higher frequency ω_2 a position in the transmission line can be found at which the admittance is $1 - jB$, where B is positive. If a resonant iris is added in shunt with the line at this point and the resonant frequency is ω_1, the combination is reflectionless at ω_1. If the Q of the iris is properly chosen, the iris susceptance can be made to equal B at the frequency ω_2. Thus the total admittance of the iris and the load is again equal to unity at ω_2. If the two frequencies do not differ by too much, the reflections may be small in the whole region from ω_1 to ω_2. The curve of standing-wave ratio as a function of ω will have the form indicated in Fig. 6·29.

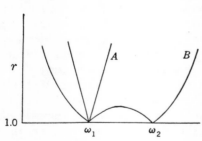

FIG. 6·29.—Standing-wave-ratio curve of a load matched at ω_1 with (Curve B) and without (Curve A) a resonant iris.

If the susceptance of the load at ω_2 is small, the separation of the resonant iris and the load is a quarter wavelength. This technique is thus equivalent to that employed in the construction of bandpass filters[1] and is analogous to a double-tuned circuit at low frequencies.

Other resonant devices can be used in a similar manner to extend the bandwidth over which a load is matched. A resonant transformer can be made by use of a length of line of high characteristic admittance. If the

[1] See Vol. 14, Chap. 3, and Vol. 9, Chap. 10, Radiation Laboratory Series.

junction effects can be neglected, the line section should be a half wavelength long. The resonant length when junction effects are present can be found from Eq. (68) of Sec. 6·17 by equating Y_{in} and Y_L. The resonant length is given by

$$\tan \beta l = \frac{2Y_1 B}{Y_L^2 - Y_1^2 + B^2}. \qquad (87)$$

Another procedure that can be used rather often to obtain a broadband match is illustrated in the admittance diagram of Fig. 6·30. The curves give the input admittance of a transition section between coaxial line and waveguide when the coaxial line is matched. By adjustment of

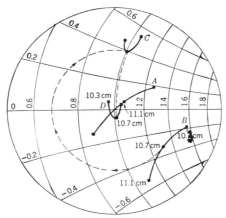

Fig. 6·30.—Admittance diagram of a transition from waveguide to coaxial line. The dimensions of the transition can be adjusted to give a match over a small band shown as Curve A. If, however, the dimensions are altered to mismatch the junction (Curve B) a broadband match (Curve D) is obtained by adding an inductive susceptance.

the dimensions of the transition a match can be obtained at one frequency without the use of any matching susceptances. The match is, however, sensitive to frequency, and the input admittance is shown as Curve A. An alteration of the dimensions results in the input admittance shown in curve B. If the transition were now matched by a capacitive susceptance placed only a short distance from the input terminal, an admittance curve very similar to Curve A would be obtained. If, however, an inductive susceptance is placed a little more than one-quarter wavelength away from the input terminal, as shown by Curves C and D, a smaller variation of input susceptance is achieved. Thus the usual rule that the matching susceptance be placed as close as possible to the junction is violated here. A more complete discussion of this example and others where a similar technique is fruitful can be found in Chap. 6 of Vol. 9 of this series.

In all the matching techniques so far described, matching elements are placed in only one transmission line. If a two-terminal-pair device is to be matched, it may be possible to achieve better results, since the matching elements can be placed in both transmission lines. The broadband stub described in Sec. 6·20 is an example of this procedure. By means of quarter-wavelength transformers in each of the two lines, it is possible to make the reflections vanish at three wavelengths, and a much broader bandwidth is obtained. The technique has been employed very little up to the present time in the design of microwave components. Since the method appears to be a fruitful one, some discussion of it is warranted. One possible general approach to the problem has been

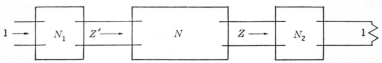

FIG. 6·31.—Procedure for broadband matching of a two-terminal-pair network N by means of the broadband transformers N_1 and N_2.

suggested by R. H. Kyhl. It can be shown[1] that for two arbitrary frequencies there exists a load impedance Z that is transformed by a given two-terminal-pair network to an impedance Z'. This is evident, since if Z exists, it is given at one frequency by

$$Z' = Z_{11} - \frac{Z_{12}^2}{Z_{22} + Z} \tag{88}$$

and at the other by

$$Z' = Z'_{11} - \frac{Z_{12}'^2}{Z'_{22} + Z}. \tag{89}$$

Equations (88) and (89) can be solved simultaneously for Z and Z' in terms of the matrix elements at the two frequencies. A given network N can then be matched over a broad band by the arrangement shown in Fig. 6·31. The networks N_1 and N_2 are broadband transformers that transform Z' to 1 and 1 to Z, respectively. The network is matched at two frequencies at one terminal pair and consequently is also matched at the other.

[1] Albert Weissfloch, "Application of the Transformer Theorem for Lossless Fourpoles to Fourpoles in Cascade Connection, *Hochf. u. Elekakus.* **61**, 19–28, 1943.

CHAPTER 7
RESONANT CAVITIES AS MICROWAVE CIRCUIT ELEMENTS

By Robert Beringer

A dielectric region completely surrounded by conducting walls is a resonant electromagnetic system. There exist electromagnetic field solutions of certain frequencies and spatial configurations that satisfy the boundary conditions and correspond to the storage of electromagnetic energy over time intervals that are long compared with the periods of the resonant frequencies. Such a system is commonly called a cavity, and the resonant solutions are the normal modes of the cavity. Each cavity has, of course, a different set of normal modes, differing both in frequency and in spatial configuration. All such sets have an infinite number of members.

If the set of normal modes is ordered with respect to increasing resonant frequency, it is found that there is always a lowest resonant frequency but, in general, no highest resonant frequency. In the direction of higher frequencies, the normal modes increase in complexity and in density, becoming infinitely dense at infinite frequencies. It is the first few members of the set that are of interest here because, for cavities of convenient size, these members are found to lie in the microwave region. In fact, for cavities of simple shape, the linear dimensions of the cavity are of the order of the wavelength of the lowest resonant frequency. For this reason, such cavities are commonly used as resonant elements in microwave circuits.

When a cavity is used as a circuit element, it is necessary to provide openings in the cavity walls for connection to the remainder of the circuit. These couplings will, of course, change the nature of the normal-mode fields. However, in most practical cases, the fields are not changed appreciably, and a knowledge of the normal-mode fields is therefore still useful.

The normal-mode fields can be found in terms of known functions for certain completely isolated cavities of simple geometrical shapes. Some of these solutions are tabulated in Chap. 4, Vol. 11, of this series. Our main concern here will be the circuit features of coupled cavities. Although the complete field solution of a coupled cavity is very difficult and may be impossible, there are many circuit features which all coupled cavities have in common.

A microwave circuit that contains cavities as resonant elements is composed of one or more transmission lines or systems, connected to the cavities by openings in the cavity walls and perhaps connected together in various ways as well. Although the general system may be of great complexity, it can be simplified by cutting each of the transmission lines near each cavity and separating each cavity from the system. Each of the separated elements consists of a cavity provided with one or more short sections of transmission line and their associated couplings to the cavity proper. Such elements are called cavity-coupling systems. In a complete circuit solution, each cavity-coupling system is treated separately, and the solutions are combined with an analysis of the transmission lines to give the complete description.

In developing this analysis, it is convenient to use the method of equivalent circuits, that is, to replace each cavity-coupling system by impedance elements of known magnitude and frequency dependence. Such a representation contains all of the facts that are pertinent to the circuit analysis of transmission systems containing cavities as elements.

There appear thus far two aspects of the problem: the complete field description of the cavity-coupling system and the equivalent-circuit representation. It is, of course, true that the two descriptions are equivalent and that a complete field description is required to give all of the details of the equivalent circuit. It is also true, however, that many of the features of the equivalent circuits can be found without a complete field description. These features are common to all cavity-coupling systems and form a basis for the circuit theory of such systems. In any particular case, a field description is required to specify completely the equivalent-circuit elements.

**EQUIVALENT CIRCUIT OF A
SINGLE-LINE, LOSSLESS CAVITY-COUPLING SYSTEM**

Consider for simplicity a cavity-coupling system without loss, containing only a single emergent transmission line. Such a system is shown in Fig. 7·1. It is possible to define, at a reference plane A, a voltage and a current that are uniquely determined by the electromagnetic fields interior to A. The voltage-to-current ratio at this plane defines an impedance, which is the input impedance of the cavity-coupling system. It is pure imaginary and a function of frequency alone.

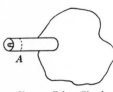

FIG. 7·1.—Single-line cavity-coupling system.

This impedance function can frequently be represented by an equivalent circuit. It is desired to find this function or representation and, in particular, its frequency dependence.

7·1. Impedance Functions of Lossless Lumped Circuits.

The problem of finding the impedance function of a cavity-coupling system is closely related to the impedance-function representation of a lossless, or purely reactive, n-mesh circuit with a single pair of exposed terminals. The n-mesh circuit is analyzed in the well-known manner of Foster's reactance theorem.[1] It is found that any such circuit can be represented by an input impedance function that is a rational fraction of the form

$$Z(\omega) = jA\omega \frac{(\omega^2 - \omega_1^2)(\omega^2 - \omega_3^2) \cdots (\omega^2 - \omega_{2n-1}^2)}{\omega^2(\omega^2 - \omega_2^2)(\omega^2 - \omega_4^2) \cdots (\omega^2 - \omega_{2n-2}^2)}. \quad (1)$$

The function is a pure imaginary. It has poles and zeros at frequencies $0, \omega_2, \omega_4, \ldots, \omega_{2n-2}$ and $\omega_1, \omega_3, \ldots, \omega_{2n-1}$, respectively. As Eq. (1) illustrates, and it is frequently proved, the poles and zeros of an imped-

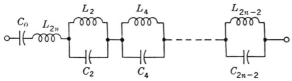

Fig. 7·2.—Input-impedance representation of an n-mesh circuit with poles at $\omega = 0$ and $\omega = \infty$.

ance function are simple, or of first order. It is seen from Eq. (1) that two networks are equivalent, that is, they have the same impedance at all frequencies, if they have the same poles and zeros and have impedances numerically equal at a single nonresonant frequency. This latter condition fixes the value of A.

An expression of the form

$$Z(\omega) = jA\omega \left(1 + \frac{a_0}{\omega^2} + \frac{a_2}{\omega^2 - \omega_2^2} + \cdots + \frac{a_{2n-2}}{\omega^2 - \omega_{2n-2}^2} \right) \quad (2)$$

is obtained by expanding Eq. (1) in partial fractions. Such a function can be represented by the series circuit of Fig. 7·2. The circuit parameters are given by

$$L_{2n} = A$$
$$C_k = -\frac{1}{a_k A} = -\left[\frac{j\omega Z^{-1}(\omega)}{\omega^2 - \omega_k^2} \right]_{\omega = \omega_k} \Bigg\} k = 0, 2, 4, \ldots, 2n - 2.$$
$$L_k = \frac{1}{\omega_k^2 C_k}$$

In Eqs. (1) and (2) the network is taken to have poles at both $\omega = 0$ and $\omega = \infty$. In special cases, one or both of these may be removed.

[1] Guillemin, *Communication Networks*, Vol. II, Wiley, New York, 1935, Chap. 5; Schelkunoff, *Electromagnetic Waves*, Van Nostrand, New York, 1943, Chap. 5.

Removal of the pole at $\omega = 0$ corresponds to putting $C_0 = \infty$ (short-circuiting C_0 in Fig. 7·2) so that $Z(\omega) \to 0$ as $\omega \to 0$. The pole at $\omega \to \infty$ is removed by putting $L_{2n} = 0$. Then $Z(\omega) \to 0$ as $\omega \to \infty$.

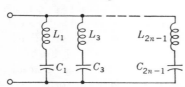

Fig. 7·3.—Input-admittance representation of n-mesh circuit with zeros at $\omega = 0$ and $\omega = \infty$.

The input admittance function analogous to Eq. (2) is obtained by expanding $Z^{-1}(\omega) = Y(\omega)$ around the poles of $Y(\omega)$, namely, $\omega_1, \omega_3, \ldots, \omega_{2n-1}$. This yields

$$Y(\omega) = Z(\omega)^{-1}$$
$$= -j\omega A^{-1}\left(\frac{b_1}{\omega^2 - \omega_1^2} + \frac{b_3}{\omega^2 - \omega_3^2} + \cdots + \frac{b_{2n-1}}{\omega^2 - \omega_{2n-1}^2}\right)\cdots. \quad (3)$$

Equation (3) may be represented by the shunt circuit of Fig. 7·3, where

$$\left.\begin{aligned} \omega_l^2 &= \frac{1}{L_l C_l} \\ b_l &= \frac{1}{L_l A^{-1}} \\ L_l &= -\left(\frac{j\omega Y^{-1}(\omega)}{\omega^2 - \omega_l^2}\right)_{\omega = \omega_l} \end{aligned}\right\} l = 1, 3, \cdots, 2n - 1.$$

In the expansion of Eq. (3) $Y(\omega)$ has been assumed to have zeros at $\omega = 0$ and at $\omega = \infty$. As in the impedance representation, these zeros may be removed in special cases, corresponding to degeneracies in one or two of the resonant elements. The zero at $\omega = 0$ is removed by letting one of the $C_l = \infty$, in which case $Y(\omega) \to \infty$ as $\omega \to 0$. The zero at $\omega = \infty$ is removed by letting one of the $L_l = 0$ so that $Y(\omega) \to \infty$ at $\omega \to \infty$.

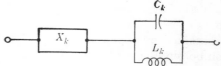

Fig. 7·4.—Input-impedance representation near a pole $\omega = \omega_k$.

At frequencies near one of the poles of $Z(\omega)$, Eq. (2) may be written as

$$Z(\omega) \approx jX_k + j\frac{\omega A \alpha_k}{\omega^2 - \omega_k^2}, \quad (4)$$

an approximation that lumps all contributions from other poles into the almost constant term X_k. This approximation is good at frequencies near the pole ω_k and far removed from any other poles. Equation (4) can be represented by the circuit of Fig. 7·4, where

$$A\alpha_k = -\frac{1}{C_k},$$

$$\omega_k^2 = \frac{1}{L_k C_k}.$$

In a similar manner the admittance function $Y(\omega)$ reduces to

$$Y(\omega) \approx -jB_l - j\frac{A^{-1}b_l\omega}{\omega^2 - \omega_l^2} \tag{5}$$

near the pole ω_l, and Eq. (5) can be represented by the circuit of Fig. 7·5, where

$$\omega_l^2 = \frac{1}{L_l C_l}$$

and

$$b_l = \frac{1}{L_l A^{-1}}.$$

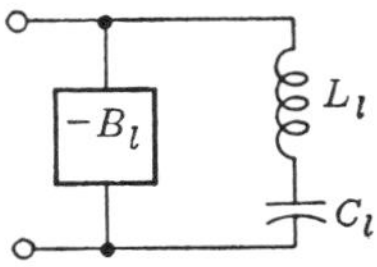

Fig. 7·5.—Input-admittance representation near a pole $\omega = \omega_l$.

In the study of cavity-coupling systems, it is often necessary to know the behavior of the system at frequencies near a particular resonance. There is therefore need for equivalent-circuit representations of the form of Figs. 7·4 and 7·5.

7·2. Impedance Functions of Lossless Distributed Circuits.—The impedance-function representation just discussed is of great generality and utility in lumped circuits. It furnishes a method for finding the equivalent circuit of any lossless, single-terminal-pair network in terms of the frequencies at which either $Z(\omega)$ or $Y(\omega)$ is infinite. An extension of the method to distributed circuits is clearly desirable.

It has been stated that a cavity has an infinite number of resonant frequencies. This is true for all distributed circuits. The impedance function, therefore, has an infinite number of poles and zeros, corresponding to an infinite number of network meshes. This suggests an extension of the foregoing method to the representation of a network with an infinite number of meshes. Such an extension is formally possible and has been carried out.[1] The expansions of Eqs. (2) and (3) are formally the same as those for the lumped-constant circuits, except for the fact that $n \to \infty$. Schelkunoff has stated that convergence difficulties sometimes arise in such series.

Although this formal extension is possible, it is more satisfactory to use the methods of Chap. 5 which make use of the field equations in defining and formulating the input impedance of a distributed circuit. In Chap. 5, a region surrounded by a perfectly conducting surface perforated by a single transmission line is discussed. Nothing is specified about the region except that the dielectric constant, permeability, and

[1] Schelkunoff, *Proc. IRE*, **32**, 83 (1944).

conductivity are independent of time and of the field quantities. The transmission line is assumed to operate in a single transmission mode. It is found possible to define voltages and currents at a reference plane normal to the axis of the transmission line which describe uniquely the field configurations in the region interior to the reference plane, except for certain degeneracies in the completely lossless case. These degeneracies will be ignored for the moment. By applying energy considerations to the interior and to the waves in the line, it is shown [see Eqs. (5·13) and (5·14)] that

and
$$\left. \begin{array}{l} Z(\omega) = \dfrac{j\omega 2(W_H - W_E)}{\frac{1}{2}ii^*}, \\[6pt] Y(\omega) = \dfrac{j\omega 2(W_E - W_H)}{\frac{1}{2}ee^*}, \end{array} \right\} \quad (6)$$

where W_H and W_E are respectively the time average of the magnetic and electric energies stored in the system interior to the reference plane and i and e are respectively the terminal current and voltage measured at that plane.

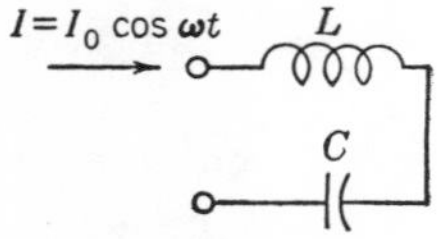

$I = I_0 \cos \omega t$

Fig. 7·6.—Simple series-resonant circuit.

The zeros of $Z(\omega)$ [poles of $Y(\omega)$] and zeros of $Y(\omega)$ [poles of $Z(\omega)$] occur at frequencies for which $W_E = W_H$. These frequencies, infinite in number, are defined as the resonant frequencies. This definition of resonance in terms of electric and magnetic energies is equivalent to the more usual definition of resonance for a simple lumped circuit. Consider, for example, the circuit of Fig. 7·6. The time average of the stored electric energy is

$$W_E = \frac{1}{2} C I_0^2 \left(\frac{1}{\omega C}\right)^2 \frac{\omega}{\pi} \int_0^{\pi/\omega} \cos^2 \omega t \, dt$$
$$= \frac{I_0^2}{4\omega^2 C},$$

and the time average of the stored magnetic energy is

$$W_H = \frac{1}{2} L I_0^2 \frac{\omega}{\pi} \int_0^{\pi/\omega} \cos^2 \omega t \, dt$$
$$= \tfrac{1}{4} L I_0^2.$$

At the resonant frequency, $\omega_0^2 = 1/LC$ and so

$$W_E = \frac{I_0^2}{4\omega_0^2 C} = \frac{1}{4} L I_0^2 = W_H.$$

Thus the frequency at which $W_E = W_H$ is the usual resonant frequency.

Referring again to the general expression [Eqs. (6)], it has been shown

[Sec. 5·24, Eq. (5·162)] that if $Y(\omega)$ has zeros at $\omega_1, \omega_2, \cdots$, then $Z(\omega)$ can be expanded as

$$Z(\omega) = -2 \sum_{n=1}^{\infty} r_n \frac{j\omega}{\omega^2 - \omega_n^2} + j\alpha_1 \omega, \qquad (7)$$

which is an obvious extension of Eq. (2) to the case of an infinite number of resonances. The question of the convergence of the series in Eq. (7), in the most general distributed case, is not always straightforward. All ordinary cavity-coupling systems, however, are free from convergence difficulties. In all practical cases, Eq. (7) reduces to Eq. (4) near a resonance.

7·3. Impedance-function Synthesis of a Short-circuited Lossless Transmission Line.—It is illustrative to consider a simple example of a distributed circuit in which Eq. (7) is evidently convergent and the poles and zeros of the impedance function are well known. A short-circuited lossless transmission line is such an example.

Let the line operate in the fundamental TEM-mode ($\lambda = \lambda_g$), and let the characteristic admittance be Y_0. Then, at terminals at a distance l from the short-circuited end, the input admittance is

$$Y = -jY_0 \cot \frac{2\pi}{\lambda} l$$
$$= -jY_0 \cot \frac{\omega l}{c}. \qquad (8)$$

The admittance has poles at

$$\omega = n\omega_1, \quad n = 0, 1, 2, \cdots,$$

where $\omega_1 = \pi c/l$, that is, $l = \lambda_1/2$.
If the substitution $l/c = \pi/\omega_1$ is made, Eq. (8) becomes

$$\frac{Y}{Y_0} = -j \cot \pi \frac{\omega}{\omega_1}. \qquad (9)$$

Expansion of $\cot \pi(\omega/\omega_1)$ in partial fractions results in

$$\frac{Y}{Y_0} = -j \frac{1}{\pi \frac{\omega}{\omega_1}} + j \frac{2}{\pi} \sum_{n=1}^{\infty} \frac{\frac{\omega}{\omega_1}}{n^2 - \left(\frac{\omega}{\omega_1}\right)^2}. \qquad (10)$$

Consider the admittance of an inductance L_0; it is $Y_0 = -j/\omega L_0$. The first term of Eq. (10) can be identified with such an inductance by putting

$$-jY_0 \frac{1}{\pi \frac{\omega}{\omega_1}} = -j \frac{1}{\omega L_0},$$

whereby

$$L_0 = \frac{\pi}{\omega_1 Y_0} = \frac{l}{cY_0}. \tag{11}$$

Each term of the sum in Eq. (10) can be identified as the admittance of a series LC-circuit. Such a series circuit has an admittance

$$Y_n = \frac{1}{j\omega L_n - j\frac{1}{\omega C_n}} = j\frac{\frac{1}{L_n \omega_n^2}}{1 - \frac{\omega^2}{\omega_n^2}} = j\frac{\omega \frac{1}{L_n \omega_1^2}}{n^2 - \frac{\omega^2}{\omega_1^2}}, \tag{12}$$

where $\omega_n^2 = n^2\omega_1^2 = 1/L_n C_n$. The values of L_n and C_n may be identified if it is noted that

$$\frac{2Y_0}{\pi} j \frac{\frac{\omega}{\omega_1}}{n^2 - \frac{\omega^2}{\omega_1^2}} = j\frac{\frac{\omega}{\omega_1^2}\frac{1}{L_n}}{n^2 - \frac{\omega^2}{\omega_1^2}},$$

from which

and

$$\left.\begin{array}{l} L_n = \dfrac{\pi}{2Y_0\omega_1} = \dfrac{1}{2}L_0 = \dfrac{l}{2cY_0} \\[2mm] C_n = \dfrac{1}{\omega_n^2 L_n} = \dfrac{2Y_0}{n^2\omega_1\pi} = \dfrac{2lY_0}{n^2\pi^2 c} \end{array}\right\} \quad n = 1, 2, \cdots. \tag{13}$$

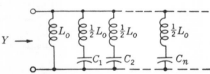

Fig. 7·7.—Representation of a short-circuited transmission line.

The short-circuited transmission line is evidently represented by the circuit of Fig. 7·7, where the circuit elements are given by Eqs. (11) and (13). The convergence of this representation is assured by Eq. (10). Near a resonance, a single term predominates and represents the admittance very well. Figure 7·8 shows the exact form of Eq. (10) and curves for two approximations[1] (one and five resonant elements).

EQUIVALENT CIRCUIT OF A SINGLE-LINE CAVITY-COUPLING SYSTEM WITH LOSS

The theory that has been given thus far is incomplete for the solution of the problem of cavity-coupling systems, since it deals only with completely lossless systems. The treatment must be extended to include

---

[1] E. A. Guillemin, "Development of Procedure for Pulse-forming Network," RL Report No. 43, Oct. 16, 1944. Guillemin has shown that a better approximation can be obtained with a finite number of resonant elements by choosing the resonant frequencies to be slightly different from $n\omega_1$.

loss. This can be done by any one of several approximate methods, each particularly useful for a certain class of problems. Although a general circuit representation,[1] valid at low frequencies, exists for an

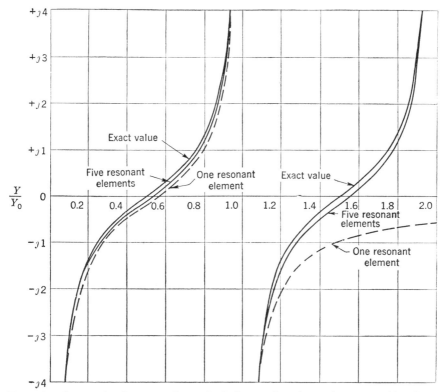

FIG. 7·8.—Curves showing exact form of admittance function and two approximations.

arbitrary two-terminal network that has lossy elements, it is difficult to handle. The generalization of the representation to microwave circuits has not been carried out, and in fact it may not prove to be feasible. Indeed, even the solution of the field equations obtained by expansion in terms of sets of orthogonal modes breaks down if the losses become too great. In practice, microwave devices do not have large losses, and it will suffice to treat only the cases where the approximate methods apply. A fruitful method of treating slightly lossy networks proceeds from Foster's theorem.

7·4. Foster's Theorem for Slightly Lossy Networks.—The extension of Foster's theorem can be carried out by introducing a complex fre-

[1] S. Darlington, "Synthesis of Reactance 4-Poles," *J. Math. Phys.*, **18**, 257–353 (1939).

quency, or oscillation constant, defined as $\lambda = j\omega + \xi$. Input impedances, being functions of this complex frequency, are written as $Z(\lambda)$. Voltages are defined as Re $(Ve^{\lambda t})$, and currents as Re $(Ie^{\lambda t})$;

$$Ve^{\lambda t} = Z(\lambda) I e^{\lambda t}.$$

In this notation, Eq. (1) for the input impedance to a single-terminal-pair network is written

$$Z(\lambda) = A \frac{(\lambda^2 - \lambda_1^2)(\lambda^2 - \lambda_3^2) \cdots}{\lambda(\lambda^2 - \lambda_2^2)(\lambda^2 - \lambda_4^2) \cdots}, \tag{14}$$

where $\lambda_1, \lambda_3, \ldots,$ and $0, \lambda_2, \lambda_4, \ldots,$ are respectively the zeros and poles of $Z(\lambda)$. For passive networks all decrements are negative, corresponding to a decrease in amplitude with time. Hence, all functions $Z(\lambda)$ have negative ξ at the poles and zeros; that is, all poles and zeros of $Z(\lambda)$ lie in the left half of the λ-plane.[1] As in the lossless case, Eq. (14) is expanded around its poles, with the result that

$$Z(\lambda) = A\left(\lambda + \frac{a_0}{\lambda} + \frac{a_2}{\lambda - \lambda_2} + \frac{a_4}{\lambda - \lambda_4} + \cdots\right)$$

$$= A\lambda + \frac{Aa_0}{\lambda} + A\sum_n \frac{a_n}{\lambda - \lambda_n}, \quad n = 0, 2, 4, \cdots. \tag{15}$$

The poles and zeros of $Z(\lambda)$ occur as conjugate pairs so that one term in the sum is of the form

$$\frac{a_n}{\lambda - \lambda_n} + \frac{a_n}{\lambda - \lambda_n^*}, \tag{16}$$

where $\lambda_n = j\omega_n + \xi_n$ and $\lambda_n^* = -j\omega_n + \xi_n$.

If, then, each mesh of a reactive network has a very small amount of loss (*i.e.*, small ξ_n) introduced into it, Expression (16) may be written as

$$\frac{a_n}{\lambda - j\omega_n - \xi_n} + \frac{a_n}{\lambda + j\omega_n - \xi_n}.$$

The values of $Z(\lambda)$ that are of interest are those for real frequencies $\lambda = j\omega$, for which Eq. (15) becomes

$$Z(\lambda) = jA\omega - j\frac{Aa_0}{\omega} + A\sum_n \frac{a_n(2j\omega - 2\xi_n)}{(\omega_n^2 - \omega^2 - 2j\omega\xi_n + \xi_n^2)},$$

which reduces to

$$Z(\lambda) = jA\omega - j\frac{Aa_0}{\omega} + A\sum_n \frac{a_n 2j\omega}{(\omega_n^2 - \omega^2 - 2j\omega\xi_n)} \tag{17}$$

for small loss. It is seen that Eq. (17) is of the same form as in the loss-

[1] Guillemin, *loc. cit.*, or S. A. Schelkunoff, *Electromagnetic Waves*, Van Nostrand, New York, 1943, Chap. V.

less case, except that the summation contains an imaginary term in the denominator. At each resonance $\omega = \omega_n$, therefore, $Z(\lambda)$ does not go to infinity but has a real contribution Aa_n/ξ_n from the summation.

The admittance case for small loss proceeds in an analogous fashion. The result is obtained, for real frequencies ($\lambda = j\omega$), that

$$Y(\lambda) = jB\omega - j\frac{Bb_0}{\omega} + B \sum_n \frac{b_n 2j\omega}{(\omega_n^2 - \omega^2 - 2j\omega\xi_n)}, \quad (18)$$

where ω_n are the poles of the admittance function, $\lambda_1, \lambda_2, \ldots$.

7·5. The Impedance Functions of Simple Series- and Parallel-resonant Circuits.—The representation of Eqs. (17) and (18) may be accomplished in a way analogous to Foster's method. Consider the simple series circuit of Fig. 7·9; its impedance is

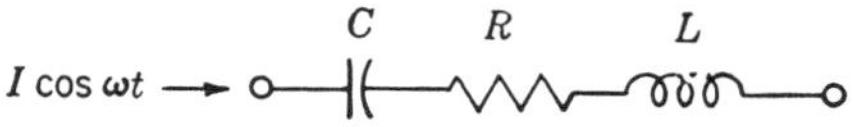

FIG. 7·9.—Simple series-resonant circuit with loss.

$$Z(\omega) = R + j\left(\omega L - \frac{1}{\omega C}\right).$$

If the resonant frequency ω_0 is defined as that frequency at which $Z(\omega)$ is pure real, $\omega_0^2 = 1/LC$ and

$$Z(\omega) = R + jL\omega_0\left(\frac{\omega}{\omega_0} - \frac{\omega_0}{\omega}\right).$$

The Q of the circuit at resonance is

$$Q = 2\pi \frac{\text{energy stored}}{\text{energy lost per cycle}}$$

$$= 2\pi f_0 \frac{\frac{1}{2}LI^2}{\frac{1}{2}I^2 R} = \frac{\omega_0 L}{R}.$$

In terms of this Q,

$$Z(\omega) = R + jQR\left(\frac{\omega}{\omega_0} - \frac{\omega_0}{\omega}\right),$$

or

$$Y(\omega) = \frac{1}{R + jQR\left(\dfrac{\omega}{\omega_0} - \dfrac{\omega_0}{\omega}\right)},$$

$$= j\frac{\dfrac{\omega\omega_0}{QR}}{\omega_0^2 - \omega^2 + j\dfrac{\omega\omega_0}{Q}}. \quad (19)$$

We see that the summation in Eq. (18) can be expressed by a sum of terms of the form of Eq. (19) corresponding to a number of simple series-resonant circuits in parallel.

Consider the admittance of the circuit of Fig. 7·10; it is

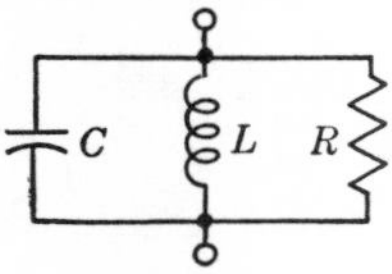

$$Y(\omega) = \frac{1}{R} + \frac{1}{j\omega L} + j\omega C$$

$$= \frac{1}{R} + j\omega_0 C \left(\frac{\omega}{\omega_0} - \frac{\omega_0}{\omega} \right),$$

Fig. 7·10.—Simple shunt-resonant circuit with loss.

where

$$\omega_0^2 = \frac{1}{LC}.$$

At resonance, the Q of this circuit is

$$Q = \omega_0 RC,$$

from which

$$Y(\omega) = \frac{1}{R} + j\frac{Q}{R}\left(\frac{\omega}{\omega_0} - \frac{\omega_0}{\omega}\right),$$

$$Z(\omega) = \frac{1}{\dfrac{1}{R} + j\dfrac{Q}{R}\left(\dfrac{\omega}{\omega_0} - \dfrac{\omega_0}{\omega}\right)}$$

$$= \frac{\dfrac{j\omega\omega_0 R}{Q}}{\omega_0^2 - \omega^2 + \dfrac{j\omega\omega_0}{Q}}. \qquad (20)$$

Comparison of Eq. (20) with Eq. (17) shows that the summation in Eq. (17) can be represented by a number of shunt-resonant circuits connected in series.

7·6. The Equivalent Circuit of a Loop-coupled Cavity.—Thus far, arbitrary cavity-coupling systems have been treated, first lossless systems and then systems having a small amount of loss introduced into each resonant mesh. The circuit representations have been derived either directly from field considerations or by analogy with lumped circuits. Although this approach is very fruitful, it is only qualitative in that there are a number of arbitrary features in the equivalent circuit. These features can be specified completely only by an actual solution of the field problem for the cavity-coupling system. This, unfortunately, has been done only in a few special cases. One of these is the case of a loop-coupled cavity.

It is instructive to study this solution. The circuits are derived from a new and more physical point of view; and in particular, the physical form of the approximations necessary to treat the dissipative system are made apparent. It has been shown that the treatment of such systems by an extension of Foster's theorem is uncertain with regard to the physi-

SEC. 7·6]   LOOP-COUPLED CAVITY   219

cal meaning of the small-loss approximation. The field approach displays this approximation more clearly.

Only in certain cases can the general field problem be solved. There are several limitations. Since such methods begin with the unperturbed normal modes of an isolated, lossless cavity, these must first be found. This is possible only for certain regular shapes. Also, the exact form of the coupling of the transmission line to the resonant mode must be found. This is very difficult, except for a cavity coupled to a coaxial line by a small loop or probe. The details of the problem of iris-coupled waveguides have not been solved.

A number of authors have treated the field aspects of the cavity problem, with various degrees of completeness. The normal-mode fields have received most of the attention. Condon[1] has treated the normal-mode fields and the effects of dissipation in the cavity walls. Slater[2] has made a very exhaustive treatment of the general problem; he treats wall losses and solves the loop-coupling case. Crout[3] and Baños[4] have applied the very elegant Lagrangian methods to the problem. This approach will be adopted here, and a brief summary of the simpler features of the Lagrangian method, as formulated by Baños, will be given.

It will be assumed that the normal modes of the lossless, unperturbed cavity have been found. These normal modes are the periodic solutions of Maxwell's equations satisfying the boundary conditions; that is, they are the solutions corresponding to standing waves in the cavity. They form a set, each member of which is characterized by a resonant frequency ω_a (wave numbers $k_a = \omega_a \sqrt{\epsilon\mu}$) and a pair of vector functions of position $\mathcal{E}_a$ and $\mathcal{3C}_a$ satisfying the wave equations

$$\nabla^2 \mathcal{E}_a + k_a^2 \mathcal{E}_a = 0,$$
$$\nabla^2 \mathcal{3C}_a + k_a^2 \mathcal{3C}_a = 0$$

and the divergence conditions,

$$\nabla \cdot \mathcal{E}_a = 0,$$
$$\nabla \cdot \mathcal{3C}_a = 0.$$

In addition, they satisfy the boundary conditions

$$\mathbf{n} \times \mathcal{E}_a = 0,$$
$$\mathbf{n} \cdot \mathcal{3C}_a = 0,$$

on the cavity walls. They form a normalized orthogonal set, in that they satisfy

[1] E. U. Condon, *Rev. Mod. Phys.*, **14**, 341 (1942); *J. Applied Phys.*, **12**, 129 (1941).
[2] J. C. Slater, "Forced Oscillations and Cavity Resonators," RL Report No. 188, Dec. 31, 1942.
[3] P. D. Crout, RL Report No. 626, Oct. 6, 1944.
[4] A. Baños RL Report No. 630, Nov. 3, 1944.

$$\int \mathcal{E}_a \cdot \mathcal{E}_b \, d\tau = \delta_a^b \mathcal{U}$$

$$\int \mathcal{H}_a \cdot \mathcal{H}_b \, d\tau = \delta_a^b \mathcal{U},$$

where $\mathcal{U}$ is the cavity volume. $\mathcal{E}_a$ and $\mathcal{H}_a$ are related by the expressions

$$k_a \mathcal{E}_a = \nabla \times \mathcal{H}_a,$$
$$-k_a \mathcal{H}_a = \nabla \times \mathcal{E}_a.$$

The electromagnetic field in the cavity can be expanded in this set of functions by

$$\mathbf{H} = \sum_a k_a \dot{q}_a \mathcal{H}_a,$$

$$\mathbf{D} = \sum_a k_a^2 q_a \mathcal{E}_a,$$

where the q_a are scalar functions of the time. The coefficients in the expansion have the dimensions of electric charge. The functions q_a corresponding to a normal mode k_a are the amplitude functions of the mode and are analogous to the coordinates of a dynamical problem. They are the so-called *normal coordinates* of the system.

The dynamical equation of the system in terms of these normal coordinates is the Lagrangian equation

$$\frac{d}{dt}\left(\frac{\partial T}{\partial \dot{q}_a}\right) + \frac{\partial V}{\partial q_a} = 0, \tag{21}$$

where T and V are the total kinetic and potential energies of the system. The quantity T is identified as the magnetic energy, and V as the electric energy of the system; that is,

$$T = \frac{1}{2}\mu \int H^2 \, d\tau$$

$$= \frac{1}{2}\mu \sum_a k_a^2 \mathcal{U} \dot{q}_a^2,$$

and

$$V = \frac{1}{2\epsilon} \int D^2 \, d\tau$$

$$= \frac{1}{2\epsilon} \sum_a k_a^4 \mathcal{U} q_a^2.$$

These have the form of stored energies in an inductance and a capacitance, where

$$\left.\begin{array}{l}L_a = \mu k_a^2 \mathcal{U}, \\ C_a = \dfrac{\epsilon}{k_a^4 \mathcal{U}}.\end{array}\right\} \quad (22)$$

In terms of these relations, T and V may be written as

$$\left.\begin{array}{l}T = \dfrac{1}{2} \sum_a L_a \dot{q}_a^2, \\ V = \dfrac{1}{2} \sum_a \dfrac{q_a^2}{C_a}.\end{array}\right\} \quad (23)$$

The quantities L_a and C_a are the equivalent inductance and capacitance respectively, of the mode k_a and correspond to the inductance and capacitance of the circuit of Fig. 7·11. From the relation $k_a = \omega_a \sqrt{\epsilon\mu}$, it may be seen at once from Eq. (22) that

$$\omega_a^2 = \frac{1}{L_a C_a}.$$

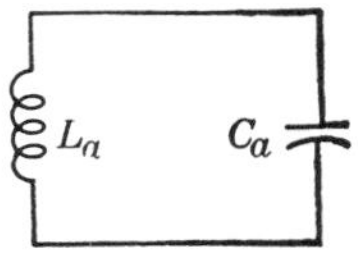

Substitution of Eq. (22) in Eq. (20) results in

$$L_a \ddot{q}_a + \frac{q_a}{C_a} = 0,$$

FIG. 7·11.—Normal-mode mesh in a lossless cavity.

as the set of dynamical equations of the system. It is seen that q_a is the charge on the capacitance C_a of the normal-mode mesh.

The frequency ω_a is the natural oscillation frequency of the mesh of Fig. 7·11. This is the frequency at which the mesh resonates, or "rings," after an exciting field has been removed. It is also the resonant frequency of the mesh, that is, the frequency at which the stored electric and magnetic energies are equal and the series reactance of the mesh vanishes.

If the lossless cavity is excited by a coupling loop and coaxial line, Lagrange's equations are of the form

$$\frac{d}{dt}\left(\frac{\partial T}{\partial \dot{q}_a}\right) + \frac{\partial V}{\partial q_a} = \theta_a,$$

where θ_a is the electromotive force induced in the normal-mode mesh by currents flowing in the loop. The loop is assumed to be so small that the current distribution is uniform along the loop. Then,

$$\theta_a = M_{0a} \frac{di_0}{dt},$$

where i_0 is the loop current and M_{0a} the mutual inductance between the loop and the normal-mode mesh, given by

$$M_{0a} = \mu k_a \int_{\text{loop}} \mathfrak{K}_a \cdot d\mathbf{s}. \tag{24}$$

The loop is seen to couple to all modes except when the integral of Eq. (24) vanishes. If $\mathfrak{K}_a$ is known, M_{0a} can be found by integrating Eq. (24) over the loop area.

The introduction of loss in the cavity walls adds another term to the dynamical equation, which then becomes

$$\frac{d}{dt}\left(\frac{\partial T}{\partial \dot{q}_a}\right) + \frac{\partial V}{\partial q_a} + \frac{\partial F}{\partial \dot{q}_a} = \theta_a, \tag{25}$$

where F is given by

$$F = \frac{1}{2} \int \frac{H^2 \, dS}{\sigma \delta}, \tag{26}$$

σ being the conductivity of the walls and δ the skin depth. It is assumed that the dissipation does not change the normal-mode fields $\mathcal{E}_a$ and $\mathfrak{K}_a$. It can be shown that Eq. (26) becomes

$$F = \tfrac{1}{2} \sum_a \sum_b R_{ab} \dot{q}_a \dot{q}_b.$$

The expansion coefficients R_{ab} may be written as

$$R_{ab} = \frac{\sqrt{\omega_a \omega_b L_a L_b}}{Q_{ab}},$$

where

$$\frac{1}{Q_{ab}} = \frac{\sqrt{\delta_a \delta_b}}{2\mathfrak{v}} \int \mathfrak{K}_a \cdot \mathfrak{K}_b \, dS. \tag{27}$$

For $b = a$,

$$\frac{1}{Q_{aa}} = \frac{1}{Q_a} = \frac{\delta_a}{2\mathfrak{v}} \int \mathfrak{K}_a^2 \, dS, \tag{28}$$

and

$$R_{aa} = R_a = \frac{\omega_a L_a}{Q_a}.$$

In terms of these relations, Eq. (25) becomes

$$L_a \ddot{q}_a + \frac{q_a}{C_a} + R_a \dot{q}_a + \sum_{b \neq a} R_{ab} \dot{q}_b = M_{0a} \frac{di_0}{dt}. \tag{29}$$

This is now interpreted as an equivalent circuit of the following form. Each normal mode is considered as a resonant mesh containing L_a, C_a, R_a, and $\sum_b R_{ab}$ in series and coupled through a mutual inductance M_{0a} to the loop. The mutual resistances R_{ab} couple all meshes for which Eq. (27) does not vanish.

It has been shown that Eq. (27) vanishes, for a cavity having the form of a right circular cylinder, for all modes that can become simultaneously resonant. It is probable that this is true also for other shapes. This fact considerably simplifies the circuit representation of Eq. (29), since the mutual resistances vanish and the meshes can be separated. Figure 7·12 is the representation of Eq. (29) under such conditions.

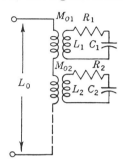

Fig. 7·12.—Circuit representation of a loop-coupled cavity.

The frequencies ω_a are the *resonant* frequencies of the meshes of Fig. 7·12 when the terminals are open-circuited, that is, the frequencies at which the series mesh reactances vanish and the stored electric and magnetic energies are equal. The natural oscillation frequencies of the meshes are somewhat different, being given by

$$\omega'_a = \omega_a \sqrt{1 - \tfrac{1}{4}Q_a^2},$$

which for high Q's reduces to

$$\omega'_a \approx \omega_a - \omega_a \frac{1}{8Q_a^2}.$$

Most cavities have such high Q's that ω'_a and ω_a differ at most by only a few parts in 10^5.

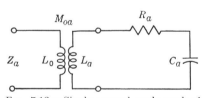

Fig. 7·13.—Single normal-mode mesh of Fig. 7·12.

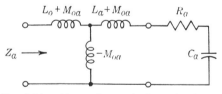

Fig. 7·14.—Alternative equivalent circuit for a single normal-mode mesh.

The circuit of Fig. 7·12 can be transformed easily into the form of Eq (17). Each normal-mode mesh is of the form of Fig. 7·13, which is equivalent to the circuit of Fig. 7·14. The circuit of Fig. 7·14 has an input impedance

$$Z_a = j\omega L_0 + \frac{\omega^2 M_{0a}^2}{j\omega L_a + R_a - j\dfrac{1}{\omega C_a}} \tag{30}$$

$$= j\omega L_0 + \frac{j\omega^3 M_{0a}^2}{L_a\left(\omega_a^2 - \omega^2 + \dfrac{jR_a\omega}{L_a}\right)}, \tag{31}$$

where
$$\omega_a = \frac{1}{L_a C_a}.$$

If Q_a is defined by
$$Q_a = \frac{\omega_a L_a}{R_a},$$

$$Z_a = j\omega L_0 + \frac{j\omega^3 \frac{M_{0a}^2}{L_a}}{\left(\omega_a^2 - \omega^2 + \frac{j\omega\omega_a}{Q_a}\right)}$$

$$= j\omega L_0 - j\omega \frac{M_{0a}^2}{L_a} + \frac{j\omega\omega_a^2 \frac{M_{0a}^2}{L_a}}{\omega_a^2 - \omega^2 + \frac{j\omega\omega_a}{Q_a}} - \frac{\omega^2 \omega_a \frac{M_{0a}^2}{Q_a L_a}}{\omega_a^2 - \omega^2 + \frac{j\omega\omega_a}{Q_a}}.$$

If the Q is high, the last term is negligible [at resonance the third term is $\omega_a(M_{0a}^2/L_a)Q_a$ and the last term is $\omega_a(M_{0a}^2/L_a)$]. Thus, to a very good approximation for even moderate Q's,

$$Z_a = j\omega L_0 - j\omega \frac{M_{0a}^2}{L_a} + \frac{j\omega\omega_a^2 \frac{M_{0a}^2}{L_a}}{\omega_a^2 - \omega^2 + \frac{j\omega\omega_a}{Q_a}}. \tag{32}$$

A comparison of this equation with Eqs. (20) and (17) shows that the last term may be represented as a shunt-resonant circuit.

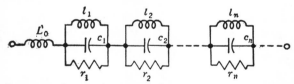

Fig. 7·15.—Representation of a loop-coupled cavity which is equivalent to Fig. 7·12.

Each of the various normal-mode meshes can be written in the form of Eq. (32). Since in an impedance representation they appear in series, the total input impedance is

$$Z = j\omega\left(L_0 - \sum_a \frac{M_{0a}^2}{L_a}\right) + j\sum_a \frac{\omega\omega_a^2 \frac{M_{0a}^2}{L_a}}{\omega_a^2 - \omega^2 + \frac{j\omega\omega_a}{Q_a}}. \tag{33}$$

A comparison of Eqs. (33) and (20) shows that the summation can be represented as a circuit of the form of Fig. 7·15, where

$$L_0' = L_0 - \sum_a \frac{M_{0a}^2}{L_a},$$

and the terms of Eqs. (32) and (19) can be identified. This leads to the following relations between the circuit elements of Figs. 7·12 and 7·15.

$$\omega_a^2 = \frac{1}{l_a c_a} = \frac{1}{L_a C_a}, \tag{34a}$$

$$Q_a = \omega_a r_a c_a = \frac{\omega_a L_a}{R_a}, \tag{34b}$$

$$r_a = \frac{\omega_a^2 M_{0a}^2}{R_a}, \tag{34c}$$

$$l_a = \frac{M_{0a}^2}{L_a}. \tag{34d}$$

The relations of Eqs. (34) do not identify the magnitudes of the circuit elements of Fig. 7·15 but only their interrelations. Clearly, there would be a loss in generality by putting each $l_a \equiv L_a$. However, l_a may be made proportional to L_a, in order to separate the magnitudes of the elements of Fig. 7·15 from the amount of coupling. This is accomplished by introducing ideal transformers and putting Fig. 7·15 into the form of Fig. 7·16, where, without loss of generality, the inductances and capacitances in Fig. 7·16 have been identified with those of Fig. 7·12. By use of Eqs. (34), it is seen that the resistances in Fig. 7·16 are given by

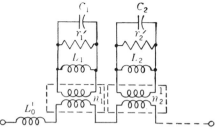

Fig. 7·16.—Alternative representation of a loop-coupled cavity.

$$r_a' = \frac{\omega_a^2 M_{0a}^2}{n_a^2 R_a} = \frac{\omega_a^2 L_a^2}{R_a}. \tag{35}$$

The representations of Fig. 7·16 are often more convenient than those of Fig. 7·15, since the circuit elements are determined by the normal-mode meshes of Fig. 7·12 alone, not by the coupling. The coupling is introduced by the ideal transformer, and the elements of Fig. 7·16 are not changed when the coupling changes as they are in Fig. 7·15.

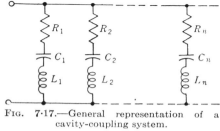

Fig. 7·17.—General representation of a cavity-coupling system.

7·7. Impedance Functions Near Resonance.—It has been shown, both by the introduction of a small loss into the Foster representation and directly from the field equations, that a single-line cavity-coupling system can be represented by an equivalent circuit of the form of Fig. 7·15 or 7·17, where in special cases one or more of the series or parallel elements may be degenerate.

At frequencies near a particular resonance, these circuits are greatly simplified, since the effects of all but the resonant mesh can be replaced by nonresonant elements. The most general representations near resonance are those of Figs. 7·18 and 7·19. In cavity-coupling systems

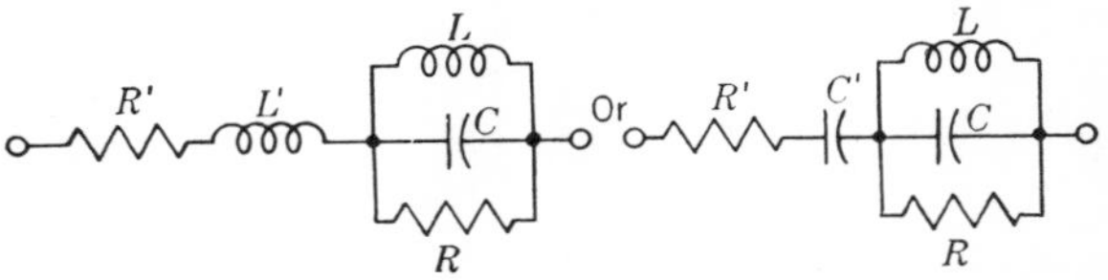

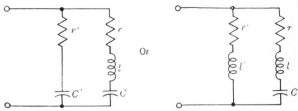

Fig. 7·18.—Impedance representation of cavity-coupling system near a resonance.

Fig. 7·19.—Admittance representation of cavity-coupling system near a resonance.

near resonance the terms R' and r' are always very small, since they correspond to off-resonance losses in the transmission line and coupling. They will be considered to vanish.

A very convenient tool in the study and design of cavity-coupling systems is the Smith impedance diagram. Consider a simple series RLC-circuit terminating a transmission line of characteristic impedance Z_0. On a Smith impedance diagram, the variation of input impedance of the circuit with frequency describes a locus like the circle (a) in Fig. 7·20. At the real axis, $\omega = \omega_0 = 1/\sqrt{LC}$, the resonant frequency of the circuit. If the circuit is shunted by a capacitor, the locus is a circle such as (b) in Fig. 7·20.[1] The new resonant frequency ω_0' is different from ω_0. The radius of the new circle is also different. If the simple RLC-circuit is shunted by an inductance, the circle will be shifted in a counterclockwise direction.

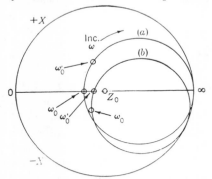

Fig. 7·20.—Loci of input impedances on a Smith impedance diagram, for a cavity-coupling system.

[1] This is obtained by transforming (a) to the admittance plane, which is accomplished by the reflection of (a) through the point Z_0. The capacitive susceptance is then added, which rotates the admittance circle. This circle is then reflected back through Z_0 to obtain (b).

It has been shown that the circuits of Fig. 7·18 are the most general impedance representations for a cavity-coupling system near resonance. Hence, loci such as (b) are the most general impedance contours for a cavity-coupling system near resonance.

Since a change of reference terminals in the transmission line corresponds, to a first approximation, to a simple rotation of locus (b) around Z_0, it is evident that by a suitable choice of reference terminals, any cavity-coupling system can be brought into the form of locus (a). Hence, any cavity-coupling system near resonance behaves as a simple series RLC-circuit at suitable terminals in the transmission line.

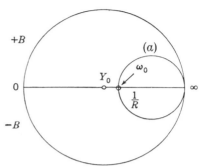

Fig. 7·21.—Locus of input admittance on a Smith admittance chart, for a cavity-coupling system.

A discussion of the representations of Fig. 7·19 in terms of the Smith admittance diagram proceeds in an exactly parallel fashion. The admit-

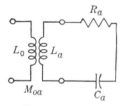

Fig. 7·22.—Loop-coupled cavity near resonance of ath mesh.

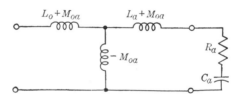

Fig. 7·23.—Alternative representation of loop-coupled cavity near resonance of ath mesh.

tance of a simple RLC shunt circuit describes a locus such as (a) in Fig. 7·21. The introduction of a series reactance rotates and changes the scale of (a). As before, a suitable choice of terminals in the transmission line can bring the locus of the most general representation of Fig. 7·19 to a form such as (a), and hence the input admittance of any cavity-coupling system near resonance behaves, at suitable terminals, as a simple shunt RLC-circuit.

In the case of the loop-coupled cavity of Fig. 7·12, it is easy to derive another near-resonance representation which is often more convenient than the Foster representations of Figs. 12·18 and 12·19. It is clear that near the resonant frequency $\omega = \omega_a$, Fig. 7·12 reduces to the circuit of Fig. 7·22, which may be represented as Fig. 7·23. It can be shown that at a single frequency any lossless two-terminal-pair network can be represented by a length of transmission line, an ideal transformer, and a series reactance. Thus, Fig. 7·23 can be represented by the circuit of

Fig. 7·24. It can be shown that the parameters of this representation are given by

$$\frac{\omega l}{c} = -\cot^{-1}\frac{\omega L_0}{Z_0},$$

$$n = \frac{\omega M_{0a}}{Z_0 \csc\left(\cot^{-1}\dfrac{\omega L_0}{Z_0}\right)},$$

and

$$L' = L_a - n^2 L_0,$$

at frequencies ω near the frequency at which the circuit of Fig. 7·24 is to be used, namely, ω_a. It is seen that the new circuit resonates at a

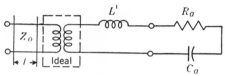

Fig. 7·24.—Another alternative representation of loop-coupled cavity near resonance of ath mesh.

frequency different from ω_a but near to it if $L_a \gg n^2 L_0$, which is the case of small coupling. It is further seen that if the length of transmission line chosen in the representation is the same kind as that physically connected to the cavity, then new reference terminals may be chosen such that the cavity can be represented by the lumped circuit of Fig. 7·25. This ideal-transformer representation is particularly convenient for cavity-coupling systems with several emerging transmission lines.

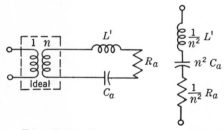

Fig. 7·25.—Representations of loop-coupled cavity near resonance ω_a at preferred terminal plane in transmission line.

7·8. Coupling Coefficients and External Loading.—Suppose that the transmission line emanating from a cavity-coupling system is terminated in its characteristic impedance (admittance). At suitable terminals, the equivalent circuit of the total system is as shown in Fig. 7·26. The Q of the cavity-coupling system is $\omega_0 l/r$, when $\omega_0^2 = 1/lc$. This is the unloaded Q of the system, Q_u, the Q when terminals A are short-circuited. The loaded Q of the circuit of Fig. 7·26 is $\omega_0 l/(r + Z_0)$. If the loaded Q is denoted by Q_L, then

$$Q_L = \frac{\omega_0 l}{r + Z_0} = \frac{1}{\dfrac{1}{Q_u} + \dfrac{Z_0}{\omega_0 l}} = \frac{1}{\dfrac{1}{Q_u} + \dfrac{Z_0}{Q_u r}},$$

and
$$Q_u = Q_L\left(1 + \frac{Z_0}{r}\right). \tag{36}$$

The coupling parameter is defined as $\beta = Z_0/r$; which is also the input conductance of the cavity-coupling system at the terminals of Fig. 7·26, normalized to Y_0, or

$$\frac{Z_c}{r} = \frac{\frac{1}{r}}{Y_0} = \beta. \tag{37}$$

Another convenient parameter is the external, or radiation, Q of the circuit. If Eq. (36) is written as

$$Q_L = \frac{1}{\frac{1}{Q_u} + \frac{\beta}{Q_u}}, \tag{38}$$

then

$$\frac{1}{Q_L} = \frac{1}{Q_u} + \frac{\beta}{Q_u} = \frac{1}{Q_u} + \frac{Z_0}{\omega_0 l}.$$

The quantity $Q_u/\beta = \omega_0 l/Z_0$ is defined as Q_r, the radiation Q.

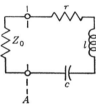

Fig. 7·26.—Equivalent circuit of cavity-coupling system terminated in Z_0 at a particular set of terminals.

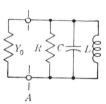

Fig. 7·27.—Equivalent circuit of cavity-coupling system terminated in Y_0 at a particular set of terminals.

A similar development exists for the shunt-resonant case as shown in Fig. 7·27. Here, $Q_u = R\omega_0 C$ where $\omega_0^2 = 1/LC$. The loaded Q is

$$Q_L = \frac{\omega_0 C}{\frac{1}{R} + Y_0} = \frac{1}{\frac{1}{R\omega_0 C} + \frac{Y_0}{\omega_0 C}} = \frac{1}{\frac{1}{Q_u} + \frac{RY_0}{Q_u}},$$

and the unloaded Q is

$$Q_u = Q_L(1 + RY_0) = Q_L\left(1 + \frac{R}{Z_0}\right).$$

Here $\beta = R/Z_0 = RY_0$. That is, β is the input impedance of the cavity-coupling system at resonance, normalized to Z_0.

It should be noted that if two cavity-coupling systems, one a series representation and the other a shunt representation, are connected to identical transmission lines and have the same coupling coefficients, then $RY_0 = Z_0/r$, or $R = Z_0^2/r = 1/rY_0^2$.

7·9. General Formulas for Q Factors.—Thus far only the Q's of simple series- and shunt-resonant circuits have been treated. Frequently, it is desired to find the Q of a more complex circuit, such as a cavity-coupling system with arbitrary terminals. A general formula for this is easily derived from the fundamental definition of Q which is

$$Q = 2\pi \frac{\text{energy stored}}{\text{energy dissipated/cycle}}.$$

It can be shown [Sec. 5.3, Eqs. (5.13) and (5.14)] that in general,

$$B = \frac{4(W_E - W_H)}{ee^*},$$
$$X = \frac{4(W_H - W_E)}{ii^*}, \qquad (39)$$

where B and X are the shunt susceptance and series reactance at the terminals, e and i are terminal voltages and currents, and W_E and W_H are the stored electric and magnetic energies in the system interior to the terminals. It can also be shown [Sec. 5·23, Eqs. (5.140) and (5.141)] that

$$\left.\begin{array}{l} \dfrac{\partial B}{\partial \omega} = \dfrac{4(W_E + W_H)}{ee^*}, \\[6pt] \dfrac{\partial X}{\partial \omega} = \dfrac{4(W_H + W_E)}{ii^*}. \end{array}\right\} \qquad (40)$$

If values for B or X at some reference terminals can be found and $\partial B/\partial \omega$ or $\partial X/\partial \omega$ can be evaluated, the total stored energy can be found from Eq. (40). The Q can then be written

$$Q = 2\pi f \frac{\frac{1}{4} ee^* \dfrac{\partial B}{\partial \omega}}{\text{energy dissipated/sec}}$$

$$= 2\pi f \frac{\frac{1}{4} ii^* \dfrac{\partial X}{\partial \omega}}{\text{energy dissipated/sec}}.$$

If, in addition, the shunt conductance G or the series resistance R at the terminals is known, the energy dissipated per second, $\frac{1}{2}ee^*G$ or $\frac{1}{2}ii^*R$, can be calculated and the Q may be written as

$$Q = \frac{\omega}{2G} \frac{\partial B}{\partial \omega}$$
$$= \frac{\omega}{2R} \frac{\partial X}{\partial \omega}. \qquad (41)$$

Equation (41) gives the general Q for a system whose impedance or admittance function is known at a given terminal pair.

7·10] IRIS-COUPLED, SHORT-CIRCUITED WAVEGUIDE

The various Q's thus far defined correspond to various choices of G or R. To find Q_u, the loss (and hence the value of G or R) is chosen to correspond to the internally dissipated energy. To determine Q_r or Q_L, the loss is chosen to correspond, respectively, to the externally dissipated energy or to the total dissipation.

7·10. Iris-coupled, Short-circuited Waveguide.—The iris-coupled, short-circuited waveguide is a particularly interesting example of a single-line cavity-coupling system, because it can be approached from several points of view and illustrates many of the features of the preceding paragraphs.

In the first place, transmission-line methods may be used to construct the equivalent circuit. Consider the device of Fig. 7·28. The broken line in plane A represents a coupling hole or inductive iris of inductive susceptance $-jb$. Assume that only one mode is propagated in the waveguide. If the losses in the short-circuiting plate are neglected in comparison with those in the waveguide of length l, then the input impedance of the short-circuited waveguide (not including the iris) at plane A is

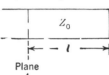

FIG. 7·28.—Iris-coupled short-circuited waveguide.

$$Z_s = Z_0 \tanh (\alpha + j\beta)l, \qquad (42)$$

where $\beta = 2\pi/\lambda_g$ and α is the attenuation constant. If the waveguide is made of a metal that is a good conductor, so that α is small, and if the length l is not excessive, the Q will be large, $\alpha l \approx \tanh \alpha l$, and from Eq. (42)

$$Z_s = Z_0 \frac{\alpha l + j \tan \beta l}{1 + j(\alpha l) \tan \beta l}. \qquad (43)$$

Near resonance, Eq. (43) is further simplified, since at resonance

$$l \approx \frac{n\lambda_g}{2}$$

where $n = 1, 2, \cdots$. Thus $\beta l \approx \pi$, and $|\tan \beta l| \ll 1$. Since $\alpha l \ll 1$ also,

$$Z_s \approx Z_0(\alpha l + j \tan \beta l). \qquad (44)$$

This may be written in another form by considering $\tan \beta l$ near resonance,

$$\tan \beta l \approx \tan\left(\pi - \pi \frac{\Delta \lambda_g}{\lambda_g}\right) = -\pi \frac{\Delta \lambda_g}{\lambda_g} \approx \frac{\lambda_g^2}{\lambda^2} \frac{\pi}{\omega_0} \cdot \delta, \qquad (45)$$

where ω_0 is the frequency at which $\tan \beta l = 0$ and $\delta = \omega - \omega_0$. Substitution of Eq. (45) into Eq. (44) yields

$$Z_s \approx Z_0\left(\alpha l + j\frac{\lambda_g^2}{\lambda^2} \frac{\pi}{\omega_0} \delta\right). \qquad (46)$$

This expression is seen to be of the same form as the input impedance of a series RLC-circuit for which

$$Z_s = R + j\left(\omega L - \frac{1}{\omega C}\right)$$
$$\approx R + j2L\delta \qquad (47)$$

near resonance. Here $\delta = \omega - \omega_0$, and $\omega_0^2 = 1/LC$. We can, therefore, identify the terms in Eq. (46) as

$$\left. \begin{array}{l} R = Z_0(\alpha l), \\ L = \dfrac{Z_0}{2}\dfrac{\lambda_g^2}{\lambda^2}\cdot\dfrac{\pi}{\omega_0}. \end{array} \right\} \qquad (48)$$

The iris appears in shunt with this circuit so that the complete equivalent circuit of the cavity-coupling system at plane A is that shown in Fig. 7·29. The iris susceptance is $-jb = -j/\omega\mathcal{L}$.

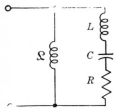 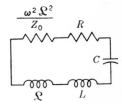

Fig. 7·29.—Equivalent circuit of an iris-coupled short-circuited waveguide near resonance. Terminals are at the plane of the iris.

Fig. 7·30.—Equivalent series circuit for an iris-coupled short-circuited waveguide cavity.

If the waveguide of Fig. 7·28 is terminated in its characteristic impedance to the left of plane A, the impedance Z_0 appears across the terminals of the equivalent circuit in Fig. 7·29. The loaded Q under these conditions may be calculated. The series impedance of the parallel $\mathcal{L}$, Z_0 combination is $j\omega\mathcal{L}Z_0/(Z_0 + j\omega\mathcal{L})$. For a high-$Q$ system $\omega\mathcal{L} \ll Z_0$, and this series impedance is $j\omega\mathcal{L} + \omega^2\mathcal{L}^2/Z_0$. The complete series circuit is that of Fig. 7·30. The radiation Q is

$$Q_r = \frac{\omega_0 L + \omega_0 \mathcal{L}}{\dfrac{\omega_0^2 \mathcal{L}^2}{Z_0}}$$
$$\approx \frac{b^2}{Y_0^2}\frac{\pi}{2}\frac{\lambda_g^2}{\lambda^2} \qquad (49)$$

for $\mathcal{L} \ll L$, from Eq. (48).

The loaded Q is given by

$$\left. \begin{array}{l} \dfrac{1}{Q_L} = \dfrac{\dfrac{\omega_0^2 \mathcal{L}^2}{Z_0} + R}{\omega_0 \mathcal{L} + \omega_0 L} \approx \dfrac{\omega_0^2 \mathcal{L}^2}{Z_0 \omega_0 L} + \dfrac{R}{\omega_0 L}. \\ \\ \dfrac{1}{Q_L} = \dfrac{Y_0^2}{b^2}\dfrac{2}{\pi}\dfrac{\lambda^2}{\lambda_g^2} + \dfrac{2(\alpha l)}{\pi}\dfrac{\lambda^2}{\lambda_g^2}. \end{array} \right\} \qquad (50)$$

or

Sec. 7·10] IRIS-COUPLED, SHORT-CIRCUITED WAVEGUIDE

The unloaded Q is

$$Q_u = \frac{\pi}{2(\alpha l)} \frac{\lambda_g^2}{\lambda^2},$$

since $\omega_0 L/R$ is the unloaded Q of the circuit when the terminals of Fig. 7·29 are short-circuited.

It is seen by examination of Fig. 7·29 that resonance occurs for frequencies somewhat less than ω_0, corresponding to $n\lambda_g > 2l$, or cavity lengths somewhat shorter than $n\lambda_g/2$. This is also seen by examining the total admittance at plane A which is

$$\frac{Y_T}{Y_0} = -j\frac{b}{Y_0} + \frac{1}{\alpha l + j \tan \beta l}$$

$$= \frac{\alpha l}{(\alpha l)^2 + \tan^2 \beta l} - j\frac{b}{Y_0} - j\frac{\tan \beta l}{(\alpha l)^2 + \tan^2 \beta l}. \quad (51)$$

In the high-Q case $|\tan \beta l| \gg \alpha l$. Thus Eq. (51) becomes

$$\frac{Y_T}{Y_0} = \frac{\alpha l}{\tan^2 \beta l} - j\frac{b}{Y_0} - j\frac{1}{\tan \beta l}. \quad (52)$$

At resonance, Y_T/Y_0 is pure real. This occurs for negative $\cot \beta l$, or l slightly less than $n\lambda_g/2$. We see that the input conductance at resonance is

$$\left(\frac{Y_T}{Y_0}\right)_{\text{res}} = \frac{\alpha l}{\tan^2 \beta l} = \alpha l \left(\frac{b}{Y_0}\right)^2. \quad (53)$$

For a cavity one-half wavelength long, in copper waveguide of dimensions 1- by $\frac{1}{2}$-in. by 0.050-in. wall, at $\lambda = 3.2$ cm, $\alpha l \approx 4(10)^{-4}$, so that an iris susceptance of $b/Y_0 = 50$ is required for critical coupling (that is, for $Y_T/Y_0 = 1$ at resonance).

We have seen in Sec. 7·9 that the Q of a circuit is given by $Q = (1/2G)\omega(\partial B/\partial \omega)$, where B is the total susceptance at the reference plane and G the conductance. Now,

$$\omega \frac{\partial B}{\partial \omega} = \omega \frac{\partial B}{\partial \lambda} \frac{\partial \lambda}{\partial \omega} = -\frac{2\pi c}{\omega} \frac{\partial B}{\partial \lambda} = -\frac{2\pi c}{\omega} \frac{\partial \lambda_g}{\partial \lambda} \frac{\partial B}{\partial \lambda_g}.$$

Since $\lambda/\lambda_g = \sqrt{1 - (\lambda/\lambda_c)^2}$ and $\partial \lambda_g/\partial \lambda = \lambda_g^3/\lambda^3$, then

$$\omega \frac{\partial B}{\partial \omega} = -\frac{2\pi c}{\omega} \frac{\lambda_g^3}{\lambda^3} \frac{\partial B}{\partial \lambda_g} = -\frac{\lambda_g^3}{\lambda^2} \frac{\partial B}{\partial \lambda_g}.$$

Since $\beta = 2\pi/\lambda_g$ and $\partial \beta/\partial \lambda_g = -2\pi/\lambda_g^2$, then

$$\omega \frac{\partial B}{\partial \omega} = 2\pi \frac{\lambda_g}{\lambda^2} \frac{\partial B}{\partial \beta}. \quad (54)$$

From Eq. (52),

$$B = -b - \frac{Y_0}{\tan \beta l},$$

$$\frac{\partial B}{\partial \beta} = -\frac{\partial b}{\partial \beta} + Y_0 l(1 + \cot^2 \beta l). \tag{55}$$

Near resonance, $\cot^2 \beta l \approx b^2/Y_0^2 \gg 1$ and $l \approx \lambda_g/2$; and from Eqs. (55) and (54),

$$\omega \frac{\partial B}{\partial \omega} = -\frac{\lambda_g^2}{\lambda^2} \beta \frac{\partial b}{\partial \beta} + \pi \frac{\lambda_g^2}{\lambda^2} \frac{b^2}{Y_0} = -\frac{\lambda_g^2}{\lambda^2} b + \pi \frac{\lambda_g^2}{\lambda^2} \frac{b^2}{Y_0} \approx \pi \frac{\lambda_g^2}{\lambda^2} \frac{b^2}{Y_0}, \tag{56}$$

since $b/Y_0 \gg 1$.

For the radiation Q the shunt conductance is Y_0, and so

$$Q_r = \frac{\pi}{2} \frac{\lambda_g^2}{\lambda^2} \left(\frac{b}{Y_0}\right)^2. \tag{57}$$

This is seen to be identical with Eq. (49). For the unloaded Q, the shunt conductance is $Y_0(\alpha l)(b/Y_0)^2$, and

$$Q_u = \frac{\pi}{2(\alpha l)} \frac{\lambda_g^2}{\lambda^2}, \tag{58}$$

which is seen to agree with Eq. (50).

CAVITY-COUPLING SYSTEMS WITH TWO EMERGENT TRANSMISSION LINES

As in the treatment of the single-line cavity-coupling system it is convenient to consider the general representation theory that has been derived for lossless n-mesh networks and to show the equivalence of this representation to that derived from field theory.

7·11. General Representation of Lossless Two-terminal-pair Networks.—It is not difficult to extend the reactance theorem to two-terminal-pair networks of n meshes.[1] It will suffice here merely to state the result that such a two-terminal pair can be represented as a series combination of T-sections or a parallel combination of Π-sections. The T-section representation is shown in Fig. 7·31, where each Z_a, Z_b, Z_c may

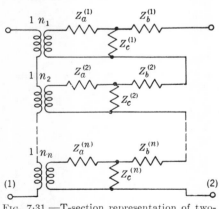

Fig. 7·31.—T-section representation of two-terminal-pair n-mesh lossless network.

[1] E. A. Guillemin, *Communication Networks*, Vol. II, Wiley, New York, 1935, Chap. 5, p. 216.

be an inductance, a capacitance, or a parallel LC-element. The ideal transformers are necessary to ensure the physical realizability of the elements. It is not necessary that the self- and transfer impedances of the component T-sections have coincident poles.

At frequencies near a particular resonant frequency the representation of Fig. 7·31 simplifies to that of Fig. 7·32, where the elements X_a, X_b, X_c represent the contributions from all other poles at frequencies near the chosen resonant frequency. They are representable as inductances or capacitances near the resonant frequency in question. The elements Z_a, Z_b, Z_c consist of L, C, or parallel LC-combinations.

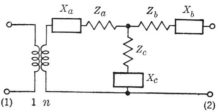

Fig. 7·32.—T-section representation of two-terminal-pair lossless network near a particular resonance.

The reactance X_c represents the contributions of the other resonances to the transfer impedance. This is essentially the direct mutual impedance of the two coupling systems and is almost always negligible; that is, only near resonance does the cavity transmit any appreciable power. In the following sections, X_c will be neglected.

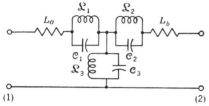

Fig. 7·33.—A special T-representation of a two-terminal-pair lossless network near resonant frequency ω_0.

Consider the special case where X_a and X_b are inductances, the ideal transformer has a ratio of one to one, $X_c = 0$, and Z_a, Z_b and Z_c are parallel-resonant elements at the frequency ω_0. This situation is shown in Fig. 7·33, where

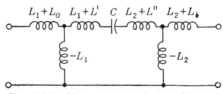

Fig. 7·34.—Network equivalent of Fig. 7·33 near the resonant frequency $\omega_0 = \sqrt{1/LC}$.

$$\omega_0^2 = \frac{1}{\mathcal{L}_1 \mathcal{C}_1} = \frac{1}{\mathcal{L}_2 \mathcal{C}_2} = \frac{1}{\mathcal{L}_3 \mathcal{C}_3}.$$

The self- and transfer impedances of this network are

$$\left.\begin{aligned}
Z_{11} &= j\omega L_a + \frac{j\omega_0^2 \omega (\mathcal{L}_1 + \mathcal{L}_3)}{\omega_0^2 - \omega^2} \approx \frac{j\omega^3(\mathcal{L}_1 + \mathcal{L}_3)}{\omega_0^2 - \omega^2} + j\omega L_a, \\
Z_{22} &= j\omega L_b + \frac{j\omega_0^2 \omega (\mathcal{L}_2 + \mathcal{L}_3)}{\omega_0^2 - \omega^2} \approx \frac{j\omega^3(\mathcal{L}_2 + \mathcal{L}_3)}{\omega_0^2 - \omega^2} + j\omega L_b. \\
Z_{12} &= \frac{j\omega_0^2 \omega \mathcal{L}_3}{\omega_0^2 - \omega^2} \approx j\omega^3 \mathcal{L}_3.
\end{aligned}\right\} \quad (59)$$

These expressions are of the same form as those for the network of Fig. 7·34. For this network,

$$Z_{11} = j\omega L_a + \frac{\dfrac{j\omega^3 L_1^2}{L}}{\omega_0^2 - \omega^2},$$

$$Z_{22} = j\omega L_b + \frac{\dfrac{j\omega^3 L_2^2}{L}}{\omega_0^2 - \omega^2}, \qquad (60)$$

$$Z_{12} = \frac{\dfrac{j\omega^3 L_1 L_2}{L}}{\omega_0^2 - \omega^2},$$

where $\omega_0^2 = 1/LC$, and $L' + L'' = L$. The equivalence is established by identifying

$$\frac{L_1^2}{L} = \mathcal{L}_1 + \mathcal{L}_3,$$
$$\frac{L_2^2}{L} = \mathcal{L}_2 + \mathcal{L}_3, \qquad (61)$$
$$\frac{L_1 L_2}{L} = \mathcal{L}_3.$$

The network of Fig. 7·34 is easily put into a convenient form by representing each T-section as a combination of a length of transmission line, an ideal transformer, and a series inductance. By this representation, it is transformed to the network of Fig. 7·35. The equivalence is established by the equations

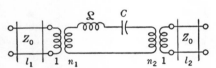

FIG. 7·35.—Network equivalent of Figs. 7·33 and 7·34 near resonance.

$$\left.\begin{array}{l} j\omega(L_1 + L_a) = -jZ_0 \cot \beta_1 l_1 - jn_1 Z_0 \csc \beta_1 l_1, \\ j\omega(L_1 + L') = j\omega \mathcal{L}' - jn_1^2 Z_0 \cot \beta_1 l_1 - jn_1 Z_0 \csc \beta_1 l_1, \\ -j\omega L_1 = jn_1 Z_0 \csc \beta_1 l_1, \end{array}\right\} \qquad (62)$$

and

$$\left.\begin{array}{l} j\omega(L_2 + L_b) = -jZ_0 \cot \beta_2 l_2 - jn_2 Z_0 \csc \beta_2 l_2, \\ j\omega(L_2 + L'') = j\omega \mathcal{L}'' - jn_2^2 Z_0 \cot \beta_2 l_2 - jn_2 Z_0 \csc \beta_2 l_2, \\ -j\omega L_2 = jn_2 Z_0 \csc \beta_2 l_2, \end{array}\right\} \qquad (63)$$

where $\mathcal{L}' + \mathcal{L}'' = \mathcal{L}$.

The solutions for $\mathcal{L}'$ and $\mathcal{L}''$ are,

$$\left.\begin{array}{l} \mathcal{L}' = L' - n_1^2 L_a, \\ \mathcal{L}'' = L'' - n_2^2 L_b. \end{array}\right\} \qquad (64)$$

In all practical cases, $L' \gg n_1^2 L_a$ and $L'' \gg n_2^2 L_b$; therefore $\mathcal{L} \approx L$, and the series circuit of Fig. 7·35 resonates at nearly the same frequency as that of Fig. 7·33 or that of Fig. 7·34.

It is clear from Fig. 7·35 that if the lines l_1 and l_2 are chosen to be identical with the physical lines connected to the cavity, new terminals may be chosen in the physical lines such that the cavity-coupling system is simply a series LC-circuit coupled by ideal transformers at the input and output terminals.

The transformation of various other forms of Fig. 7·32 to the form of Fig. 7·35 will not be treated in detail. This transformation can be performed in all cases for which $X_c = 0$.

7·12. Introduction of Loss.—Just as in the case of the single line or single terminal pair, it is possible to treat the small-loss approximation, in which loss is introduced into each mesh of a purely reactive network. It is clear that the general representation of such a network is of the form of Fig. 7·31 with resistive elements added to each Z_a, Z_b, and Z_c. Near a particular resonance this reduces to the form of Fig. 7·33 and finally to that of Fig. 7·35, where a resistive element now appears in the resonant mesh.

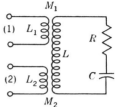

Fig. 7·36.—Cavity with two loop-coupled lines near $\omega_0^2 = 1/LC$.

This can be verified by a straightforward analysis following that shown for the single-terminal-pair network.

7·13. Representation of a Cavity with Two Loop-coupled Lines.—The loop-coupled cavity, like the single-line cavity-coupling system discussed previously, can be treated rather exactly by field methods. A simple extension of the single-loop case of Fig. 7·12 shows that Fig. 7·36 is the

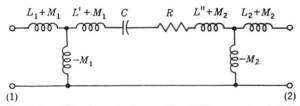

Fig. 7·37.—Circuit equivalent to Fig. 7·36 near $\omega_0^2 = 1/LC$.

equivalent circuit of a cavity with two loop-coupled lines at frequencies near $\omega_0^2 = 1/LC$. No direct coupling between the loops L_1 and L_2 is assumed. The circuit of Fig. 7·36 is easily transformed to that of Fig. 7·37 near the resonance $\omega_0^2 = 1/LC$, where $L = L' + L''$. This in turn can be transformed to the form of Fig. 7·35 (including series resistance R), where as before $\mathcal{L} \approx L$.

7·14. Transmission through a Two-line Cavity-coupling System.—Let us consider a transmission system that includes a two-line cavity. Let us suppose the terminals to be so chosen as to simplify Fig. 7·35, and assume the generator and load impedances to be real and to be given by R_G and R_L at these terminals. This circuit is shown in Fig. 7·38. Transforming the load and generator into the resonant mesh results in the circuit of Fig. 7·39. The unloaded Q, which is obtained by putting

$$R_G = R_L = 0,$$

is $Q_u = \omega_0 L/R$. The loaded Q is

$$Q_L = \frac{\omega_0 L}{R + n_1^2 R_G + n_2^2 R_L},$$

from which

$$Q_u = Q_L \left(1 + n_1^2 \frac{R_G}{R} + n_2^2 \frac{R_L}{R}\right).$$

The input- and output-coupling parameters are defined, respectively, as

$$\beta_1 = \frac{n_1^2 Z_0}{R},$$

$$\beta_2 = \frac{n_2^2 Z_0}{R}.$$

It is customary to define the transmission through the cavity in terms of a matched generator and load (i.e., $R_G = R_L = Z_0$). Under these

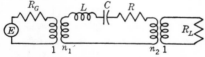

FIG. 7·38.—Equivalent circuit of a two-line cavity-coupling system at particular reference planes. The generator and load impedances R_G and R_L are real at these planes.

FIG. 7·39.—Alternative form for the circuit of Fig. 7·38.

conditions, the impedance of the mesh of Fig. 7·39 is

$$Z = R\left[(1 + \beta_1 + \beta_2) + jQ_u\left(\frac{\omega}{\omega_0} - \frac{\omega_0}{\omega}\right)\right],$$

and the power into the load impedance is

$$P_L = n_2^2 Z_0 |I|^2 = \beta_2 R |I|^2$$

$$= \frac{\beta_1 \beta_2 \dfrac{E^2}{Z_0}}{(1 + \beta_1 + \beta_2)^2 + Q_u^2 \left(\dfrac{\omega}{\omega_0} - \dfrac{\omega_0}{\omega}\right)^2}.$$

The available power from the generator line is $\frac{1}{4} E^2/Z_0 = P_0$.

The transmission-loss function $T(\omega)$ is defined as P_L/P_0, or

$$T(\omega) = \frac{4\beta_1 \beta_2}{(1 + \beta_1 + \beta_2)^2 + Q_u^2 \left(\dfrac{\omega}{\omega_0} - \dfrac{\omega_0}{\omega}\right)^2}. \tag{65}$$

At resonance the transmission loss is

$$T(\omega_0) = \frac{4\beta_1\beta_2}{(1 + \beta_1 + \beta_2)^2}. \tag{66}$$

Dividing Eq. (65) by Eq. (66) has the result

$$T(\omega) = \frac{T(\omega_0)}{1 + Q_L^2\left(\dfrac{\omega}{\omega_0} - \dfrac{\omega_0}{\omega}\right)^2}$$

or, putting $\omega = \omega_0 + (\Delta\omega/2)$,

$$T(\omega) = \frac{T(\omega_0)}{1 + Q_L^2\left(\dfrac{\Delta\omega}{\omega_0}\right)^2}. \tag{67}$$

It is noted that $T(\omega) = \tfrac{1}{2}T(\omega_0)$ (*i.e.*, half-power points of transmission occur) for $\Delta\omega/\omega_0 = 1/Q_L$. The quantity $\Delta\omega$ is frequently called the "bandwidth" of the cavity at the resonant frequency ω_0.

CHAPTER 8

RADIAL TRANSMISSION LINES

By N. Marcuvitz

8·1. The Equivalent-circuit Point of View.—The description of the electromagnetic fields within a region of space enclosed by conducting walls may be made in either of two ways. On the one hand, the given region may be considered as a whole and its electromagnetic behavior described by the indication of its resonant properties and field behavior as a function of frequency. In this "cavity" method of description, the problem of determination of the fields within the region under consideration is treated as an individual one, and no effort is made to apply the results of its solution to other similar problems. On the other hand, if the given region possesses a certain regularity of geometrical structure in some direction—the direction of energy transmission—an alternative, equally rigorous and more systematic treatment is possible. The given region is regarded as a composite structure whose constituent subregions are of two principal types. The fields in each of these subregions can be described as a superposition of an infinite number of wave types or modes characteristic of the cross-sectional shape of the subregion. In the frequency range of interest it is found that in one type of subregion—the transmission-line region—only a single dominant mode is necessary to characterize completely the field behavior whereas in the other type of subregion—the discontinuity region—the entire infinite set of modes is necessary for the field description. The resulting complication in the description of the discontinuity regions is not serious because as a consequence of the rapid damping out of the higher modes the discontinuity structure may be regarded as effectively lumped as far as all modes but the dominant one are concerned. It is usually necessary, therefore, to indicate only the dominant-mode discontinuity effects that are introduced by such regions. The justification for such a simple procedure lies in the fact that it is the dominant mode which determines the energy transmission and the interaction characteristics of the over-all system. Hence, only a knowledge of this mode is of interest.

The description of a composite structure in terms of transmission regions and discontinuity or junction regions can be put into ordinary electrical-network form. This is accomplished by the introduction of voltage and current as measures of the transverse electric and magnetic fields, respectively, of the dominant mode. As a consequence, the field behavior of the dominant mode in each of the constituent transmission regions may be represented schematically by the voltage and current behavior on a corresponding transmission line. The dominant-mode discontinuity effects introduced by the junction regions are given by specification of the relations between the voltages and currents at the terminals of each junction. Such relations may be schematically represented by means of an equivalent circuit. In this manner it is seen that the energy-transmission characteristics of the original region are described by the electrical characteristics of a system of distributed transmission lines interconnected by lumped circuits. A knowledge of the propagation constant and characteristic impedance of each transmission line as well as the values of the lumped circuit parameters thus suffices to give a complete description of the electromagnetic properties of interest for the over-all system. In contradistinction to the cavity description, it is to be noted that the results of many comparatively simple transmission-line and junction problems are combined to give this composite description.

8·2. Differences between Uniform and Nonuniform Regions.—The circuit point of view for electromagnetic problems has been only sketched, since it has been amply discussed elsewhere in this book for uniform waveguide structures. Its application is not limited to uniform structures, which are characterized almost everywhere by uniformity of cross sections transverse to the direction of energy transmission. The same point of view may be applied as well to certain nonuniform structures that possess almost everywhere a type of symmetry in which cross sections transverse to the transmission direction are geometrically similar to one another rather than identical. Several examples of structures of this sort are illustrated in Fig. 8·1.

The loaded cavity in Fig. 8·1a is an example of a structure containing two nonuniform transmission regions, from 0 to r and from r to R, separated by the junction region at r. The cross sections of the transmission regions are cylindrical surfaces of differing radii. The tapered waveguide section in Fig. 8·1b is a structure composed of two uniform transmission regions separated by a nonuniform transmission region from r to R with junction regions at r and R. In the nonuniform region the cross sections perpendicular to the direction of energy flow are segments of cylindrical surfaces of variable radii. In Fig. 8·1c is represented a spherical cavity containing a dielectric and composed of two nonuniform transmission regions from 0 to r and from r to R. The cross sections in each of these regions are spherical surfaces. In Fig. 8·1d is represented a conical antenna that is composed of nonuniform transmission regions from r to R and R to ∞ with junction regions at

242 RADIAL TRANSMISSION LINES [SEC. 8·2

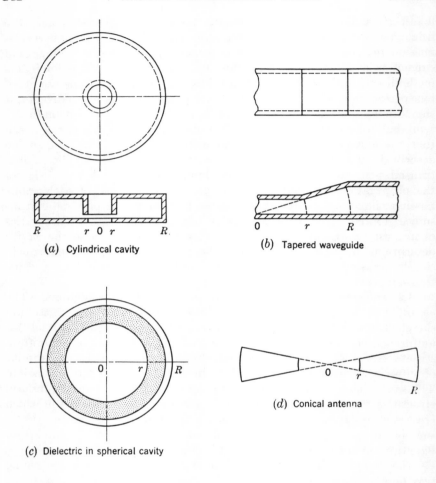

(a) Cylindrical cavity

(b) Tapered waveguide

(c) Dielectric in spherical cavity

(d) Conical antenna

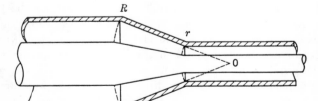

(e) Tapered coaxial line

FIG. 8·1.—Structures containing nonuniform transmission lines.

r and R. Figure 8·1e represents a tapered section that is the coaxial-line analogue of the waveguide taper in Fig. 8·1b. The cross sections in the nonuniform transmission regions represented in Fig. 8·1c and d are either complete spherical surfaces or segments thereof.

The subject matter of this chapter will be devoted principally to the description of the electromagnetic fields that can exist within the nonuniform transmission regions represented in Fig. 8·1a and b. This field description will be carried through from an impedance point of view in close analogy with the corresponding description of uniform transmission regions, which for the sake of the analogy will also be briefly treated. The impedance point of view stems from the existence of traveling and standing waves that characterize the field behavior in the transmission regions. The nature of these characteristic waves may be found by investigation of solutions to the field equations or, better, the wave equation, in a coordinate system appropriate to the geometry of the transmission region in question.

For example, in a rectangular xyz coordinate system the wave equation determining the steady-state behavior of a typical field component may be written in the form

$$\frac{\partial^2 u}{\partial z^2} + \left(\frac{\partial^2}{\partial x^2} + \frac{\partial^2}{\partial y^2} + k^2\right) u = 0, \tag{1}$$

where, as is customary, the complex amplitude u of the field variable is assumed to have a time dependence $\exp(j\omega t)$ with $\omega = kc$ denoting the complex angular frequency and c the speed of wave propagation. Let the z-axis be chosen as the direction of energy transmission. In the associated uniform transmission system the cross-sectional surfaces are parallel to the xy-plane and identical with one another. Characteristic modes or field patterns exist for each of which the operator shown within parentheses in Eq. (1) is the square of a constant κ called the mode propagation constant or wave number (cf. Chap. 2). For such modes the wave equation may be rewritten as a one-dimensional transmission-line equation,

$$\frac{\partial^2 u}{\partial z^2} + \kappa^2 u = 0, \tag{2}$$

which determines the variation of the mode amplitude along the transmission system. The two independent mathematical solutions to this equation may be expressed as

$$\cos \kappa z, \quad \sin \kappa z$$

and interpreted physically as standing waves; or alternatively, they may be written in exponential form as

$$e^{+j\kappa z}, \quad e^{-j\kappa z}$$

and interpreted as traveling waves (*cf.* Chap. 3). Since either set of solutions suffices to specify completely the propagation in the z-direction, the impedance description of uniform transmission lines must be expressed in terms of these trigonometric functions.

Similarly, in a cylindrical $r\phi z$ coordinate system the wave equation determining the complex amplitude u of one of the field components may be written as

$$\frac{1}{r}\frac{\partial}{\partial r}\left(r\frac{\partial u}{\partial r}\right) + \left(\frac{1}{r^2}\frac{\partial^2}{\partial \phi^2} + \frac{\partial^2}{\partial z^2} + k^2\right)u = 0. \tag{3}$$

In contradistinction to the rectangular case, every cylindrical field component does not obey the same wave equation [Eq. (3)]; however, only this equation will be considered for the moment. Let the direction of energy transmission be chosen along the r-coordinate. In a transmission system of this type the cross sections, which are ϕz cylindrical surfaces, are no longer uniform but only similar. Examples of such nonuniform transmission regions are shown in Fig. 8·1a and b. Characteristic modes still exist for such regions, but the operator within parentheses in Eq. (3) associated with each of the modes is a constant only over the cross-sectional surfaces and varies along the direction of propagation. The functional dependence of the mode constant on r has been given in Sec. 2·13. Substitution of this result into the wave equation leads to the one-dimensional radial-transmission-line equation

$$\frac{1}{r}\frac{\partial}{\partial r}\left(r\frac{\partial u}{\partial r}\right) + \left(\kappa^2 - \frac{m^2}{r^2}\right)u = 0, \qquad m = 0, 1, 2, \cdots, \tag{4}$$

which determines the variation of the mth mode amplitude along the direction of energy propagation. As before, two independent mathematical solutions to this equation exist; these are the Bessel and Neumann functions of order m (Bessel functions of first and second kind, respectively)

$$J_m(\kappa r) \qquad N_m(\kappa r)$$

and, in analogy with the trigonometric functions encountered in the rectangular geometry, may be interpreted physically as standing waves. Alternatively a set of solutions may be written in terms of the two types of mth-order Hankel functions (Bessel functions of the third kind)

$$H_m^{(1)}(\kappa r), \qquad H_m^{(2)}(\kappa r),$$

and these similarly may be interpreted as ingoing and outgoing traveling waves. Since the impedance description of radial transmission lines must necessarily be based on these wave solutions, it is desirable to list a few properties of the cylinder functions. These properties bear a strong resemblance to those of the trigonometric functions. For example, the relation, or rather identity, among the three different kinds of Bessel functions,

$$H_m^{(1)}(x) = J_m(x) \pm jN_m(x), \qquad (5)$$
$$H_m^{(2)}$$

is analogous to the exponential and trigonometric relation

$$e^{\pm jx} = \cos x \pm j \sin x.$$

The resemblance is particularly close in the range $x \gg m$ as can be inferred from the asymptotic identity of Eqs. (2) and (4) in this range. In fact for $x \gg m$ there obtains the asymptotic relation

$$H_m^{(1)}(x) \approx \sqrt{\frac{2}{\pi x}}\, e^{\pm j\left(x - \frac{2m+1}{4}\pi\right)} = (\mp j)^m \sqrt{\frac{2}{\pi x}}\, e^{\pm j\left(x - \frac{\pi}{4}\right)} \qquad (6a)$$

or, equivalently,

$$J_m(x) \approx \sqrt{\frac{2}{\pi x}} \cos\left(x - \frac{2m+1}{4}\pi\right),$$
$$N_m(x) \approx \sqrt{\frac{2}{\pi x}} \sin\left(x - \frac{2m+1}{4}\pi\right), \qquad (6b)$$

Physically these approximations imply that at large distances traveling radial waves are identical with traveling plane waves save for the decrease in amplitude of radial waves along the direction of propagation. This

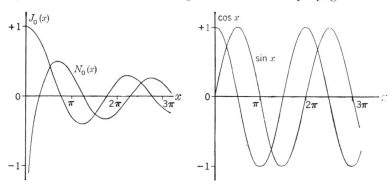

Bessel functions   Trigonometric functions
FIG. 8·2.—Comparison of zero-order Bessel functions and trigonometric functions.

decrease is to be expected from the spreading of radial (cylindrical) waves as they propagate, in contrast to the nonspreading of the constant-amplitude plane waves. The variable amplitude as well as the quasi periodicity of the Bessel functions is perhaps best illustrated in the graphical comparison in Fig. 8·2 where the zero-order Bessel functions are compared with the corresponding trigonometric functions. The similarity of Bessel functions to trigonometric functions and their interpretation as traveling and standing waves are emphasized by writing them in the polar form

$$H_m^{(1)}_{(2)}(x) = h_m e^{\pm j\left(\eta_m - \frac{m\pi}{2}\right)} = (\mp j)^m h_m e^{\pm j\eta_m}, \tag{7a}$$

or, equivalently,

$$J_m(x) = h_m \cos\left(\eta_m - \frac{m\pi}{2}\right), \qquad N_m(x) = h_m \sin\left(\eta_m - \frac{m\pi}{2}\right), \tag{7b}$$

from which it follows that

$$h_m(x) = \sqrt{J_m^2(x) + N_m^2(x)}, \qquad h_m(x) \approx \sqrt{\frac{2}{\pi x}}, \qquad \text{for } x \gg 1,$$

$$\eta_m(x) = \frac{m\pi}{2} + \tan^{-1}\frac{N_m(x)}{J_m(x)}, \qquad \eta_m(x) \approx x - \frac{\pi}{4}, \qquad \text{for } x \gg 1.$$

The amplitude $h(x)$ and phase $\eta(x)$ of the Hankel functions of orders zero and one are shown graphically in Fig. 8·3 and are tabulated in

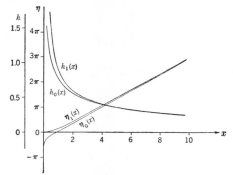

Fig. 8·3.—The amplitude and phase of the Hankel function of order zero and one.

Table 8·1. The values of the Bessel functions $J_0(x)$ and $J_1(x)$ are also included in this table.[1]

Since in a radial transmission line the point $r = 0$ is a singular point, it is not to be expected that the asymptotic small-argument approximations (for $x \ll m, m \geqq 1$),

$$\left.\begin{aligned} J_m(x) &\approx \frac{1}{m!}\left(\frac{x}{2}\right)^m, \qquad N_m(x) \approx -\frac{(m-1)!}{\pi}\left(\frac{2}{x}\right)^m, \qquad \text{for } m \geqq 1, \\ J_0(x) &\approx 1 - \left(\frac{x}{2}\right)^2, \qquad N_0(x) \approx -\frac{2}{\pi}\ln\frac{2}{\gamma x}, \qquad\qquad\qquad m = 0, \\ & \gamma = 1.781 \ldots, \end{aligned}\right\} \tag{8}$$

are closely related to those of the trigonometric functions, although there is a qualitative resemblance. The corresponding small-argument rela-

[1] *Cf.* the tables in G. N. Watson, *Theory of Bessel Functions*, Cambridge, London, 1944.

SEC. 8·2]   UNIFORM AND NONUNIFORM REGIONS   247

tions for the Hankel function may be obtained by use of the identity of Eq. (5).

TABLE 8·1.—VALUES OF THE BESSEL FUNCTIONS

x	$h_0(x)$	$h_1(x)$	$\eta_0(x)$	$\eta_1(x)$	$J_0(x)$	$J_1(x)$
0	∞	∞	−90°	0°	1.0000	0.000
0.1	1.830	6.459	−57.0	0.5	0.9975	0.0499
0.2	1.466	3.325	−47.5	1.7	0.9900	0.0995
0.3	1.268	2.298	−39.5	3.7	0.9776	0.1483
0.4	1.136	1.792	−32.3	6.3	0.9604	0.1960
0.5	1.038	1.491	−25.4	9.4	0.9385	0.2423
0.6	0.9628	1.293	−18.7	12.8	0.9120	0.2867
0.7	0.9016	1.151	−12.2	16.6	0.8812	0.3290
0.8	0.8507	1.045	−5.9	20.7	0.8463	0.3688
0.9	0.8075	0.9629	+0.4	24.9	0.8075	0.4060
1.0	0.7703	0.8966	6.6	29.4	0.7652	0.4401
1.1	0.7377	0.8421	12.7	34.0	0.7196	0.4709
1.2	0.7088	0.7963	18.8	38.7	0.6711	0.4983
1.3	0.6831	0.7572	24.8	43.6	0.6201	0.5220
1.4	0.6599	0.7234	30.8	48.5	0.5669	0.5420
1.5	0.6389	0.6938	36.8	53.5	0.5118	0.5579
1.6	0.6198	0.6675	42.7	58.6	0.4554	0.5699
1.7	0.6023	0.6441	48.6	63.8	0.3980	0.5778
1.8	0.5861	0.6230	54.6	69.0	0.3400	0.5815
1.9	0.5712	0.6040	60.4	74.2	0.2818	0.5812
2.0	0.5573	0.5866	66.3	79.5	0.2239	0.5767
2.2	0.5323	0.5560	78.0	89.1	0.1104	0.5560
2.4	0.5104	0.5298	89.7	101.0	0.0025	0.5202
2.6	0.4910	0.5071	101.4	111.8	−0.0968	0.4708
2.8	0.4736	0.4872	113.0	122.8	−0.1850	0.4097
3.0	0.4579	0.4694	124.6	133.8	−0.2601	0.3391
3.5	0.4245	0.4326	153.6	161.5	−0.3801	0.1374
4.0	0.3975	0.4034	182.4	189.4	−0.3972	−0.0660
5.0	0.3560	0.3594	240.1	245.7	−0.1776	−0.3276
6.0	0.3252	0.3274	297.6	302.3	0.1507	−0.2767
7.0	0.3012	0.3027	355.1	359.1	0.3001	−0.0047
8.0	0.2818	0.2829	412.5	416.0	0.1717	0.2346
9.0	0.2658	0.2666	469.9	473.0	−0.0903	0.2453
10.0	0.2522	0.2528	527.2	530.1	−0.2459	0.0435

Several important differential properties that will prove useful later are

$$\left. \begin{array}{l} \dfrac{1}{x^{-m}} \dfrac{d}{dx}[x^{-m}J_m(x)] = -J_{m+1}(x), \quad \dfrac{1}{x^{-m}} \dfrac{d}{dx}[x^{-m}N_m(x)] = -N_{m+1}(x), \\ \dfrac{1}{x^m} \dfrac{d}{dx}[x^m J_m(x)] = J_{m-1}(x), \quad \dfrac{1}{x^m} \dfrac{d}{dx}[x^m N_m(x)] = N_{m-1}(x), \\ J_m(x)N'_m(x) - N_m(x)J'_m(x) = \dfrac{2}{\pi x}, \end{array} \right\} \quad (9)$$

where the prime denotes the derivative with respect to the argument. Equations (9) are also quite similar to the corresponding trigonometric relations

$$\frac{d}{dx} \cos x = - \sin x,$$

$$\frac{d}{dx} \sin x = \cos x,$$

$$\cos x \frac{d}{dx} \sin x - \sin x \frac{d}{dx} \cos x = 1,$$

particularly for large arguments where the x^m and x^{-m} factors of Eqs. (9) may be omitted in first approximation.

With this brief discussion of the characteristic waves in uniform and radial transmission lines it is now desirable to turn to the impedance description of such transmission systems. An impedance description exists for every characteristic mode. In the following however, this will be carried out only for the lowest or dominant mode, since this mode is usually the most important for practical applications. The treatment of any other mode in terms of impedances is carried through in an exactly similar manner. As a preliminary to the impedance description of nonuniform radial transmission systems, that of the uniform line will be reviewed briefly (*cf.* Chap. 3). The corresponding treatment of radial lines is developed in close analogy thereto.

8·3. Impedance Description of Uniform Lines.—In Chaps. 2 and 3 the electromagnetic field within nondissipative uniform transmission systems, such as linear waveguides, was described in terms of a superposition of characteristic modes. The introduction of a voltage V and a current I as measures of the transverse electric field $\mathbf{E}_t$ and magnetic field $\mathbf{H}_t$ associated with the dominant mode was made quantitative by the definitions

$$\mathbf{E}_t(x,y,z) = V(z)\mathbf{\Phi}(x,y),$$
$$\mathbf{H}_t(x,y,z) = I(z)\boldsymbol{\zeta} \times \mathbf{\Phi}(x,y),$$

with $\boldsymbol{\zeta}$ a unit vector in the direction of propagation z, and $\mathbf{\Phi}$ a transverse vector function characteristic of the mode. In the steady state, the voltage and current were shown to satisfy the transmission-line equations

$$\left. \begin{aligned} \frac{dV}{dz} &= -j\kappa Z_0 I, \\ \frac{dI}{dz} &= -j\kappa Y_0 V, \\ \kappa^2 &= -\gamma^2, \end{aligned} \right\} \qquad (10)$$

where κ was defined as the propagation constant and $Y_0 = 1/Z_0$ as the characteristic admittance of the dominant-mode transmission line. A

Sec. 8·3] IMPEDANCE DESCRIPTION OF UNIFORM LINES 249

section of line is represented schematically in Fig. 8·4 which also shows the positive directions of V and I. The constant κ is expressed in terms of the angular frequency ω and cutoff wave number k_c of the dominant mode by

$$\kappa^2 = \left(\frac{\omega}{c}\right)^2 - k_c^2 = k^2 - k_c^2. \tag{11}$$

Since both the voltage and current satisfy equations similar to Eq. (4), solutions to Eq. (10) can be written in terms of standing waves as

$$\begin{aligned} V(z) &= A \cos \kappa z + B \sin \kappa z, \\ jZ_0 I(z) &= A \sin \kappa z - B \cos \kappa z. \end{aligned} \tag{12}$$

The application of the boundary conditions that the voltage and current at $z = z_0$ are $V(z_0)$ and $I(z_0)$ leads to

$$\begin{aligned} V(z) &= V(z_0) \cos \kappa(z_0 - z) + jZ_0 I(z_0) \sin \kappa (z_0 - z), \\ I(z) &= I(z_0) \cos \kappa(z_0 - z) + jY_0 V(z_0) \sin \kappa(z_0 - z) \end{aligned} \tag{13}$$

as the complete solution for the behavior of the dominant mode. Since many of the quantities of physical interest depend only on the ratio of

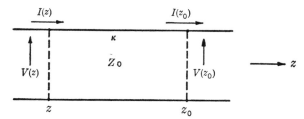

FIG. 8·4.—Section of a uniform transmission line showing positive directions of I and V.

the voltage to the current, it is expedient to define relative admittances in the positive direction by

$$Y'(z) = \frac{1}{Y_0} \frac{I(z)}{V(z)} = \frac{Y(z)}{Y_0}, \quad \text{and} \quad Y'(z_0) = \frac{1}{Y_0} \frac{I(z_0)}{V(z_0)} = \frac{Y(z_0)}{Y_0}, \tag{14}$$

and by division to convert the solution, Eqs. (13), into the fundamental admittance relation

$$Y'(z) = \frac{j + Y'(z_0) \cot \kappa(z_0 - z)}{\cot \kappa(z_0 - z) + jY'(z_0)}. \tag{15}$$

Equation (15) may be employed for admittance calculations in either of two ways: analytically with the aid of cotangent tables or graphically with the aid of transmission-line charts (cf. Sec. 3·6).

An alternative solution to Eqs. (10) may be given on a traveling-wave basis as

$$V(z) = Ce^{-j\kappa z} + De^{+j\kappa z}, \quad Z_0 I(z) = Ce^{-j\kappa z} - De^{+j\kappa z}, \quad (16)$$

where C and D represent the complex amplitudes of the incident and reflected voltage waves at $z = 0$. Again it is desirable to introduce a ratio, in this case the ratio of the amplitudes of the reflected and incident voltage waves at any point z (or z_0). This quantity is called the reflection coefficient

$$\Gamma(z) = \frac{D}{C} e^{j2\kappa z} \quad \text{or} \quad \Gamma(z_0) = \frac{D}{C} e^{j2\kappa z_0}.$$

By the elimination of D/C there is obtained the fundamental transmission-line relation

$$\Gamma(z) = \Gamma(z_0) e^{j2\kappa(z-z_0)}. \quad (17)$$

From Eq. (16) it is seen that the relation between the reflection coefficient and relative admittance at any point z is

$$\Gamma(z) = \frac{1 - Y'(z)}{1 + Y'(z)}. \quad (18)$$

Equations (17) and (18) provide a method alternative to that of Eq. (15) of relating admittances at two different points on a transmission line.

With this brief review of the description of the uniform transmission regions of an electromagnetic system, it is now appropriate to turn to the treatment of the discontinuity regions. A discontinuity region is described by indication of the relations between the voltages and currents at the transmission lines connected to its terminals. For the case of two such terminal lines, distinguished by subscripts 1 and 2, the voltage-current relations are linear and of the form

$$I_1 = Y_{11}V_1 + Y_{12}V_2, \quad I_2 = Y_{12}V_1 + Y_{22}V_2. \quad (19)$$

The positive directions of voltage and current are chosen as in the equivalent circuit of Fig. 8·5 which is a schematic representation of Eqs. (19). Since the principal interest in this chapter is in transmission systems, the discussion of the properties of the circuit parameters Y_{11}, Y_{12}, Y_{22} will be omitted. It is assumed that these parameters are known either from theoretical computations or from experimental measurements on the discontinuity.

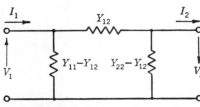

FIG. 8·5.—Equivalent circuit of a discontinuity region.

In connection with circuit descriptions of discontinuities, it is useful to observe that equivalent-circuit descriptions exist for transmission

Sec. 8·3] IMPEDANCE DESCRIPTION OF UNIFORM LINES

regions as well. This may be seen by rewriting Eqs. (13) as

$$\left.\begin{array}{l}I(z) = -jY_0 \cot \kappa(z_0 - z)V(z) - jY_0 \csc \kappa(z_0 - z)[-V(z_0)],\\ I(z_0) = -jY_0 \csc \kappa(z_0 - z)V(z) - jY_0 \cot \kappa(z_0 - z)[-V(z_0)],\end{array}\right\} \quad (20)$$

where $-V(z_0)$ is chosen as the positive voltage at z_0 in accord with the sign convention of Eqs. (19). Comparison of Eqs. (19) and (20) then shows that the equivalent-circuit parameters of a length $(z_0 - z)$ of transmission line are

$$\left.\begin{array}{c}Y_{11} = Y_{22} = -jY_0 \cot \kappa(z_0 - z),\\ Y_{12} = -jY_0 \csc \kappa(z_0 - z),\\ Y_{11} - Y_{12} = Y_{22} - Y_{12} = -jY_0 \tan \dfrac{\kappa(z_0 - z)}{2}.\end{array}\right\} \quad (21)$$

In addition to the knowledge of the relation between the admittances at two different points of a transmission line, that of the frequency derivative of the relative admittance is important. This relation may be obtained simply for a uniform line by forming first the differential of the logarithm of Eqs. (17) and (18) as

$$\frac{\Delta\Gamma(z)}{\Gamma(z)} = \frac{\Delta\Gamma(z_0)}{\Gamma(z_0)} + j2(z - z_0)\Delta\kappa, \quad (22a)$$

and

$$\frac{\Delta\Gamma(z)}{\Gamma(z)} = -\frac{2\Delta Y'(z)}{1 + [jY'(z)]^2}. \quad (22b)$$

Equation (22a) states that on a change in frequency the resulting relative change in the input reflection coefficient of a nondissipative uniform line differs from that of the output reflection coefficient only by a phase change of value twice the change in the electrical length of the line. Rewriting Eq. (22a) in terms of admittance with the aid of Eq. (22b), one finds that

$$\frac{\kappa \dfrac{dY'(z)}{d\kappa}}{1 + [jY'(z)]^2} = j\kappa(z_0 - z) + \frac{\kappa \dfrac{dY'(z_0)}{d\kappa}}{1 + [jY'(z_0)]^2}, \quad (23a)$$

and the desired relation is finally obtained by observing from Eq. (11) that

$$\frac{d\kappa}{\kappa} = \left(\frac{k}{\kappa}\right)^2 \frac{d\omega}{\omega} = \left(\frac{k}{\kappa}\right)^2 d\ln\omega,$$

and therefore Eq. (23a) may be rewritten as

$$\frac{\dfrac{dY'(z)}{d\ln\omega}}{1 + [jY'(z)]^2} = j\left(\frac{k}{\kappa}\right)^2 \kappa(z_0 - z) + \frac{\dfrac{dY'(z_0)}{d\ln\omega}}{1 + [jY'(z_0)]^2}. \quad (23b)$$

It should be emphasized that Eq. (23b) determines the frequency derivative of the *relative* admittance. If the characteristic admittance Y_0 varies with frequency, the following relation should be employed to distinguish between the frequency derivatives of the admittance $Y(z)$ and relative admittance $Y'(z)$

$$\frac{dY(z)}{d \ln \omega} = Y_0 \frac{dY'(z)}{d \ln \omega} + Y'(z) \frac{dY_0}{d \ln \omega}. \qquad (24)$$

8·4. Field Representation by Characteristic Modes.—In the following sections the electromagnetic fields within radial transmission regions[1] of cylindrical shape will be described on an impedance—transmission-line basis in a manner similar to that employed in the preceding sections. A

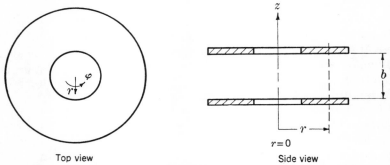

Top view   Side view
Fig. 8·6.—Coordinates for the cylindrical region between two disks.

typical example of a radial transmission system is the cylindrical region between two annular disks as shown in Fig. 8·6. The discussion starts from a transmission-line form of the fundamental Maxwell equations and proceeds to a characteristic-mode representation of the fields and thence to a detailed treatment of the impedance (or admittance) properties of several of the lowest modes. As indicated in Fig. 8·6, an $r\phi z$ polar coordinate system is most suitable for the description of the field. In this coordinate system the electric field **E** and magnetic field **H** are given implicitly, for the steady state of angular frequency ω, by

$$\left. \begin{aligned} \frac{1}{r}\frac{\partial E_z}{\partial \phi} - \frac{\partial E_\phi}{\partial z} &= -j\omega\mu H_r, & \frac{1}{r}\frac{\partial H_z}{\partial \phi} - \frac{\partial H_\phi}{\partial z} &= j\omega\epsilon E_r, \\ \frac{\partial E_r}{\partial z} - \frac{\partial E_z}{\partial r} &= -j\omega\mu H_\phi, & \frac{\partial H_r}{\partial z} - \frac{\partial H_z}{\partial r} &= j\omega\epsilon E_\phi, \\ \frac{1}{r}\frac{\partial (rE_\phi)}{\partial r} - \frac{1}{r}\frac{\partial E_r}{\partial \phi} &= -j\omega\mu H_z, & \frac{1}{r}\frac{\partial (rH_\phi)}{\partial r} - \frac{1}{r}\frac{\partial H_r}{\partial \phi} &= j\omega\epsilon E_z. \end{aligned} \right\} \qquad (25)$$

These equations may be converted to a form that emphasizes the radial transmission character of a cylindrical region by eliminating the com-

[1] *Cf* S. A. Shelkunoff, *Electromagnetic Waves*, Van Nostrand, New York, 1943; Ramo and Whinnery, *Fields and Waves in Modern Radio*, Wiley, New York, 1944.

SEC. 8·4]   CHARACTERISTIC MODES   253

ponents of **E** and **H** in the direction of propagation r. The resulting transmission-line equations for the remaining transverse components then become

$$\frac{\partial E_z}{\partial r} = -jk\zeta \left[-H_\phi + \frac{1}{k^2}\left(\frac{1}{r}\frac{\partial^2 H_z}{\partial \phi\, \partial z} - \frac{\partial^2 H_\phi}{\partial z^2}\right) \right],$$
$$\frac{1}{r}\frac{\partial}{\partial r}(rE_\phi) = -jk\zeta \left[H_z + \frac{1}{k^2}\left(\frac{1}{r^2}\frac{\partial^2 H_z}{\partial \phi^2} - \frac{1}{r}\frac{\partial^2 H_\phi}{\partial \phi\, \partial z}\right) \right],$$
(26a)

and

$$\frac{\partial H_z}{\partial r} = -jk\eta \left[E_\phi + \frac{1}{k^2}\left(\frac{\partial^2 E_\phi}{\partial z^2} - \frac{1}{r}\frac{\partial^2 E_z}{\partial \phi\, \partial z}\right) \right],$$
$$\frac{1}{r}\frac{\partial(rH_\phi)}{\partial r} = -jk\eta \left[-E_z + \frac{1}{k^2}\left(\frac{1}{r}\frac{\partial^2 E_\phi}{\partial \phi\, \partial z} - \frac{1}{r^2}\frac{\partial^2 E_z}{\partial \phi^2}\right) \right],$$
(26b)

where $k = \omega\sqrt{\mu\epsilon}$ and $\zeta = 1/\eta = \sqrt{\mu/\epsilon}$ is the intrinsic impedance of the medium. It is evident from the first of Eqs. (25) that the knowledge of the transverse field components suffices to determine the radial components; the latter will therefore not be considered in what follows.

Since the boundary conditions on the periphery of the cross-sectional ϕz-surfaces of the transmission region are known, the transverse behavior of the field can be found. This immediately suggests the possibility of finding a transverse-field representation, in terms of characteristic modes, that replaces the transverse derivatives in Eqs. (26) by known expressions. The virtue of this procedure is that the field problem is thereby reduced to a one-dimensional transmission-line problem. For the case of uniform regions, the appropriate field representation is expressed in terms of transverse vector modes of two types: the E-modes derivable from a single component of **E** in the direction of propagation and the H-modes derivable from a single component of **H** in the direction of propagation. For radial transmission regions, on the other hand, no similar vector decomposition into E- and H-modes exists. A scalar mode representation does however exist, and in fact it is related to that employed in Sec. 2·13 for uniform lines. The representation applicable to a radial transmission region of height b (Fig. 8·6) consists, for the case in which H_z vanishes, of a superposition of E-type modes

$$E_z = -\frac{\epsilon_n V'_l}{b} \cos\frac{n\pi}{b} z \cos m\phi,$$
$$E_\phi = -\frac{\epsilon_n V'_l}{\kappa^2 b}\left(\frac{m}{r}\right)\left(\frac{n\pi}{b}\right)\sin\frac{n\pi}{b} z \sin m\phi,$$
$$H_\phi = \epsilon_m \frac{I'_l}{2\pi r}\cos\frac{n\pi}{b} z \cos m\phi,$$
(27a)

and for the case in which E_z vanishes, of a superposition of H-type modes

$$H_z = \frac{\epsilon_n I_l''}{b} \sin \frac{n\pi}{b} z \sin m\phi,$$

$$H_\phi = \frac{\epsilon_n I_l''}{\kappa^2 b} \left(\frac{m}{r}\right)\left(\frac{n\pi}{b}\right) \cos \frac{n\pi}{b} z \cos m\phi, \qquad (27b)$$

$$E_\phi = \epsilon_m \frac{V_l''}{2\pi r} \sin \frac{n\pi}{b} z \sin m\phi,$$

together with the corresponding modes of opposite ϕ-polarization. In the above,

$$l = (m,n),$$
$$m = 0, 1, 2, \cdots,$$
$$n = 0, 1, 2, \cdots,$$

$$\epsilon_n \begin{cases} = 1, & \text{for } n = 0, \\ = 2, & \text{for } n \geqq 1, \end{cases}$$

$$\kappa^2 = k^2 - \left(\frac{n\pi}{b}\right)^2.$$

The mode amplitudes V_l and I_l, called the mode voltage and current, respectively, are functions only of r and have been so defined that the average power flow in the positive r-direction is $\frac{1}{2}$ Re $(V_l I_l^*)$. After substitution of the mode forms (27a) or (27b) into the Eqs. (26) and evaluation of the transverse derivatives therein, there are obtained the transmission-line equations

$$\left.\begin{array}{l} \dfrac{dV_l}{dr} = -j\kappa_l Z_l I_l, \\ \dfrac{dI_l}{dr} = -j\kappa Y_l V_l, \end{array}\right\} \qquad (28a)$$

where

$$\kappa_l^2 = \kappa^2 - \left(\frac{m}{r}\right)^2 = k^2 - \left(\frac{n\pi}{b}\right)^2 - \left(\frac{m}{r}\right)^2, \qquad (28b)$$

and

$$Z_l = \frac{1}{Y_l} = \zeta \frac{\kappa^2}{k \kappa_l} \frac{b \epsilon_m}{2\pi r \epsilon_n} \quad \text{for the } E\text{-type modes,}$$

$$Z_l = \frac{1}{Y_l} = \zeta \frac{k \kappa_l}{\kappa^2} \frac{2\pi r \epsilon_n}{b \epsilon_m} \quad \text{for the } H\text{-type modes.} \qquad (28c)$$

The superscript distinguishing the mode type has been omitted, since the equations for both mode types are of the same form. A field representation similar to this one can also be given if the region has an angular aperture less than 2π. Such a representation differs from that in Eqs. (27) and (28) only in the dependence on ϕ and in the value of the amplitude normalization required to maintain the power definition.

If neither E_z nor H_z vanishes, the transverse field may be represented as a superposition of the mode fields in Eqs. (27a) and (27b). In this case the two types of modes can no longer be distinguished on the basis of vector orthogonality as in the uniform line. Nevertheless, the four voltage and current amplitudes of the mixed nmth mode satisfy Eqs. (28)

and can be determined at any point r from the four scalar components of the total transverse field at r.

As in the corresponding case of the uniform line, Eqs. (28) constitute the basis for the designation of V_l and I_l as the mode voltage and current. Concomitantly they also provide the basis for the introduction of a transmission line of propagation constant κ_l and characteristic impedance Z_l to represent the variation of voltage and current along the direction of propagation. The transmission line so defined differs from that for a uniform region in that both the propagation constant and characteristic impedance are variable. This variability of the line parameters must be taken into account when eliminating I'_l from Eqs. (28a) by differentiation in order to obtain the wave equation

$$\frac{1}{r}\frac{d}{dr}\left(r\frac{dV'_l}{dr}\right) + \kappa_l^2 V'_l = 0, \qquad (29)$$

obeyed by the voltage of an E-type mode. The current I''_l of an H-type mode satisfies the same equation. However both I'_l and V''_l obey more complicated wave equations and are best obtained from V'_l or I''_l respectively, by use of Eqs. (28).

The waves defined by Eq. (29) have already been treated in Sec. 8·2. There it was shown that outgoing traveling waves were of the form

$$H_m^{(2)}(\kappa r), \qquad \text{where } \kappa^2 = k^2 - \left(\frac{n\pi}{b}\right)^2.$$

In a nondissipative uniform line the exponential waves corresponding to these are classified as propagating or nonpropagating depending on whether the wave amplitudes remain constant or decrease rapidly with the distance traveled. The nature of a wave is determined simply from the sign of the square of the corresponding propagation constant; a positive sign indicates a propagating wave, and a negative sign an attenuating wave. For a radial line the situation is somewhat more complex. (1) There is an over-all decrease in amplitude of radial waves due to cylindrical spreading, but this decrease will not serve as a basis for distinguishing the various waves. (2) The square of the propagation constant κ is really not a constant, as can be seen from Eq. (28b). The sign may be positive for large r and negative for small r. No difficulty as to classification arises if κ_l^2 is either everywhere positive or everywhere negative, for this implies that κ is either positive real or negative imaginary. Hence from Eqs. (6a) the corresponding waves are either propagating or nonpropagating. On the other hand, if κ_l^2 has a variable sign, there is an apparent difficulty that is, however, easily resolvable. Since for this case κ is necessarily positive and m is equal to or greater than 1, it follows from Eq. (6a) that in the range $\kappa r > m$ (i.e., κ^2 positive) the wave is propagating. Conversely it is seen from Eq. (8) that for $\kappa r < m$ the waves damp out as $(\kappa r)^{-m}$. Thus the propagating or nonpropagating nature of a radial wave is determined from the sign of the square of the propagation constant κ_l exactly as in the case of a uniform wave. In the case of radial waves it should be noted that the same mode may be both propagating and nonpropagating, the propagation being characteristic of the wave behavior at large distances.

Uniform regions are most suitable as transmission systems if only one mode is capable of propagating. This is likewise true for radial regions, and therefore the following section will be concerned with the detailed discussion of the transmission properties of a radial region in which only one E- or H-type mode is propagating.

8·5. Impedance Description of a Radial Line.—A typical radial region to be considered is that shown in Fig. 8·6. The applicability of a simple transmission-line description to such a region is subject to the restrictions that only one mode propagate and that no higher-mode interactions exist between any geometrical discontinuities in the region. These restrictions are not essential and may be taken into account by employing a multiple transmission-line description although this will not be done in this chapter. The simple transmission-line description for the case of only the lowest E-type mode propagating is based on the line equations

$$\left.\begin{aligned}\frac{dV}{dr} &= -jkZ_0I, \\ \frac{dI}{dr} &= -jkY_0V, \\ Z_0 &= \frac{1}{Y_0} = \sqrt{\frac{\mu}{\epsilon}}\frac{b}{2\pi r},\end{aligned}\right\} \tag{30}$$

obtained by setting $m = n = 0$ in Eqs. (28). The field structure of this mode is circularly symmetric about the z-axis; the electric field has only one component parallel to the z-axis; and the magnetic field lines are circles concentric with the z-axis.

The transmission-line description corresponding to the case where only the lowest symmetrical H-type mode can propagate is closely related to that for the E-type mode. The field structure of this mode consists, for an infinite guide height b, of a magnetic field with only a z-component and a circular electric field concentric with the z-axis. A duality thus exists between the E- and H-type modes, since the (negative) electric field and the magnetic field of one are replaced, respectively, by the magnetic field and electric field of the other. This correspondence or duality between the two mode types is an illustration of Babinet's principle discussed in Chap. 2. The point of adducing such a principle is that a field situation for one type of mode can be deduced from that of the other type merely by the duality replacements. In the transmission-line description, duality is manifested by the replacement of V, I, μ, ϵ, Z, and Y of the one mode by I, V, ϵ, μ, Y, and Z of the other. The line equations for the lowest H-type mode are therefore identical with the line equations [Eqs. (30)] for the E-type mode. The duality is, however, an idealization that is possible only if the height of the transmission structure is infinite. For a finite height b the z-component of the magnetic field of the H-type mode must vanish on the planes that constitute the cross-sectional boundaries. Since the magnetic field is divergenceless, this implies the existence of a radial component of the magnetic field. Duality between the fields of the two modes is thus no longer possible. Moreover, because of the variability of the magnetic field along the z-direction, n cannot be set equal to zero in the general line equations [Eqs. (28)]. As a result the transmission-line equations for the lowest symmetrical H-type mode (i.e., $m = 0$) become

$$\frac{dV}{dr} = -j\kappa Z_0 I, \qquad Z_0 = \frac{1}{Y_0} = \sqrt{\frac{\mu}{\epsilon}}\frac{k}{\kappa}\frac{4\pi r}{b}, \\ \frac{dI}{dr} = -j\kappa Y_0 V, \qquad \kappa^2 = k^2 - \left(\frac{n\pi}{b}\right)^2. \tag{31}$$

From a comparison of Eqs. (30) and (31) it is again apparent that duality, in the above-mentioned sense, no longer exists because of the different characteristic impedances and propagation constants of the two modes. A modified and useful form of duality, however, still obtains. If V, $Z_0 I$, and k of the E-type mode are replaced by I, $Y_0 V$, and κ of the H-type mode, the line Eqs. (30) go over into Eqs. (31) and conversely. This is easily seen if the line equations are rewritten in the forms

E-type $\qquad\qquad\qquad H$-type

$$\frac{dV}{dr} = -jkZ_0 I, \qquad \frac{d(Y_0 V)}{dr} + \frac{(Y_0 V)}{r} = -j\kappa I, \\ \frac{d}{dr}(Z_0 I) + \frac{(Z_0 I)}{r} = -jkV, \qquad \frac{dI}{dr} = -j\kappa Y_0 V. \tag{32}$$

As a consequence of this modified duality, all relative impedance relations of the one mode become identical with the relative admittance relations of the other mode provided the propagation constant k is associated with the E-type and κ with the H-type relations.

In both the E- and H-type modes the voltage V and current I are measures of the intensities of the electric and magnetic fields associated with the propagating mode. This fact is indicated quantitatively in Eqs. (27a) and (27b). The positive directions of V and I may be shown schematically by a transmission-line diagram of the usual type as in Fig. 8·7. This schematic representation of the behavior of the lowest

E- (or H-) type mode differs from the corresponding representation for the uniform line in that the characteristic impedance, being variable, must be specified at each reference plane

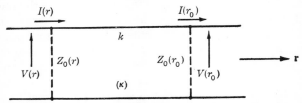

FIG. 8·7.—A portion of a radial transmission line with positive directions indicated for I and V.

With this preliminary discussion of the interrelations between the dominant E- and H-type modes, it is now appropriate to treat the transmission and impedance properties of the individual modes. The modified duality existing between these modes makes necessary the consideration of only one mode—the lowest E-type is chosen in the following—since the properties of the other are easily obtained by the duality replacements. Transmission and impedance properties are readily deduced by consideration of the wave-equation form of the line equations. The wave equation for the lowest E-type mode, obtained by eliminating I from Eqs. (30), is

$$\frac{1}{r}\frac{d}{dr}\left(r\frac{dV}{dr}\right) + k^2 V = 0. \tag{33}$$

The waves defined by this equation have been discussed in Sec. 8·2. The standing-wave solution to Eq. (33) was there shown to be of the form

$$V(r) = A J_0(kr) + B N_0(kr). \tag{34a}$$

With the aid of Eqs. (30) and the differential properties of the Bessel functions, the solution for the current I can be written as

$$j Z_0(r) I(r) = A J_1(kr) + B N_1(kr). \tag{34b}$$

The arbitrary constants A and B can be evaluated from the boundary condition that at $r = r_0$, the voltage and current are $V(r_0)$ and $I(r_0)$. In terms of these quantities Eqs. (34) become

$$V(r) = V(r_0) \left[\frac{J_{10} N_0 - N_{10} J_0}{\frac{2}{\pi k r_0}} \right] + j Z_0(r_0) I(r_0) \left[\frac{N_{00} J_0 - J_{00} N_0}{\frac{2}{\pi k r_0}} \right],$$

$$Z_0(r) I(r) = Z_0(r_0) I(r_0) \left[\frac{N_{00} J_1 - J_{00} N_1}{\frac{2}{\pi k r_0}} \right] - j V(r_0) \left[\frac{J_{10} N_1 - N_{10} J_1}{\frac{2}{\pi k r_0}} \right],$$

(35)

IMPEDANCE DESCRIPTION OF A RADIAL LINE

where
$$J_0(kr) = J_0, \quad J_0(kr_0) = J_{00}, \quad J_1(kr_0) = J_{10},$$
and similarly for the N's.

As in the case of the uniform line it is convenient to define relative admittances at the radii r and r_0 as

$$Y'(r) = \frac{Z_0(r)I(r)}{V(r)} = \frac{Y(r)}{Y_0(r)}, \quad \text{and} \quad Y'(r_0) = \frac{Z_0(r_0)I(r_0)}{V(r_0)} = \frac{Y(r_0)}{Y_0(r_0)}. \tag{36}$$

Because of the lack of a unique characteristic impedance for a radial line, it is important to emphasize that relative admittance is here defined as the ratio of the absolute admittance at a radius to the characteristic admittance at the same radius. Incidentally, the choice of sign convention for V and I implies that relative admittance is positive in the direction of increasing radius.

The fundamental wave solutions [Eqs. (35)] can be rewritten as an admittance relation that gives the relative admittance at any radius r in terms of that at any other radius r_0. This relation, obtained by division of Eqs. (35) by one another, may be expressed as

$$Y'(r) = \frac{j + Y'(r_0)\zeta(x,y)\,\text{ct}(x,y)}{\text{Ct}(x,y) + jY'(r_0)\zeta(x,y)}, \tag{37}$$

where

$$\left.\begin{aligned}
\text{ct}(x,y) &= \frac{J_1 N_{00} - N_1 J_{00}}{J_0 N_{00} - N_0 J_{00}} = \frac{1}{\text{tn}\,(x,y)}, \\
\text{Ct}(x,y) &= \frac{J_{10} N_0 - N_{10} J_0}{J_1 N_{10} - N_1 J_{10}} = \frac{1}{\text{Tn}\,(x,y)}, \\
\zeta(x,y) &= \frac{J_0 N_{00} - N_0 J_{00}}{J_1 N_{10} - N_1 J_{10}} = \zeta(y,x),
\end{aligned}\right\} \tag{38}$$

and[1]

$$x = kr, \quad y = kr_0.$$

The ct and Ct functions are asymmetrical in x and y and may be termed the small and large radial cotangents, respectively; their reciprocals tn and Tn may be called correspondingly the small and large radial tangent functions. The nature of the asymmetry of these functions is made evident in the relation

$$\text{ct}(x,y)\zeta(x,y) = -\text{Ct}(y,x), \tag{39}$$

which, in addition, may be employed to obtain alternative forms for Eq. (37). In fact this relation implies that Eq. (37) can be expressed in terms of only two of the radial functions rather than in terms of the three employed. Several practical applications appear simpler, however, if the three functions ct, Ct, and ζ are used as above. These functions are plotted in the graphs of Figs. 8·8 to 8·10 with the electrical length $(y - x)$

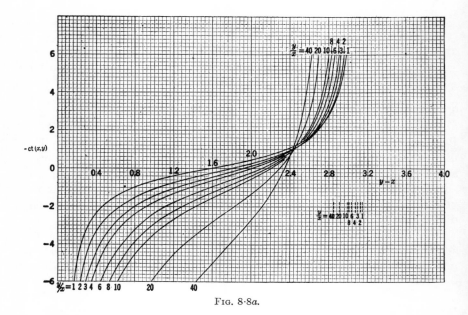

Fig. 8·8a.

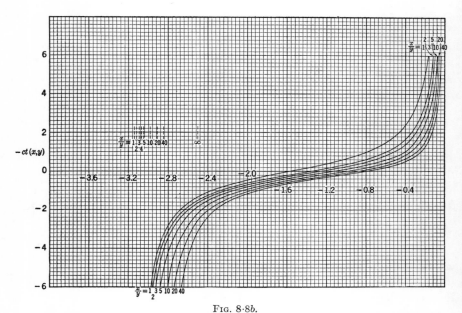

Fig. 8·8b.

Fig. 8·8.—(a) Values of ct (x, y) for $y/x > 1$. (b) Values of ct (x, y) for $y/x < 1$.

Sec. 8·5] IMPEDANCE DESCRIPTION OF A RADIAL LINE

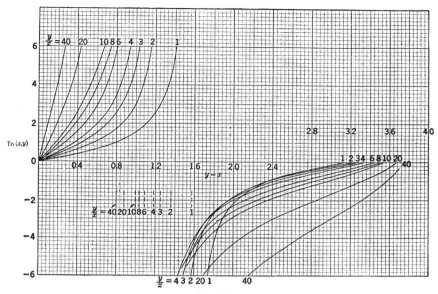

Fig. 8·9a.

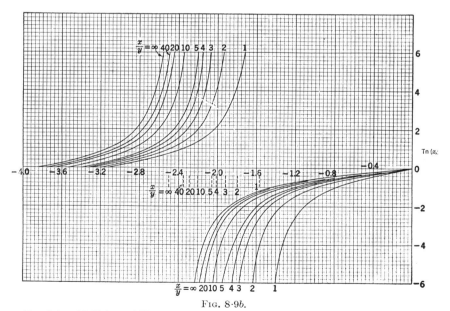

Fig. 8·9b.

Fig. 8·9.—(a) Values of Tn (x, y) for $y/x > 1$. (b) Values of Tn (x, y) for $y/x < 1$.

as the independent variable and the curvature ratio y/x as a parameter. The curves of Figs. 8·8a and 8·9a are for y/x greater than unity and apply to the case where the input terminals are at a smaller radius than the output terminals of the transmission line. Conversely Figs. 8·8b and 8·9b apply to the case where the input terminals are at a larger radius than the output terminals. The symmetry in x and y of the ζ function permits the use of the single Fig. 8·10 for its representation.

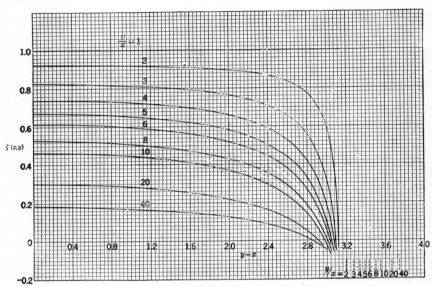

Fig. 8·10.—Values of $\zeta(x, y) = \zeta(y, x)$.

Since the radial functions become infinite at certain values of their arguments, it is difficult to plot a complete range of their functional values. Thus in the graphs of Figs. 8·8 to 8·10 it is to be noted that values above 6 are not shown. The abscissas, however, corresponding to infinite functional values are indicated by short dashed vertical lines in the various figures. The hiatus in plotted values constitutes a definite restriction on the use of the graphs, particularly for impedance calculations on radial lines of approximately one-half wavelength. In this connection it should be pointed out that by the use of the identity of Eq. (39), large values of the Tn function may be found from the graphs of the ct and ζ functions. A more complete table of values of the radial functions will be found in the *Waveguide Handbook*.

The extension of the graphs to include large values of line length $y - x$ is rendered unnecessary because of the asymptotic identity of the radial and trigonometric functions in the range of large x and y.

IMPEDANCE DESCRIPTION OF A RADIAL LINE

This identity may be seen by employing the asymptotic expressions [Eq. (6b)] for the Bessel functions in the definitions [Eq. (38)] with the result that for x and $y \gg 1$

$$\text{ct}(x,y) \approx \frac{\cos\left(x - \frac{3\pi}{4}\right)\sin\left(y - \frac{\pi}{4}\right) - \sin\left(x - \frac{3\pi}{4}\right)\cos\left(y - \frac{\pi}{4}\right)}{\cos\left(x - \frac{\pi}{4}\right)\sin\left(y - \frac{\pi}{4}\right) - \sin\left(x - \frac{\pi}{4}\right)\cos\left(y - \frac{\pi}{4}\right)}$$
$$\approx \cot(y - x),$$

$$\text{Ct}(x,y) \approx \frac{\cos\left(y - \frac{3\pi}{4}\right)\sin\left(x - \frac{\pi}{4}\right) - \sin\left(y - \frac{3\pi}{4}\right)\cos\left(x - \frac{\pi}{4}\right)}{\cos\left(x - \frac{3\pi}{4}\right)\sin\left(y - \frac{3\pi}{4}\right) - \sin\left(x - \frac{3\pi}{4}\right)\cos\left(y - \frac{3\pi}{4}\right)} \quad (40)$$
$$\approx \cot(y - x),$$

$$\zeta(x,y) \approx \frac{\cos\left(x - \frac{\pi}{4}\right)\sin\left(y - \frac{\pi}{4}\right) - \sin\left(x - \frac{\pi}{4}\right)\cos\left(y - \frac{\pi}{4}\right)}{\cos\left(x - \frac{3\pi}{4}\right)\sin\left(y - \frac{3\pi}{4}\right) - \sin\left(x - \frac{3\pi}{4}\right)\cos\left(y - \frac{3\pi}{4}\right)}$$
$$\approx 1.$$

The insertion of these asymptotic values of the radial functions into the radial-transmission-line relation [Eq. (37)] immediately yields the uniform-transmission-line relation [Eq. (15)]. This is not an unexpected result; for in the range of large radii, the cylindrical cross-sectional surfaces of a radial guide tend to become parallel planes and thus the radial geometry approaches the geometry of a uniform transmission line. As a consequence, impedance computations on radial lines for which x and y are sufficiently large may be performed with the aid of the uniform-line relation [Eq. (15)] rather than of Eq. (37). The accuracy of such a computation depends, of course, on the values of x and y; in fact, for reasonable accuracy it will be found that both the input and output terminals of the line must be located at least one wavelength from the axis of radial symmetry.

For impedance computations on long radial lines in which either the input or output terminals are located at a small radius (x or $y < 2\pi$), a stepwise method of calculation is necessary because of the limited range of $y - x$ over which the radial functions are plotted. The procedure is to divide the long line arbitrarily into a number of radial lines of length such that the charts of Figs. 8·8 to 8·10 can be employed. In most cases a division into only two lines is necessary. The output terminals of the first line are chosen at some convenient radius such that its length $y - x$ is less than about π. For such a line length the range of the radial functions plotted in the charts is adequate. The same procèdure is repeated for the second line. The input and output terminals of the remaining portion of the long radial line will then usually be at sufficiently large radii to permit the application either of the uniform-line relation [Eq. (15)] or, equivalently, of the circle diagram with relatively small error.

Although applications of the radial-transmission relation [Eq. (37)] will be made in the next sections, it is desirable to consider a few preliminary illustrations of its use with a view toward obtaining some idea of the physical significance of the radial functions. As a first illustration let us find the relative input admittance $Y'(r)$ of an E-type radial line with a short circuit at its output terminals located at $r = r_0$. Setting $Y'(r_0) = \infty$ in Eq. (37), one obtains for the input admittance

$$Y'(r) = -j \text{ ct } (x,y). \tag{41}$$

The radial function $-\text{ct}(x,y)$ is thus the relative input susceptance of a short-circuited E-type line or, by duality, the relative input reactance of an open-circuited H-type line. It is to be noted from Figs. 8·8a and b that the length of short-circuited line required to produce a given relative input susceptance is greater in the case of a radial E-type line than for the uniform line if the relative input susceptance is less than unity and if $y > x$; the radial line is shorter, however, for relative susceptances greater than unity if $y > x$ and for all susceptances if $y < x$.

The relative input admittance $Y'(r)$ of an E-type line with an open circuit at its output terminals is obtained by placing $Y'(r_0) = 0$ in Eq. (37) with the result that

$$Y'(r) = j \frac{1}{\text{Ct } (x,y)} = j \text{ Tn } (x,y). \tag{42}$$

The radial function Tn (x,y) is thus the relative input susceptance of an open-circuited E-type line or by duality the input reactance of a short-circuited H-type line. From Fig. 8·9a and b it is apparent that the relative lengths of open-circuited radial E-type and uniform lines have, for a given relative input susceptance, a behavior almost inverse to that of the short-circuited lines.

The case of an infinite radial line, that is, one extending from $r = 0$ to $r = \infty$, is of interest and will now be considered. The relative input admittance at r of an E-type infinite line looking in the direction of increasing radius is obtained from Eq. (37) by setting both $Y'(r_0) = 1$ and $y \gg 1$. With the aid of the relations of Eqs. (5), (6b), and (38), one finds for the relative input admittance

$$Y'(r) = -j \frac{H_1^{(2)}(kr)}{H_0^{(2)}(kr)}, \tag{43a}$$

REFLECTION COEFFICIENTS

$$Y'(r) \approx -j \frac{1}{kr \ln \frac{2}{\gamma kr}}, \quad \text{for } kr \ll 1, \tag{43b}$$

$$Y'(r) \approx 1, \quad \text{for } kr \gg 1. \tag{43c}$$

By use of the definition [Eq. (7a)], Eq. (43a) may be put into the form

$$Y'(r) = \frac{h_1}{h_0} [\cos(\eta_1 - \eta_0) - j \sin(\eta_1 - \eta_0)]. \tag{43d}$$

As is apparent from Eqs. (43a) and (43d) the input admittance of an infinite radial line is in general not equal to the characteristic admittance $Y_0(r)$! Moreover, the relative input admittance is complex with a negative imaginary (that is, inductive) part. Correspondingly, by duality, the relative input impedance of an infinite H-type line is complex with a negative imaginary (capacitive) part.

The relative input admitance at r of the same infinite E-type line but now looking in the direction of decreasing radius is obtained by setting $r_0 = 0$ in Eq. (37). With the use of Eqs. (8) and (38), it is now found that for all finite $Y'(0)$ the relative input admittance is

$$Y'(r) = -j \frac{J_1(kr)}{J_0(kr)}, \tag{44a}$$

$$Y'(r) = -j \frac{kr}{2}, \quad kr \ll 1, \tag{44b}$$

$$Y'(r) = -j \tan\left(kr - \frac{\pi}{4}\right), \quad kr \gg 1. \tag{44c}$$

The input admittance is thus seen to be positive imaginary (capacitive) if it is remembered that admittances are counted negative when looking in the direction of decreasing radius. Correspondingly, by duality, the relative input impedance of an infinite H-type line is positive imaginary (inductive). This asymmetric behavior of the admittance of an infinite radial line is in marked contrast to that of an infinite uniform line where the relative input admittance is always real and equal to unity looking in either direction along the line.

8·6. Reflection Coefficients in Radial Lines.—In addition to the standing-wave or admittance description of the fields in radial transmission regions, there exists, as in uniform regions, an alternative description on a reflection or scattering basis. This latter description is based on the traveling-wave solution to the wave equation [Eq. (33)]:

$$V(r) = A H_0^{(2)}(kr) + B H_0^{(1)}(kr), \tag{45a}$$

where A and B are the complex amplitudes of the waves traveling in the directions of increasing and decreasing radius, respectively. The solution for the current I follows from Eq. (45a) by use of Eq. (31) and the differential properties of the Hankel functions as

$$jZ_0(r)I(r) = AH_1^{(2)}(kr) + BH_1^{(1)}(kr), \qquad (45b)$$

and this again may be interpreted in terms of waves traveling in both directions. Reflection coefficients may now be introduced at arbitrary radii r and r_0 and defined as ratios of waves traveling in each direction at these points. Both current and voltage reflection coefficients may be defined. The voltage reflection coefficients are

$$\Gamma_v(r) = \frac{B}{A}\frac{H_0^{(1)}(kr)}{H_0^{(2)}(kr)}, \qquad \Gamma_v(r_0) = \frac{B}{A}\frac{H_0^{(1)}(kr_0)}{H_0^{(2)}(kr_0)}, \qquad (46a)$$

and the current reflection coefficients are correspondingly

$$\Gamma_I(r) = \frac{B}{A}\frac{H_1^{(1)}(kr)}{H_1^{(2)}(kr)}, \qquad \Gamma_I(r_0) = \frac{B}{A}\frac{H_1^{(1)}(kr_0)}{H_1^{(2)}(kr_0)}. \qquad (46b)$$

On the elimination of the factor B/A there are obtained the fundamental transmission-line relations

$$\Gamma_v(r) = \Gamma_v(r_0)e^{j2[\eta_0(kr)-\eta_0(kr_0)]}, \qquad (47a)$$
$$\Gamma_I(r) = \Gamma_I(r_0)e^{j2[\eta_1(kr)-\eta_1(kr_0)]}, \qquad (47b)$$

that relate the values of the reflection coefficients at the radii r and r_0. The quantities η_0 and η_1 are the phases of the Hankel functions as defined in Eq. (7a). The transmission Eqs. (47) also provide a means of relating the admittances at two points on a radial line. This relation may be obtained from the wave solutions by division of Eq. (45a) by Eq. (45b) with the result that at the radius r

$$\frac{I(r)}{V(r)} = -jY_0(r)\frac{H_1^{(2)}(kr)}{H_0^{(2)}(kr)}\frac{1+\Gamma_I}{1+\Gamma_v}. \qquad (48)$$

This equation may be transformed to the more familiar form of a relation between the reflection coefficient and the admittance as

$$Y'_+(r) = \frac{1+\Gamma_I}{1+\Gamma_v} = \frac{1-\Gamma_v\left(\dfrac{Y_0^-}{Y_0^+}\right)}{1+\Gamma_v}, \qquad (49a)$$

or conversely as

$$\Gamma_v(r) = \frac{Z'_+ - 1}{Z'_- + 1}, \qquad \Gamma_I = -\frac{1 - Y'_+}{1 + Y'_-}, \qquad (49b)$$

if the relative admittance (or impedance) ratios are introduced as

$$Y'_+(r) = \frac{1}{Z'_+(r)} = \frac{Y(r)}{Y_0^+(r)}, \qquad Y'_-(r) = \frac{1}{Z'_-(r)} = \frac{Y(r)}{Y_0^-(r)}, \qquad (50)$$

where

$$Y_0^+(r) = -jY_0(r)\frac{H_1^{(2)}(kr)}{H_0^{(2)}(kr)}, \qquad Y_0^-(r) = jY_0(r)\frac{H_1^{(1)}(kr)}{H_0^{(1)}(kr)} \qquad (51)$$

are the admittances associated with waves traveling on an infinite E-type radial line in the directions of increasing and decreasing radius, respectively [cf. Eq. (43a)]. It is instructive to note that the radial-line relations Eqs. (47) to (50) go over into the corresponding uniform-line relations Eqs. (17) and (18) in the limit of large kr, for in this limiting region

$$\eta_0(kr) = \eta_1(kr) = kr - \frac{\pi}{4},$$
$$\Gamma_v = -\Gamma_I,$$

and

$$Y_0^+ = Y_0^- = Y_0.$$

Equations (47) and (49) provide a method alternative to that of Eq. (37) of relating the admittance at two different points of an E-type radial line. The corresponding relations for the case of an H-type line are obtained by the duality replacements discussed above.

8·7. Equivalent Circuits in Radial Lines.—The radial-transmission-line relations so far developed permit the determination of the dominant-mode voltage and current or, alternatively, the admittance at any point on a radial line from a knowledge of the corresponding quantities at any other point. These relations assume that no geometrical discontinuities exist between the two points in question. The existence of such a non-uniformity in geometrical structure implies that relations like Eqs. (35), for example, must be modified. The form of the modification follows directly from the linearity and reciprocity of the electromagnetic field equations as

$$\begin{aligned} I_1 &= Y_{11}V_1 + Y_{12}V_2, \\ I_2 &= Y_{12}V_1 + Y_{22}V_2, \end{aligned} \qquad (52)$$

where I_1 and V_1 are the dominant-mode current and voltage at a reference point on one side of the discontinuity and I_2, V_2 are the same quantities at a reference point on the other side. These equations are identical with Eqs. (19) which describe a discontinuity in a uniform line. The discontinuity can likewise be represented schematically by the equivalent circuit indicated in Fig. 8·5 which shows the choice of positive directions for the voltage and the current. The equivalent-circuit parameters Y_{11}, Y_{12}, and Y_{22} depend on the geometrical form of the discontinuity as well as on the choice of reference points. As in the case of the uniform line, the explicit evaluation of the circuit parameters involves the solution of a boundary-value problem and will not be treated here. The customary assumption that they are known either from measurement or from computation will be made. Even when the reference points designated by the subscripts 1 and 2 in the circuit equations are chosen as coincident at some radius, the equations are still valid. Equations (52) then indicate in general an effective discontinuity in the dominant-mode voltage and current at the reference radius in contrast to the continuity that exists at every point in a smooth radial line.

A particularly instructive example of an equivalent circuit is that corresponding to a length $r_0 - r$ of an E-type radial line with no discontinuities between r and r_0. The relations between the currents and voltages at the reference radii r and r_0 have already been derived in Eqs. (35). They can readily be put into the form of Eqs. (52) by a simple algebraic manipulation that yields

$$I_1 = -jY_0(r)\,\mathrm{ct}(x,y)V_1 - j\sqrt{Y_0(r)Y_0(r_0)}\,\mathrm{cst}(x,y)V_2, \\ I_2 = -j\sqrt{Y_0(r)Y_0(r_0)}\,\mathrm{cst}(x,y)V_1 + jY_0(r_0)\,\mathrm{ct}(y,x)V_2, \quad (53)$$

where

$$I_1 = I(r), \quad V_1 = V(r), \quad x = kr, \\ I_2 = I(r_0), \quad V_2 = -V(r_0), \quad y = kr_0,$$

and

$$\mathrm{cst}(x,y) = \frac{\dfrac{2}{\pi}\dfrac{1}{\sqrt{xy}}}{J_0 N_{00} - N_0 J_{00}} = -\mathrm{cst}(y,x).$$

The function cst (x,y) is termed the radial cosecant function, since it becomes asymptotically identical with the trigonometric cosecant function for sufficiently large x and y. The value of the radial cosecant function may be computed from the tabulated values of the radial cotangent functions by use of the identity

$$\mathrm{cst}^2(x,y) = \frac{1 + \mathrm{ct}(x,y)\,\mathrm{Ct}(x,y)}{\zeta(x,y)}. \quad (54)$$

This identity goes over into the corresponding trigonometric identity

$$\csc^2(y - x) = 1 + \cot^2(y - x)$$

for sufficiently large x and y. From Eqs. (53) it is to be noted that the equivalent circuit of a length of radial line is unsymmetrical in contrast to the case of the uniform line. The shunt and series parameters of the π-circuit representation (*cf.* Fig. 8·5) for the radial line are seen to be

$$Y_{11} - Y_{12} = -jY_0(r)\left[\mathrm{ct}(x,y) + \sqrt{\frac{y}{x}}\,\mathrm{cst}(x,y)\right], \\ Y_{22} - Y_{12} = -jY_0(r_0)\left[-\mathrm{ct}(y,x) + \sqrt{\frac{x}{y}}\,\mathrm{cst}(y,x)\right], \quad (55) \\ Y_{12} = -j\sqrt{Y_0(r)Y_0(r_0)}\,\mathrm{cst}(x,y).$$

Since the corresponding equivalent-circuit representation of a finite length of an H-type radial line differs somewhat from that of the E-type line just considered, it is perhaps desirable to indicate explicitly how the duality principle may be employed to obtain the H-type representation. On use of the duality replacements in Eqs. (53), the circuit equations are obtained corresponding to a length $r_0 - r$ of H-type radial line in which there are no discontinuities; thus

$$V_1 = -jZ_0(r)\,\text{ct}(x,y)\,I_1 - j\,\sqrt{Z_0(r)Z_0(r_0)}\,\text{cst}(x,y)I_2,$$
$$V_2 = -j\,\sqrt{Z_0(r)Z_0(r_0)}\,\text{cst}(x,y)\,I_1 + jZ_0(r_0)\,\text{ct}(y,x)I_2, \quad (56)$$

where the various quantities are defined as in Eq. (53). These circuit equations are of the form

$$V_1 = Z_{11}I_1 + Z_{12}I_2,$$
$$V_2 = Z_{12}I_1 + Z_{22}I_2$$

and may be schematically represented as the T-circuit shown in Fig. 8·11.

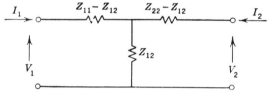

Fig. 8·11.—T-circuit for a discontinuity with positive directions of V and I.

The series and shunt elements of the T are given by

$$\begin{aligned}
Z_{11} - Z_{12} &= -jZ_0(r)\left[\text{ct}(x,y) + \sqrt{\frac{y}{x}}\,\text{cst}(x,y)\right], \\
Z_{22} - Z_{12} &= -jZ_0(r_0)\left[-\text{ct}(y,x) + \sqrt{\frac{x}{y}}\,\text{cst}(y,x)\right], \\
Z_{12} &= -j\,\sqrt{Z_0(r)Z(r_0)}\,\text{cst}(x,y).
\end{aligned} \quad (57)$$

It should be emphasized that equivalent-circuit representations of the type described in Eqs. (53) apply to radial lines on which only one mode is being propagated, or at least on which all higher modes are effectively terminated by their characteristic impedance. If such is not the case, the equivalent circuits must be modified to include the description of more than one propagating mode.

The admittance description of radial lines will be completed with the derivation of the differential form of the relation between the relative admittances at two points r and r_0 on a radial line. To derive this relation in a manner analogous to that employed for the uniform line, one first forms the differential of the logarithm of the transmission-line relation [Eq. (47a)] and obtains

or
$$\frac{\Delta\Gamma_v(r)}{\Gamma_v(r)} = \frac{\Delta\Gamma_v(r_0)}{\Gamma_v(r_0)} + j2[\Delta\eta_0(x) - \Delta\eta_0(y)]$$

$$\frac{\Delta\Gamma_v(r)}{\Gamma_v(r)} = \frac{\Delta\Gamma_v(r_0)}{\Gamma_v(r_0)} + j\frac{4}{\pi}\left[\frac{1}{h_0^2(kr)} - \frac{1}{h_0^2(kr_0)}\right]\frac{\Delta k}{k}. \quad (58)$$

This differential relation indicates on one hand that the change in phase of the voltage reflection coefficient at the input end of an E-type radial line is the sum of a phase change associated with the output reflection coefficient plus a phase change associated with the electrical length of the line. On the other hand, the relative change in amplitude of the input voltage reflection coefficient is identical with the relative change in amplitude of the output reflection coefficient. This behavior is analogous to that of a uniform line, and in fact Eq. (58) becomes identical with the uniform-line relation [Eq. (22)] for large kr and kr_0. By means of Eq. (58) it is possible to compute the change in the input voltage reflection coefficient due either to a frequency change $\Delta k/k$ or to a change in output reflection coefficient or to a simultaneous change of both these factors. In order to express Eq. (58) in the admittance form it is necessary to find the differential relation between the voltage reflection coefficient and the relative admittance. Although this relation can be obtained by differentiation of Eq. (49b), such a procedure is not too simple. It is somewhat more desirable to derive the desired differential relation between the relative admittances at any two points r and r_0 by starting from the expression for the amplitude ratio

$$\frac{-B}{A} = \frac{J_1 - jY'(r)J_0}{N_1 - jY'(r)N_0} = \frac{J_{10} - jY'(r_0)J_{00}}{N_{10} - jY'(r_0)N_{00}},$$

obtained from Eqs. (34a) and (34b) and the definitions of Eq. (36). Forming the differential of the logarithm of this ratio at $x = kr$ and $y = kr_0$ and equating the results, one gets, with the aid of the Wronskian relation, Eq. (9c),

$$\left(\frac{\Delta Y'(r)}{1 + [jY'(r)]^2} + \left\{jx + \frac{Y'(r)}{1 + [jY'(r)]^2}\right\}\frac{\Delta k}{k}\right)\alpha(r)$$
$$= \left(\frac{\Delta Y'(r_0)}{1 + [jY'(r_0)]^2} + \left\{jy + \frac{Y'(r_0)}{1 + [jY'(r_0)]^2}\right\}\frac{\Delta k}{k}\right)\alpha(r_0), \quad (59)$$

where
$$\alpha(r) = \frac{1 + [jY'(r)]^2}{[J_1(kr) - jY'(r)J_0(kr)]^2}\frac{2}{\pi kr}.$$

The reason for the introduction of a coefficient $\alpha(r)$ lies in the fact that $\alpha(r)/\alpha(r_0)$ approaches unity as kr and kr_0 become sufficiently large. In this far region Eq. (59) becomes asymptotically identical with the corresponding uniform-line relation [Eq. (23a)]. It should be pointed out that Eq. (59) can be employed to compute the change in relative input admittance due to a change either in output admittance $\Delta Y'(r_0)$ or in frequency $\Delta k/k = \Delta\omega/\omega$ or in both. A particular case of Eq. (59) which is often of use occurs when a short circuit exists at the output terminals. For this case

$$Y'(r_0) = \infty, \qquad \Delta Y'(r_0) = \Delta\left(\frac{1}{Z'(r_0)}\right) = -\frac{\Delta Z'(r_0)}{[Z'(r_0)]^2},$$

and therefore Eq. (59) reduces to

$$\left[\frac{\Delta Y'(r)}{1 + [jY'(r)]^2} + \left\{jx + \frac{Y'(r)}{1 + [jY'(r)]^2}\right\}\frac{\Delta k}{k}\right]\alpha(r)$$
$$= \left[\Delta Z'(r_0) + jy\,\frac{\Delta k}{k}\right]\frac{2}{\pi y J_0^2(y)}. \quad (60)$$

The importance of differential relations of the above type stems from the fact that they provide a relatively simple admittance means for rigorously computing Q's of cavities and other parameters, as will be illustrated in the following sections.

The corresponding differential relation for an H-type radial line may be obtained by the usual duality replacements. The fundamental differential relative admittance relation [Eq. (59)] becomes, under the duality transformation, a relative impedance relation of exactly the same form. It is important to note that the characteristic impedance $Z_0(r)$ of an H-type line is a function of k in contradistinction to an E-type line.

8·8. Applications.—In the preceeding sections a variety of methods employed to describe the electromagnetic fields within radial transmission systems have been investigated. These investigations indicate that complex radial systems of the sort often encountered in practice may be regarded as composite structures consisting of transmission regions and discontinuity regions. The associated descriptions of the fields within these component regions fall naturally into two distinct categories: the transmission-line description and the equivalent-circuit description. It is thereby implied that the electrical properties of such composite systems may be computed by straightforward engineering methods involving only impedance calculations on the transmission-line and circuit equivalents of these systems. As illustrations of such computations we shall first consider a class of resonant-cavity problems with particular consideration of some cases associated with the design and operation of certain high-frequency electronic oscillators.

The electrical properties of a resonant system are determined by specification of the three fundamental parameters:
1. The frequency of resonance.
2. The Q of the resonance.
3. The resonant conductance.

The meaning as well as computation of these parameters may perhaps be clarified by consideration of a general resonant system schematically represented in the vicinity of the resonant frequency by an equivalent circuit of the form shown in Fig. 8·12. The representation is recognized to be that of a lumped-constant low-frequency shunt-resonant circuit. It is also representative at some reference point for the admittance description of the behavior of the electromagnetic fields in spatially extended resonant systems of many types.

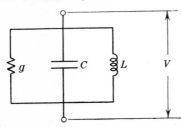

Fig. 8·12.—The equivalent circuit of a system near resonance.

The total admittance associated with the resonant system may be written

$$Y_t = g_t + j\left(\omega C - \frac{1}{\omega L}\right) = g_t(\omega) + jB_t(\omega), \tag{61}$$

where the first form pertains to the lumped-constant circuit and the second to the extended structure. The fact that in the second case the circuit parameters may depend on the angular frequency has been explicitly emphasized in the expressions for the total conductance g_t and total susceptance B_t.

A resonant frequency of an electromagnetic system is defined as a frequency at which the average electric and magnetic energies within the system are equal. Since by an energy theorem (cf. Chap. 5) the total susceptance of an electromagnetic system is proportional to the difference between the average electric and magnetic energies stored in the system, it follows that an angular frequency ω_0 of resonance is identical with the frequency for which

$$\omega_0 C - \frac{1}{\omega_0 L} = B_t(\omega_0) = 0. \tag{62}$$

Since the total susceptance may vanish at more than one frequency, Eq. (62) determines, in general, a series of resonant frequencies.

The Q of a resonant electromagnetic system is defined as the product, at resonance, of the angular frequency ω and the ratio of the total energy stored in the system to the power dissipated or otherwise coupled out of the system. The total energy stored within an electromagnetic system can be expressed in terms of the frequency derivative of the total susceptance and the rms voltage V associated with the equivalent circuit describing the system. From the energy theorem discussed in Chap. 5 this expression for the total energy is

$$\frac{1}{2}\left(C + \frac{1}{\omega^2 L}\right)V^2 = \frac{1}{2}\frac{dB_t}{d\omega}V^2.$$

A COAXIAL CAVITY

Correspondingly, the total power lost from the system is $g_t V^2$. As a consequence, the desired expression for Q at the resonant angular frequency ω_0 is

$$Q = \frac{1}{2g_t}\left(\frac{\omega dB_t}{d\omega}\right)_{\omega=\omega_0} = \frac{\omega_0 C}{g_t} \qquad (63)$$

and is seen to be independent of the voltage V associated with the reference plane to which the equivalent circuit pertains. The fact that both the resonant frequency and the Q of an enclosed system are invariant with respect to choice of reference point should be evident from their original definitions in terms of energy. This independence with respect to reference point does not apply to the last of the three aforementioned parameters—the resonant conductance. As has already been implied the resonant conductance $g_t(\omega_0)$ is defined as the ratio, at resonance, of the total power lost from the system to the square of the rms voltage at the reference plane.

As an application of the developments in Sec. 8·3 to 8·7, we shall now consider a few examples of the computation of the resonant parameters defined in Eqs. (61) to (63).

8·9. A Coaxial Cavity.—The first case to be considered is that of a nondissipative cylindrical cavity oscillating in the lowest symmetrical E-type mode. The cavity dimensions are indicated in Fig. 8·13. As also

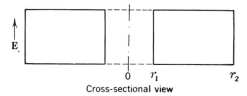

Fig. 8·13.—A coaxial cavity.

shown in the figure the equivalent electrical network for this structure is an E-type transmission line of length $r_2 - r_1$ and propagation constant $k = 2\pi/\lambda = \omega/c$ with infinite-admittance terminations at r_1 and r_2. With the choice of r_1 as the reference radius, the total admittance at r_1 is seen to be the sum of the infinite admittance of the termination plus the input admittance of a short-circuited E-type radial line of electrical length $y - x$. By Eqs. (41) and (63) the resonant wave number k (or angular frequency ω_0) is therefore determined from the

resonance condition
$$\text{ct}(x,y) = -\infty. \tag{64}$$

Reference to the plot of the radial cotangent function in Fig. 8·8a shows that for
$$\frac{y}{x} = \frac{r_2}{r_1} = 5.75,$$
the solution to Eq. (64) is
$$y_0 - x_0 = k_0(r_2 - r_1) = 3.04,$$
or
$$\lambda_0 = \frac{2\pi}{k_0} = 2\pi \frac{(0.095)}{3.04} = 0.196 \text{ in.} = 0.50 \text{ cm.}$$

The resonant wavelength is thus about 3 per cent greater than that of a corresponding uniform cavity of equal length. The remaining resonant parameters Q and g are, for this nondissipative case, infinite and zero, respectively.

8·10. Capacitively Loaded Cavity.—As a second illustration let us consider the calculation of the resonant frequency of a loaded cylindrical

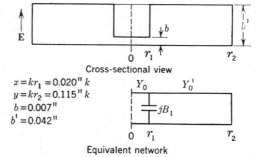

Fig. 8·14.—A capacitively loaded cavity.

cavity oscillating in the lowest angularly symmetric E-type mode. As indicated in Fig. 8·14, the equivalent network that describes the fields within such a structure consists of a junction of an open- and a short-circuited E-type radial transmission line of unequal characteristic admittances $Y_0(r)$ and $Y'_0(r)$ but identical propagation constants $k = 2\pi/\lambda$ and a junction admittance of value jB_1. The network parameters are obtained from Eq. (30) and the *Waveguide Handbook* as

$$\frac{Y_0(r)}{Y'_0(r)} = \frac{b'}{b}, \tag{65a}$$

$$\frac{B_1}{Y'_0(r_1)} = \frac{2kb'}{\pi} \ln \frac{eb'}{4b}, \tag{65b}$$

where ln is the logarithm to the base e. Equation (65b) is an approximation that is valid to within a few per cent for the cavity shape under consideration. At the reference plane $r = r_1$, the total susceptance relative to the characteristic admittance $Y_0'(r_1)$ is, from Eqs. (41) and (44a),

$$\frac{B_t(k)}{Y_0'(r_1)} = \frac{B_1}{Y_0'(r_1)} + \frac{Y_0(r_1)}{Y_0'(r_1)} \frac{J_1(x)}{J_0(x)} - \operatorname{ct}(x,y),$$

or to a good approximation by Eqs. (44b) and (65a)

$$\frac{B_t(k)}{Y_0'(r_1)} = \frac{B_1}{Y_0'(r_1)} + \frac{b'x}{b2} - \operatorname{ct}(x,y). \tag{66}$$

The resonant frequency of the lowest E-type mode is obtained from that value k_0 for which the total susceptance vanishes. With the insertion of the numerical values into Eq. (66) the vanishing of the total susceptance leads to the transcendental resonance condition

$$-0.0379k_0 - 0.060k_0 = -\operatorname{ct}(0.020k_0, 0.115k_0) \tag{67}$$

which can be solved graphically for k_0 with the aid of Fig. 8·8a. The left-hand side of Eq. (67) is expressed as a function of

$$y_0 - x_0 = k_0(r_2 - r_1) = 0.095k_0,$$

and as such it is representable in Fig. 8·8a as a straight line of slope $-(0.098/0.095) = -1.03$. The intersection of this line with the function $-\operatorname{ct}(x,y)$, with $y/x = 5.75$, then yields for the solution of Eq. (67) the value

$$y_0 - x_0 = 0.095k_0 = 1.35,$$

and consequently the resonant wavelength is

$$\lambda_0 = \frac{2\pi}{k_0} = 0.441 \text{ in.} = 1.12 \text{ cm.}$$

In many applications it is necessary to know the dependence of the resonant wavelength on the gap height b. This dependence may be computed exactly as in the above case and will be found to yield the values shown in Table 8·2.

TABLE 8·2.—VARIATION OF RESONANT WAVELENGTH FOR THE CAVITY OF FIG. 8·14

b, in.	λ_0, cm
0.005	1.24
0.006	1.17
0.007	1.12
0.008	1.08
0.009	1.04

8·11. Capacitively Loaded Cavity with Change in Height.—As a variant of the two preceding examples let us consider the problem of finding the gap height b to make the cylindrical cavity illustrated in Fig. 8·15 resonate at the wavelength $\lambda = 1.25$ cm in the lowest angularly symmetric E-type mode. In this specification of the problem the electrical lengths,

$$x_1 = kr_1 = 0.255,$$
$$x_2 = kr_2 = 0.638,$$
$$x_3 = kr_3 = 1.468$$

are explicity known, and hence the values of the radial functions may be obtained directly from the plotted curves. The electrical network that describes the lowest E-type mode in the above cavity consists of three E-type radial lines of characteristic admittances $Y_0(r)$, $Y_0'(r)$, and $Y_0''(r)$

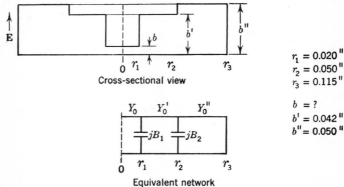

Fig. 8·15.—Loaded coaxial cavity with a change in height.

joined at the radii r_1 and r_2, with the first and last lines being open- and short-circuited, respectively. The junction admittances jB_1 and jB_2 at r_1 and r_2 are capacitances whose susceptance values may be determined to within a few per cent from the *Waveguide Handbook*. Higher-mode interaction effects between the discontinuities at r_1 and r_2 are assumed to be of negligible importance. For the cavity indicated in Fig. 8·15, the values of the network parameters are

$$\frac{Y_0'(r)}{Y_0''(r)} = \frac{b''}{b'} = 1.19, \qquad \frac{Y_0(r)}{Y_0'(r)} = \frac{b'}{b} = \frac{0.042}{b}, \qquad (68a)$$

$$\frac{B_2}{Y_0'(r_2)} = \frac{2kb'}{\pi} \ln\left(\frac{1-\alpha^2}{4\alpha}\right)\left(\frac{1+\alpha}{1-\alpha}\right)^{\frac{1}{2}\left(\alpha+\frac{1}{\alpha}\right)}, \qquad \alpha = \frac{b'}{b''}, \qquad (68b)$$

$$\frac{B_2}{Y_0'(r_2)} = 0.016,$$

and B_1 is determined from Eq. (65b) which incidentally is a limiting form of Eq. (68b) for $\alpha \ll 1$. To determine the gap height b, it is first necessary to compute the total susceptance at some reference point, say $r = r_1 + 0$. The total susceptance at this reference point is the sum of the susceptances looking in the directions of increasing and decreasing radius. The susceptance in the direction of increasing radius is computed by a stepwise

SEC. 8·12] OSCILLATOR COUPLED TO WAVEGUIDE 277

procedure. As a first step, the sum of the junction susceptance at r_2 and the susceptance of the short-circuited line of electrical length $x_3 - x_2$ relative to $Y_0'(r_2)$ is computed to be

$$\frac{B_2}{Y_0'(r_2)} - \frac{Y_0''(r_2)}{Y_0'(r_2)} \operatorname{ct}(x_2, x_3) = 0.016 - 1.334 = -1.32$$

with the aid of Eqs. (68) and Fig. 8·8a. It is noted that the change in height at r_2 has a relatively minor effect in terms of the junction susceptance introduced thereby but a major effect in terms of the change in characteristic admittance. With this knowledge of the relative susceptance at x_2, the relative susceptance at $x_1 + 0$ in the direction of increasing radius may be calculated with the aid of the radial-transmission-line relation [Eq. (37)] as

$$\frac{\vec{B}}{Y_0'(r_1)} = \frac{1 - 1.32 \operatorname{ct}(x_1,x_2)\varsigma(x_1,x_2)}{\operatorname{Ct}(x_1,x_2) + (1.32)\varsigma(x_1,x_2)}, \tag{69a}$$

or explicitly by use of Figs. 8·8 to 8·10 (with $x_2/x_1 = 2.5$)

$$\frac{\vec{B}}{Y_0'(r_1)} = \frac{1 - (1.32)(4.04)(0.872)}{1.36 + (1.32)(0.872)} = -1.45. \tag{69b}$$

The relative susceptance at $x_1 + 0$ looking in the direction of decreasing radius has already been found in Sec. 8·10 to be

$$\frac{\overleftarrow{B}}{Y_0'(r_1)} = \frac{2kb'}{\pi} \ln \frac{eb'}{4b} + \frac{b'}{b}\frac{x_1}{2}, \tag{70a}$$

$$\frac{\overleftarrow{B}}{Y_0'(r_1)} = 0.341 \ln \frac{0.0285}{b} + \frac{0.00535}{b}. \tag{70b}$$

The resonance condition [Eq. (62)] that the total susceptance $\vec{B} + \overleftarrow{B}$ vanish then leads to

$$0.341 \ln \frac{0.0285}{b} + \frac{0.00535}{b} - 1.45 = 0,$$

which can be solved graphically for b to give

$$b = 0.00590 \text{ in.}$$

For the case in which the height b is known, the problem of finding the resonant frequency is solved by a method similar to the one just carried out. Values of k are assumed, and by a trial-and-error method the resonance value k_0 is found as that for which Eqs. (69a) and (70a) are equal in magnitude but opposite in sign.

8·12. Oscillator Cavity Coupled To Rectangular Waveguide.—As a final illustration of the computation of the electrical properties of cylindrical cavities we shall consider the case of the radial cavity of Fig. 8·16 oscillating in the lowest angularly symmetric E-type mode and coupled to a matched rectangular waveguide. Such structures, frequently encountered in high-frequency oscillator tubes, are excited by an electronic beam along the symmetry axis. The case illustrated in Fig.

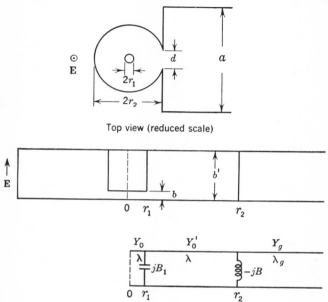

FIG. 8·16.—Cavity and equivalent circuit of an osillator coupled to waveguide.

8·16 resembles closely the cavity of the Neher tube designed for operation in the 1-cm wavelength range. The calculation of the over-all electrical characteristics of such an electronic-electromagnetic system requires a knowledge of the interaction of the electronic beam and the electromagnetic field as summed up in the expression for the electronic admittance at some reference plane. This electronic problem has been considered elsewhere in the Radiation Laboratory Series[1] and will be omitted in the following discussion, since the modification thereby introduced is taken into account simply by inclusion of the electronic admittance in the expression for the total admittance.

The computations as always are based on the network equivalent of the oscillating system. As shown in Fig. 8·16, the radial-transmission-line description of the system to the left of the radius r_2 is identical with that employed in Sec. 8·10, and hence the values of the associated circuit parameters need not be indicated again. The only additional values necessary are those associated with the coupling network at r_2 and with

[1] *Klystrons and Microwave Triodes*, Vol. 7.

the H_{01}-mode uniform transmission line describing the rectangular waveguide. In the case under consideration the coupling is accomplished by a small slit of width d in a wall of infinitesimal thickness. The coupling network is a simple shunt inductive element whose relative susceptance is given by (cf. *Waveguide Handbook*)

$$\frac{B}{Y_0'(r_2)} = \frac{\lambda}{\pi r_2}\left(\frac{4r_2}{d}\right)^2. \qquad (71a)$$

The effective ratio of the characteristic admittance Y_g of the rectangular guide to the characteristic admittance $Y_0'(r_2)$ of the radial guide at r_2 is

$$\frac{Y_g}{Y_0'(r_2)} = \frac{\lambda}{\lambda_g}\frac{4\pi r_2}{a}, \qquad (71b)$$

where the wavelength λ_g in the rectangular guide of width a is related to the free-space wavelength λ by

$$\lambda_g = \frac{\lambda}{\sqrt{1 - \left(\frac{\lambda}{2a}\right)^2}}.$$

With the knowledge of the pertinent circuit parameters as contained in Eqs. (65) and (71), it is now possible to compute in a straightforward transmission-line manner the total admittance at some arbitrary reference point and thence the desired resonant properties of the system. However, rather than proceed in this manner there can be employed a simpler and more convenient perturbation method of computation that is based on the smallness of the impedance coupled into the cavity system at $r = r_2$. If this impedance were identically zero, the total admittance at $r = r_1$ would be zero and correspondingly the resonant wavelength would be $\lambda_0 = 1.12$ cm as computed in Sec. 8·10. The effect of the coupling is to introduce at $r = r_2$ a small impedance ΔZ_2 and thereby bring about a small relative change in the unperturbed resonant frequency of amount

$$\frac{\Delta\omega}{\omega} = \frac{\Delta k}{k} = -\frac{\Delta\lambda}{\lambda}.$$

Concomitant to these changes there is produced a change $\Delta Y_t'$ in the total relative admittance given by

$$\Delta Y_t' = k\frac{\partial Y_t'}{\partial k}\frac{\Delta k}{k} + \frac{\partial Y_t'}{\partial Z_2'}\Delta Z_2', \qquad (72)$$

which merely expresses the fact that the total admittance is a function both of the frequency and the terminating impedance at $r = r_2$. Equation (72) provides the basis for the computation of the frequency shift caused by the perturbation $\Delta Z_2'$. The imposition of the equilibrium condition [Eq. (62)] that $\Delta B_t' = 0$ leads to the desired value for $\Delta k/k$.

To perform the calculation it is first necessary to know the numerical values of the partial derivatives in Eq. (72) that indicate the rate of change of the total admittance with respect to both the frequency and the output impedance Z_2.

The total admittance at $r = r_1 + 0$ is the sum of the relative input admittance $\vec{Y'}$ of the radial line to the right of $r_1 + 0$ plus the relative admittance $\overleftarrow{Y'}$ looking to the left of $r_1 + 0$. From Eq. (67) of Sec. 8·10, the unperturbed values at $k_0 = 14.21$ in^{-1} ($\lambda_0 = 1.12$ cm) of these admittances are

$$\vec{Y'}(r_1) = -j1.39,$$
$$\overleftarrow{Y'}(r_1) = +j(0.098)(14.21) = +j1.39.$$

The partial derivative with respect to frequency of $\vec{Y'}$ for the unperturbed cavity is obtained by means of Eq. (60) and Table 8·1 as

$$k\frac{\partial \vec{Y'}}{\partial k} = j\left\{j\vec{Y'} - x(1 - \vec{Y'}^2) + x\left[\frac{J_1(x) - j\vec{Y'}J_0(x)}{J_0(y)}\right]^2\right\},$$

$$k\frac{\partial \vec{Y'}}{\partial k} = j\left\{1.39 - 0.284(1 + 1.93) + 0.284\left[\frac{0.141 - 1.39(0.980)}{0.437}\right]^2\right\}$$
$$= j2.78.$$

The corresponding frequency derivative of $\overleftarrow{Y'}$ may be computed in a similar manner, but it is simpler to differentiate $\overleftarrow{Y'}$ directly, since it is a linear function of k, and obtain

$$k\frac{\partial \overleftarrow{Y'}}{\partial k} = j(0.098)k_0 = j1.39,$$

a value identical with $\overleftarrow{Y'}$ itself as is to be expected, since $\overleftarrow{Y'}$ is effectively a lumped admittance. The frequency derivative of the total admittance at r_1 is thus, in the unperturbed case, imaginary and of the value

$$k\frac{\partial Y'_t}{\partial k} = k\left(\frac{\partial \overleftarrow{Y'}}{\partial k} + \frac{\partial \vec{Y'}}{\partial k}\right) = j4.17. \tag{73}$$

Likewise from Eq. (60) the variation of the total admittance with the output impedance at the unperturbed frequency k_0 is

$$\frac{\partial Y'_t}{\partial Z'_2} = \frac{\partial \vec{Y'}}{\partial Z'_2} = \frac{x}{y}\left[\frac{J_1(x) - j\vec{Y'}J_0(x)}{J_0(y)}\right]^2$$
$$= \frac{0.020}{0.115}\left[\frac{0.141 - (1.39)(0.098)}{0.437}\right]^2 = 1.35.$$

With the knowledge of the numerical values of the partial derivatives in Eq. (72), it is now possible to compute the frequency shift caused by the introduction of a coupled impedance whose value relative to $Z_0'(r_2)$ is

$$\Delta Z_2' = \Delta R_2' + j\Delta X_2' = \cfrac{1}{\cfrac{Y_g}{Y_0'(r_2)} - j\cfrac{B}{Y_0'(r_2)}}$$

or

$$\Delta Z_2' = \frac{1}{2.97 - j45.9} = (1.41 + j21.8)10^{-3},$$

the numerical values being obtained by use of Eqs. (71a) and (71b). The frequency perturbation necessary to maintain the equilibrium condition of vanishing total susceptance is obtained by setting the imaginary part of Eq. (72) equal to zero

$$\Delta B_t' = k\frac{\partial B_t'}{\partial k}\frac{\Delta k}{k} + \frac{\partial Y_t'}{\partial Z_2'}\Delta X_2' = 0,$$

and therefore

$$\frac{\Delta k}{k} = -\frac{\dfrac{\partial Y_t'}{\partial Z_2'}}{k\dfrac{\partial B_t'}{\partial k}}\Delta X_2' = -\frac{(1.35)(21.8)10^{-3}}{4.17} = -7.05 \times 10^{-3}. \quad (74)$$

The actual resonant frequency of the coupled system is thus 0.705 per cent less than the unperturbed resonant frequency, and the smallness of this frequency shift justifies the perturbation method of calculation.

The remaining characteristics of the system can be computed with the aid of the results obtained above. The resonant relative conductance, for example, is the real part of the change in the total admittance at $r = r_1$ due to the coupled perturbation $\Delta Z_2'$ and is given by the real part of Eq. (72) as

$$g' = \Delta g_t' = \frac{\partial Y_t'}{\partial Z_2'}\Delta R_2' = (1.35)(1.41)10^{-3} = 1.90 \times 10^{-3}. \quad (75)$$

In connection with these calculations it is of importance to note that the coefficients of $\Delta X_2'$ and $\Delta R_2'$ in Eqs. (74) and (75) indicate respectively the frequency and power-pulling factors of the output load on the cavity (the electronic effects being neglected). It is a straightforward procedure to further compute the pulling factors with respect to the load in the rectangular guide.

The loaded Q of the system is given by Eq. (63) as

$$Q = \frac{1}{2g'}k\frac{\partial B_t'}{\partial k},$$

or with the insertion of the numerical values from Eqs. (73) and (75)

$$Q = \frac{1}{2(1.90)10^{-3}} \times (4.17) = 1100.$$

For the computation of the unloaded Q the effects of dissipation on the metallic boundaries of the system must be taken into account. The modification required in the over-all equivalent-circuit—transmission-line picture consists in the introduction of resistive elements into the circuit parameters and in the introduction of a complex propagation constant k and characteristic admittance (with variable imaginary parts) to describe the radial transmission lines. These effects, as well as finite-thickness effects, will not be taken into account in this chapter.

CHAPTER 9

WAVEGUIDE JUNCTIONS WITH SEVERAL ARMS

By C. G. Montgomery and R. H. Dicke

Low-frequency communication circuits usually consist of networks composed of interconnected single-terminal-pair circuit elements such as resistors, condensers, and inductors. Occasionally, circuit elements with two terminal pairs, such as transformers, are used. In microwave circuits, waveguide junctions with several arms are often employed. This is necessary because lengths of line long compared with the wavelength must be used to connect different components together. The simple connection of a number of components in series or in shunt is not possible because of the finite size of the components. Thus there must be a length of transmission line associated with each component, and the individual transmission lines must be connected together at a microwave junction.

The description of the behavior of a junction increases in complexity as the number of arms increases. A convenient classification of junctions may be made on this basis. Thus it is convenient to speak of a junction with four interconnecting arms as a member of a class that might be called "fourth-order junctions" or "four-junctions." In low-frequency terminology, such a junction would be termed a network or transducer with four pairs of terminals. As usual, only linear passive lossless junctions are to be considered.

T-JUNCTIONS

If a waveguide junction has three arms, it will be designated as a T-junction. Such a junction is completely characterized by a matrix of the third order containing six independent elements.

9.1. General Theorems about T-junctions.—Three fundamental statements that are simple and useful may be made about a T-junction. By the arguments of Chap. 5, the behavior of the T is identical with the behavior of an equivalent circuit. Circuits that contain the required number of independent parameters and are therefore suitable are discussed in Sec. 4·11. From the properties of these circuits the general theorems can be proved.

Theorem I.—It is always possible to place a short circuit in one arm of a T-junction in such a position that there is no transmission of power between the other two arms. The proof is simple. In Sec. 4·12 the expression was derived for the resulting elements of the impedance matrix when a load is placed on one terminal pair of a network. If the short circuit is placed in arm (3), a load of impedance Z_3 is placed across the terminals and the impedance matrix of the resulting two-arm junction is given by

$$Z = \begin{pmatrix} Z_{11} - \dfrac{Z_{13}^2}{Z_{33} + Z_3} & Z_{12} - \dfrac{Z_{13}^2}{Z_{33} + Z_3} \\ Z_{12} - \dfrac{Z_{13}^2}{Z_{33} + Z_3} & Z_{22} - \dfrac{Z_{13}^2}{Z_{33} + Z_3} \end{pmatrix}. \quad (1)$$

The plunger stops all power transmission if the mutual element of this matrix vanishes, or if

$$Z_{12} = \frac{Z_{13}^2}{Z_{33} + Z_3}. \quad (2)$$

Since Z_3 can take any value from $-\infty$ to $+\infty$, this equation can always be satisfied, and the theorem is proved. The input impedances of arms (1) and (2) under these conditions are given by

$$\left. \begin{aligned} Z_{\text{in}}^{(1)} &= Z_{11} - \frac{Z_{12} Z_{13}}{Z_{23}}, \\ Z_{\text{in}}^{(2)} &= Z_{22} - \frac{Z_{12} Z_{13}}{Z_{23}}. \end{aligned} \right\} \quad (3)$$

An even simpler proof of this theorem may be given by means of the circuits of Fig. 4·35. In the series circuit of Fig. 4·35, if the distance of the short circuit from the terminal plane is such that the total line length from the short circuit to the terminals of the transformer is an odd number of quarter wavelengths, the circuit is open at that point and no transmission occurs. In the shunt case, the transformer must be short-circuited to prevent transmission.

The determination of the position of the short circuit to stop transmission is a convenient method of determining experimentally the parametric lengths of the lines in Fig. 4·35. The measurement can be accurately made, and the indication is positive. If a small amount of loss is present, either in the junction or in the movable short-circuiting plunger, the power transmission does not become exactly zero but passes through a minimum at the proper position.

Theorem II.—If the T-junction is symmetrical about arm (3), a second theorem is true. A short circuit in the arm of symmetry can be so placed that transmission between the other two arms is possible without reflection. A three-junction with a short circuit in one arm is equivalent to a transmission line whose characteristic impedance can be found from Eq. (1) if the subscripts 1 and 2 are made equal. The value is

GENERAL THEOREMS ABOUT T-JUNCTIONS

$$Z_0^2 = \left(Z_{11} - \frac{Z_{13}^2}{Z_{33} + Z_3}\right)^2 - \left(Z_{12} - \frac{Z_{13}^2}{Z_{33} + Z_3}\right)^2. \quad (4)$$

Since Z_3 may have any value either positive or negative, Z_0 may also have any value, and in particular Z_0 may be unity. This demonstrates the second theorem.

The theorem may also be proved from the equivalent circuits. For a symmetrical T the circuits of Fig. 4·35 reduce to those shown in Fig. 9·1. Perfect transmission from terminals (1) to (2) results if the proper impedance is placed on arm (3) to resonate the impedance Z. The cir-

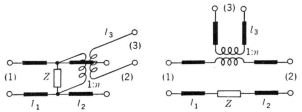

FIG. 9·1.—Shunt and series equivalents for a symmetrical T-junction.

cuit then reduces to a transmission line with a characteristic impedance of unity and a length $l_1 + l_2$.

Theorem III.—A third theorem applicable to a general T-junction is that it is impossible to match such a junction completely. A junction is said to be completely matched if the input impedance at any arm is unity when matched loads are connected to the other two arms. The proof of this theorem follows most easily from the unitary character of the scattering matrix. If a matched junction is possible, the scattering matrix must have diagonal elements equal to zero; thus

$$\mathsf{S} = \begin{pmatrix} 0 & S_{12} & S_{13} \\ S_{12} & 0 & S_{23} \\ S_{13} & S_{23} & 0 \end{pmatrix}. \quad (5)$$

Since S is unitary, the off-diagonal elements of SS^* are zero. From the (1,2) element

$$S_{11}S_{12}^* + S_{12}S_{22}^* + S_{13}S_{32}^* = 0$$

or
$$S_{13}S_{23}^* = 0. \quad (6)$$

Thus either S_{13} or S_{23} must be zero. However, neither S_{13} nor S_{23} can vanish if the diagonal elements of $\mathsf{S}^*\mathsf{S}$ are to be unity. Thus the diagonal elements of S may not all vanish.

It can also be seen, by the use of similar arguments, that any two of the diagonal elements of S can vanish only if the third arm of the junction is completely decoupled from the junction.

These three theorems are often useful in understanding the operation of microwave devices, since the theorems are completely general. Individual varieties of T-junctions often have other special properties which are best described by the equivalent circuit.

9·2. The Choice of an Equivalent Circuit. Transformation of Reference Planes.—Although any one of a number of forms of circuit may be used to represent a junction, certain circuits are particularly suitable for a given junction. To establish a criterion of suitability is difficult. Perhaps the choice could be made on the ease with which the numerical computation can be performed for a particular application of the junction. The simplicity of the computations depends upon the application for which the junction is intended as well as upon the nature of the junction. For example, since the lengths of transmission line of the circuits of Fig. 4·35 are easily determined experimentally, it might well be argued that the terminal planes of the junction should be chosen so as to make the line lengths all zero. The circuits would reduce to simple series and shunt circuits whose behavior is easy to visualize. This criterion does not offer any basis for choice between the shunt and series circuits; both are true equivalent circuits. For a given junction, however, the positions of the terminal or reference planes are different for the two circuits. Thus an additional criterion is suggested. If the reference planes fall close to a plane in the junction that has some physical significance, the plane of the junction of the center conductors of a coaxial-line T, for example, then this choice of reference planes seems a natural one. How this may be applied to a particular junction is illustrated in the examples to be discussed.

A second method for choosing a correct equivalent circuit depends on the manner of the variation of the elements of the circuit with frequency. Thus if a positive reactance occurs in an equivalent circuit, it would be satisfying if the variation with frequency were just that of an inductance varying as ωL. It is somewhat more likely that the reactance would be proportional to $2\pi c/\lambda_g$ rather than $2\pi c/\lambda$, since relative values of the reactance are usually employed. Whether or not this criterion can be applied to all cases is not known. If it were necessary to resort to a very complex equivalent circuit in order to have the amount of the frequency dependence of all the circuit elements correct, then the usefulness of the equivalent circuit might be largely destroyed. A simpler circuit with few elements is valuable, even if the dependence on frequency is wrong. An examination of special cases will aid in the understanding of this point.

Since transformations from one set of terminal planes to another are often made, it is desirable to give the general relations that express the change in the impedance-matrix elements. A change in the position of a terminal plane can be regarded as the addition of a length of transmission line to the equivalent circuit. The length may be, of course, either positive or negative, since a line of length $-l$ is equivalent to one of $(\lambda_g/2) - l$ in so far as impedance calculations are concerned. If the phases of the currents are also to be preserved, the equivalent length is $n\lambda_g - l$, where n is an integer. Since the calculations are easy, a somewhat more general case will be considered. Instead of a line, suppose that a general two-terminal-pair network is added to the original junction.

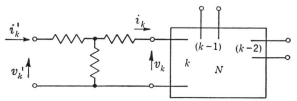

Fig. 9·2.—The addition of a T-network to a general network N.

In Fig. 9·2 is shown a network N to which has been added a T-network on the kth terminals. The current i_k and the voltage v_k at the original network are to be eliminated and replaced by the new voltage v'_k and current i'_k. The new and old values are related by

$$\left. \begin{array}{l} v'_k = z_{11} i'_k - z_{12} i_k, \\ v_k = z_{12} i'_k - z_{22} i_k. \end{array} \right\} \quad (7)$$

The negative signs result from the convention that positive currents flow into the network at the positive terminal of each pair. The network N is described by the set of equations

$$\left. \begin{array}{l} v_1 = Z_{11} i_1 + \cdots + Z_{1k} i_k + \cdots, \\ \cdots \\ v_k = Z_{1k} i_1 + \cdots + Z_{kk} i_k + \cdots. \\ \cdots \end{array} \right\} \quad (8)$$

It is a simple matter of algebra to eliminate v_k and i_k from Eqs. (7) and (8). The resulting set of equations will have coefficients that may be denoted by primes. The values are

$$Z'_{ij} = Z_{ij} - \frac{Z_{ik} Z_{jk}}{Z_{kk} + z_{22}}, \quad \text{if } i,j \neq k, \quad (9)$$

$$Z'_{ik} = Z_{ik} \frac{z_{12}}{Z_{kk} + z_{22}} = Z'_{ki}, \quad (10)$$

$$Z'_{kk} = \frac{Z_{kk} z_{11}}{Z_{kk} + z_{22}} + \frac{z_{11} z_{22} - z_{12}^2}{Z_{kk} + z_{22}}. \quad (11)$$

If the T-network that was added is a length of transmission line of unity characteristic impedance and length l, the impedance elements are

$$z_{11} = z_{22} = -j \cot \frac{2\pi l}{\lambda_g},$$
$$z_{12} = -j \csc \frac{2\pi l}{\lambda_g}, \quad (12)$$
$$z_{11}z_{22} - z_{12}^2 = 1.$$

By successive applications of this transformation, lines may be added to all the terminals of the network N.

9·3. The E-plane T-junction at Long Wavelengths.—The properties of branched transmission lines were discussed in Chap. 6 under the simple assumption that the lines behaved as though they were connected in series. This approximation is valid for low-frequency transmission lines

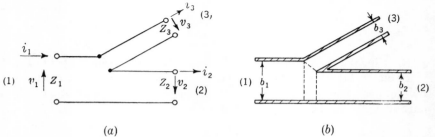

(a) (b)

FIG. 9·3.—A series junction. (a) Equivalent circuit of an E-plane T-junction; (b) cross-sectional view of an E-plane T-junction.

of all forms. As the frequency increases, the properties of the junction become more complicated and depend upon the shape of the junction and the kind of transmission line. Figure 9·3 illustrates diagrammatically a series junction and defines the convention of positive directions of the voltages and the currents flowing into the junction. The impedance and admittance matrices of this system both contain infinite elements and are consequently of little use. The linear equations that relate the voltages and currents are

$$\begin{matrix} v_1 + v_2 + v_3 = 0, \\ i_1 = i_2 = i_3. \end{matrix} \quad (13)$$

A rectangular waveguide, operated in the dominant mode, that has a branch waveguide joining it on the broad face behaves, at long wavelengths, as a pure series junction of three waveguides. Such a junction is called an E-plane T-junction, since the change in structure at the branch occurs in the plane of the electric field. Figure 9·3b shows a cross section of a waveguide with an E-plane branch. The equivalent circuit of this E-plane junction is shown in Fig. 9·3a. The proper choice of the characteristic impedance of the three lines is immediately obvious. At long wavelengths the integral of the electric field taken around the path shown by the dotted lines is zero;

$$b_1 E_1 + b_2 E_2 + b_3 E_3 = 0. \quad (14)$$

Thus if the voltages are taken proportional to the heights of the waveguides, Eq. (14) is equivalent to the first of Eqs. (12). Since the currents in the three lines are equal, the characteristic impedance of each line must be chosen proportional to b. In Fig. 9·3a the characteristic impedances are denoted by Z_1, Z_2, Z_3.

Although the impedance matrix of this system does not exist, the scattering matrix does. By the application of elementary circuit theory, the scattering matrix S may be shown to be

$$S = \begin{Bmatrix} \dfrac{-Z_1 + Z_2 + Z_3}{Z_1 + Z_2 + Z_3} & \dfrac{Z\sqrt{Z_1 Z_2}}{Z_1 + Z_2 + Z_3} & \dfrac{Z\sqrt{Z_1 Z_3}}{Z_1 + Z_2 + Z_3} \\ \cdots & \dfrac{Z_1 - Z_2 + Z_3}{Z_1 + Z_2 + Z_3} & \dfrac{Z\sqrt{Z_2 Z_3}}{Z_1 + Z_2 + Z_3} \\ \cdots & \cdots & \dfrac{Z_1 + Z_2 - Z_3}{Z_1 + Z_2 + Z_3} \end{Bmatrix}. \quad (15)$$

This equation as well as Eqs. (12) is independent of the angle of the branched line with respect to the main line.

As the frequency is increased, the line integral of the electric field along the path in Fig. 9·3b begins to have an appreciable value, and the behavior of the junction is no longer represented by a pure series circuit. To investigate the proper equivalent circuit it is useful to restrict the branching angle to 90° and further to impose the conditions that

$$b_3 \ll b_1, \qquad b_1 = b_2.$$

To simplify the notation, it is convenient to write

$$b_1 = b_2 = b, \qquad b_3 = b'.$$

If a wave is transmitted down the branch line, it is reflected at the junction with a large reflection coefficient. For small values of b', the reflection coefficient is given by S_{33} of Eq. (15):

$$S_{33} = \frac{2b - b'}{2b + b'}.$$

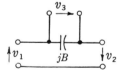

Fig. 9·4.—Equivalent circuit of an E-plane T-junction with junction susceptance.

If the lines (2) and (3) are terminated in open circuits, the reflection coefficient is unity. At high frequencies the field configuration in the neighborhood of the junction must be similar to that produced by the symmetrical change in height of a waveguide described in Sec. 6·16. The electric field is distorted near the branch line and concentrated to produce an excess storage of electrical energy. Therefore a capacitive susceptance must appear at the junction. The equivalent circuit becomes that shown in Fig. 9·4.

The value of the susceptance B is

$$B = \frac{2b}{\lambda_g}\left(1 + \ln \frac{b}{2b'}\right), \tag{16}$$

relative to the characteristic admittance of the main waveguide. Equation (16) is valid if both b/λ_g and b'/b are small compared with unity. This value B may be compared with the junction susceptance B' at a step change of height in a waveguide. This junction susceptance was found to be

$$B' = \frac{2b}{\lambda_g}\left(1 + \ln \frac{b}{4b'}\right), \tag{17}$$

which is valid under the same conditions.

It is interesting to compare these two junction effects by calculating the input admittance of the waveguide structure shown in Fig. 9·5. This waveguide termination can be regarded as a change in height of the guide and a short circuit a distance b away. The input admittance is therefore

$$Y'_{in} = jB' - j\frac{b'}{b}\cot\frac{2\pi b}{\lambda_g}. \tag{18}$$

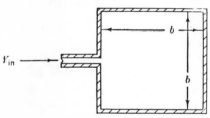

Fig. 9·5.—Structure for comparing the junction effects of a change in cross section and of an E-plane T.

The structure may also be regarded as an E-plane T-junction with a short circuit in each of the two main arms a distance $b/2$ away from the branch. The input admittance on this basis is

$$Y_{in} = jB - j\frac{1}{2}\frac{b'}{b}\cot\frac{\pi b}{\lambda_g}. \tag{19}$$

Since $b/\lambda_g \ll 1$, the difference between Eqs. (18) and (19) may be shown to be

$$Y_{in} - Y'_{in} = j\frac{2b'}{\lambda_g}\left(\ln 2 - \frac{\pi}{4}\right) = j0.09\frac{2b'}{\lambda_g}. \tag{20}$$

Thus the two methods of calculation give the same results to a good approximation.

This comparison is a critical test of the reliability of both formulas, since the short circuits are placed close to the junction ($b \ll \lambda_g$) and consequently interaction effects are to be expected.

It is often convenient to transform the equivalent circuit of Fig.

9·4 to a new reference plane in the branch arm. Since the impedance matrix is infinite, the general form of the transformation given in Sec. 9·2 cannot be used. It is easy, however, to proceed from first principles. The junction equations are

$$v_1 + v_2 + v_3 = 0, \\ i_1 = i_2 = i_3 - jBv_3. \quad (21)$$

If a Π-network is connected to arm (3), the new currents and voltages are given by

$$i'_3 = y_{11}v'_3 - y_{12}v_3, \\ i'_3 = y_{12}v_3 - y_{22}v_3.$$

The network relations become

$$v_1 + v_2 + \frac{y_{11}}{y_{12}}v'_3 + \frac{i'_3}{y_{12}} = 0, \\ i_1 = i_2 = i'_3\left(\frac{y_{22}}{y_{12}} + \frac{jB}{y_{12}}\right) + v'_3\left(\frac{y_{12}^2 - y_{11}y_{22}}{y_{12}} - jB\frac{y_{11}}{y_{12}}\right). \quad (22)$$

If the added line is short,

$$y_{11} = y_{22} = \frac{y_0}{j\beta l} = y_{12}.$$

The junction becomes a pure series junction if the coefficient of v'_3 in the second of Eqs. (22) vanishes or if

$$\beta l = -\frac{B}{y_0}. \quad (23)$$

For this length, Eqs. (22) reduce to Eqs. (13). Since l is negative, the reference plane is within the junction as shown in Fig. 9·6.

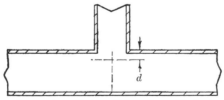

Fig. 9·6.—Reference planes of an E-plane T-junction with first-order end correction.

9·4. E-plane T-junction at High Frequencies.—At high frequencies for which the condition $b \ll \lambda_g$ is no longer fulfilled, the equivalent circuit of an E-plane junction becomes more complicated. For the symmetrical case, there must be four circuit parameters. An equivalent circuit which is convenient to use and which reduces directly to the simple series circuits for long wavelengths is shown in Fig. 9·7. The elements of the admittance matrix are indicated on the left-hand side of the figure. If the Π-network in arm (3) is replaced by the T-network that is equivalent to it, the circuit becomes that shown in Fig. 9·7b. The impedances are related to the admittances of Fig. 9·7a by the equations

$$Z_a = \frac{1}{y_{11} - y_{12}},$$
$$Z_b = \frac{y_{33} - y_{13}}{y_{12}y_{33} - y_{13}^2},$$
$$Z_c = \frac{y_{13}}{y_{12}y_{33} - y_{13}^2},$$
$$Z_d = \frac{y_{12} - y_{13}}{y_{12}y_{33} - y_{13}^2}.$$
(24)

As b' becomes small compared with b, Z_b and Z_d approach zero, Z_a becomes very large, and the only remaining element is the shunt capacitance Z_c. The circuit elements are shown as inductances or capacitances

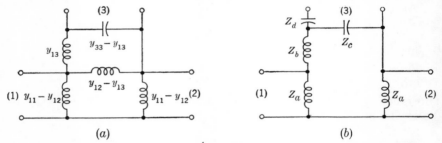

Fig. 9·7.—Exact equivalent circuits for an E-plane T-junction at high frequencies.

in Fig. 9·7 according to the sign of the admittances when $b' = b$. Typical experimental values, for $b/\lambda_g = b'/\lambda_g = 0.227$, are

$$\begin{aligned} Z_a &= -j10.4, \\ Z_b &= j0.50, \\ Z_c &= -j4.85, \\ Z_d &= -j0.57. \end{aligned}$$
(25)

Additional values of the circuit parameters are to be found in Vol. 10 of this series.

The circuits of Fig. 9·1 are, of course, equally valid representations of the symmetrical E-plane T-junction. The values of the circuit parameters of the series representation are

$$\frac{l_1}{\lambda_g} = \frac{l_2}{\lambda_g} = 0.014, \quad \frac{l_3}{\lambda_g} = 0.028,$$
$$Z = j0.01, \quad n^2 = 0.829,$$
(26)

for $b' = b = 0.2\lambda_g$.

If the three arms of the T-junction have equal heights and branch at equal angles of 120°, the junction possesses a higher degree of symmetry than the 90° junction, and the number of circuit parameters necessary to describe the behavior is reduced to two. The impedance matrix is of the form

$$Z = \begin{pmatrix} Z_{11} & Z_{12} & Z_{12} \\ Z_{12} & Z_{11} & Z_{12} \\ Z_{12} & Z_{12} & Z_{11} \end{pmatrix}. \quad (27)$$

A circuit that exhibits the same symmetry as this junction is shown in Fig. 9·8. The elements Z_1 and Z_2 are related to the elements of Z by the equations

$$Z_{11} = \frac{Z_1(2Z_1 + 3Z_2)}{3(Z_1 + Z_2)}, \quad Z_{12} = \frac{Z_1^2}{3(Z_1 + Z_2)},$$
$$Z_1 = Z_{11} + Z_{12}, \quad Z_2 = \frac{(Z_{11} + Z_{12})(Z_{11} - 2Z_{12})}{3Z_{12}}. \quad \Big\} \quad (28)$$

Another special case of considerable interest is the 180° E-plane junction or the bifurcation of a waveguide in the E-plane. Such a junction is shown schematically in Fig. 9·9a. If the dividing wall has negligible thickness, certain special properties are manifest. The equivalent circuit of Fig. 9·7b may be drawn as in Fig. 9·9b for the reference planes indicated in Fig. 9·9a. Since the reference planes of the three arms coincide, the voltage $v_3' = v_1 + v_2$. The T-network in Fig. 9·9b must reduce to a shunt element, and the equivalent circuit becomes that of Fig. 9·9a. If the heights of the smaller guides are not equal, an additional element must be added to the circuit. This additional ele-

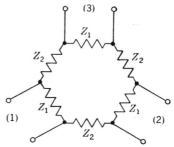

Fig. 9·8.—Equivalent circuit of a 120° E-plane junction.

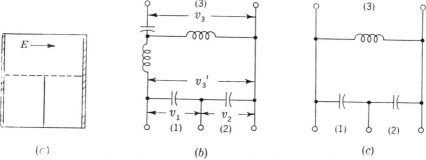

Fig. 9·9.—E-plane bifurcation of a waveguide and the equivalent circuit.

ment may be an impedance in series with the terminal common to arms (1) and (2).

Another equivalent circuit which may be useful in some situations is one that contains a three-winding ideal transformer. This circuit is shown in Fig. 9·10. The six parameters are the three line lengths, two turn ratios of the transformer, and Z. The circuit equations are

$$\left.\begin{array}{l} v_1 + v_2 + v_3 = 0, \\ n_1 i_1 + n_2 i_2 - n_3 i_3 = 0, \\ i_1 - i_2 = \dfrac{1}{Z}\left(\dfrac{n_1 n_2}{n_3}\right)\left(\dfrac{v_1}{n_1} - \dfrac{v_2}{n_2}\right), \end{array}\right\} \qquad (29)$$

when the positive directions for the currents and voltages are those shown on the figure.

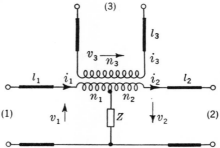

Fig. 9·10.—Equivalent circuit of an E-plane T-junction with a three-winding transformer.

The series impedance Z in Fig. 9·1b may also be placed in shunt with the transformer. If the characteristic impedance of line (3) is chosen as $n^2 b'/b$, the circuit reduces to that shown in Fig. 9·4 at the proper reference planes in the three lines.

9·5. H-plane T-junctions.—Junctions with three arms in which the branching takes place in the H-plane may be discussed in a similar fashion to E-plane junctions. For very long wavelengths, the junction is a pure

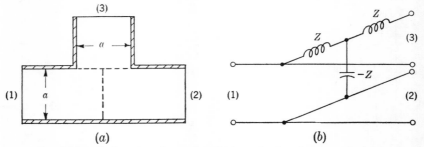

Fig. 9·11.—H-plane junction and equivalent circuit at long wavelengths.

shunt junction. The coupling from the main waveguide to the branch guide is by means of the magnetic fields. If the branch arm is at 90° to the main line, the coupling will be from the longitudinal magnetic field in the main line to the transverse magnetic field in the branch line. Since these fields are in quadrature with respect to the transverse electric field, the junction behaves as a pure shunt junction only if a quarter-wavelength line is inserted between the branch line and the main line. Figure 9·11a shows the junction, with the reference plane indicated. The equivalent circuit is shown in Fig. 9·11b. The proper ratio of impedances of the main line and the quarter-wavelength line is the ratio of the transverse magnetic field to the longitudinal magnetic field. The characteristic impedance of the quarter-wavelength

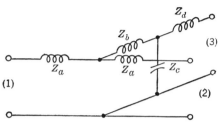

Fig. 9·12.—Exact equivalent circuit for an H-plane T-junction.

line relative to that of lines (1) and (2) is therefore $2a/\lambda_g$. As the frequency is increased, additional circuit elements must be employed to express the effect of higher modes at the junction, and the circuit becomes that shown in Fig. 9·12. The values of the elements on the side arm will depart from those of a quarter-wavelength line. Experimental values of these parameters for $\lambda = 3.20$ cm and $a = 0.902$ in. are

$$Z_a = j0.17, \quad Z_b = j0.19, \\ Z_c = -j1.04, \quad Z_d = j1.00. \quad (30)$$

The value of $2a/\lambda_g$ is 1.002 for these conditions.

9·6. A Coaxial-line T-junction.—Stubs or T-junctions in coaxial line have equivalent circuits that are similar to those of waveguide junctions.

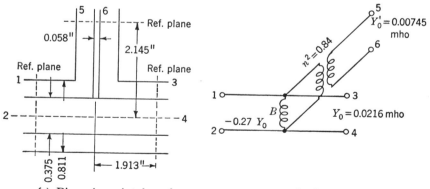

(a) Dimensions of stub used.   (b) Circuit.

Fig. 9·13.—Equivalent circuit of a coaxial T-junction.

Since the coaxial-line junction is physically like a shunt junction, it is natural to employ a shunt equivalent circuit. The coupling is magnetic, and no phase shift between the branch line and the main line would be expected for low-frequency waves. At high frequencies a junction effect becomes important, which may be represented most suitably as shown in Fig. 9·1 or 9·12. As an example of the circuit of a coaxial-line T-junction, parameters measured[1] at a wavelength of 10 cm are presented in Fig. 9·13. As expected, the reference planes for which the circuit has a simple form are close to the junction of the inner conductors. The reference planes in the main line overlap, as for the *H*-plane T. The junction effect is represented by the shunt inductance and the ideal transformer, and the effects of these elements are not large.

9·7. The T-junction with a Small Hole.—The theory of diffraction by a small hole, discussed in Chap. 6, may be applied to two waveguides joined together to form a T-junction whose arms are connected by a small hole. It will be recalled that the theory proceeds from the assumption that the field in the hole depends on the incident field only. Radiation takes place from the hole into waveguide or into free space as though the hole were an electric or magnetic dipole or both, depending on whether the incident field has an electric or magnetic component. For the irises discussed in Chap. 6, a large reflection is produced by the wall containing the hole and a small amount of power is transmitted through the hole. If a hole is located in the side wall of a waveguide as shown in Fig. 9 14, power incident upon it from guide (3) is almost totally reflected, but some power leaks through the guides (1) and (2). Guide (3) appears to be terminated in a large susceptance given by

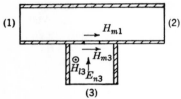

FIG. 9·14.—T-junction coupled with a small hole.

$$\frac{1}{B} = \frac{2\pi}{S\lambda}(PE_n^2 - M_1 H_l^2 - M_2 H_m^2),$$

where E_n is the electric field in guide (3) normal to the hole, H_l and H_m are the tangential magnetic fields in the directions of the principal axes of the hole, P is the electric polarizability, and M_1 and M_2 are the two magnetic polarizabilities. The quantity S is a normalizing factor that is equal to $\frac{1}{2}(\lambda_g/\lambda)ab$ for unit transverse magnetic field and for the dominant mode in rectangular waveguide. It will be recalled that this expression neglects the reaction of the load upon the matched generator, which is small if B is large.

The power leaking into guides (1) and (2) is given by similar expressions. If the amplitudes of the waves into guides (1) and (2) are denoted by A_1 and A_2, respectively, then

[1] J. R. Harrison, "Design Considerations for Directional Couplers," RL Report No. 724, Dec. 31, 1945, Fig. 56.

$$A_1 = -\frac{\pi j}{\lambda S_1}(PE_{3n}E_{1n} - M_1H_{3l}H_{1l} - M_2H_{3m}H_{1m}),$$

$$A_2 = -\frac{\pi j}{\lambda S_1}(PE_{3n}E_{1n} + M_1H_{3l}H_{1l} - M_2H_{3m}H_{1m}),$$

where the difference in sign results from the fact that the waves A_1 and A_2 are traveling in opposite directions.

If a wave is incident in guide (1) of the junction, a somewhat different situation arises. The incident field excites a field in the hole which radiates waves outward from all three arms of the T-junction. The wave from arm (3) is given by

$$A_3 = A_1 \frac{S_1}{S_3}.$$

The wave radiated away from the hole in arm (1) is a wave reflected from the hole. The amplitude is

$$B_1 = \frac{-j\pi}{\lambda S_1}(PR_1^2 + M_1H_{l1}^2 - M_2H_{m1}^2).$$

The wave radiated away from the hole in arm (2) produces only a small change in phase in the incident wave. The amplitude is

$$A_2 = -\frac{j\pi}{\lambda S_1}(PE_1^2 - M_1H_{l1}^2 - M_2H_{m1}^2).$$

The hole in the side of guide (1) is therefore equivalent to a two-arm junction with a scattering matrix having the elements

$$S_{12} = 1 + A_2, \qquad S_{11} = B_1.$$

The scattering matrix may be replaced by an equivalent circuit in the usual way. If a T-network is chosen for the circuit representation, the elements are

$$Z_{12} = \frac{2(1 + A_2)}{(1 - B_1)^2 - (1 + A_2)^2},$$

$$Z_{11} - Z_{12} = \frac{A_2^2 - B_1^2}{(1 - B_1)^2 - (1 + A_2)^2}.$$

Since $|A_2|$ and $|B_1|$ are small compared with unity, Z_{12} is large and $Z_{11} - Z_{12}$ is small.

If the T-junction of Fig. 9·14 represents an H-plane junction, then

$$B_1 = \frac{j2\pi}{\lambda_g ab}\left(\frac{\lambda_g}{2a}\right)^2 M_2 = A_2,$$

$$Z_{11} - Z_{12} = 0,$$

$$Z_{12} = j\frac{a^2 b}{\pi M_2}\left(\frac{a}{\lambda_g}\right).$$

WAVEGUIDE JUNCTIONS WITH FOUR ARMS

Junctions in waveguide which have four arms have many special properties that have proved to be particularly useful in microwave circuits. Such junctions are related to the familiar bridge circuits at low frequencies. These properites are fundamentally the result of the high degree of symmetry that a four-junction may possess.

9·8. The Equivalent Circuit of a Four-junction.—It is easy to find many equivalent circuits for a junction with four arms by proceding

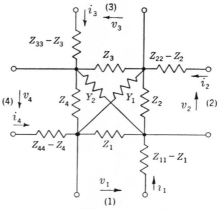

Fig. 9·15.—Equivalent circuit of a four-junction.

according to the general methods outlined in Chap. 4. One equivalent circuit is shown in Fig. 4·37 and consists of six transmission lines connected in shunt with four shunt admittances. A more useful form results from the application of the general methods. Figure 4·38 illustrates a four-terminal-pair network with a core consisting of a four-terminal network. If the four-terminal network of Fig. 4·34 is used for the core, a general form for the equivalent circuit of a four-junction results. The circuit is shown in Fig. 9·15. The number of independent circuit parameters is 10—the number necessary to determine the impedance matrix completely. For the voltages and currents defined in the figure, the impedance matrix can be found from elementary principles, and the result is

$$Z = \begin{pmatrix} Z_{11} & Z_1Z_2Y_2 & Z_1Z_3(Y_1+Y_2) & Z_1Z_4Y_1 \\ Z_1Z_2Y_2 & Z_{22} & Z_2Z_3Y_1 & Z_2Z_4(Y_1+Y_2) \\ Z_1Z_3(Y_1+Y_2) & Z_2Z_3Y_1 & Z_{33} & Z_3Z_4Y_2 \\ Z_1Z_4Y_1 & Z_2Z_4(Y_1+Y_2) & Z_3Z_4Y_2 & Z_{44} \end{pmatrix} \quad (31)$$

where Y_1 and Y_2 are the admittances of the crossed circuit elements.

From the form of the circuit or from the impedance matrix, certain properties are evident. (1) With respect to a given arm, there is one opposite arm and two adjacent arms. (2) The mutual impedance between the opposite arms (1) and (3) can be made to vanish either by making Z_1 or Z_3 equal to zero, a trivial case, or by making $Y_1 + Y_2 = 0$. For this condition, however, the mutual impedance between arms (2) and (4) also vanishes. (3) If the mutual impedance vanishes between two adjacent arms, for example, (2) and (3), then $Y_1 = 0$, and the mutual impedance between the other pair of adjacent arms (1) and (4) also vanishes if the trivial cases are neglected.

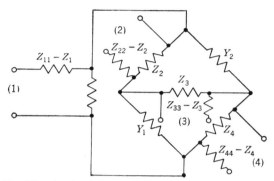

Fig. 9·16.—The circuit of the four-junction of Fig. 9·15 redrawn.

The analogy of the circuit of Fig. 9·15 to that of a Wheatstone bridge may be made obvious by rearranging the components as shown in Fig. 9·16. The choice of arm (1) to represent the "battery" of the bridge is, of course, arbitrary. The circuit could have been drawn with any arm in this position. The relation that is maintained is that arms (1) and (3) are opposite arms, as are arms (2) and (4). It would be possible to proceed from this equivalent circuit to develop the interesting properties of four-arm junctions. The method that will be adopted, however, is to consider the scattering of waves falling upon the junction and to derive the behavior in these terms.

9·9. Directional Couplers.—It is evident from the circuit of Fig. 9·16 that by altering the impedances connected to arms (2) and (4) it is possible to balance the bridge. The impedances on arms (2) and (4) may consist of matched loads together with some reflecting irises. In a similar fashion, if proper irises and matched loads are connected to arms (1) and (3), no power is coupled between these two arms. The waveguide junction may now be extended to include the irises. Furthermore it is possible to choose the irises in such a way that no reflections are produced when waves are incident in arms (1) or (2). Under these conditions the device is called a directional coupler.

A directional coupler is defined as junction of four transmission lines (1), (2), (3), and (4) such that with all lines terminated in their characteristic impedances, the terminals (1) and (2) are matched and there is no

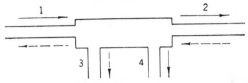

FIG. 9·17.—A directional coupler.

coupling between (1) and (3) and (2) and (4). The suitability of the name can be seen from Fig. 9·17. A wave incident on the junction in line (1) leaves by lines (2) and (4). A wave incident in line (2) leaves

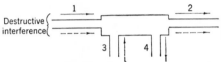

FIG. 9·18.—Directional coupler with two incident waves, first case.

by lines (3) and (1). Thus the powers absorbed by matched loads on arms (3) and (4) are indicative of the powers traversing the junction in lines (1) and (2) in the two directions. The ratio of the power emerging from line (4) to that incident in line (1) is called the coupling coefficient of the directional coupler.

A directional coupler has several interesting properties. One of these is that all the terminals are matched. The waves indicated by dotted lines in Fig. 9·18 may be reversed in time and combined with the waves indicated by solid arrows. If the amplitudes and phases of the two waves relative to each other are properly adjusted, the two waves entering the junction at arm (1) can be made to cancel each other. All that remains is a wave entering the junction at arm (3) and leaving it at arms (2) and (4). Thus a wave incident on the junction at arm (3) is transmitted to arms (2) and (4) without reflection, and the terminal (3) is matched. In a similar way it can be shown that terminal (4) is matched. Thus all directional couplers are completely matched.

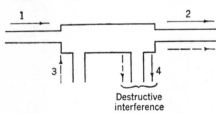

FIG. 9·19.—Directional coupler with two incident waves, second case.

Another theorem of importance is that a junction such that two noncoupling terminals are matched is a directional coupler. It may be assumed that terminals (1) and (3) of Fig. 9·19 do not couple with each other and are matched. Waves can fall upon the junction in lines (1) and (3) as shown by the solid and dotted arrows. If the phase and amplitude of one of these waves are adjusted with respect to those of the other wave, the two waves in line (4) can be made to cancel each other. A reversal of time causes a wave to enter line (2) and leave at lines (1) and (3). This is the same situation which is indicated by the dotted arrows in Fig. 9·17, and therefore the junction is a directional coupler.

9·10. The Scattering Matrix of a Directional Coupler.—The scattering matrix of a directional coupler is of the form

$$\mathbf{S} = \begin{pmatrix} 0 & S_{12} & 0 & S_{14} \\ S_{21} & 0 & S_{23} & 0 \\ \hline 0 & S_{32} & 0 & S_{34} \\ S_{41} & 0 & S_{43} & 0 \end{pmatrix}. \quad (32)$$

The zero elements on the diagonal indicate that the junction is completely matched, and the remainder of the zero elements indicate that there is zero coupling between arms (1) and (3) and also between arms (2) and (4). The remainder of the elements of **S** are not completely independent but must be such as to make the matrix symmetrical and unitary. The conditions thus imposed are that

and
$$\left. \begin{array}{l} S_{jk} = S_{kj}, \\ |S_{12}|^2 + |S_{14}|^2 = 1, \end{array} \right\} \quad (33)$$

$$\left. \begin{array}{l} S_{21}S_{23}^* + S_{41}S_{43}^* = 0, \\ S_{12}S_{14}^* + S_{32}S_{34}^* = 0. \end{array} \right\} \quad (34)$$

From Eqs. (33) and (34),

$$\left. \begin{array}{l} |S_{12}||S_{23}| = |S_{14}||S_{34}|, \\ |S_{12}||S_{14}| = |S_{23}||S_{34}|, \end{array} \right\} \quad (35)$$

and hence

$$\left. \begin{array}{l} |S_{12}| = |S_{34}|, \\ |S_{14}| = |S_{23}|. \end{array} \right\} \quad (36)$$

Equations (36) state that the coupling from arm (1) to arm (2) is equal to that from arm (3) to arm (4) and also the coupling from arm (1) to arm (4) is equal to that from arm (2) to arm (3). Thus a wave incident in arm (1) couples the same fraction of its power into arm (4) that a wave in arm (2) couples into arm (3).

There is still a great deal of arbitrariness in the phases of the elements of **S**. This indeterminateness can be eliminated by the correct choice of the locations of the reference planes. As an example, the location of the terminal plane (2) may be chosen in such a way that S_{12} is real and positive. Similarly the location of the plane of arm (4) may be chosen in such a way that S_{14} is positive imaginary, and the location of the plane in arm (3) may be chosen so that S_{34} is positive real. From Eqs. (36),

$$S_{12} = S_{34} = \alpha, \tag{37}$$

a positive real number. From Eq. (34),

$$\alpha S_{23}^* + S_{41}\alpha = 0; \tag{38}$$

and from Eq. (38),

$$S_{23} = S_{41} = j\beta, \tag{39}$$

where β is positive real.
Thus the scattering matrix becomes

$$\mathbf{S} = \begin{pmatrix} 0 & \alpha & 0 & j\beta \\ \alpha & 0 & j\beta & 0 \\ \hline 0 & j\beta & 0 & \alpha \\ j\beta & 0 & \alpha & 0 \end{pmatrix}. \tag{40}$$

The scattering matrix (40) will be regarded as a standard form for a directional coupler. From the first of Eqs. (33),

$$\alpha^2 + \beta^2 = 1.$$

A theorem that is of considerable importance may now be proved. It has been shown previously that every directional coupler is completely matched. It will now be shown that any completely matched junction of four transmission lines is a directional coupler. If the junction is matched, the scattering matrix is

$$\mathbf{S} = \begin{pmatrix} 0 & S_{12} & S_{13} & S_{14} \\ S_{21} & 0 & S_{23} & S_{24} \\ \hline S_{31} & S_{32} & 0 & S_{34} \\ S_{41} & S_{42} & S_{43} & 0 \end{pmatrix}. \tag{41}$$

The location of the terminal planes may be chosen in such a way that S_{21} and S_{43} are pure real and positive and that S_{41} is pure imaginary. Then, since **S** is unitary,

$$\left. \begin{array}{c} S_{21}S_{23}^* - S_{41}S_{43}^* = 0, \\ -S_{21}S_{41}^* + S_{23}S_{43}^* = 0. \end{array} \right\} \tag{42}$$

If the first of Eqs. (42) is multiplied by S_{21}, the second by S_{43}, and the difference is taken,

$$(S_{21}^2 - S_{43}^2)S_{23} = 0. \tag{43}$$

Hence either S_{23} vanishes and the junction is a directional coupler as

already proved, or
$$S_{21} = S_{43} = \alpha, \tag{44}$$
where α is a real positive number.
If α is substituted in Eqs. (42),
$$S_{23} = S_{41} = j\beta, \tag{45}$$
where β is a pure real number.
If these relations are substituted in Eq. (41), S takes the form

$$\mathsf{S} = \begin{pmatrix} 0 & \alpha & S_{13} & j\beta \\ \alpha & 0 & j\beta & S_{24} \\ \hline S_{31} & j\beta & 0 & \alpha \\ j\beta & S_{42} & \alpha & 0 \end{pmatrix}. \tag{46}$$

If the first row of the matrix (46) is multiplied by the complex conjugate of the second column, and if the first column is multiplied by the complex conjugate of the fourth row, then because S is unitary,
$$\left.\begin{array}{r} -j\beta S_{13} + j\beta S_{42}^* = 0, \\ \alpha S_{31} + \alpha S_{42}^* = 0. \end{array}\right\} \tag{47}$$

If neither α nor β vanishes, Eqs. (47) imply that both S_{31} and S_{42} vanish. This puts Eq. (46) into the same form as Eq. (40), and the junction is a directional coupler. If either α or β vanishes, the junction is also a directional coupler. Thus the result has been obtained that any junction of four transmission lines which is completely matched is a directional coupler.

9·11. The Arbitrary Junction of Four Transmission Lines.—It might be thought that any junction of four transmission lines could be completely matched by a transformer in each of the four transmission lines and therefore could be made into a directional coupler. This cannot be done in general, however, as will now be shown.[1]

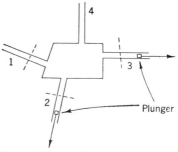

Fig. 9·20.—Arbitrary junction of four transmission lines.

Let us consider the arbitrary junction of four transmission lines shown in Fig. 9·20. If plungers are inserted in lines (2) and (3) as shown in the figure, then for any position of the plunger in arm (2) there exists a position of the plunger in line (3) such that no coupling exists between lines (1) and (4). The junction with the plunger in line (2) at some definite position is

[1] This proof is due to R. L. Kyhl.

reduced to a three-terminal-pair junction for which a plunger in one of the remaining arms can always be used to decouple the remaining two transmission lines. For such locations of the plungers, there exist standing waves in each of the transmission lines (1), (2), and (3). If now the plunger in line (2) is moved to a new location, there will again be a new

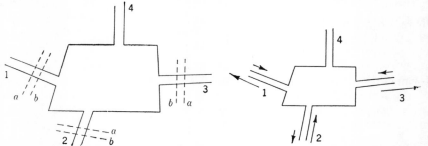

Fig. 9·21.—Positions of nodes in arms of four-junction for two positions of the plungers.

Fig. 9·22.—Linear combination of solutions.

position of the plunger in line (3) which causes line (4) to be decoupled from line (1), and there are pure standing waves in lines (1), (2), and (3). The nodes of the standing waves in line (2) must now be in a new position. However, the nodes in lines (1) and (3) may or may not be in the same position. In Fig. 9·21, the dotted lines marked a represent one position of the nodes, those marked b represent the other position.

Since these solutions represent pure standing waves, they are characterized by the fact that the electric fields are everywhere in phase or 180° out of phase, as shown in Chap. 5. If none of the positions a and b coincide, a linear combination of these two solutions, taken with different time phases, corresponds to running and standing waves in the lines (1), (2), and (3). Only if the nodes of the two solutions coincide in one of the guides will the linear combination be a pure standing wave in that guide. The linear combination of solutions is shown in Fig. 9·22. Since the waves in the guides are not pure standing waves, the amplitude of a wave running one way is different from that running the other way. This is indicated in Fig. 9·22 by the arrows of different length. It will be assumed that one long arrow points into the network and two point out. If the converse condition were obtained, a time reversal would lead to the desired condition. By the addition of matching transformers to the lines (1), (2), and (3)

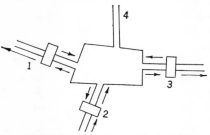

Fig. 9·23.—Four-junction with matching transformers.

the conditions of Fig. 9·23 are obtained. The transformers set up the standing-wave pattern of Fig. 9·22.

From a consideration of the terminals located on the transmission-line side of the transformers in Fig. 9·23, it is seen that the new modified junction is matched looking into line (2). Moreover, there is no coupling between lines (2) and (4). The terminals of line (4) can now be matched by the inclusion of a transformer in line (4). Thus both lines (2) and (4) are matched, and there is no coupling between them. Hence, by an earlier theorem, the junction is a directional coupler. Conversely, the junction may be considered as a perfect directional coupler with transformers that mismatch it in the four transmission lines.

Four transformers are actually not required; it is easily seen that three are sufficient. If the linear combination of the two standing-wave solu-

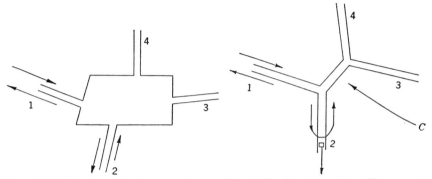

Fig. 9·24.—Behavior of a degenerate four-junction.

Fig. 9·25.—Junction of two T's connected together.

tions is taken correctly, one of the lines, line (1), for example, can be made to contain a pure running wave. Thus the transformer on line (1) is not essential, and three transformers are sufficient to match the junction of four transmission lines.

It should be remembered that the foregoing derivation hinges upon the assumption that none of the nodes a and b in Fig. 9·21 coincide. Since the plunger in line (2) is moved to a new position in going from case a to case b, it is clear that the resulting nodes also move. Hence if the nodes do coincide in one of the transmission lines, it will have to be either in line (1) or in line (3). If the nodes a and b of Fig. 9·21 coincide in line (3) and nowhere else, a linear combination of the two solutions may be taken such that the waves are completely canceled in line (3). The combination must be one with equal time phases. The resulting linear combination is one with standing waves in lines (1) and (2) only, as in Fig. 9·24. Figure 9·25 is an example of a junction in which standing waves are set up in lines (1) and (2) by a plunger in arm (2). It is to be noted that the junction of Fig. 9·25 is composed of two T-junctions connected together by a transmission line c. The plunger in arm (2) is in such a position as to decouple c from arm (1).

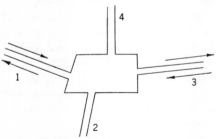

Fig. 9·26.—Behavior of a degenerate four-junction.

It may turn out that the nodal planes a and b coincide in both lines (1) and (3). This is possible only if the waves are completely absent from line (2). This implies, however, that there are pure standing waves in lines (1) and (3) with no waves in the other lines. This condition is indicated in Fig 9·26.

The procedures just described could be repeated for a wave introduced into line (4). A standing-wave solution analogous to Fig. 9·24 or 9·26 would be obtained. In each case the network is equivalent to a junction of the type shown in Fig. 9·25 (the lines need not be numbered in this order) provided all degenerate forms of Fig. 9·25, such as those shown in Figs. 9·27 and 9·28, are included.

To recapitulate, any arbitrary junction of four transmission lines can be represented either as a directional coupler with transformers in three of the lines or as a junction consisting of two interconnected T-junctions.

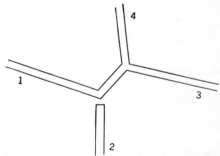

Fig. 9·27.—Equivalent form of a degenerate four-junction.

9·12. The Magic T.—A matched directional coupler with a coupling coefficient of $\frac{1}{2}$ has proved to be an extremely useful device for many microwave applications and has become known as a magic T. The low-frequency analogue of a magic T is the well-known hybrid coil used in telephone repeater circuits. Such a device is indicated in Fig. 9·29.

Among the many applications of the magic T, the more important ones are impedance bridges,[1] balanced mixers,[2] balanced duplexers,[3] and microwave discriminators.[4]

A magic T can be realized in any one of a number of forms, in wave-

---

[1] Vol. 11, Chap. 9, Radiation Laboratory Series.
[2] Vol. 16, Chap. 6, Radiation Laboratory Series.
[3] Vol. 14, Chap. 8, Radiation Laboratory Series.
[4] Vol. 11, Chap. 2, Radiation Laboratory Series.

guides or in coaxial lines. One of the simplest of these forms is the combination of an E-plane and an H-plane T-junction which is shown in Fig. 9·30. A wave incident on the junction in arm (4) has even symmetry about the symmetry plane, and the transmitted power is divided with even symmetry between arms (1) and (2). No power is coupled to arm (3), since no mode that has even symmetry can propagate in

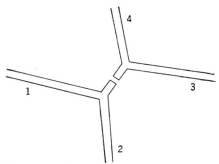

Fig. 9·28.—Another form of a degenerate four-junction.

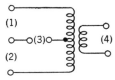

Fig. 9·29.—Circuit of a hybrid coil.

arm (3). There is also a reflected wave in arm (4). The reflected wave can be matched out, however, by adding to the junction some post or iris that does not destroy the symmetry of the junction. A wave incident in arm (3) possesses odd symmetry and therefore excites fields in arms (1) and (2) which have odd sym-

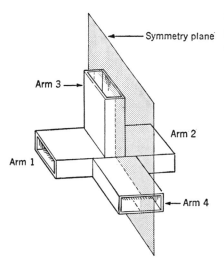

Fig. 9·30.—A magic T.

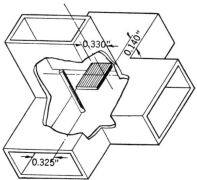

Fig. 9·31.—Positions of post and iris for matching a $\frac{1}{2}$- by 1-in. waveguide T.

metry. No power is transmitted to arm (4), since arm (4) will not support a mode with odd symmetry. To eliminate the reflected wave in arm (3), a second matching device must be added to the junction. Figure 9·31 shows one method for matching the junction. The dimensions are given for a wavelength of 3.33 cm in $\frac{1}{2}$- by 1-in. waveguide. The post is 0.125 in. in diameter, and the iris $\frac{1}{32}$ in. thick.

The argument of the preceding section can be repeated with coherent generators connected to arms (3) and (4) to show that in the matched junction, arms (1) and (2) are also decoupled. It is easy to see that the impedance matrix must have the simple form

$$Z = \frac{j}{\sqrt{2}} \begin{pmatrix} 0 & 0 & 1 & 1 \\ 0 & 0 & -1 & 1 \\ \hline 1 & -1 & 0 & 0 \\ 1 & 1 & 0 & 0 \end{pmatrix}, \qquad (48)$$

for the proper choice of reference planes. It may be shown that

$$Y = -Z = -S,$$

since $Z^2 = -I$. An equivalent circuit of a magic T is shown in Fig. 9·32.

Another form of a magic T-junction is shown in Fig. 9·33. Three of the arms are coaxial lines, and the fourth arm is rectangular waveguide.

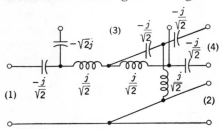

FIG. 9·32.—Equivalent circuit of a magic T.

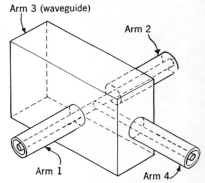

FIG. 9·33.—A magic T in coaxial line and waveguide.

The symmetry of the junction ensures that there is no coupling between arms (3) and (4). If the junction were matched, there would be no coupling between arms (1) and (2). Many other[1] magic T's can be made from ring circuits, which are discussed in the next section.

9·13. Ring Circuits.—A four-arm junction may be equivalent to four three-arm junctions connected together to form a ring circuit as in Fig. 9·34. The lines l_1, l_2, l_3, l_4 which interconnect the networks may have arbitrary lengths and arbitrary characteristic impedances. If the series transformer representation of Fig. 4·35 is used for each of the three-arm junctions, the equivalent circuit may be reduced to a simpler form. The series impedance elements of each three-junction may be combined with the lines that interconnect the networks, with the possible exception of one remaining series element. If the network possesses some symmetry,

[1] See also Sec. 12·24.

this remaining element is also absent. The turn ratios of the transformers may also be adjusted if corresponding changes are made in the line impedances.

A ring circuit may be easily realized in waveguide by combining four E-plane T-junctions as shown in Fig. 9·35 for a symmetrical case. If

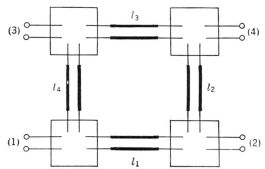

Fig. 9·34.—A ring circuit made up of four three-arm junctions.

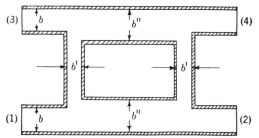

Fig. 9·35.—A symmetrical ring circuit made of four E-plane T-junctions.

the distances between the reference planes are so chosen that each T-junction is a pure series junction and that each is separated by $\lambda_g/4$ from the adjacent junctions, the device becomes a directional coupler of a common form, sometimes called a double-stub coupler. If $b' \ll b$, the reference planes may be chosen as shown in Fig. 9·6. The directional coupler is matched if b'' has the value

$$b'' = b\sqrt{1 + \left(\frac{b'}{b}\right)^2},$$

and the coupling coefficient C is given by

$$C = \frac{\left(\dfrac{b'}{b}\right)^2}{1 + \left(\dfrac{b'}{b}\right)^2}.$$

If b' is very small compared with b, b'' may be made equal to b and the coupling coefficient is simply $(b'/b)^2$. A simple equivalent circuit for this directional coupler is shown in Fig. 9·36.

If the coupling coefficient of a directional coupler is equal to $\frac{1}{2}$, the device becomes a magic T. At long wavelengths, a magic T could be made as shown in Fig. 9·37b from E-plane T-junctions. Shunt junctions, which are the duals of the series junctions, may also be used. A synthesis of a magic T in coaxial line is shown in Fig. 9·37a.

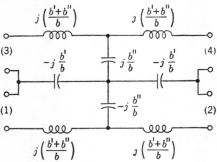

Fig. 9·36.—A simple equivalent circuit for the directional coupler of the kind shown in Fig. 9·35.

A still more general form[1] of directional coupler in which each line has a different characteristic impedance can be constructed. Such a device is illustrated schematically in Fig. 9·38 where only the characteristic impedances of the lines are indicated. There are four parameters: the coupling coefficient C and three arbitrary quantities K, L, M. If $C = 0.5$, a magic T is obtained.

The configurations so far discussed are composed of T-junctions separated by quarter-wavelength sections of line. Another series of

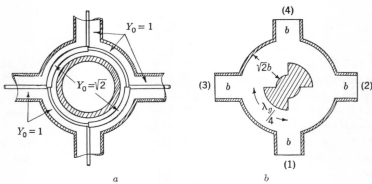

a b

Fig. 9·37a.—A magic T in coaxial line.
Fig. 9·37b.—A magic T made from E-plane T-junctions.

junctions results if one of the lines is chosen to be three-quarters of a wavelength long. For example, a magic T composed of E-plane waveguide junctions is shown in Fig. 9·39. The four lines may also have different characteristic impedances, and Fig. 9·40 shows the values that

[1] B. A. Lippmann, "The Theory of Directional Couplers," RL Report No. 860, Dec. 28, 1945.

these impedances must have for the junction to be a matched directional coupler.

9·14. Four-junctions with Small Holes.—The theory of the diffraction by small holes is useful in the discussion of several important types of directional couplers. One example is the two-hole directional coupler shown in Fig. 9·41, where the holes are spaced by a quarter of a guide wavelength. The equivalent circuit is a special case of the ring circuit

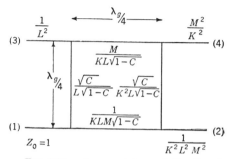
Fig. 9·38.—A general form of directional coupler in which each line has a different characteristic impedance.

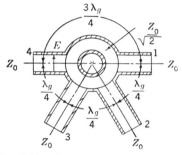
Fig. 9·39.—A magic T composed of E-plane waveguide junctions.

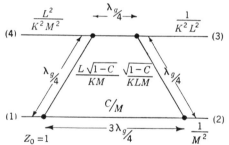
Fig. 9·40.—Values of impedances necessary to make the junction of Fig. 9·39 a matched directional coupler.

of Fig. 9·34. The line lengths l_2 and l_4 are effectively zero, and l_1 and l_3 are one-quarter wavelength. The two waveguides are coupled by the longitudinal magnetic fields, which excite magnetic dipoles within the holes. If the holes are small, the dipoles are of equal strength and 90° out of phase. The dipoles radiate into the second waveguide; but because of spacing between the holes, the radiation is reinforced in the direction of the original wave in the first waveguide and is canceled in the opposite direction. Thus one arm of the junction (that containing the termination in the figure) is decoupled from the arm in which the power is incident. The amount of radiation from the holes may be computed by the formulas given in Sec. 9·7. The interactions between the holes are, of course, neglected in the formulas, whereas in practical directional couplers such as the one shown in the figure, the interactions may not be negligible.

It is also possible to couple from one waveguide to another by two holes so that the wave in the auxiliary line travels in the opposite direction

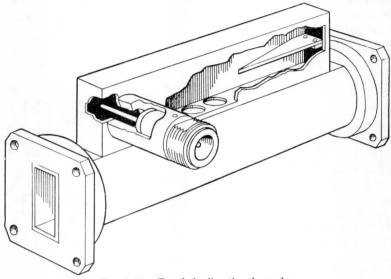

FIG. 9·41.—Two-hole directional coupler.

to that in the main line. Such a device is shown in Fig. 9·42. Directional couplers of this type are called reverse couplers. The two coupling holes are located on opposite sides of the center line, on the broad side of one waveguide and on the narrow side of the other. The longitudinal

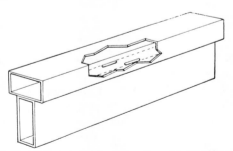

FIG. 9·42.—Schwinger reverse-coupling directional coupler.

magnetic fields that excite the holes are in the opposite directions. The holes are spaced along the waveguide by a quarter of a wavelength, and therefore radiation from one hole is reinforced by radiation from the other for the wave traveling in the backward direction. The coupling coefficient for this coupler also may be calculated from the small-hole formulas if the proper relative values of the fields are inserted.

Another directional coupler, which operates on a different principle, is shown in Fig. 9·43. This device is known as a Bethe-hole coupler. Since the hole is in the center of the broad face of the waveguide, it is excited both by the normal electric field and by the transverse magnetic field in the waveguide. Both an electric dipole and a magnetic dipole are produced in the hole, and both dipoles radiate in both directions into the second waveguide. The electric coupling is an even coupling about the axis of the hole, wheras the magnetic coupling produces fields with odd symmetry. The strength of the magnetic coupling can be adjusted, therefore, to be equal to the electric coupling, by the rotation of one waveguide with respect to the other. The phases of the fields are such that if power is incident on the coupler in the lower waveguide from the left side of the figure, the power coupled into the upper waveguide proceeds in the direction of the arrow labeled "To detector."

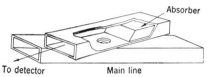

FIG. 9·43.—Bethe hole coupler.

Design formulas for Bethe-hole directional couplers and for other types as well are given in Chap. 14 of Vol. 11 of this series.

9·15. Degenerate Four-junctions.—In Sec. 9·11 it was shown that a degenerate form of a four-junction that did not have the properties of a directional coupler consisted of two three-junctions connected together as in Fig. 9·25. To find the conditions on the elements of the impedance matrix that must be satisfied in order that the junction be degenerate, it is most convenient to find the impedance matrix for two T-junctions connected together. If the terminals of one T-junction are designated by (1), (2), and (3) and those of the other T-junction by (4), (5), and (6), and if terminals (3) and (4) are connected together, then

$$v_3 = v_4, \qquad i_3 = -i_4.$$

These relations may be used to eliminate v_3, v_4, i_3, and i_4 from the network equations. The result is the impedance matrix

$$Z = \begin{pmatrix} Z_{11} - \dfrac{Z_{13}^2}{Z_{33}+Z_{44}}, & Z_{12} - \dfrac{Z_{13}Z_{23}}{Z_{33}+Z_{44}} & \vdots & \dfrac{Z_{13}Z_{45}}{Z_{33}+Z_{44}} & \dfrac{Z_{13}Z_{46}}{Z_{33}+Z_{44}} \\ \cdots & Z_{22} - \dfrac{Z_{23}^2}{Z_{33}+Z_{44}} & \vdots & \dfrac{Z_{23}Z_{45}}{Z_{33}+Z_{44}} & \dfrac{Z_{23}Z_{46}}{Z_{33}+Z_{44}} \\ \hline & & \vdots & & \\ & \cdots & \vdots & Z_{55} - \dfrac{Z_{45}^2}{Z_{33}+Z_{44}} & Z_{56} - \dfrac{Z_{45}Z_{46}}{Z_{33}+Z_{44}} \\ & \cdots & \vdots & \cdots & Z_{66} - \dfrac{Z_{46}^2}{Z_{33}+Z_{44}} \end{pmatrix}.$$

The characteristic feature of this impedance matrix is that the determinant of the upper left-hand portion which is partitioned off is equal to zero. The corresponding determinant of the general impedance matrix of Eq. (31) is

$$\begin{vmatrix} Z_1Z_3(Y_1+Y_2) & Z_1Z_4Y_1 \\ Z_2Z_3Y_1 & Z_2Z_4(Y_1+Y_2) \end{vmatrix}.$$

This determinant is equal to zero if

$$(Y_1 + Y_2)^2 = Y_1^2$$

or

$$Y_2 = 0, \quad -2Y_1.$$

When the admittance $Y_2 = 0$, the general equivalent circuit can be redrawn in a way that makes evident the two three-junctions. This circuit is shown in Fig. 9·44.

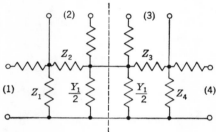

FIG. 9·44.—Equivalent circuit for two T-junctions connected together.

Another degenerate case of particular importance is that for which Y_2 becomes infinite. Again the circuit may be redrawn to make the relationships between the lines obvious. Figure 9·45 shows this circuit. The admittance Y' consists of Z_1 and Z_4 in parallel, or

$$Y' = \frac{1}{Z_1} + \frac{1}{Z_4},$$

and similarly

$$Y'' = \frac{1}{Z_2} + \frac{1}{Z_3}.$$

It is more customary to replace the Π-network formed by Y', Y'', and Y_1 by the equivalent T-network. For the symmetrical case, when lines (1) and (4) are identical as are lines (2) and (3), the circuit becomes that of Fig. 9·46, where the network elements are labeled with the values of the impedance elements. The circuit of Fig. 9·46 is useful for symmetrical structures such as four waveguides joined together in the H-plane, the H-plane cross (Fig. 9·47a), or the coupling of a coaxial line to a waveguide (Fig. 9·47b). Some values of the circuit parameters for the H-plane

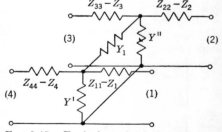

FIG. 9·45.—Equivalent circuit for two symmetrical T-junctions connected together.

cross are to be found in the *Waveguide Handbook*.[1] Unfortunately few data are available for the transition from coaxial line to waveguide.

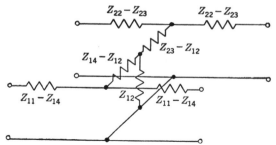

FIG. 9·46.—Second form of equivalent circuit of Fig. 9·45.

The problem has, however, received considerable experimental[2] and theoretical[3] investigation. If suitable reference planes are chosen, it is possible to reduce the values of the series elements ($Z_{11} - Z_{14}$, for example) to zero.

9·16. A Generalization of the Theory of Four-terminal Networks to Four-terminal-pair Networks.—By partioning the impedance matrix of a four-terminal-pair network, a formal extension[4] to the usual four terminal network can be made.

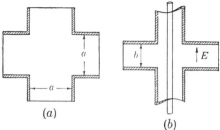

FIG. 9·47.—Examples of junctions for which the circuit of Fig. 9·46 is valid: (a) H-plane cross; (b) coupling of coaxial line to waveguide.

The network equations may be written

$$v' = Z_{11}i' + Z_{12}i'',$$
$$v'' = Z_{21}i' + Z_{22}i'',$$

where the Z_{ij}'s are portions of the impedance matrix Z, thus

$$Z = \begin{pmatrix} Z_{11} & Z_{12} & Z_{13} & Z_{14} \\ Z_{12} & Z_{22} & Z_{23} & Z_{24} \\ \hline Z_{13} & Z_{23} & Z_{33} & Z_{34} \\ Z_{14} & Z_{24} & Z_{34} & Z_{44} \end{pmatrix} = \begin{pmatrix} Z_{11} & Z_{12} \\ Z_{21} & Z_{22} \end{pmatrix},$$

[1] *Waveguide Handbook*, Vol. 10, Radiation Laboratory Series.
[2] *Microwave Transmission Circuits*, Vol. 9, Chap. 6, Radiation Laboratory Series.
[3] J. C. Slater, *Microwave Transmission*, McGraw-Hill, New York, 1942, Chap. VII; "Properties of the Coaxial-wave Guide Junction in the 725A and 2J51 Output," BTL Report No. MM-44-180-4, Nov. 20, 1944; S. Kuhn, "The Coupling between a Rectangular Wave Guide Carrying an H_{01} Wave and a Concentric Line," ASE Report No. M439, September 1942; L. B. Turner, "Impedance Transformer and Junction Box for Cm-wave Coaxial Circuit," ASE Report No. M522, June 1943.
[4] Lippmann, *op. cit.*

and

$$\mathbf{v}' = \begin{pmatrix} v_1 \\ v_2 \end{pmatrix}, \qquad \mathbf{v}'' = \begin{pmatrix} v_3 \\ v_4 \end{pmatrix},$$

$$\mathbf{i}' = \begin{pmatrix} i_1 \\ i_2 \end{pmatrix} \qquad \mathbf{i}'' = \begin{pmatrix} i_3 \\ i_4 \end{pmatrix}.$$

It should be noted that

$$Z_{11} = \tilde{Z}_{11}, \qquad Z_{12} = \tilde{Z}_{21}, \qquad Z_{22} = \tilde{Z}_{22}.$$

As in Fig. 9·48, a four-terminal network Z_1 may be connected between terminal pairs (1) and (2) of Z and a second network Z_2 between ter-

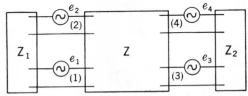

FIG. 9·48.—Circuit for matrix generalization of image impedance.

minal pairs (3) and (4), together with series generators e_i. The network equation then becomes

$$e' = (Z_{11} + Z_1)\mathbf{i}' + Z_{12}\mathbf{i}'',$$
$$e'' = Z_{21}\mathbf{i}' + (Z_{22} + Z_2)\mathbf{i}'',$$

where

$$\mathbf{e}' = \begin{pmatrix} e_1 \\ e_2 \end{pmatrix}, \qquad \mathbf{e}'' = \begin{pmatrix} e_3 \\ e_4 \end{pmatrix},$$

$$Z_1 = \begin{pmatrix} Z'_{11} & Z'_{12} \\ Z'_{12} & Z'_{22} \end{pmatrix}, \qquad Z_2 = \begin{pmatrix} Z''_{11} & Z''_{12} \\ Z''_{12} & Z''_{22} \end{pmatrix}.$$

The solutions obtained for the currents are

$$\mathbf{i}' = (Z_1 + Z_1^{\text{in}})^{-1}\mathbf{e}' - (Z_1 + Z_1^{\text{in}})^{-1}Z_{12}(Z_{22} + Z_2)^{-1}\mathbf{e}'',$$
$$\mathbf{i}'' = -(Z_2 + Z_2^{\text{in}})^{-1}Z_{21}(Z_{11} + Z_1)^{-1}\mathbf{e}' + (Z_2 + Z_2^{\text{in}})^{-1}\mathbf{e}'',$$

where

$$Z_1^{\text{in}} = Z_{11} - Z_{12}(Z_{22} + Z_2)^{-1}Z_{21},$$
$$Z_2^{\text{in}} = Z_{22} - Z_{21}(Z_{11} + Z_1)Z_{12}$$

are the generalized input impedances. The image impedance of the four-junction may be similarly generalized, and the image impedances Z' and Z'' are defined by

$$Z' = Z_{11} - Z_{12}(Z_{22} + Z'')^{-1}Z_{21},$$
$$Z'' = Z_{22} - Z_{21}(Z_{11} + Z')^{-1}Z_{12}.$$

If a four-terminal-pair network is terminated by the image impedances on each side, then

$$i' = \tfrac{1}{2}(Z')^{-1}e', \quad \text{for } e'' = 0,$$
$$i'' = \tfrac{1}{2}(Z'')^{-1}e'', \quad \text{for } e' = 0.$$

From this equation, it is evident that if terminals (1) and (2) are to be decoupled and terminals (3) and (4) are also to be decoupled, Z' and Z'' must be diagonal matrices,

$$Z' = \begin{pmatrix} Z_1 & 0 \\ 0 & Z_2 \end{pmatrix}, \quad Z'' = \begin{pmatrix} Z_3 & 0 \\ 0 & Z_4 \end{pmatrix}.$$

The currents are therefore

$$i_1 = \frac{e_1}{2Z_1}, \quad i_2 = \frac{e_2}{2Z_2}, \quad \text{for } e'' = 0,$$

$$i_3 = \frac{e_3}{2Z_3}, \quad i_4 = \frac{e_4}{2Z_4}, \quad \text{for } e'' = 0.$$

Lippmann[1] has extended the development of the theory to the treatment of chains of four-terminal-pair networks. The method runs parallel to the corresponding treatment of chains of four-terminal networks. An image transfer constant and a propagation constant may be defined, and the generalization is essentially complete. A somewhat similar problem arises in the consideration of a chain of cavities connected in a ring that forms the resonator system of a strapped magnetron.[2]

RADIATION AND SCATTERING BY ANTENNAS

The concept of a generalized waveguide junction which was established in Chap. 5 may be extended to antenna problems. One terminal of the junction is the transmission line connected to the antenna. The other terminals are infinite in number and may be chosen in any of several ways. The currents and voltages at these terminals or, alternatively, the wave amplitudes form a representation of the electromagnetic fields in the vicinity of the antenna.

9·17. Representation in Terms of Plane Waves.—The transmission-line terminal of an antenna differs in no way from a terminal line of any waveguide junction. A wave incident on the antenna through the transmission line may be partially reflected, or the terminal may be matched, and no reflection takes place. The amplitude of the reflected wave may be expressed in terms of a scattering coefficient or in terms of a shunt or series impedance terminating the waveguide. If large reflecting objects are near the antenna, the magnitude and phase of the reflection coefficient in the transmission line depends upon the position and nature of the reflector. In nearly all cases, however, such reflected

---
[1] Lippmann, *op. cit.*
[2] See *Microwave Magnetrons*, Vol. 6, Chap. 4, Radiation Laboratory Series.

power is negligible compared with the reflected power from the antenna itself. As an example, the particular case of an antenna consisting of the open end of a rectangular waveguide may be considered. At the end of the waveguide the fields are distorted to satisfy the boundary conditions, and both electric and magnetic energy is stored near the end of the pipe. This stored energy produces effects equivalent to those produced by a lumped susceptance. Energy is also radiated, and this energy may be represented by a lumped conductance. The equivalent circuit of the transmission line is therefore that shown in Fig. 9·49a. In Fig. 9·49b the field near the end of the pipe is indicated. In the E-plane shown, E-modes are excited and produce the equivalent of a capacitive susceptance. For a rectangular waveguide, H-modes are also excited. The net effect is capacitive, however, for waveguides of the dimensions usually encountered in practice.

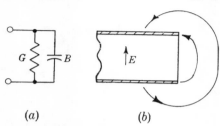

Fig. 9·49.—The termination of a waveguide or a parallel-plate transmission line in free space.

Several cases of the abrupt termination of a transmission line in space have been rigorously solved,[1] and the results are collected in the *Waveguide Handbook*.[2] The conductance G may be regarded as the relative conductance of space; its magnitude is of the order of $(ka)^2$, where k is the wave number and a a dimension of the waveguide.

The remaining terminals of the antenna are usually chosen to be located a very large distance away from the antenna, and the wave amplitudes of plane waves proceeding outward are taken as a representation of the field. Thus the radiation pattern of the antenna is commonly used to describe the antenna properties. The maximum radiation intensity relative to an isotropic radiator that radiates the same total power is called the gain of the antenna. The radiation patterns of certain simple cases have also been rigorously calculated.

It is also possible to calculate the radiation from a small hole. For a small hole at the end of a waveguide the value of B in Fig. 9·49 is the same as that of an iris placed across the guide, and G is given by

$$G = \frac{2\pi}{3} \frac{ab\lambda_g}{\lambda^3}.$$

Certain general theorems may be proved about the radiation and scattering from antennas, but it is more convenient to choose a different representation for the terminals for this purpose.

[1] J. Schwinger, unpublished work.
[2] *Waveguide Handbook*, Vol. 10, Radiation Laboratory Series.

9·18. Representation in Terms of Spherical Waves.

—It is convenient, in dealing with infinities, to replace a nondenumerable set by a denumberable one. One way in which this can be done in the case of the radiation problem is to expand the field quantities in spherical waves whose angular dependence is simply related to the various tesseral harmonics. The total electromagnetic field is determined by the amplitudes of these various waves. This representation in terms of spherical waves is equivalent to the introduction of terminals with incident and reflected wave amplitudes for each of the spherical waves.

In a vacuum, the electric and magnetic field vectors satisfy the same differential equation, namely, the wave equation,

$$\nabla^2 \mathbf{E} - c^2 \frac{\partial^2 \mathbf{E}}{\partial t^2} = 0. \tag{49}$$

By ∇^2 is meant the vector operator

$$\nabla^2 = \nabla\nabla \cdot - \nabla \times \nabla \times. \tag{50}$$

In terms of cartesian components, Eq. (49) reduces to the three scalar equations

$$\Delta E_j - c^2 \frac{\partial^2 E_j}{\partial t^2} = 0, \tag{51}$$

where Δ is the Laplacian operator. It is seldom useful to solve Eq. (51), as the three solutions are not independent, being coupled by the condition

$$\nabla \cdot \mathbf{E} = 0. \tag{52}$$

A solution of the scalar equation [Eq. (51)], however, can be used to construct permissible solutions of Eq. (49) for certain special coordinate systems. One such coordinate system is that of spherical coordinates. The solutions are systems of spherical waves about the center of the system.

It is convenient to assume a sinusoidal time dependence. Equation (49) becomes, in the usual way,

$$\nabla^2 \mathbf{E} + k^2 \mathbf{E} = 0. \tag{53}$$

The scalar wave equation becomes

$$\Delta u + k^2 u = 0. \tag{54}$$

For any solution of Eq. (54), there is a vector solution of Eq. (53) given by

$$\mathbf{L} = \nabla u. \tag{55}$$

It should be noted that Eq. (52) is not satisfied by $\mathbf{L}$. However, $\mathbf{L}$ can be used to construct solenoidal vector fields. Let $\mathbf{R}$ be a radius vector from some fixed point. Then

$$\mathbf{M} = \mathbf{R} \times \mathbf{L} \tag{56a}$$

and

$$\mathbf{N} = \nabla \times (\mathbf{R} \times \mathbf{L}) \tag{56b}$$

are solutions of Eqs. (52) and (53). It should be noted that $\mathbf{M}$ is normal to $\mathbf{R}$. Because of the occurrence of $\mathbf{R}$ in Eq. (56), the coordinate system in which this equation has the most simple form is the system of spherical coordinates shown in Fig. 9·50.

Scalar Wave Equation.—In the spherical coordinate system Eq. (54) becomes

$$\frac{1}{r^2}\frac{\partial}{\partial r}\left(r^2 \frac{\partial u}{\partial r}\right) + \frac{1}{r^2 \sin \theta} \frac{\partial}{\partial \theta}\left(\sin \theta \frac{\partial u}{\partial \theta}\right) + \frac{1}{r^2 \sin^2 \theta} \frac{\partial^2 u}{\partial \phi^2} + k^2 u = 0. \tag{57}$$

A set of single-valued solutions to this equation is

$$u_{mn} = P_n^{|m|}(\cos \theta) e^{jm\phi} \frac{1}{\sqrt{kr}} Z_{n+\frac{1}{2}}(kr), \tag{58}$$

where n and m are integers such that

$$n \geqq 0, \qquad |m| \leqq n,$$

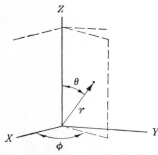

Fig. 9·50.—System of spherical coordinates.

$P_n^{|m|}$ (cos θ) is the associated Legendre polynomial, and Z is a cylinder function. If the origin lies inside the region under consideration, the field quantities must be finite at this point and $Z_{n+\frac{1}{2}}$ must be the Bessel function $J_{n+\frac{1}{2}}$. If the point $r = 0$ is excluded from the region, $Z_{n+\frac{1}{2}}$ may be a Bessel, Neumann, or Hankel function. The details of the solution of this equation, together with the properties of the various functions, may be found in many places[1] and will not be discussed here.

The associated Legendre polynomials have the form

$$P_n^m(x) = \frac{(1 - x^2)^{m/2}}{2^n n!} \frac{d^{m+n}(x^2 - 1)^n}{dx^{n+m}}, \tag{59}$$

where $x = \cos \theta$. Some examples of these polynomials are

$$\left.\begin{array}{l} P_0^0 = 1, \qquad P_1^0 = \cos \theta, \qquad P_1^1 = \sin \theta, \\ P_2^0 = \frac{1}{2}(3 \cos^2 \theta - 1), \\ P_2^1 = 3 \sin \theta \cos \theta, \\ P_2^2 = 3 \sin^2 \theta. \end{array}\right\} \tag{60}$$

[1] H. Margenau and G. M. Murphy, *The Mathematics of Physics and Chemistry*, Van Nostrand, New York, 1943.

The functions $e^{jm}P_n^{|m|}$ (cos θ) are called tesseral harmonics and have the important property of orthogonality. It may be shown that

$$\int_0^\pi \int_0^{2\pi} e^{jm\phi} P_n^m(\cos\theta) e^{jp\phi} P_q^{|p|}(\cos\theta) \sin\theta \, d\theta \, d\phi = 0 \tag{61}$$

for $p \neq m$ and/or $q \neq n$.

The behavior of the various cylinder functions for large values of kr will shed some light on the physical meaning of Eq. (58). It is easily shown that in the limit $x \to \infty$,

$$\left.\begin{aligned}
J_{n+\frac{1}{2}}(x) &= \sqrt{\frac{2}{\pi x}} \cos\left(x - \frac{n+1}{2}\pi\right), \\
N_{n+\frac{1}{2}}(x) &= \sqrt{\frac{2}{\pi x}} \sin\left(x - \frac{n+1}{2}\pi\right), \\
H_{n+\frac{1}{2}}^{(1)}(x) &= \sqrt{\frac{2}{\pi x}} (-j)^{n+1} e^{jx}, \\
H_{n+\frac{1}{2}}^{(2)}(x) &= \sqrt{\frac{2}{\pi x}} (j)^{n+1} e^{-jx}.
\end{aligned}\right\} \tag{62}$$

It is clear that a solution with a Bessel or Neumann function for its radial dependence represents a spherical standing wave. The two Hankel functions represent running waves that either diverge from the origin or converge upon it. It is clear that a wave running out from the origin requires a source, and this is a physical reason for the necessity of excluding the origin in the case where the radial dependence is that of a Hankel function.

Vector Wave Equation.—Each of the solutions [Eq. (58)] of the scalar wave equation can be used to generate solutions of the vector wave equation as in Eqs. (56a) and (56b). Let

$$\mathbf{M}_{mn} = \mathbf{R} \times \boldsymbol{\nabla} u_{mn}, \tag{63}$$
$$\mathbf{N}_{mn} = \boldsymbol{\nabla} \times (\mathbf{R} \times \boldsymbol{\nabla} u_{mn}). \tag{64}$$

These are solenoidal solutions of the vector wave equation [Eq. (53)]. It is to be noticed that $\mathbf{M}_{mn}$ is normal to $\mathbf{R}$. For this reason, a solution for which $\mathbf{M}_{mn}$ is the electric or magnetic field is called transverse-electric or transverse-magnetic respectively. It should be noted also that if the electric or magnetic field is $\mathbf{M}_{mn}$, the corresponding magnetic or electric field is of the form $\mathbf{N}_{mn}$.

It may be demonstrated that the set of waves of Eqs. (63) and (64) forms a complete set of solutions of Maxwell's equations which, for empty space, vanish at infinity.

In addition to this completeness property, the set of solutions given in Eqs. (63) and (64) have the important property of orthogonality. This is the same orthogonality property which was invoked in the case of a waveguide with several propagating modes. Because of the orthogonality of the functions in Eqs. (63) and (64), the stored electric or magnetic energy inside a simply connected charge-free region is equal to the sum of the energies computed for the various modes. Moreover, the power entering or leaving the bounding surface is equal to the sum of the powers computed for each of the spherical waves.

9·19. Solutions of the Vector Wave Equations.—Since the magnetic field may always be determined from the electric field, it is necessary to consider only the electric field in a description of the electromagnetic field in a charge-free space. It is convenient to use Hankel functions in the description of the field. Let us consider the description of one of the spherical waves of the set of Eqs. (63) and (64),

$$\mathbf{M}_{mn} = \mathbf{R} \times \boldsymbol{\nabla} u_{mn}. \tag{65}$$

The occurence of the complex function $e^{jm\phi}$ in Eq. (58) is inconvenient. Therefore a new set of waves may be defined,

$$\left.\begin{aligned}\mathbf{M}'_{mn} &= \frac{1}{2}(\mathbf{M}_{mn} + \mathbf{M}_{-m,n}), \\ \mathbf{M}''_{mn} &= \frac{1}{2j}(\mathbf{M}_{mn} - \mathbf{M}_{-m,n}).\end{aligned}\right\} \tag{66}$$

Consider the wave

$$\mathbf{M}'_{mn} = \mathbf{R} \times \boldsymbol{\nabla} u'_{mn},$$
$$u'_{mn} = P'^{|m|}_{n}(\cos\theta)\cos m\phi \frac{1}{\sqrt{kr}} Z_{n+\frac{1}{2}}(kr). \tag{67}$$

The solutions $\mathbf{M}'$ are even in ϕ, and the solutions $\mathbf{M}''$ are odd in ϕ. The function $Z_{n+\frac{1}{2}}$ will be assumed to be a linear combination of the two Hankel functions,

$$Z_{n+\frac{1}{2}} = aH_{n+\frac{1}{2}}^{(1)} + bH_{n+\frac{1}{2}}^{(2)}. \tag{68a}$$

In case the region includes the origin, the field at the origin must be finite, $Z_{n+\frac{1}{2}}$ must be a Bessel function, and

$$a = b. \tag{68b}$$

If there is an antenna at the center of the region, the origin must be excluded and then there is no simple relation between a and b. However, the various a's and b's are linearly related to each other, and it is possible to define a scattering matrix that relates the various incident waves to the scattered waves. This is completely analogous to the scattering matrix that was defined in Chap. 5 for generalized waveguide junctions.

It is convenient to introduce a single index instead of the double index of $\mathbf{M}'_{mn}$. The way in which this is done is of no particular importance. However, for purposes of definiteness let the new subscripts be assigned according to Table 9·1.

TABLE 9·1.—REPLACEMENT OF DOUBLE SUBSCRIPTS BY SINGLE SUBSCRIPTS

Subscript	Mode	Mode character
2	$\mathbf{M}'_{01}$	
3	$\mathbf{M}'_{11}$	Electric dipole
4	$\mathbf{M}''_{11}$	
5	$\mathbf{N}'_{01}$	
6	$\mathbf{N}'_{11}$	Magnetic dipole
7	$\mathbf{N}''_{11}$	
.	.	
.	.	

The subscript 1 will refer to the terminals in the transmission line that excites the antenna. As in Chap. 5, column vectors will be introduced to represent the incident and scattered waves. Let

$$\mathbf{a} = \begin{pmatrix} a_1 \\ \cdot \\ \cdot \\ \cdot \\ a_m \\ \cdot \\ \cdot \end{pmatrix}, \quad \mathbf{b} = \begin{pmatrix} b_1 \\ \cdot \\ \cdot \\ \cdot \\ b_m \\ \cdot \\ \cdot \end{pmatrix} \quad (69)$$

These vectors have an infinite number of dimensions. It is convenient to normalize the a's and b's in such a way that $\frac{1}{2}a_n^* a_n$ represents the power incident on the antenna in the nth mode. Since the a's and b's are linearly related, a scattering matrix $\mathbf{S}$ may be defined such that

$$\mathbf{Sa} = \mathbf{b}. \quad (70)$$

The methods of Chap. 5 can be used to show that $\mathbf{S}$ is unitary and symmetrical. However, it is worth while to arrive at these results in a simpler way. Since

$$\widetilde{\mathbf{S}^*\mathbf{a}^*} = \tilde{\mathbf{b}}^* = \tilde{\mathbf{a}}^*\tilde{\mathbf{S}}^*,$$
then (71)
$$\tilde{\mathbf{a}}^*\tilde{\mathbf{S}}^*\mathbf{S}^*\mathbf{a} = \tilde{\mathbf{b}}^*\mathbf{b}.$$

But $\frac{1}{2}\tilde{\mathbf{b}}^*\mathbf{b}$ is the total power scattered by the antenna, and it must equal the total incident power $\frac{1}{2}\tilde{\mathbf{a}}^*\mathbf{a}$. Therefore

$$\tilde{\mathbf{a}}^*\tilde{\mathbf{S}}^*\mathbf{Sa} = \tilde{\mathbf{a}}^*\mathbf{a}. \quad (72)$$

Since Eq. (72) applies independently of the vector **a**,

$$\tilde{S}*S = I, \tag{73}$$

or

$$\tilde{S}* = S^{-1}. \tag{74}$$

Thus S is a unitary matrix.

To show that S is symmetrical, it should be noticed that for any solution of Maxwell's equations for zero loss there is another one with time reversed. For example, if

$$\mathbf{E}(r, \theta, \phi, t) = \sum_{m,n} \text{Re} \, (\mathbf{M}'_{mn} e^{j\omega t}) \tag{75}$$

is a solution,

$$\mathbf{E}(r, \theta, \phi, -t) = \sum_{m,n} \text{Re} \, (\mathbf{M}'_{mn} e^{-j\omega t}) = \sum_{m,n} \text{Re} \, (\mathbf{M}'^{*}_{m,n} e^{j\omega t})$$

is also a solution. However,

$$\left.\begin{aligned}
\mathbf{M}'_{mn} &= \\
&\mathbf{R} \times \nabla \left\{ P_n^{|m|}(\cos \theta) \cos m\phi \, \frac{1}{\sqrt{kr}} [a_s H^{(1)}(kr) + b_s H^{(2)}(kr)] \right\}, \\
\text{and} & \\
\mathbf{M}^{*\prime}_{mn} &= \\
&\mathbf{R} \times \nabla \left\{ P_n^{|m|}(\cos \theta) \cos m\phi \, \frac{1}{\sqrt{kr}} [a_s^* H^{(2)}(kr) + b_s^* H^{(1)}(kr)] \right\}.
\end{aligned}\right\} \tag{76}$$

Thus from Eq. (76), if S**a** = **b** represents a solution of Maxwell's equations, then another allowable solution is

$$S\mathbf{b}^* = \mathbf{a}^*. \tag{77}$$

The complex conjugate of Eq. (77) is

$$S^*\mathbf{b} = \mathbf{a}$$

or

$$S^{*-1}\mathbf{a} = \mathbf{b}; \tag{78}$$

therefore

$$S^{*-1} = S. \tag{79}$$

If Eq. (79) is compared with Eq. (74), it can be seen that

$$\tilde{S} = S \tag{80}$$

and S is symmetrical. It should be noticed that the symmetry property depends upon the fact that the angular dependence of **M**' is pure real.

9·20. Scattering Matrix of Free Space.—When there is no antenna, the wave amplitudes a_1 and b_1 are meaningless and consequently the scattering matrix has its first row and column omitted. The relation Eq. (68b) requires that the remainder of the scattering matrix be just the identity matrix

$$S_0 = \begin{pmatrix} & & & & & \\ & 1 & 0 & 0 & .. & .. \\ & 0 & 1 & 0 & .. & .. \\ & 0 & 0 & 1 & .. & .. \\ & .. & .. & .. & .. & .. \\ & .. & .. & .. & .. & .. \end{pmatrix}. \tag{81}$$

9·21. Scattering Matrix of a Simple Electric Dipole.—Assume that an electric dipole radiates the mode M'_{01}, and that it is matched to its transmission line. It may be assumed also that the antenna does not affect any of the other modes. The scattering matrix is

$$S = \begin{pmatrix} 0 & e^{j\alpha} & 0 & 0 & 0 & .. & .. \\ e^{j\alpha} & 0 & 0 & 0 & 0 & .. & .. \\ 0 & 0 & 1 & 0 & 0 & .. & .. \\ 0 & 0 & 0 & 1 & 0 & .. & .. \\ 0 & 0 & 0 & 0 & 1 & .. & .. \\ .. & .. & .. & .. & .. & .. & .. \end{pmatrix}. \tag{82}$$

It should be noted that the assumption that the antenna is matched and radiates only mode (2) determines the first column of matrix (82). The second column follows from the symmetry and unitary property of S. The first two rows are fixed by the symmetry of S. The remainder of the matrix is fixed by the assumption that the antenna does not disturb the other radiation modes. It has the same form as Eq. (81).

Such a dipole antenna will absorb power only from the one dipole mode that it radiates. It also absorbs all the power incident on the antenna in this mode. If a plane wave falls on the dipole antenna, the antenna absorbs the one dipole mode and leaves the rest of the wave undisturbed. The absorption of the dipole mode may be described as a dipole wave radiating out from the antenna and having the right amplitude and phase to cancel the dipole component in the plane wave. This negative wave will be called the scattered wave. Scattering of this type may be described by the matrix

$$S' = S - S_0 = \begin{pmatrix} 0 & e^{j\alpha} & & \\ e^{j\alpha} & -1 & & 0 \\ & & & \\ 0 & & & 0 \end{pmatrix}. \tag{83}$$

It should be noted that the power scattered by the antenna is equal to the power absorbed by the antenna, independently of the type of wave falling on the dipole.

9·22. The General Antenna.—The scattering matrix of a general antenna that is assumed to be matched to its transmission line is

$$\mathbf{S} = \begin{pmatrix} 0 & S_{12} & \cdots & \cdots & \cdots \\ S_{21} & S_{22} & \cdots & \cdots & \cdots \\ \cdots & \cdots & \cdots & \cdots & \cdots \\ \cdots & \cdots & \cdots & \cdots & \cdots \end{pmatrix}. \tag{84}$$

There are a few interesting things to be seen about matrix (84). The first of these is the reciprocity property. Since

$$S_{n1} = S_{1n}, \tag{85}$$

an antenna that absorbs $|S_{1n}|^2$ of the power incident on the antenna in the nth mode also radiates $|S_{1n}|^2$ of the total radiated power in this mode. Similarly, there is a reciprocity for the scattering of one mode into another by the antenna.

For a unit incident wave in the antenna line, the wave radiated by the antenna is

$$\mathbf{b}_r = \begin{pmatrix} 0 \\ S_{21} \\ S_{31} \\ \cdot \\ \cdot \\ \cdot \end{pmatrix}. \tag{86}$$

Because of the unitary property of $\mathbf{S}$, a wave incident on the antenna such that

$$\mathbf{a}_i = \begin{pmatrix} 0 \\ S_{21}^* \\ S_{31}^* \\ \cdot \\ \cdot \\ \cdot \end{pmatrix} = \mathbf{b}_r^* \tag{87}$$

will be completely absorbed. It will be recognized that $\mathbf{a}_i$ is just $\mathbf{b}_r$ with a time reversal.

The general antenna is not of great interest. Its generality is so great that any pile of tin with a transmission line exciting it may be called an antenna. It is evident on physical grounds that such a pile of tin does not make a good antenna, and it is worth while to search for some distinguishing characteristics that can be used to differentiate between an ordinary pile of tin and one that makes a good antenna. When the properties of a good antenna are considered, the only one that stands out is the general economy of metal. A good antenna does not have an ensemble of metallic ears, flaps, and springs that play no useful role in the business of radiating. In other words, there may be two antennas that have identical radiation patterns. One of them may have miscellaneous structures attached to it that are not necessary. Since the radiation patterns are identical, the patterns cannot be used to distinguish between the antennas. It might be expected, however, that one antenna would scatter more than the other. It is worth while therefore to see what can be done in the way of differentiating between a good and bad antenna on the basis of scattering.

9·23. The General Scattering Problem.—It is convenient to break the scattering matrix of a general antenna into two parts

$$\mathbf{S} = \mathbf{S}_1 + \mathbf{S}_2, \tag{88}$$

where

$$\mathbf{S}_1 = \begin{pmatrix} 0 & S_{12} & \cdots & \cdots \\ \cdots & \cdots & \cdots & \cdots \\ S_{21} & & & \\ \vdots & & 0 & \end{pmatrix}, \tag{89}$$

$$\mathbf{S}_2 = \begin{pmatrix} 0 & & 0 & \\ \cdots & \cdots & \cdots & \cdots \\ & S_{22} & \cdots & \cdots \\ 0 & 0 & & \\ \vdots & \vdots & & \end{pmatrix}. \tag{90}$$

Similarly, it is convenient to break the column vectors a and b into two parts

$$\mathbf{a} = \mathbf{a}^{(1)} + \mathbf{a}^{(2)}, \tag{91}$$

$$\mathbf{a}^{(1)} = \begin{pmatrix} a_1 \\ 0 \\ 0 \\ 0 \\ 0 \\ \cdot \end{pmatrix}, \quad \mathbf{a}^{(2)} = \begin{pmatrix} 0 \\ a_2 \\ a_3 \\ \cdots \\ \cdots \\ \cdots \end{pmatrix}. \tag{92}$$

The quantities of interest are

$$\left. \begin{array}{l} \text{Power radiated by the antenna} = \tfrac{1}{2}\tilde{\mathbf{a}}^{(1)*}\mathbf{a}^{(1)}, \\ \text{Power absorbed by the antenna} = \tfrac{1}{2}\tilde{\mathbf{a}}^{(2)*}\mathbf{S}_1^*\mathbf{S}_1\mathbf{a}^{(2)}, \\ \text{Power scattered by the antenna} = \tfrac{1}{2}[(\mathbf{S}_2 - \mathbf{I})\mathbf{a}^{(2)}]^*[(\mathbf{S}_2 - \mathbf{I})\mathbf{a}^{(2)}]. \end{array} \right\} \tag{93}$$

The eigenvalue solutions[1] are convenient for the calculation of scattering. Let

$$Sa_k = s_k a_k, \qquad (94)$$

where the a_k are a complete, orthogonal, and real set of eigenvectors. The column vectors a_k are normalized in such a way that

$$\tilde{a}_k a_k = 1. \qquad (95)$$

Thus for each of these eigensolutions, the power radiated by the antenna is

$$\tfrac{1}{2}\tilde{a}_k^{(1)*}a_k^{(1)} = \tfrac{1}{2}a_k^2, \qquad (96)$$

where a_k is the first component of the kth eigenvector. Likewise, the power absorbed by the antenna is

$$\tfrac{1}{2}\tilde{a}_k^{(2)*}S_1^*S_1 a_k^{(2)} = \tfrac{1}{2}s_k^* s_k a_k^2. \qquad (97)$$

If

$$f_k = (S_2 - 1)a_k^{(2)}, \qquad (98)$$

the scattered power is

$$\tfrac{1}{2}\tilde{f}_k^* f_k. \qquad (99)$$

However,

$$f_k = (S - 1)a_k - S_1 a_k^{(2)} - (S - 1)a_k^{(1)}, \qquad (100)$$
$$f_k = (s_k - 1)a_k - S_1 a_k^{(2)} - (S - 1)a_k^{(1)}, \qquad (101)$$

but

$$S_1 a_k^{(2)} = s_k a_k^{(1)},$$

and

$$S_2 a_k^{(1)} = 0.$$

Therefore

$$f_k = (s_k - 1)a_k - s_k a_k^{(1)} + a_k^{(1)} - S_1 a_k^{(1)}$$
$$= (s_k - 1)a_k^{(2)} - S_1 a_k^{(1)}. \qquad (102)$$

If the first column of S is designated by the column vector

$$r = \begin{pmatrix} 0 \\ S_{21} \\ \cdot \\ \cdot \\ \cdot \end{pmatrix}, \qquad (103)$$

then $S_1 a_k^{(1)} = a_k r$, and therefore

$$f_k = (s_k - 1)a_k^{(2)} - a_k r. \qquad (104)$$

The scattered power is $\tfrac{1}{2}$ of

$$\tilde{f}_k^* f_k = (s_k^* - 1)(s_k - 1)\tilde{a}_k^{*(2)}a_k^{(2)} + a_k^2 \tilde{r}^* r - (s_k^* - 1)a_k \tilde{a}_k^{(2)*}r$$
$$\quad - (s_k - 1)a_k \tilde{r}^* a_k^{(2)}. \qquad (105)$$

[1] See Chap. 12 for the mathematical formulation of eigenvalue solutions.

Since a_k is pure real,
$$\tilde{a}_k^{(2)*}r = \tilde{r}a_k^{(2)} = s_k a_k,$$
$$\tilde{r}^* a_k^{(2)} = s_k^* a_k. \tag{106}$$
Also
$$\tilde{r}^* r = 1,$$
$$\tilde{f}_k^* f_k = (s_k^* - 1)(s_k - 1)\tilde{a}_k^{(2)} a_k^{(2)} + [1 - (s_k^* - 1)s_k - (s_k - 1)s_k^*] a_k^2. \tag{107}$$
It should be noted that
$$[\tilde{a}_k^{(2)} a_k^{(2)}](\tilde{r}^* r) \geq [\tilde{a}_k^{(2)} r][\tilde{r}^* a_k^{(2)}]. \tag{108}$$
From Eqs. (106) and (108)
$$\tilde{a}_k^{(2)} a_k^{(2)} \geq s_k^* s_k a_k^2 = a_k^2;$$
therefore
$$\tilde{f}_k^* f_k \geq [(s_k^* - 1)(s_k - 1) + 1 - (s_k^* - 1)s_k - (s_k - 1)s_k^*] a_k^2,$$
$$\tilde{f}_k^* f_k \geq [2 - s_k^* s_k] a_k^2 = a_k^2. \tag{109}$$

Thus the scattered power for an eigensolution is always equal to or greater than the absorbed power.

9·24. Minimum-scattering Antenna.—A minimum-scattering antenna will be defined as an antenna for which the scattering is a minimum for each eigensolution. Under this condition, Eq. (109) becomes an equality for all k. It is to be expected that of all antennas producing a given radiation pattern, any minimum-scattering antenna will be the least like a pile of tin. However, it remains to be seen if any arbitrary antenna pattern can be obtained with a minimum-scattering antenna.

In order for $\tilde{f}_k^* f_k = a_k^2$ in Eq. (107), either
$$\tilde{a}_k^{(2)} a_k^{(2)} = a_k^2 \tag{110}$$
or
$$s_k = 1. \tag{111}$$

It is clear that Eq. (111) cannot be satisfied for all k, as this would yield the identity matrix for the scattering matrix and this corresponds to no antenna at all. In order for Eqs. (106), (106a), and (110) to be satisfied for some value of k ($k = 1$, for example) it is necessary, however, that
$$a_1^{(2)} = e^{j\gamma} a_1 r^*, \tag{112}$$
where the phase factor γ is real.
If Eq. (112) is multiplied by S_1,
$$S_1 a_1^{(2)} = e^{j\gamma} a_1 S_1 r^* = e^{j\gamma} a_1^{(1)};$$
but
$$S_1 a_1^{(2)} = s_1 a_1^{(1)}.$$
Therefore
$$s_1 = e^{j\gamma}. \tag{113}$$

Now either the remainder of the eigenvalues are equal to unity, or there is another eigenvalue

$$s_2 \neq 1, \tag{114}$$

but then

$$\tilde{a}_2^{*(2)} a_2^{(2)} = a_2^2. \tag{115}$$

Because of the orthogonality of the eigenvectors,

$$\tilde{a}_2^* a_1 = 0, \tag{116}$$
$$\tilde{a}_2^{*(2)} a_1^{(2)} = a_2 a_1, \tag{117}$$

and

$$a_2^{(2)} = -e^{-j\gamma} a_2 r^*; \tag{118}$$

also

$$s_2 = -s_1 = -e^{j\gamma}. \tag{119}$$

This exhausts the class of all eigenvectors with eigenvalues not equal to unity. This follows because Eqs. (112) and (118) are the only pure real vectors of this form. The remaining alternative, with

$$s_2 = 1, \tag{120}$$

will be considered later.

The scattering matrix may be computed as follows:

$$\mathsf{S} = \mathsf{T}\mathsf{S}_d\mathsf{T}^{-1}, \tag{121}$$

where T is a matrix with the eigenvectors normalized to unity as columns and S_d is a diagonal matrix with eigenvalues as entries. From Eq. (110) it can be seen that for the eigenvectors normalized to unity which satisfy that equation,

$$|a_1| = |a_2| = \frac{1}{\sqrt{2}}.$$

It can be assumed, without loss in generality, that $a_1 = a_2 = \frac{1}{\sqrt{2}}$. In order to simplify the notation let

$$\mathsf{a}_1 = \frac{1}{\sqrt{2}} \begin{pmatrix} 1 \\ g_2 \\ g_3 \\ \cdot \\ \cdot \\ \cdot \end{pmatrix}. \tag{122}$$

Then

$$\mathsf{T} = \frac{1}{\sqrt{2}} \begin{pmatrix} 1 & 1 & \cdot & \cdot & \cdot \\ g_2 & -g_2 & \cdot & \cdot & \cdot \\ g_3 & -g_3 & \cdot & \cdot & \cdot \\ \cdot & \cdot & \cdot & \cdot & \cdot \\ \cdot & \cdot & \cdot & \cdot & \cdot \\ \cdot & \cdot & \cdot & \cdot & \cdot \end{pmatrix} \tag{123}$$

Sec. 9·24] MINIMUM-SCATTERING ANTENNA

$$S_d = \begin{pmatrix} s_1 & 0 & 0 & 0 & \cdots \\ 0 & -s_1 & 0 & 0 & \cdots \\ 0 & 0 & 1 & 0 & \cdots \\ 0 & 0 & 0 & 1 & \cdots \\ \cdot & \cdot & \cdot & & \\ \cdot & \cdot & \cdot & & \\ \cdot & \cdot & \cdot & & \end{pmatrix} \qquad (124)$$

Equation (121) can be written as

$$S = T(S_d - I)T^{-1} - I. \qquad (125)$$

Since

$$T^{-1} = \tilde{T}, \qquad (126)$$

Eq. (125) is easily evaluated.

$$S = \tfrac{1}{2} \begin{pmatrix} (s_1 - 1) & -(s_1 + 1) & 0 & 0 & \cdots \\ (s_1 - 1)g_2 & (s_1 + 1)g_2 & 0 & 0 & \cdots \\ (s_1 - 1)g_3 & (s_1 + 1)g_3 & 0 & 0 & \cdots \\ \cdot & \cdot & & & \\ \cdot & \cdot & & & \\ \cdot & \cdot & & & \end{pmatrix} \cdot \begin{pmatrix} 1 & g_2 & g_3 & \cdots \\ 1 & -g_2 & -g_3 & \cdots \\ & & & \\ & & & \\ & & & \end{pmatrix} + I, \qquad (127)$$

$$S = \begin{pmatrix} -1 & s_1 g_2 & s_1 g_3 & \cdots \\ s_1 g_2 & -g_2^2 & -g_2 g_3 & \cdots \\ s_1 g_3 & -g_3 g_2 & -g_3^2 & \cdots \\ \cdot & \cdot & & \\ \cdot & \cdot & & \end{pmatrix} + I. \qquad (128)$$

Equation (128) can be simplified somewhat by choosing the terminals in the antenna transmission line in a new place. It is possible to choose this location so that

$$s_1 = +1; \qquad (129)$$

then

$$S = \begin{pmatrix} 0 & g_2 & g_3 & \cdots \\ g_2 & (1 - g_2^2) & -g_2 g_3 & \cdots \\ g_3 & -g_3 g_2 & (1 - g_3^2) & \cdots \\ \cdot & \cdot & \cdot & \cdots \\ \cdot & \cdot & \cdot & \cdots \end{pmatrix}. \qquad (130)$$

It will be recalled that there was a second alternative to Eq. (114) which is that

$$s_2 = 1. \qquad (131)$$

By following through a procedure analogous to that leading to Eq. (128), it may be shown that the condition

$$S_{11} = 0 \tag{132}$$

requires that

$$s_1 = -1, \tag{133}$$

and this leads to a scattering matrix of the form Eq. (130).

The scattering matrix (130) has some interesting properties. It is, for one thing, pure real, and this introduces a symmetry in the antenna pattern. To see this, note that if **a** is the column vector of a plane wave incident on the antenna, then **a*** is the column vector of a wave incident from the opposite direction. Thus if

$$\mathbf{Sa} = \mathbf{b}, \tag{134}$$

then

$$\mathbf{Sa}^* = \mathbf{b}^*.$$

Thus the components of **b*** differ only in phase from those of **b**, and the gain of the antenna in one direction is identical with that in the opposite direction.

Another property of antennas to which the scattering matrix (130) applies is that they scatter radiation with the same pattern that they normally radiate. To see this, it should be noted that the radiation pattern of the antenna is given by the first column of matrix (130). On the other hand, the matrix that represents the scattering of the antenna is

$$\mathbf{S}_2 - \mathbf{S}_0 = \begin{pmatrix} 0 & 0 & \cdots \\ 0 & \begin{matrix} -g_2^2 & -g_2 g_3 \\ -g_2 g_3 & -g_3^2 \\ \cdot & \cdot \\ \cdot & \cdot \\ \cdot & \cdot \end{matrix} & \begin{matrix} \cdots \\ \cdots \\ \cdots \\ \cdots \\ \cdots \end{matrix} \end{pmatrix}. \tag{135}$$

This matrix is of such a form that any column vector operating on it yields a column vector of the form

$$c \begin{pmatrix} 0 \\ g_2 \\ g_3 \\ \cdot \\ \cdot \\ \cdot \end{pmatrix},$$

where c is some number.

It can be easily verified that the scattered power is always equal to the absorbed power, independently of the mode of excitation. If the incident wave is represented by

$$\mathbf{a} = \begin{pmatrix} 0 \\ a_2 \\ a_3 \\ . \\ . \\ . \end{pmatrix}, \qquad (136)$$

the power absorbed by the antenna is

$$P_a = \tfrac{1}{2}(\tilde{r}\mathbf{a})^*(\tilde{r}\mathbf{a}), \qquad (137)$$

where r has the same significance as in Eq. (103). The power scattered by the antenna is

$$\begin{aligned}
P_s &= \tfrac{1}{2}[\overbrace{(\mathbf{S}_2 - \mathbf{S}_0)\mathbf{a}}]^*[(\mathbf{S}_2 - \mathbf{S}_0)\mathbf{a}] \\
&= \tfrac{1}{2}[(\tilde{r}\mathbf{a})\tilde{r}]^*[(\tilde{r}\mathbf{a})r] \\
&= \tfrac{1}{2}(\tilde{r}\mathbf{a})^*(\tilde{r}\mathbf{a}).
\end{aligned} \qquad (138)$$

Thus the scattered power is always equal to the absorbed power.

To recapitulate, a minimum-scattering antenna has identical gain in opposite directions. It scatters with the same antenna pattern that it radiates. It always scatters the same power it absorbs. A dipole antenna is an example of such a minimum-scattering antenna.

CHAPTER 10

MODE TRANSFORMATIONS

By E. M. Purcell and R. H. Dicke

Waveguides are usually designed to have only one propagating mode over the frequency range within which operation is intended. In the neighborhood of discontinuities in the guide, to be sure, the electromagnetic field cannot be described by a single mode, but the higher, nonpropagating modes that are there excited die out quickly with distance from the region of the discontinuity, and only the dominant mode, represented by a wave running in each direction, remains. As has been seen, this circumstance greatly simplifies the analysis of waveguide circuits, for it allows any length of guide of uniform cross section to be treated as a simple transmission line. In many instances, however, the waveguide-circuit designer must concern himself with the properties of guides that admit more than one propagating mode at the frequency used. Some of the practical reasons for interest in such questions are:

1. The special types of *symmetry* that some of the higher modes possess can sometimes be utilized to advantage. The waveguide rotary joint based on the axial symmetry of the TM_{01}-mode was the first such application and remains one of the most important.
2. The attenuation of a running wave in a waveguide, which is the result of the finite conductivity of the walls, decreases as the waveguide is enlarged in cross section for a given frequency and a given mode. When attenuation is unusually costly, it may be desirable to use for transmission either the lowest mode, in a guide too large to prevent propagation of other modes, or a higher mode such as the TE_{01}-mode for which the decrease of attenuation with increasing guide dimensions is exceptionally rapid.
3. In waveguides of circular cross section there is no single lowest mode. Rather, for any arbitrary choice of a direction perpendicular to the axis of the guide, there are two modes with identical cutoff wavelengths distinguished by the polarization which may be parallel or perpendicular to the reference direction.[1] Propagation of this type provides the guided-wave counterparts of the

[1] The lowest mode is likewise double, or degenerate, in this sense for a guide whose cross section is a square or any figure of $4n$-fold symmetry, but the choice of axis is then not wholly arbitrary.

multitude of effects associated with elliptically polarized light and lends itself to a variety of uses.

From a theoretical point of view the propagation of two modes, rather than just one mode, in a uniform guide is a subject of limited interest. It was pointed out in Sec. 2·18 that the normal modes are orthogonal and should therefore be treated entirely independently, at least in regions remote from discontinuities or nonuniformities in the guide. Thus, the problem of a uniform guide carrying power in, for example, n modes is formally identical with the problem of n separate guides each permitting propagation in its fundamental mode only, these n guides being connected together in various ways at certain junction points. The latter may be thought of as corresponding to discontinuities or irregularities in the guide carrying many modes simultaneously, since they serve to transfer energy from one mode to another. It is these junctions which will demand most of our attention in this chapter. Such a junction will be referred to as a mode transducer. This is a broad and rather loose use of a term which in circuit theory is customarily applied only to two-terminal-pair networks.

10·1. Mode Transducers.—The early literature on waveguides[1] contains many descriptions of means by which a particular waveguide mode can be excited. Usually the means suggested involve one or more antennas projecting into the guide and driven in appropriate phase relation to one another. This technique is illustrated by the elementary examples of Fig. 10·1, which shows how the TE_{20}-mode in a rectangular waveguide might be excited from a coaxial line.

In Fig. 10·1a the coaxial line is made to branch at A into two lines that are carried above and below the guide, respectively, to drive antennas B and C. These antennas are displaced by the same distance $S/2$ on either side of the center of the guide; and because they are driven *in phase*, the total length back to the junction A being the same for each branch, there is no excitation of the fundamental TE_{10}-mode. The TE_{20}-mode, however, is excited. If the dimensions of the guide are so chosen as to permit propagation of the TE_{10}- and TE_{20}-modes only, transmission will actually be effected only by the TE_{20}-mode if it is assumed that perfect geometrical symmetry has been realized. Several adjustable parameters are available, notably d, h, and S in the figure, by which one might hope to satisfy some additional electrical requirement,

---

[1] R. I. Sarbacher and W. A. Edson, *Hyper and Ultrahigh Frequency Engineering*, Wiley, New York, 1944, Sec. 6·14, p. 201, Sec. 8·8, p. 309; J. G. Brainerd, G. Koehler, H. J. Reich, and L. F. Woodruff, *Ultra-high-frequency Techniques*, Van Nostrand, New York, 1942, Chap. 14, p. 479; F. E. Terman, *Radio Engineers' Handbook*, McGraw-Hill, New York, 1943, Sec. 3, p. 261.

such as that of an impedance match looking into the coaxial input line when the waveguide has a reflectionless termination.

In Fig. 10·1b a similar transformation is effected in a different way. The antennas B and C are brought in from the same side of the guide and symmetrically disposed but are driven exactly out of phase, the line AC being made just one-half wavelength longer than AB. Again the TE_{20}-mode is excited, and the TE_{10}-mode is not. Whereas in the device of Fig. 10·1a the achievement of "mode purity," that is, the freedom from excitation of the TE_{10}-mode, depends on geometrical symmetry only,

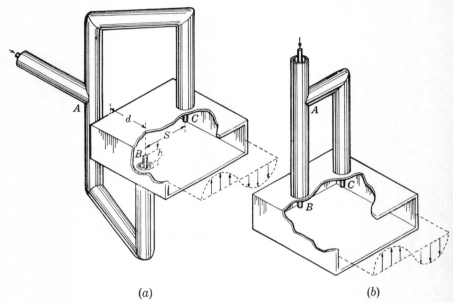

Fig. 10·1.—Excitation of the TE_{20}-mode from a coaxial line.

this is not true of the circuit of Fig. 10·1b. The extra section of line, nominally one-half wavelength long, will vary in electrical length if the frequency is changed. Thus, if extreme mode purity over a band of frequencies is a requirement, the scheme b is not a good one. This illustrates one of the practical considerations that influence the design of mode transducers.

There is another difference between the circuits of Fig. 10·1a and b which should be noted in passing. If the waveguide were large enough to allow still higher modes to propagate, these would, in general, not respond in the same way to excitation by the two probe arrangements a and b. The TE_{21}-mode, for example, the field configuration of which may be found in Fig. 10·2, would be excited by the system of Fig. 10·1b but not by that of Fig. 10·1a, whereas the converse is true for the TE_{11}-mode.

The essential difference between the antenna configurations a and b of Fig. 10·1 is that the former is *symmetrical* about a line or axis running longitudinally down the guide whereas the latter is *antisymmetrical* about a vertical plane through this axis.

Schemes of this sort can be elaborated in nearly endless variety. Also, use could be made of suitably placed current loops rather than the antennas of the electric type shown in Fig. 10·1. There are, however, difficulties associated with loop antennas if strong coupling—for example, an impedance match—between transmission line and waveguide is desired. In this case the loops are necessarily large, and large loops do not behave as simple magnetic dipoles.

The methods thus far discussed have not actually found wide use in microwave engineering for the reason that their application is confined to transitions between coaxial line and hollow waveguides. If it is desirable to use hollow waveguides for transmission because of the advantage in power-handling capacity or because of their convenient size, it is usually desirable for the same reasons to exclude even short sections of coaxial line from the circuit.

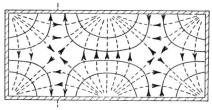

FIG. 10·2.—Field configurations for the TE_{21}-mode.

Waveguide-to-waveguide transitions, with a change in mode, are sometimes effected in a manner that resembles the multiantenna method already described. Figure 10·3 shows a transition from the TE_{10}- to the TE_{20}-mode in rectangular guide. The coupling between the guides is

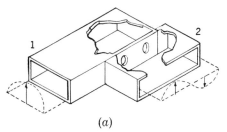
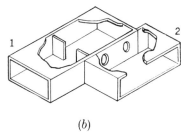

(a) (b)

FIG. 10·3.—Coupling from the TE_{10}-mode in one waveguide to the TE_{20}-mode in another by means of small holes.

effected by two small holes separated by a distance equal to one-half the wavelength of the TE_{10}-mode in guide (1). If the holes are not too large, the magnetic field in one hole due to a wave in guide (1) will be just opposite to the field in the other, which is the condition required for excitation of the TE_{20}-mode in guide (2), without simultaneous excitation of the TE_{10}-mode. The coupling is weak, however; that is, a wave entering guide (1) will be almost totally reflected, only a small fraction of the power being transferred to guide (2) when guide (2) is provided with a reflectionless termination.

If strong coupling is desired—for definiteness, let us say a matched condition—two courses are open. The nearly total reflection could be compensated by an impedance-matching element, which could be placed in either guide. For example, an inductive iris might be included in guide (1), as in Fig. 10·3b. In effect, this creates a resonant cavity at the end of guide (1). The matched condition would be attained only over a very narrow frequency range, and the resistive losses in the part of the guide that forms the cavity would be greatly increased, as would the

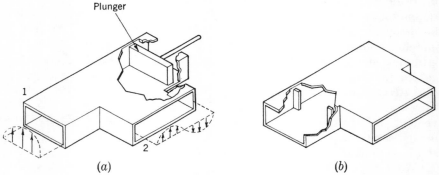

Fig. 10·4.—Matched mode transducer from the TE_{10}- to the TE_{20}-mode.

electric field strength, for a given power transfer. For some purposes, however, this might not be objectionable.

The original transducer can be altered in another way, by enlarging the coupling holes. If this course is pursued in an effort to obtain a matched mode transducer, the elementary explanation of the action of the two coupling holes soon loses its validity. Instead, there exists a complicated field configuration not easy to analyze, and there is no longer any assurance that only one mode will be excited in guide (2). Nevertheless, it may be possible to achieve the desired result by a cut-and-try procedure, in which one or more geometrical parameters are systematically varied and their effect upon the impedance match and upon the excitation of the unwanted mode in guide (2) is examined. Perhaps the simplest example of this treatment is shown in Fig. 10·4. Here the entire wall common to the two guides has been removed. A plunger in guide (a) provides an adjustable parameter which is found, experimentally, to control the excitation of the TE_{10}-mode to a considerable degree. In one specimen of this type, the fraction of the power incident on the junction from guide (1) which was transferred to the TE_{10}-mode in guide (2) could be reduced to 0.001 by adjustment of the plunger. At this plunger setting, moreover, the reflection of incident power was not large and could be eliminated by a relatively minor impedance transformation introduced ahead of the junction in guide (1).

The final junction, shown in Fig. 10·4b, belongs to a class of mode transducers of considerable importance in microwave engineering. Such transducers are characterized by an abrupt change of mode effected in a space that is closely coupled to both input and output pipes. The stored energy associated with this region is small, and as a result the

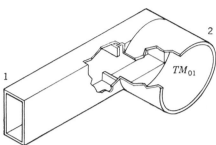

Fig. 10·5.—Transducer from the TE_{10}- to the TM_{01}-mode.

properties of the junction do not change rapidly with frequency (Chap. 5). The transducer shown in Fig. 10·5 belongs to this class also; it effects a transformation from the TE_{10}-mode in rectangular guide to the TM_{01}-mode in round guide.

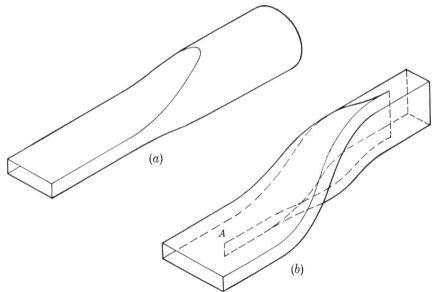

Fig. 10·6.—Mode transducers that employ a taper from one mode to another. (a) Rectangular-to-round transition, (b) transition from the TE_{10}- to the TE_{20}-mode.

A different approach to mode-transducer design is suggested by the tapered transmission line. It has already been pointed out (Chap. 6) that waveguides of different characteristic impedance can be joined in an almost reflectionless manner by means of an intermediate section in which a gradual change from the dimensions of one guide to those of the other takes place. Provided this tapered section is at least a few wavelengths long or, more precisely, if the change of cross section per wavelength is slight, propagation through the composite guide occurs with very little reflection over a wide frequency band. This principle is applied in the "rectangular-to-round" taper frequently used to connect a rectangular guide operating in the TE_{10}-mode to a guide of circular cross section operating in the TE_{11}-mode (Fig. 10·6a). Since the two modes involved have essentially the same character, this device scarcely

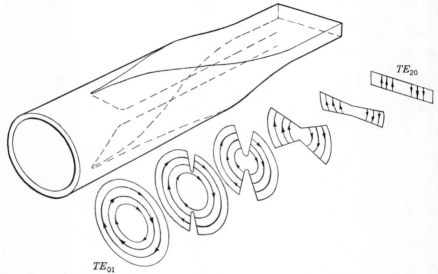

Fig. 10·7.—Transducer for converting from TE_{20}-mode in rectangular guide to TE_{01}-mode in round guide.

deserves to be called a mode transducer. Fig. 10·6b shows a "taper" by which the transformation from the TE_{10}- to the TE_{20}-mode can be effected in a gradual manner. The operation of this taper should be self-evident if the reader will notice that the abrupt cessation of the dividing wall at A can have no effect on the propagation of the TE_{20}-mode, because the currents that flow on the two sides of the wall are equal and opposite.

The TE_{20}-mode in rectangular guide can, in turn, be converted into the TE_{01}-mode in round guide by the transducer sketched in Fig. 10·7. Successive sections through the transition section are shown, and the field configuration is indicated.

10·2. General Properties of Mode Transducers.—The general statements that can be made about mode transducers are few and follow immediately from the observation that a waveguide in which n modes can propagate is equivalent to n guides, in each of which only one propagating mode can exist. Thus, the properties of a junction of m guides supporting n_1, n_2, . . . n_m modes, respectively, can be described by an N-by-N impedance matrix, N being the total number of propagating modes, or $N = \sum_{k=1}^{m} n_k$. The basis for this assertion is the orthogonality of the waveguide modes, which ensures the absence of cross terms involving the fields of two or more modes from the expression for the power flow normal to a boundary surface drawn across a guide.

To such a junction, then, all of the general theorems of Chap. 5 may be applied. The symmetry of the impedance matrix is again an expression of the reciprocity theorem or, in this case, of the *reversibility* of the mode transformation. To be sure, this result is fairly obvious, and it was not stressed in the preceding section in which each transducer was described arbitrarily in terms of a wave incident from one side only.

The transducer of Fig. 10·4 can be described by a 3-by-3 matrix,

$$Z = \begin{pmatrix} Z_{11} & Z_{12} & Z_{13} \\ Z_{12} & Z_{22} & Z_{23} \\ Z_{13} & Z_{23} & Z_{33} \end{pmatrix},$$

in which the subscript 1 is identified with the TE_{10}-mode in the narrow guide and the subscripts 2 and 3 with the TE_{20}- and TE_{10}-modes, respectively, in the wide guide. If the junction is lossless, an allowable assumption for junctions of this type, the elements of Z are all imaginary. An ideal transducer of this sort would be characterized by

$$Z_{23} = Z_{13} = 0, \tag{1}$$

and
$$\left. \begin{array}{r} Z_{11} - Z_{22} = 0, \\ Z_{12}^2 - Z_{11}Z_{22} - 1 = 0. \end{array} \right\} \tag{2}$$

Equation (1) expresses the fact that the TE_{10}-wave in the wide guide is not excited by the other waves. Equations (2) state the condition for a matched transition between terminals (1) and (3). Since the 3-by-3 impedance matrix for the lossless case originally contained six unknown quantities, the application of Eqs. (1) and (2) leaves only two quantities to be specified. One of these can be identified at once as Z_{33}; the remaining quantity is to be associated with the equivalent length of line between terminals (1) and (2). The result that has been obtained is perhaps so obvious as to command little respect. It amounts to the statement that an ideal mode transducer is equivalent to a lossless matched two-terminal-pair network and hence to some length of uniform line.

The third terminal pair is completely disconnected from the remainder of the system, by virtue of Eqs. (1). A wave incident *from* (3) will be totally reflected as if, for this mode, there existed a short circuit somewhere within the transducer. The apparent position of the short circuit is specified by Z_{33}.

In practice mode transducers are not quite perfect, either because they are not perfectly matched or because an appreciable amount of energy is transferred to the unwanted mode or for both reasons. A slight mismatch is usually tolerable and in any case presents no new problem. Excitation of one or more unwanted modes, on the other hand, often proves very troublesome. It is in the examination of this problem that the description of a transducer as an n-terminal-pair device is most helpful.

If a transducer is considered such as the one in Fig. 10·4, it is seen to involve only one unwanted mode and can be regarded as a three-terminal-pair device. The general three-arm lossless junction has the property (Chap. 9) that a short circuit suitably located in one arm, say (3), entirely decouples the other two arms from each other. Hence, no matter how weak the excitation of the unwanted mode, if it happens that the energy transferred to that mode is subsequently reflected back, without loss, to the junction and in precisely the most unfavorable phase, there will be no transfer of energy between guides (1) and (2) in the steady state. A wave incident upon the junction from guide (1) or from guide (2) will be totally reflected. A situation like this can arise when two mode transducers are connected together by joining the multimode pipe of one to the corresponding pipe of the other. This is equivalent to connecting two three-terminal-pair networks as in Fig. 10·8. The behavior of the combined network may be expected to depend critically on the length of the connecting pipe.

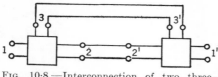

Fig. 10·8.—Interconnection of two three-terminal-pair networks.

Little can be said, in general, about junctions involving more than three modes. In special cases symmetry conditions may simplify the solution. An example is provided by the junction shown in Fig. 10·4a which, if the plunger is removed, becomes a four-terminal-pair device. It is a very special four-terminal-pair device, however, for inspection discloses that in so far as the four propagating modes are involved, it has precisely the symmetry of the magic-T four-junction. The TE_{20}-mode in arm (2) plays the role of the series branch of the magic T. Thus any result derived for the latter junction applies at once to Fig. 10·4a. It is interesting, although perhaps disappointing, to note that there is *not* obtained thereby a theoretical prediction that the plunger in Fig. 10·4 can be adjusted to eliminate excitation of the TE_{10}-mode in the wide guide. That is to say, the magic-T four-junction does not belong to the class of four-junctions that are equivalent to two cascaded three-junctions (see Chap. 9).

If a mode transducer cannot be assumed to be lossless, the only general statement that can be made about it is the statement of the reciprocity theorem.

10·3. The Problem of Measurement.—When two or more modes are excited in the same waveguide, it is not easy to determine by measurement the amplitude and phase of each. The difficulties are mainly practical ones, which would not arise if the experimenter had at his disposal infinitesimal probes that could be moved about within the guide at will without disturbing the existing fields.

Only in exceptional cases is the usual slotted-line technique applicable, for it is usually impossible to cut in a guide a slot that will be everywhere parallel to the lines of current flow in each of the two modes and will therefore seriously disturb the field of neither. One such exception is found in the case of a guide of circular cross section carrying the TE_{11}- and TM_{01}-modes, neither of which is modified by the presence of a narrow longitudinal slot. A probe in the form of a small antenna can be moved along such a slot, as in the usual standing-wave experiment, and a quantity can be measured that is some function of a linear combination of the amplitudes of the four traveling waves, two of which are associated with each mode. The interpretation of the "standing-wave" pattern so obtained is a more complicated affair than in the case of a single mode. For example, if the guide is matched for both modes so that there are two waves only, running in the same direction, the power picked up by the probe will nevertheless go through maxima and minima as the probe is moved along the guide. The reason for this is the difference between the guide wavelengths λ_{g1} and λ_{g2} of the two waves. Suppose that I_p, the rectified probe current, is proportional to the square of the electric field amplitude at the probe and that x measures the position of the probe along the guide. Then,

$$I_p \approx |e^{j(2\pi x/\lambda_{g1})} + be^{j\theta}e^{j(2\pi x/\lambda_{g2})}|^2, \qquad (3)$$

where $be^{j\theta}$ accounts for the relative amplitude of the second wave and its phase relative to the first wave at $x = 0$. The distance between successive positions of maximum I_p is

$$\frac{\lambda_{g1}\lambda_{g2}}{(\lambda_{g1} - \lambda_{g2})},$$

and the apparent "standing-wave ratio" is

$$\frac{I_{p\,\text{max}}}{I_{p\,\text{min}}} = \frac{(1+b)^2}{(1-b)^2}. \tag{4}$$

An analysis of the fields within such a guide, even when several waves are present, can be made on the basis of a number of probe readings taken at points distributed along the guide. In principle, six probe readings should suffice to determine the six quantities of interest: the relative amplitudes of the four traveling waves (three numbers) and their relative phases at some reference point (three numbers). This, of course, assumes that the guide wavelength for each mode is previously known as well as the relative degree of coupling to the probe, that is, the field at the probe in each mode when equal power is flowing in the two modes. If slots are ruled out, as they often are, fixed probes may be used. These may take the form of small holes in the guide walls coupling to external guides to which a detector can be connected. In any case, the deduction of the desired quantities from several probe readings is a process discouragingly tedious and probably inaccurate unless the circumstances permit simplifying approximations.

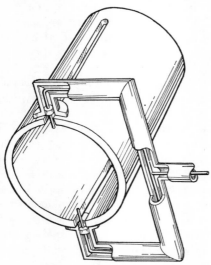

Fig. 10·9.—Probes for detecting the TM_{01}-mode.

A difficulty that may arise from the use of a probe is the coupling between the two modes caused by the probe itself. This is especially troublesome when one of the modes present has a low intensity. The difficulty can be avoided, at some cost in complexity, by a scheme that also furnishes a method for measuring the amplitude of one mode alone in the presence of the other. Let the circular guide supporting the TE_{11}- and TM_{01}-modes be provided with two diametrically opposite longitudinal slots and a traveling probe for each slot, as shown in Fig. 10·9. If the transmission lines coming from the probes are joined together externally at a point equidistant from the two probes, the voltage at this junction point will be a measure of the amplitude of the TM_{01}-mode only. Of course, this is merely an application of the multi-antenna mode transducer described at the beginning of Sec. 10·1. The principle can obviously be extended to any mode configuration if the complexity of the resulting apparatus can be tolerated.

When there are more than two allowed modes, the difficulty of quantitative determination of the fields within the guide becomes indeed formidable. In many cases, however, it is possible to ascertain the properties of the mode-transducing junction through measurements made in the single-mode guide that usually forms one arm of the transducer. Consider such a transducer, exemplified by the devices shown in Figs. 10·1 and 10·5, which serves to couple a guide carrying only one mode to another guide in which more than one mode can propagate. For brevity, the single-mode guide will be referred to as A and the multimode guide as B. It is assumed further, for simplicity, that only two modes are excited; the desired mode (2) is strongly excited, whereas the mode (3) is excited only weakly. The following remarks thus apply to the TE_{10}- to TM_{01}-mode transducer of Fig. 10·5 if the junction is free from irregularities that would introduce asymmetry and result in the excitation of the TE_{11}-mode as an elliptically polarized wave in pipe B, which is in this case the circular guide. Let us see how the electrical properties of the junction can be determined from standing-wave measurements made in guide A only. While the procedure to be described is a practical one, it is discussed here chiefly as an application of the ideas of the preceding section.

It should be noted at the outset that one operation which can be performed on the waves in pipe B in a nearly ideal manner is that of reflecting them from a short-circuiting plunger, which fits snugly into pipe B and the face of which is a flat metal surface normal to the axis of B. Such a plunger introduces no coupling between modes. Moreover, because the phase velocities of the two modes in B are different, motion of the plunger varies the terminations of the two modes at different rates.

It was shown in the preceding chapter that an equivalent circuit of the form shown in Fig. 10·10 can be drawn for a lossless three-terminal-pair network, and this representation is convenient for the present purpose. Three of the parameters in this general description are the line lengths l_1, l_2, and l_3, which are of no immediate concern. The other parameters are the turn ratios n_{12} and n_{13} of the two ideal transformers, and the convention is adopted that $n_{13} < 1$ for a voltage stepup from line (1) to line (3), which is the situation suggested in Fig. 10·10. The ideal transducer is then characterized by $n_{12} = 1$, $n_{13} = 0$. The case of interest, however, is $n_{12} = 1$, $0 < n_{13} \ll 1$.

The plunger in pipe B now serves to connect to lines (2) and (3) impedances Z_2 and Z_3 which, but for losses in pipe B, would be pure reactances jX_1 and jX_2. Any combination of X_1 and X_2 can be realized if pipe B is long enough and if the guide wavelengths of the modes in B are incommensurate. Measurements of the impedance of the junction, as seen from pipe A, for a sufficient number of plunger positions provides, in principle, information from which the parameters of the equivalent circuit can be determined. If $n_{13} \ll 1$, as assumed, it is well to examine the situation more closely. In particular, suppose that the most significant piece of information required is the value of n_{13}, which is a measure of the degree of coupling to the unwanted or parasitic mode. The effect of n_{13} will be most pronounced at plunger positions for which the line (3) is terminated in a very high impedance. In fact, if there were no loss in line (3), total reflection of power incident from A would result when the distance of the plunger from the transformer, which, of course, includes the distance l_3, was $(2n + 1)\lambda_{g_3}/4$, n being an integer and λ_{g_3} denoting the guide wavelength for the mode in question. Actually, because of attenuation in B, Z_3 will not become infinite but will be very large when the plunger is near the position just specified.

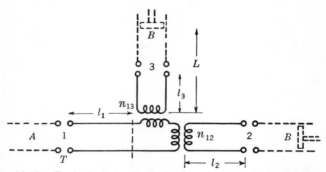

Fig. 10·10.—Equivalent circuit of a lossless three-terminal-pair network.

The behavior of Z_2 and the resulting variation of the impedance Z_1 observed at the terminals (1) can be analyzed by applying the transmission-line formulas. Suppose that the total distance L between the transformer n_{13} and the plunger is close to $m\lambda_{g_3}/4$ ($m = 1, 3, \cdots$) so that $kL = 2\pi L/\lambda_g = m\pi/2 + \theta$, where $\theta \ll 1$. Let the attenuation constant for this mode in B be denoted by α (as in Sec. 2·16, for example). If $\alpha\lambda \ll 1$, as is usually true, an approximate formula for Z_2 is easily derived

$$Z_2 \approx \frac{\dfrac{m\lambda_g\alpha}{4} - j\theta}{\left(\dfrac{m\lambda_g\alpha}{4}\right)^2 + \theta^2}. \tag{5}$$

The impedance connected to line (2) can be treated as a reactance jX_2, even though the attenuation constant of mode (2) is comparable to that of mode (3) if plunger positions in the neighborhood of $L_2 = m\lambda_{g_2}/4$ are excluded. The impedance seen looking to the right, at T, can then be written

MODE FILTERS AND MODE ABSORBERS

$$Z_T = n_{13}^2 \frac{\frac{m\lambda_{g_3}\alpha}{4} - j\theta}{\left(\frac{m\lambda_{g_3}\alpha}{4}\right)^2 + \theta^2} + jX_2. \qquad (6)$$

A measurement of the standing-wave ratio in A, as a function of plunger position, will disclose sharp minima, and it is in these regions that Eq. (6) applies. If the variation of X_2 through such a narrow region is neglected, the impedance Z_T will traverse a curve in the RX-plane similar to that drawn as Curve I in Fig. 10·11a, as the plunger is moved through a region $m = m_1$.

For m_2, some other value of m, the impedance Z_T will trace out Curve II, which is similar to Curve I but displaced by a different value of X_2. Now if m_1 and m_2 are both fairly large, and if $|m_1 - m_2| \ll m_1$, the maximum resistance observed at T will be approximately the same. Curves I and II, if transformed to the Smith chart of Fig. 10·11b, will then come close to the same resistance circle, provided the reference plane T has been properly selected. Since the position of T (governed by the length l_1) is not known in advance, the procedure is to plot the observed standing-wave ratio and the corresponding plunger position, measured in units of λ_{g2}, on the Smith chart. The whole diagram can then be rotated until the circles so obtained touch the same resistance circle. Thus, with some approximations, the position of T has been found, and if α is known n_{13} can be immediately computed.

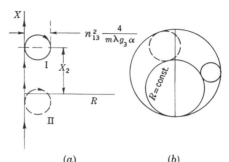

Fig. 10·11.—Impedance loci in the neighborhood of resonances.

The degree of coupling between mode (3) and the rest of the system can be estimated in another way if it can be arranged to excite in guide B this mode *only*. Such a wave, incident on a perfect transducer, will be totally reflected, as has already been noted in Sec. 10·2. If $n_{13} > 0$, however, there will be transferred to the other modes a small fraction of the incident power, a measurement of which will at once determine n_{13}.

10·4. Mode Filters and Mode Absorbers.—The example just discussed in Sec. 10·3 could have been analyzed much more easily had it been possible to assume that guide B was terminated in a *matched load* for mode (2) at the same time that mode (3) was totally reflected. The term jX_2 in Eq. (6) could then be omitted. In other words, a terminating element capable of reflecting one mode totally while absorbing another or transmitting it without reflection would be useful. In any given case, the electromagnetic field configuration of the modes in question will suggest at once means by which, at least in principle, this can be achieved. Figure 10·12a and b shows two such selective devices. Each consists of a screen of conducting wires (supported, if necessary, by a thin dielectric sheet). The wires are, respectively, perpendicular to the electric fields of the TM_{01}-mode (a) and the TE_{01}-mode (b). Hence the screen a will transmit TM_{01}-waves without reflection but will reflect, at least partially, TE_{01}-waves and others such as those of the TM_{11}-mode. Conversely, the screen b transmits TE_{01}-waves.

Modes that are reflected by such a screen, or *mode filter*, will not, in general, be *totally reflected*. The reflection coefficient will depend on the diameter and the number of the wires. In fact, such a screen can be regarded simply as a shunt susceptance, similar to a metal post in a

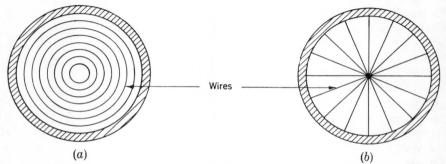

(a)   (b)

FIG. 10·12.—Wire screens for selective transmission of (a) TM_{01}-mode and (b) TE_{01}-mode.

rectangular guide. As a practical matter, it is difficult to achieve a very large reflection coefficient for one mode without causing appreciable reflection of the mode to which the screen is intended to be transparent, since complete transparency presupposes wires of infinitesimal diameter.

In certain special cases, one of which is represented by Fig. 10·12b, conducting sheets parallel to the guide axis may be employed. If each wire of the screen of Fig. 10·12b is translated along the guide, it generates a septum, and through these septa the TE_{01}-waves will pass undisturbed. Such a scheme is obviously limited to TE-modes, since the equivalent of a "magnetic wall" does not exist.

An important variant of Fig. 10·12a makes use of a single resonant ring, instead of the concentric rings of wires, to reflect TE_{11}-waves. Such a closed resonant ring would not totally reflect TE_{01}-waves.

The application of waveguide-circuit analysis to mode filters themselves is straightforward and not particularly interesting. If the filter in question does not couple one mode to another, it is necessary to insert in each of the equivalent single-mode transmission lines the appropriate impedance element.

If the wires shown in Fig. 10·12 are not perfectly conducting, the resulting device is a selective mode *absorber*. It is not easy in practice to construct a mode absorber that presents a matched load to one mode and that transmits another freely. This is not always necessary, however. Often it is required merely that sufficient selective attenuation be introduced in a multimode guide that connects two mode transducers to avoid difficulty arising from a resonance. This problem was suggested in the preceding section, the results of which are immediately applicable.

Let us suppose that the transducer to which Fig. 10·10 applies is provided with a matched termination for mode (2). This might, in practice, take the form of a similar transducer connected to transfer back to the single-mode guide. If the term jX_2 in Eq. (6) is replaced by unity, it is noted that the maximum reflection of a wave incident from pipe A will occur when $\theta = 0$, and the standing-wave ratio ρ_A in A can then be written, if ρ is not large, as

$$\rho_A - 1 \approx \frac{4n_{13}^2}{m\lambda_{g_3}\alpha}. \tag{7}$$

Now $e^{\frac{-m\lambda_g\alpha}{4}}$ is the attenuation of the amplitude of a wave running one way between the junction and the plunger in line (3). This might be written more generally as $e^{-\gamma}$, including the effect of any mode absorber introduced into pipe B. Then

$$\rho_A - 1 \approx \frac{(n_{13})^2}{\gamma}. \tag{8}$$

It should be noted that the quantity $(n_{13})^2$, which has been used to specify the coupling of the unwanted mode, is approximately $[(n_{13})^2 \ll 1]$ the ratio of the power transferred into mode (3) to that transferred into mode (2) when both are terminated without reflection. If it is required for a given application that $\rho_A - 1$ shall not exceed a specified number, regardless of the distance between the junction and the reflecting termination of the parasitic mode (3), Eq. (8) prescribes the amount of attenuation that must be provided in pipe B by a selective mode absorber.

10·5. The TE_{11}-mode in Round Guide.—The TE_{11}-mode in waveguide of circular cross section, which was mentioned at the beginning of this chapter, deserves special attention. Propagation can occur in two modes simultaneously in such a guide, but there is an essential arbitrariness in the description because the two modes have identical cutoff frequencies and because there is no unique direction to fix the coordinate system in which the field configuration is to be described.

First, two mutually perpendicular planes through the axis of the guide may be selected, and modes (1) and (2) may be defined as the two TE_{11}-modes with electric fields parallel and perpendicular respectively to each of these planes. Any wave in this pipe, for example a wave polarized in a plane making some angle with the two chosen planes, can be represented as the sum of waves associated with modes (1) and (2). The wave, in other words, can be resolved into what will be called two *basic polarizations*. The basic polarizations need not be perpendicular to each other, but this choice is usually convenient; and if it is made, the modes (1) and (2) will be said to be orthogonal.

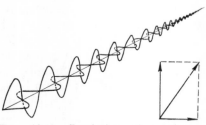

Fig. 10·13.—Resolution of a plane-polarized wave into two basic waves.

Figure 10·13 shows how a plane-polarized wave at some arbitrary polarization angle may be resolved into two basic polarizations. It is clear that there is nothing fundamental about the particular directions chosen for the basic polarizations. If these basic polarizations are rotated to a new position, an alternative representation is obtained for the plane wave.

In a similar way two plane waves may be superposed to make a circularly polarized wave as shown in Fig. 10·14. Thus, a circularly polarized wave may be said to be resolvable into these two basic polarizations. Basic waves need not be taken as plane polarized. For instance, it is well known that two circularly polarized waves can be combined to form a plane-polarized wave. Thus the two circular polarizations form a possible set of basic waves.

Fig. 10·14.—Resolution of a circularly polarized wave into two plane polarized waves.

For the analysis of circuits and circuit elements employing the TE_{11}-modes, a description by means of scattering matrices, which were introduced in Sec. 5·13, is helpful. As before, a waveguide carrying two modes is to be regarded as equivalent to two transmission lines, and the terminal of such a guide is represented by two terminal pairs, each pair being associated with one of the basic polarizations. The new element in the situation arises from the possibility of shifting, at will, from one set of basic polarizations to another. Such a change in representation involves a transformation of the scattering matrix that "mixes" the original terminal pairs, so to speak. Although this is a procedure that can be carried out formally for any guide carrying two modes, it usually lacks the useful physical interpretation that can be given in this special case.

10·6. Permissible Transformations of a Scattering Matrix.—Consider the system shown in Fig. 10·15. The two planes of polarization are designated by the numbers (1) and (2). The remainder of the terminals are located in rectangular waveguides and are designated by the numbers (3) and (4). As was shown in Chap. 5, if the incident waves are represented by the column vector

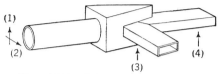

FIG. 10·15.—Waveguide junction with four terminals. Two terminals are in rectangular guide; the other two terminals are two basic polarizations in the round waveguide.

$$\mathbf{a} = \begin{pmatrix} a_1 \\ a_2 \\ a_3 \\ a_4 \end{pmatrix} \tag{9}$$

and the scattered waves by

$$\mathbf{b} = \begin{pmatrix} b_1 \\ b_2 \\ b_3 \\ b_4 \end{pmatrix}, \tag{10}$$

then

$$\mathbf{Sa} = \mathbf{b}, \tag{11}$$

where the matrix $\mathbf{S}$ is symmetrical and unitary. It is implicitly assumed that the polarizations (1) and (2) are normal to each other.

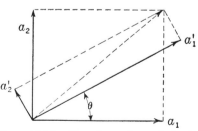

FIG. 10·16.—Transformation of basic polarizations by a rotation.

It is evident that the planes of polarization chosen in Fig. 10·15 are not unique and a rotation of (1) and (2) to a new position should be permissible. A rotation can be characterized by a linear transformation of the wave amplitudes. A rotation of the basic polarizations by an angle θ changes the wave amplitudes a_1 and a_2, in Fig. 10·16, into the corresponding primed quantities; thus

$$\left. \begin{array}{l} a_1 = a_1' \cos \theta - a_2' \sin \theta, \\ a_2 = a_1' \sin \theta + a_2' \cos \theta. \end{array} \right\} \tag{12}$$

Equations (12) may be written, using the matrix notation, as

$$a = Ta', \qquad (13)$$

where

$$T = \begin{pmatrix} \cos\theta & -\sin\theta & & \\ \sin\theta & \cos\theta & & 0 \\ \hline & & 1 & 0 \\ & 0 & & \\ & & 0 & 1 \end{pmatrix}. \qquad (14)$$

In a completely analogous way,

$$b = Tb'; \qquad (15)$$

and if these are substituted in Eq. (11),

$$b' = T^{-1}STa'. \qquad (16)$$

The matrix

$$S' = T^{-1}ST \qquad (17)$$

is the scattering matrix for the new terminals. It may be seen by inspection of Eq. (14) that

$$T^{-1} = \tilde{T}, \qquad (18)$$

and thus T is orthogonal. It was found in Sec. 5·12 that a sufficient condition that Eq. (16) provide a permissible transformation is that the matrix T be real and orthogonal. A permissible transformation is one that leads to a symmetrical and unitary scattering matrix.

Many transformations T that are real and satisfy Eq. (18) and are therefore permissible are not particularly useful, for, in general, these transformations mix all the components of a with one another. If the transformation is restricted to a mixing of the terminals (1) and (2), it is found that the resulting transformation is either the rotation described by Eq. (14) or a reflection in some plane containing the axis of the guide. There are cases in which the more general transformation (having no simple geometrical significance) is useful, but they will not be considered here.

In addition to the rotation or reflection of the reference polarizations, there is another important type of transformation. This is a transformation which is equivalent to choosing the reference plane of each of the terminals in a new place. It is effected by the diagonal matrix P discussed in Sec. 5·16. If the kth terminal reference plane is shifted out a distance d_k, then

SEC. 10·6] PERMISSIBLE TRANSFORMATIONS 353

$$P = \begin{pmatrix} e^{j\phi_1} & 0 & 0 & 0 \\ 0 & e^{j\phi_2} & 0 & 0 \\ 0 & 0 & e^{j\phi_3} & 0 \\ 0 & 0 & 0 & e^{j\phi_4} \end{pmatrix}, \quad (19)$$

where $\phi_k = 2\pi(d_k/\lambda_k)$. It should be noted that $P^* = P^{-1}$. The matrix S' obtained by the transformation

$$S' = PSP \quad (20)$$

is both symmetrical and unitary.

If transformations of the type given by Eqs. (17) and (19) are induced successively, it can be shown that any such sequence is equivalent to a single transformation $S' = \tilde{U}SU$, where U is a unitary matrix.

Consider the following example. Let

$$U = \frac{1}{\sqrt{2}} \begin{pmatrix} 1 & -1 & & \\ & & 0 & \\ j & j & & \\ \hline & & \sqrt{2} & 0 \\ & 0 & & \\ & & 0 & \sqrt{2} \end{pmatrix}. \quad (21)$$

The matrix U may be represented as the product

$$U = PT,$$

where T is a rotation [see Eq. (14)] of 45° and

$$P = \begin{pmatrix} 1 & 0 & & \\ & & 0 & \\ 0 & j & & \\ \hline & & 1 & 0 \\ & 0 & & \\ & & 0 & 1 \end{pmatrix}. \quad (22)$$

It may easily be seen that transformation by U changes the reference polarizations into circular polarizations. To see this, note that a particular case of $a = Ua'$ is

$$\frac{1}{\sqrt{2}} \begin{pmatrix} 1 \\ j \\ 0 \\ 0 \end{pmatrix} = U \begin{pmatrix} 1 \\ 0 \\ 0 \\ 0 \end{pmatrix}. \quad (23)$$

The left-hand side of Eq. (23) represents a circularly polarized wave.

The matrix given by Eq. (21) is a special case of a transformation that, in the more general case, changes the basic polarizations from plane to elliptical polarizations.

10·7. Quarter-wave Pipe.—Perhaps the simplest device that displays the characteristic features of TE_{11}-mode propagation is the *quarter-wave pipe*. It is a length of guide having a circular cross section at either end but departing from circular symmetry in the intermediate section, where it is symmetrical about a plane only. The departure from circular symmetry in this region is just great enough to introduce a net difference of $\pi/2$ in the electrical length of the pipe with respect to basic polarizations chosen respectively parallel to and perpendicular to the symmetry

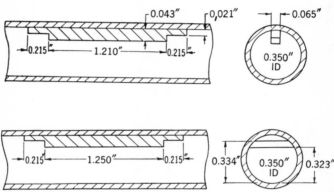

Fig. 10·17.—Quarter-wave pipes. The steps at each end of the fin are quarter-wavelength transformers for matching; the dimensions given are for $\lambda = 1.25$ cm.

plane. In other words, if two waves polarized in these directions are passed through the pipe, a quarter-wave retardation of one with respect to the other results. This device is the exact analogue of the optical quarter-wave plate made of a doubly refracting crystal, and it is employed in the same manner, that is, to convert a linearly polarized wave into a circularly polarized wave and vice versa.

The requisite retardation can be effected in many ways, with negligible reflection of the incident wave. It suffices, for example, to make the guide slightly elliptical in cross section for some distance, or a dielectric slab may be introduced or a metal fin, as in Fig. 10·17. In any case, if reflections can be neglected, the scattering matrix for the quarter-wave pipe can be written

$$S = e^{j\phi} \begin{pmatrix} 0 & 0 & 1 & 0 \\ 0 & 0 & 0 & j \\ 1 & 0 & 0 & 0 \\ 0 & j & 0 & 0 \end{pmatrix}. \tag{24}$$

An incident linearly polarized wave

$$\begin{pmatrix} 1 \\ 1 \\ 0 \\ 0 \end{pmatrix}$$

emerges as a circularly polarized wave

$$e^{j\phi}\begin{pmatrix} 0 \\ 0 \\ 1 \\ j \end{pmatrix}.$$

The angle ϕ is simply a measure of the electrical length of the pipe for a wave polarized along (1). If S_{24} and S_{42} in Eq. (24) are replaced by -1, the scattering matrix is obtained for a *half-wave pipe*, a device that can be used, for example, to convert a right-hand circularly polarized wave into a left-hand circularly polarized wave.

10·8. Rotary Phase Shifter.—The rotary phase shifter[1] may be cited to illustrate the use of the transformation discussed in the previous section to facilitate analysis. Figure 10·18 is an exploded view of such a phase shifter. It consists of two sections of round guide separated by a sandwich of a half-wave pipe between two quarter-wave pipes. The heavy arrows on these sections indicate the planes of polarization that are retarded by the pipes. The half-wave pipe is mounted in such a way as to be rotatable. If the incident plane wave has a plane of polarization (1) or (2), it will be shown that the device acts as a phase shifter as the angle θ is changed.

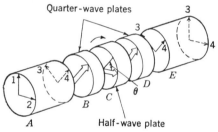

Fig. 10·18.—Rotary phase shifter.

Consider first the section marked A in Fig. 10·18. This is just a section of uniform waveguide with arrows indicating the directions of polarization for the four terminal pairs. It is easily seen that the scattering matrix for this section is

$$S_1 = \frac{1}{\sqrt{2}} e^{j\phi_1} \begin{pmatrix} 0 & 0 & 1 & 1 \\ 0 & 0 & -1 & 1 \\ \hline 1 & -1 & 0 & 0 \\ 1 & 1 & 0 & 0 \end{pmatrix}, \quad (25)$$

[1] A. G. Fox, "Waveguide Filters and Transformers," *BTL Memorandum* MM-41-160-25.

where ϕ_1 is a phase angle depending on the length of the guide. After passing through the quarter-wave pipe, the scattering matrix becomes

$$\mathsf{S}_2 = \frac{1}{\sqrt{2}} e^{j\phi_2} \begin{pmatrix} 0 & 0 & 1 & -j \\ 0 & 0 & -1 & -j \\ \hline 1 & -1 & 0 & 0 \\ -j & -j & 0 & 0 \end{pmatrix}. \qquad (26)$$

The axis of the half-wave pipe is at an angle θ with respect to the direction (4). Therefore, the problem is simplified if the matrix of Eq. (26) is transformed by a rotation through an angle θ by the matrix

$$\mathsf{T} = \begin{pmatrix} 1 & 0 & & \\ 0 & 1 & & 0 \\ \hline & & \cos\theta & -\sin\theta \\ & 0 & \sin\theta & \cos\theta \end{pmatrix}. \qquad (27)$$

Then

$$\mathsf{S}'_2 = \tilde{\mathsf{T}}\mathsf{S}_2\mathsf{T} = \frac{1}{\sqrt{2}} e^{j\phi_2} \begin{pmatrix} & & e^{-j\theta} & -je^{-j\theta} \\ & 0 & -e^{j\theta} & -je^{j\theta} \\ \hline e^{-j\theta} & -e^{j\theta} & & \\ -je^{-j\theta} & -je^{j\theta} & & 0 \end{pmatrix}. \qquad (28)$$

Now passage through the half-wave pipe has the effect of changing the sign of the fourth row and column. Therefore,

$$\mathsf{S}_3 = \frac{1}{\sqrt{2}} e^{j\phi_3} \begin{pmatrix} & & e^{-j\theta} & +je^{-j\theta} \\ & 0 & -e^{j\theta} & +je^{j\theta} \\ \hline e^{-j\theta} & -e^{j\theta} & & \\ je^{-j\theta} & +je^{j\theta} & & 0 \end{pmatrix}. \qquad (29)$$

A rotation by $-\theta$ is now indicated, in order that the scattering matrix of the quarter-wave pipe should be simple. This rotation is effected by $\mathsf{T}' = \mathsf{T}(-\theta) = \tilde{\mathsf{T}}(\theta)$. Thus,

$$S_4 = \tilde{T}'S_3\tilde{T}' = \frac{1}{\sqrt{2}} e^{j\phi_3} \begin{pmatrix} 0 & \begin{array}{cc} e^{-2j\theta} & je^{-2j\theta} \\ -e^{2j\theta} & je^{2j\theta} \end{array} \\ \begin{array}{cc} e^{-2j\theta} & -e^{2j\theta} \\ je^{-2j\theta} & +je^{2j\theta} \end{array} & 0 \end{pmatrix}. \quad (30)$$

Passage through the final quarter-wave pipe changes Eq. (30) to

$$S_5 = \frac{1}{\sqrt{2}} e^{j\phi_5} \begin{pmatrix} 0 & \begin{array}{cc} e^{-2j\theta} & e^{-2j\theta} \\ -e^{2j\theta} & e^{2j\theta} \end{array} \\ \begin{array}{cc} e^{-2j\theta} & -e^{2j\theta} \\ e^{-2j\theta} & +e^{2j\theta} \end{array} & 0 \end{pmatrix}. \quad (31)$$

A rotation by 45° brings the basic polarization planes into coincidence with (1) and (2) and is accomplished by setting $S_6 = \tilde{T}''ST''$, in which T'' is given by Eq. (27) with $\theta = 45°$. When this transformation is carried out and the terminals subsequently moved to the end of the guide, the final result is

$$S_7 = e^{j\phi_7} \begin{pmatrix} 0 & \begin{array}{cc} e^{-2j\theta} & 0 \\ 0 & e^{2j\theta} \end{array} \\ \begin{array}{cc} e^{-2j\theta} & 0 \\ 0 & e^{2\theta j} \end{array} & 0 \end{pmatrix}. \quad (32)$$

It should be noted that for an incident wave

$$\mathbf{a} = \begin{pmatrix} 1 \\ 0 \\ 0 \\ 0 \end{pmatrix}, \quad (33)$$

the scattered wave is

$$\mathbf{b} = e^{j(\phi_7 - 2\theta)} \begin{pmatrix} 0 \\ 0 \\ 1 \\ 0 \end{pmatrix}. \quad (34)$$

The outgoing wave contains θ as a phase angle. Thus, as θ is increased, the equivalent line length between input and output terminals is increased. For an incident wave the scattered wave is

$$a = \begin{pmatrix} 0 \\ 1 \\ 0 \\ 0 \end{pmatrix}, \quad (35)$$

$$b = e^{j(\phi_7 + 2\theta)} \begin{pmatrix} 0 \\ 0 \\ 0 \\ 1 \end{pmatrix}, \quad (36)$$

and the line length is decreased as θ is increased.

10·9. A Rectangular-to-round Transducer.—Figure 10·19 shows a rectangular-to-round mode transducer. The junction between the two guides may be symmetrical as shown (i.e., have one or more symmetry planes), or it may be completely unsymmetrical. The terminals (2) and (3) may be plane polarizations located symmetrically with respect to a symmetry plane as shown, or they may be elliptical polarizations in any position.

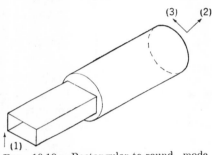

FIG. 10·19.—Rectangular-to-round mode transducer.

It is evident that since the junction has three terminal pairs and negligible loss, all the results obtained for lossless T-junctions of various kinds are also applicable here. These results are mainly the following:

A. It is impossible to make a T-junction that is completely matched.
B. If a T-junction is matched at two terminal pairs, the third terminal pair is completely decoupled from the other two.
C. A plunger in any arm of a T-junction can be so placed as to decouple the remaining two arms.
D. If the T-junction is symmetrical, a plunger in the arm containing the symmetry plane can be so placed that the remaining two junctions are matched.

Each of these results is applicable to Fig. 10·19. For instance, assume that the E-plane of the figure is a symmetry plane of the junction. Assume that the terminals (2) and (3) are elliptical polarizations whose major axes make angles of 45° with respect to the symmetry plane. The plane polarization indicated in the figure is clearly a limiting case. Two circular polarizations form the other limiting case. From the result (D) it is clear that there is a position of a plunger in the rectangular guide such that the terminals (2) and (3) are matched. For example, there is a position of the plunger such that a linearly polarized wave, polarized as (2) in the figure, is reflected back, polarized as (3).

The statements (A), (B), and (C) apply regardless of the symmetry of the junction or orientation of the polarization. The interpretation of the statement (A) is obvious. Theorem B states that if terminal (1) is matched and (2) is the polarization excited in the round guide by a wave incident in (1), then the polarization (3) orthogonal to (2) is completely decoupled from (1) and (2). Alternatively, if a polarization (2) is completely reflected polarized as (3), then (1) does not couple with the round guide.

The theorem (C) has an interesting application to the junction of Fig. 10·19. If a plunger is inserted in (1), then for any polarization (2) there is a position of the plunger such that this polarization is reflected back unchanged. Also, for this same position the orthogonal polarization (3) is unchanged by reflection.

10·10. Discontinuity in Round Guide.—A round guide with a discontinuity in the middle is an example of a four-terminal-pair network. As such, it satisfies the conditions applicable to directional couplers. One of the most interesting of these is the following: Any four-terminal-pair junction that is completely matched[1] is a directional coupler. To state it in other words, if a junction of four waveguides is matched at each waveguide, then for any guide there is one other guide to which it does not couple (see Chap. 9).

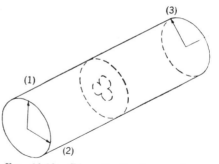

FIG. 10·20.—Discontinuity in round waveguide.

This theorem may be taken over completely in the case of the round-guide junction. If, for a given set of basic polarizations, the junction is matched, then it has the properties of a directional coupler. Referring to Fig. 10·20 this implies that the polarization (1) (shown as a plane polarization, whereas it may be elliptical or circular) does not couple to one of the polarizations (2), (3), or (4).

The case of no coupling between guides (1) and (2) is rather trivial. In this case no power is reflected in either (1) or (2) for incident power in (2) or (1). This is also true for any linear combination of (1) and (2).

[1] As usual the term "matched" is intended to mean that if all terminal pairs but one are provided with reflectionless terminations, a wave incident at the one terminal pair is not reflected.

For any polarization incident on one side, the power is transmitted wholly through to the other side.

If (1) and (2) couple, then there is zero coupling between (1) and (3) and between (2) and (4). Thus, all los less matched junctions of two round guides may be divided into two classes:

1. Those in which power entering one guide in any polarization is wholly transmitted out the other guide.
2. Those in which each of the basic polarizations of one guide couples with one and only one basic polarization of the other guide.

10·11. Principal Axes in Round Guide.—In Fig. 10·21 is shown a round guide having a lossless termination. Plane polarizations (1) and (2) are chosen as basic polarizations for the device. Since the directions of (1) and (2) were chosen at random, the resulting scattering matrix is complicated. It will now be shown that it is possible to choose directions for the polarizations (1) and (2) such that these polarizations are not changed by the reflection at the termination.

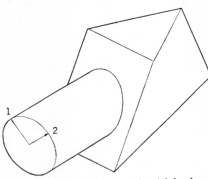

FIG. 10·21.—Round waveguide with lossless termination.

Let S be the scattering matrix (a 2-by-2 matrix, of course) of the junction referred to the original terminals. Now assume that there is some incident wave **a** which is reflected back unchanged, except perhaps. for a change in phase. Assume, in other words, that

$$S\mathbf{a} = s\mathbf{a}, \qquad (37)$$

where $s = e^{j\phi}$.

In order for Eq. (37) to be satisfied by a nonvanishing **a**, the determinant

$$\begin{vmatrix} S_{11} - s & S_{12} \\ S_{21} & S_{22} - s \end{vmatrix} = 0. \qquad (38)$$

This leads to a quadratic equation in s (called the characteristic equation of S), the two roots of which are, in general, unequal. The roots will be of the form $e^{j\phi}$ because the box is lossless.

Provided s has one of the roots of Eq. (38) as its value, there will be nontrivial solutions of Eq. (37). As yet, however, there is no guarantee that the resulting solutions for **a** do not represent elliptical polarizations. It will now be shown that the **a**'s do indeed represent linearly polarized waves.

If Eq. (37) is multiplied by s^*S^*, and if it is remembered that

then
$$S^* = S^{-1},$$
$$S^*a = s^*a. \qquad (39)$$
If the complex conjugate of Eq. (39) is added to Eq. (37), the result is
$$S(a + a^*) = s(a + a^*). \qquad (40)$$
But $a + a^*$ represents a pure real column vector, or a *linearly* polarized wave. Hence, linearly polarized waves can be found that satisfy Eq. (37). These particular planes of polarization afford convenient basic polarizations because the resulting scattering matrix has only diagonal elements.

If the termination of the waveguide is lossy, the scattering matrix is no longer unitary and Eq. (40) is no longer valid. It is still possible to obtain solutions of Eq. (37), but these solutions, in general, represent elliptically polarized waves.

10·12. Resonance in a Closed Circular Guide.—The problem to be discussed in this section derives its importance from the use of waveguide rotary joints employing the TM_{01}-mode. It has already been pointed out in Sec. 10·3 that even very weak excitation of one of the TE_{11}-modes by a transducer such as that shown in Fig. 10·5 may have a pronounced effect on the behavior of a rotary joint made from two such transducers connected in cascade in the event of a TE_{11}-*resonance* in the pipe connecting the two. To find the conditions under which such a resonance can occur, it is necessary to take into account the fact that a TE_{11}-wave incident on one of the transducers is reflected with a phase that depends, in general, on the polarization of the wave. That is to say, the transducer, as regards TE_{11}-waves, acts very

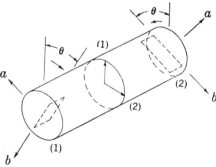

FIG. 10·22.—Round waveguide closed at both ends.

much like the junction of Fig. 10·21. There is, of course, some "loss" representing the leakage of power through the transducer into the rectangular guide, but this can be disregarded in the search for the resonance condition, as can the presence of the TM_{01}-wave in the pipe. The problem can thus be reduced to that of the section of round guide closed at both ends, which is shown in Fig. 10·22.

The waveguide is cut in the middle to allow the two sections to be rotated with respect to each other. The two halves of the guide are identical. Their terminations are totally reflecting and are identical also, although of some irregular shape. The principal axes of the ends are indicated by the directions a and b. The median plane is taken as a common terminal plane for the two halves of the system, and the polarizations (1) and (2) are common to the two halves of the system. The problem is to find the conditions for which, at a fixed frequency, a wave can exist in this closed lossless system. This problem was first treated by H. K. Farr.[1]

The two ends are designated by the numbers 1 and 2. If the scattering matrices for the two ends are S_1 and S_2, then

$$S_1 a_1 = b_1,$$
$$S_2 a_2 = b_2. \qquad (41)$$

Since the incident wave for one end is the scattered wave coming from the other end, it is clear that

$$\begin{aligned} a_1 &= b_2, \\ a_2 &= b_1. \end{aligned} \qquad (42)$$

From Eqs. (41) and (42),

$$S_1 S_2 a_2 = S_1 b_2 = S_1 a_1 = b_1 = a_2. \qquad (43)$$

From this result it may be inferred, as in the preceding section, that if solutions are to exist with $a_2 \neq 0$, we must have

$$\det (S_1 S_2 - I) = 0. \qquad (44)$$

Now S_1 and S_2 are not unrelated, for the two ends are the same, except for rotation of one with respect to the other. It is convenient to take as reference the situation in which the ends are oriented so that their principal axes are parallel and coincide with the basic polarizations (1) and (2) of Fig. 10·22. In this case $\theta = 0$,

$$S_1 = S_2 = \begin{pmatrix} e^{j\phi_1} & 0 \\ 0 & e^{j\phi_2} \end{pmatrix}. \qquad (45)$$

If now the two halves of the guide are rotated by θ and $-\theta$, respectively, we have

$$S_1 = S(\theta)$$

and

$$S_2 = S(-\theta),$$

where $S(\theta)$ is the matrix of Eq. (45) transformed by a rotation of $T(\theta)$, according to Eq. (14). Thus S_1 becomes

$$S_1 = \begin{pmatrix} \cos^2 \theta e^{j\phi_1} + \sin^2 \theta e^{j\phi_2} & \cos\theta \sin\theta (e^{j\phi_1} - e^{j\phi_2}) \\ \cos\theta \sin\theta (e^{j\phi_1} - e^{j\phi_2}) & \cos^2 \theta e^{j\phi_2} + \sin^2 \theta e^{j\phi_1} \end{pmatrix}, \qquad (46)$$

[1] H. K. Farr, "A Theory of Resonance in Rotary Joints of the TM_{01} Type," RL Report No. 993, January 1946.

and S_2 is the same except for a change in sign of the off-diagonal elements. It is merely a matter of straightforward algebra to substitute these matrices in Eq. (44) and thus obtain a relation connecting θ, ϕ_1, and ϕ_2. If, instead of ϕ_1 and ϕ_2, the angles ϵ and δ are introduced, these angles being defined by

$$\epsilon = \phi_1 + \phi_2, \quad \delta = \phi_1 - \phi_2, \tag{47}$$

the result assumes a particularly compact form

$$\sin^2 2\theta = \frac{\cos \delta - \cos \epsilon}{\cos \delta - 1}. \tag{48}$$

The angles δ and ϵ have a simple interpretation. If the termination is represented by two selective reflectors or short circuits, one acting on a wave polarized along one of the principal axes and the other effective for

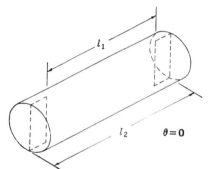

Fig. 10·23.—The quantities l_1 and l_2 in resonant round waveguide.

a wave polarized along the other principal axis, then $\epsilon \lambda_g/4\pi$ is the sum of the distances from the midplane to each of the two reflectors and $\delta \lambda_g/4\pi$ is the distance between the two reflectors. For the structure shown in Fig. 10·23,

$$\epsilon = \frac{2\pi}{\lambda_g} (l_1 + l_2)$$

and

$$\delta = \frac{2\pi}{\lambda_g} (l_1 - l_2),$$

assuming that the edge of the fin acts as a short circuit *at that point* for a wave polarized parallel to the fin.

If the guide were terminated at each end by a flat plunger, δ would be zero and solutions of Eq. (48) would exist only for $\epsilon = 2n\pi$, ($n = 1$, 2, . . .). This means that the total length of the pipe between plungers must be $n\lambda_g/2$, the familiar condition for resonances in a closed uniform guide.

If δ is not zero, Eq. (48) shows that resonance can occur at some value of θ for any ϵ within a limited range. The situation is best illustrated by curves of constant δ on the $(\theta - \epsilon)$-plane, as in Fig. 10·24. If $\delta = \pi/4$, for example, resonance will occur for some value of θ if ϵ lies within the

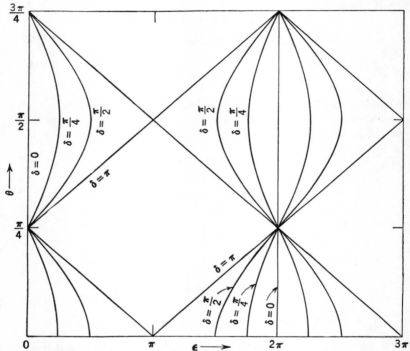

Fig. 10·24.—The angle of rotation θ for the occurrence of resonance of a closed round waveguide.

range $2n\pi \pm \pi/4$, $(n = 1, 2, \cdots)$; and if $\delta = \pi$, a resonance will occur for some value of θ, no matter what value of ϵ is chosen.

If resonances are to be avoided, it is clearly desirable to have δ small, thereby increasing the range of lengths for a given frequency (or, at fixed length, increasing the range of frequencies) over which resonances are excluded.

CHAPTER 11

DIELECTRICS IN WAVEGUIDES

By C. G. Montgomery

11·1. Waveguides Filled with Dielectric Materials.—In the preceding portions of this book, it has usually been assumed that the medium in which the phenomena take place is either free space or a material in which the energy losses can be neglected. It is now desired to investigate more closely what happens when the energy losses must be taken into account as well as to indicate the mechanism of these energy losses. Dielectric loss can be formally described by considering the dielectric constant to be a complex number which can be written as

$$\epsilon = \epsilon' - j\epsilon''.$$

In a condenser made with a lossy dielectric, the power loss for a given applied voltage will be proportional to the real part of the admittance. The conductance is

$$G = \mathrm{Re}\,(Y) = \mathrm{Re}\,(j\omega C) = \mathrm{Re}\left(j\omega \frac{\epsilon}{\epsilon_0} C_0\right)$$
$$= \mathrm{Re}\left(j\omega \frac{\epsilon'}{\epsilon_0} C_0 + \omega \frac{\epsilon''}{\epsilon_0} C_0\right) = \omega \frac{\epsilon''}{\epsilon_0} C_0.$$

It is evident therefore, that the amount of energy loss is proportional to ϵ''. The phase angle φ of the dielectric constant is often used as a measure of the loss, and

$$\tan \varphi = \frac{\epsilon''}{\epsilon'}.$$

If ϵ is a complex number, then the wave impedance becomes complex, and the transverse electric and magnetic fields become out of phase. The propagation constant γ also becomes complex. If frequencies above the cutoff frequency are considered, then

$$\gamma = \alpha + j\beta = \alpha + j\left(\frac{2\pi}{\lambda_g}\right),$$

and

$$\alpha = \lambda_g \frac{\omega^2 \epsilon'' \mu}{4\pi} = \pi \frac{\lambda_g}{\lambda^2} \frac{\epsilon''}{\epsilon'} = \pi \frac{\lambda_g}{\lambda_0^2} \frac{\epsilon''}{\epsilon_0} = \pi \frac{\lambda_g}{\lambda^2} \tan \varphi, \qquad (1)$$

where λ is the wavelength corresponding to the frequency ω of uniform plane waves in the medium and λ_0 is the corresponding wavelength of uniform plane waves in free space. The guide wavelength λ_g is given by

$$\lambda_g = \lambda_0 \left(\frac{\epsilon'}{\epsilon_0} - \frac{\lambda_0^2}{\lambda_c^2}\right)^{-\frac{1}{2}} \left\{\frac{1}{2} + \frac{1}{2}\left[1 + \left(\frac{\dfrac{\epsilon''}{\epsilon_0}}{\dfrac{\epsilon'}{\epsilon_0} - \dfrac{\lambda_0^2}{\lambda_c^2}}\right)^2\right]^{\frac{1}{2}}\right\}^{-\frac{1}{2}}. \quad (2)$$

If it may be assumed that ϵ''/ϵ_0 is small compared with unity, λ_g may be expanded in powers of ϵ''/ϵ_0

$$\lambda_g = \lambda_0 \left(\frac{\epsilon'}{\epsilon_0} - \frac{\lambda_0^2}{\lambda_c^2}\right)^{-\frac{1}{2}} \left[1 - \frac{1}{8}\left(\frac{\dfrac{\epsilon''}{\epsilon_0}}{\dfrac{\epsilon'}{\epsilon_0} - \dfrac{\lambda_0^2}{\lambda_c^2}}\right)^2 + \cdots\right] \quad (3)$$

Thus, to the first order, the wavelength is unchanged by the presence of the ϵ'' term in the dielectric constant. These formulas are also valid, of course, for TEM-waves. It is necessary only to set $\lambda_g = \lambda$ and $\lambda_c = \infty$. The equations become

$$\begin{aligned}
\lambda &= \lambda_0 \sqrt{\frac{\epsilon_0}{\epsilon'}} \left\{\frac{1}{2} + \frac{1}{2}\left[1 + \left(\frac{\epsilon''}{\epsilon'}\right)^2\right]^{\frac{1}{2}}\right\}^{-\frac{1}{2}} \\
&= \lambda_0 \sqrt{\frac{2\epsilon_0}{\epsilon'(1 + \sec \varphi)}}, \\
\lambda &\approx \lambda_0 \sqrt{\frac{\epsilon_0}{\epsilon'}} \left(1 - \frac{\varphi^2}{8} + \cdots\right).
\end{aligned} \quad (4)$$

If it is desired to express the losses of the medium in terms of an equivalent conductivity of the medium, then the substitution

$$\sigma = \omega \epsilon''$$

gives the results

$$\alpha = \frac{\lambda_g}{\lambda_0} \frac{\sigma}{2} \sqrt{\frac{\mu_0}{\epsilon_0}} \quad (5)$$

and

$$\lambda_g = \lambda_0 \left(\frac{\epsilon'}{\epsilon_0} - \frac{\lambda_0^2}{\lambda_c^2}\right)^{-\frac{1}{2}} \left[1 - \frac{1}{8}\left(\frac{\dfrac{\sigma}{2\pi} \dfrac{\mu_0}{\epsilon_0} \lambda_0}{\dfrac{\epsilon'}{\epsilon_0} - \dfrac{\lambda_0^2}{\lambda_c^2}}\right)^2 + \cdots\right]. \quad (6)$$

Still another set of parameters is sometimes used to describe a lossy medium. The index of refraction is assumed to be complex, and quantities n and k are defined by

$$\frac{\epsilon'}{\epsilon_0} - j\frac{\epsilon''}{\epsilon_0} = (n - jk)^2. \quad (7)$$

SEC. 11·1] WAVEGUIDES FILLED WITH DIELECTRICS 367

The quantity k is called the absorption index. It can be seen that

$$\begin{aligned}\frac{\epsilon'}{\epsilon_0} &= n^2 - k^2, \\ \frac{\epsilon''}{\epsilon_0} &= 2nk.\end{aligned}\quad\quad(8)$$

The parameters n and k are most often used when dealing with optical frequencies but occasionally are used in the microwave region.

When a dielectric is introduced into a waveguide, the transmission loss is increased because of the losses in the dielectric and the increase is measured by α as calculated above. The Joule heating losses in the walls of the waveguide are also altered. The change in the metal losses caused by a change in the real part of the dielectric constant is a simple one. Since the propagation constant is given by

$$\gamma^2 = k^2 - \omega^2\epsilon\mu,$$

we see that a fractional change in ϵ produces a change in γ of the same amount as would be produced by the same fractional change in ω^2. Thus, it can be seen from Fig. 2·13 that an increase in ϵ will decrease the metal losses at a frequency near the cutoff frequency but will increase these losses at higher frequencies. Since the only metals considered here are those which have conductivities so high that the form of the fields in the waveguide is unchanged by the losses that result, the total loss can be computed by adding directly the values of α for the metal and for the dielectric.

To find the effect on the metal losses of an imaginary term in ϵ, a more detailed examination must be made. In Chap. 2 it was shown that the value of the attenuation constant for the metal, α_m, is given by the expression

$$\alpha_m = \frac{1}{2\gamma\sigma\ \mathrm{Re}\ (Z_w)}\frac{\int_{\mathrm{walls}}|H_{\mathrm{tan}}|^2\,dS}{\int_{\mathrm{cross\ section}}|H_t|^2\,dS},$$

where H_{tan} is the magnetic field tangential to the metal walls and H_t is the transverse magnetic field. Two cases must now be treated. First, E-modes may be considered, where no longitudinal magnetic field exists, and $H_t = H_{\mathrm{tan}}$ on the boundary of the waveguide. The ratio of the two integrals in the expression for α_m depends only on the shape and size of the guide, and the dielectric constant occurs only in $\mathrm{Re}\ (Z_E)$. Thus

$$\frac{1}{\mathrm{Re}\ (Z_E)} = \frac{1}{\mathrm{Re}\left(\dfrac{\gamma}{j\pi\epsilon}\right)} = \frac{(\omega\epsilon')^2 + (\omega\epsilon'')^2}{\alpha\omega\epsilon'' + \beta\omega\epsilon'}.$$

For a lossless dielectric,

$$\frac{1}{\text{Re}(Z'_E)} = \frac{\omega \epsilon'}{\beta}.$$

Hence, the fractional change in the metal losses resulting from an imaginary term in the dielectric constant can be written as

$$\frac{\Delta \alpha_m}{\alpha_m} = \frac{\dfrac{1}{\text{Re}(Z_E)} - \dfrac{1}{\text{Re}(Z'_E)}}{\dfrac{1}{\text{Re}(Z_E)}}$$

or

$$\frac{\Delta \alpha_m}{\alpha_m} = \sin^2 \varphi \left[1 - \frac{1}{2}\left(\frac{\lambda_g}{\lambda}\right)^2\right]. \qquad (9)$$

It is seen from this equation that the effect is of the second order in ϵ''/ϵ'. Moreover, the change can be either positive or negative, depending on whether λ_g is less or greater than $\sqrt{2}\,\lambda$.

For H-modes, the situation is somewhat more complicated, because there is a longitudinal magnetic field. On the boundary of the waveguide

$$|H_{\tan}|^2 = |H_t|^2 + |H_l|^2,$$

where H_l is the longitudinal magnetic field. However, H_l is proportional to H_t, and from Eq. (2·89) it is seen that

$$H_l = \text{const.}\,\frac{H_t}{\gamma}.$$

Thus α_m can be split into two parts and written as

$$\alpha_m = \frac{1}{\text{Re}(Z_H)}\left(A + \frac{B}{|\gamma|^2}\right),$$

where A and B are geometrical constants. However,

$$\text{Re}(Z_H) = \text{Re}\left(\frac{j\omega\mu}{\gamma}\right) = \frac{\beta\omega\mu}{\alpha^2 + \beta^2},$$

since

$$|\gamma|^2 = \alpha^2 + \beta^2.$$

Hence

$$\frac{\Delta \alpha_m}{\alpha_m} = \frac{\alpha^2}{\beta^2}\,\frac{A}{A + \dfrac{B}{\beta^2}}.$$

A and B are positive quantities, and therefore $\Delta \alpha_m / \alpha_m$ is always positive. For high frequencies, β becomes small and $\Delta \alpha_m / \alpha_m$ approaches $\alpha^2 \dfrac{A}{B}$ and becomes independent of frequency. It is seen that the expression

$$\frac{\Delta\alpha_m}{\alpha_m} = \frac{\tan^2\varphi}{4}\left(\frac{\lambda_g}{\lambda}\right)^4 \frac{A}{A + \left(\frac{\lambda_g}{2\pi}\right)^2 B} \tag{10}$$

is of the second order in ϵ''/ϵ'. Here again, because α_m has been calculated only under the assumption that $\alpha/\beta \ll 1$, the total attenuation is given by $\alpha + \alpha_m$.

11·2. Reflection from a Change in Dielectric Constant.—The boundary conditions that are to be satisfied at the surface of a discontinuity in the properties of the medium are that the tangential components of the electric and magnetic fields must be continuous. Suppose that there is such a discontinuity in a waveguide and that the plane of the discontinuity is perpendicular to the axis of the guide. The power flow down the guide is $P = \dfrac{\text{Re}(Z_w)}{2} \int |H_t|^2 \, ds$, where the integral is taken over the cross section of the guide. Now H_t must be continuous across the interface of the two dielectrics, whereas the wave impedance Z_w will change discontinuously. Therefore, it is clear that there must be a reflected wave at the interface, and this will be sufficient to satisfy the conditions at the boundary. Moreover, the equivalent impedance that will correctly describe this reflection should be chosen proportional to the wave impedance. Hence, the equivalent circuit representing the discontinuity is simply that shown in Fig. 11·1.

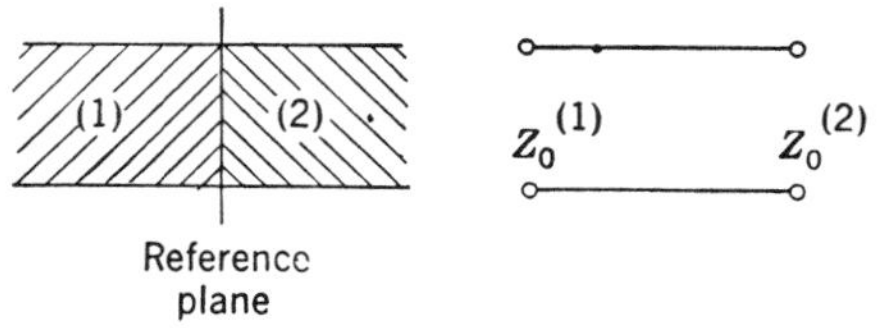

Fig. 11·1.—Equivalent circuit for a change in dielectric constant.

The standing-wave ratio r is the ratio of the impedances taken in such a way that $r > 1$. The position of the minimum is either at the junction or a quarter wavelength away from it, depending on the location of the observation point with respect to the reference plane.

The value of the wave impedance, as derived in Chap. 2, differs for E- and H-modes, and the two values are

$$Z_H = \frac{j\omega\mu}{\gamma}, \qquad Z_E = \frac{\gamma}{j\omega\mu}. \tag{2·91}$$

For the case of no loss, these reduce to

$$Z_H = \frac{\lambda_g}{\lambda}\zeta, \tag{2·35}$$

$$Z_E = \frac{\lambda}{\lambda_g}\zeta, \tag{2·39}$$

where λ is the wavelength in the medium and ζ is $\sqrt{\mu/\epsilon}$. In terms of the wavelength in free space, the impedance of guide (2) relative to that of guide (1) may be written

$$\frac{Z_H^{(2)}}{Z_H^{(1)}} = \sqrt{\frac{\dfrac{\epsilon^{(1)}}{\epsilon_0} - \left(\dfrac{\lambda_0}{\lambda_c}\right)^2}{\dfrac{\epsilon^{(2)}}{\epsilon_0} - \left(\dfrac{\lambda_0}{\lambda_c}\right)^2}}, \tag{11}$$

and

$$\frac{Z_E^{(2)}}{Z_E^{(1)}} = \frac{\epsilon^{(1)}}{\epsilon^{(2)}} \sqrt{\frac{\dfrac{\epsilon^{(2)}}{\epsilon_0} - \left(\dfrac{\lambda_0}{\lambda_c}\right)^2}{\dfrac{\epsilon^{(1)}}{\epsilon_0} - \left(\dfrac{\lambda_0}{\lambda_c}\right)^2}}. \tag{12}$$

If $\epsilon^{(2)} > \epsilon^{(1)}$, then the relative H-mode impedance decreases monotonically as $\epsilon^{(2)}$ increases. The relative E-mode impedance is not a monotonic function but may increase, pass through a maximum, and then decrease. It is easy to show that the maximum value occurs when

$$\frac{\epsilon^{(2)}}{\epsilon_0} = 2\left(\frac{\lambda_0}{\lambda_c}\right)^2, \tag{13}$$

and the maximum value is

$$\frac{Z_E^{(2)}}{Z_E^{(1)}} = \frac{\epsilon^{(1)}}{2\epsilon_0 \left(\dfrac{\lambda_0}{\lambda_c}\right)} \frac{1}{\sqrt{\dfrac{\epsilon^{(1)}}{\epsilon_0} - \left(\dfrac{\lambda_0}{\lambda_c}\right)^2}}. \tag{14}$$

Thus, for sufficiently large values of λ_0/λ_c, there are two values of $\epsilon^{(2)}$ that make $Z_E^{(2)}/Z_E^{(1)}$ equal to unity, namely,

$$\epsilon^{(2)} = \epsilon^{(1)}$$

and

$$\epsilon^{(2)} = \frac{\epsilon_0}{\left(\dfrac{\lambda_c}{\lambda_0}\right)^2 - \dfrac{\epsilon_0}{\epsilon^{(1)}}}$$

The second value corresponds to the case of Brewster's angle for plane waves, that is, the angle at which, when the electric vector lies in the plane of incidence, there is no reflected ray from the boundary between two mediums. Figure 11·2 shows values of the relative wave impedance for $(\lambda_0/\lambda_c)^2 = 0.8$ when $\epsilon^{(1)} = \epsilon_0$.

When the dielectric material is lossy, the wave impedance becomes complex. Its value can be expressed in terms of the complex dielectric constant as follows. From Eq. (2·91),

$$Z_H = \frac{j\omega\mu}{\alpha + j\beta}$$

$$= \frac{\omega\mu\beta}{\alpha^2 + \beta^2}\left(1 + j\frac{\alpha}{\beta}\right)$$

$$= \frac{\omega\mu}{\beta + \dfrac{\omega^4 \epsilon''^2 \mu^2}{4\beta^3}}\left(1 + j\frac{\omega\epsilon''\mu}{2\beta^2}\right), \quad (15)$$

and

$$Z_E = \frac{\beta}{\omega\epsilon' + \omega\dfrac{\epsilon''^2}{\epsilon'}}\left[1 + \frac{\omega^2\epsilon''^2\mu}{2\beta^2\epsilon'} - j\left(\frac{\omega^2\epsilon''\mu}{2\beta^2} - \frac{\epsilon''}{\epsilon'}\right)\right]. \quad (16)$$

It is seen from these expressions that the change in the real part of Z_H or Z_E is of the second order in ϵ''. For TE-modes, the presence of

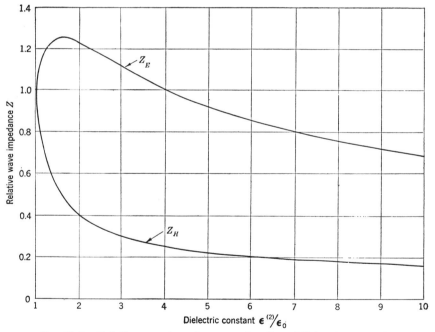

Fig. 11·2.—Relative wave impedance as a function of dielectric constant.

loss makes the wave impedance slightly inductive; for TM-modes, the reactive part is either positive or negative depending on the frequency. Near cutoff Z_E is capacitive; at higher frequencies it becomes inductive. The expressions for Z_H and Z_E can be transformed to a somewhat more useful form in the following way:

$$Z_H = \frac{\lambda_g}{\lambda}\sqrt{\frac{\mu}{\epsilon'}}\frac{1 + j\frac{1}{2}\tan\varphi\left(\frac{\lambda_g}{\lambda}\right)^2}{1 + \frac{1}{4}\tan^2\varphi\left(\frac{\lambda_g}{\lambda}\right)^2}, \tag{17}$$

$$Z_E = \frac{\lambda}{\lambda_g}\sqrt{\frac{\mu}{\epsilon'}}\cos^2\varphi\left\{1 + \frac{1}{2}\tan^2\varphi\left(\frac{\lambda_g}{\lambda}\right)^2 - j\tan\varphi\left[\frac{1}{2}\left(\frac{\lambda_g}{\lambda}\right)^2 - 1\right]\right\}. \tag{18}$$

The wave impedance Z_E is thus purely resistive when

$$\left(\frac{\lambda_g}{\lambda}\right)^2 = 2,$$

and its value is

$$Z_E = \sqrt{\frac{\mu}{2\epsilon'}}.$$

To obtain corresponding expressions for transmission in a dielectric medium but not in a waveguide, it is necessary only to set $\lambda_g = \lambda$.

If it be assumed that guide (1) is empty and that guide (2) is filled with a lossy dielectric, then the reflection produced at the junction can be easily calculated. Let the dielectric-filled guide be so terminated that there are no reflected waves from the termination. If the dielectric is very lossy, then the use of a sufficient length of the material is the obvious solution. If the dielectric has low loss, the use of a tapered section from the dielectric back to empty guide is a relatively easy method of obtaining a matched termination. The reflection will be equal to that in a line terminated by the relative impedance $Z = R + jX$. The quantities R and X may be obtained from the impedance given by Eq. (17) or (18), divided by the value of the wave impedance when no dielectric is present. If an H-mode is present, and if λ_g' denotes the guide wavelength in the dielectric and λ_g the guide wavelength without dielectric, then

$$\left.\begin{array}{l} R = \dfrac{\dfrac{\lambda_g'}{\lambda_g}}{1 + \dfrac{1}{4}\tan^2\varphi\left(\dfrac{\lambda_g'}{\lambda'}\right)^2}, \\[1em] X = \dfrac{1}{2}R\tan\varphi\left(\dfrac{\lambda_g'}{\lambda'}\right)^2, \end{array}\right\} \tag{19}$$

where λ' is the wavelength of uniform plane waves in the dielectric medium. Equations (2) and (4) may be used to obtain λ_g' and λ' in terms of ϵ' and ϵ''. Thus, expressions for ϵ' and ϵ'' in terms of R and X may be written. These expressions are useful in the determination of dielectric constants. If $X/R = \tan\theta$, then ϵ'/ϵ_0 is given by the positive solution of

$$\left(\frac{\epsilon'}{\epsilon_0}\right)^2 - \frac{\epsilon'}{\epsilon_0}\left[\nu^2(1 + \tan^2\theta) + (1 - \tan^2\theta)^2\frac{1-\nu^2}{R^2}\right] + \nu^4\tan^2\theta = 0, \quad (20)$$

where $\nu = \lambda_0/\lambda_c$. When ϵ'/ϵ_0 has been found, it is possible to calculate

$$\tan\varphi = \frac{\epsilon''}{\epsilon'} = \tan 2\theta\,\frac{\dfrac{\epsilon'}{\epsilon_0} - \nu^2}{\dfrac{\epsilon'}{\epsilon_0}}. \quad (21)$$

These expressions can be simplified considerably in most cases. Thus if $\varphi \ll 1$ and $\theta \ll 1$,

$$\left.\begin{array}{l}\dfrac{\epsilon'}{\epsilon_0} = \nu^2 + \dfrac{1-\nu^2}{R^2} - \theta^2\dfrac{1-\nu^2}{R^2}\dfrac{\nu^2 + 2\dfrac{1-\nu^2}{R^2}}{\nu^2 + \dfrac{1-\nu^2}{R^2}}, \\[2em] \text{and} \\[1em] \varphi = 2\theta\,\dfrac{\dfrac{1-\nu^2}{R^2}}{\nu^2 + \dfrac{1-\nu^2}{R^2}}.\end{array}\right\} \quad (22)$$

If θ is not small but ν^2 may be neglected compared with ϵ'/ϵ_0, then

$$\left.\begin{array}{l}\tan\varphi = \tan 2\theta, \\[0.5em] \dfrac{\epsilon'}{\epsilon_0} = \left|\dfrac{\epsilon}{\epsilon_0}\right|\cos\varphi, \\[0.5em] \dfrac{\epsilon''}{\epsilon_0} = \left|\dfrac{\epsilon}{\epsilon_0}\right|\sin\varphi, \\[0.5em] \left|\dfrac{\epsilon}{\epsilon_0}\right| = \dfrac{1-\nu^2}{R^2}(1 - \tan^4\theta).\end{array}\right\} \quad (23)$$

Suppose that the dielectric-filled waveguide is terminated not in a matched impedance but in a short circuit. The reflection may be most conveniently expressed by the equivalent circuit shown in Fig. 11·3. The value of the admittance Y_{11} of a transmission line of length d is

$$Y_{11} = Y_0 \coth \gamma d. \quad (24)$$

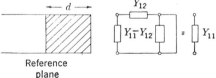

Fig. 11·3.—Equivalent circuit of short-circuited dielectric-filled waveguide.

The characteristic admittance Y_0 is given by Eq. (15) or (16), and γ is given by Eqs. (1) and (2). The values of Y_{11} have often been used to determine the values of dielectric constants at microwave frequencies.[1] For this purpose Eq. (24) must be expressed in terms of ϵ' and $\tan \varphi$ and solved. Since the equation is a complex transcendental one, it is necessary to resort to graphical methods.

11·3. Dielectric Plates in Waveguides.—Elementary transmission-line theory may be applied to calculate the impedance of dielectric plates perpendicular to the axis of the waveguide. A few simple cases will be discussed to show certain applications of dielectric materials to microwave techniques. Let it be assumed, for the moment, that losses may be neglected. The impedance at the face of the plate is

$$Z_{\text{in}} = Z \frac{1 + jZ \tan \beta t}{Z + j \tan \beta t},$$

where Z is the relative impedance and t the thickness of the plate. When $\beta t = n\pi$, where n is an integer, $Z_{\text{in}} = 1$. Thus a plate n half-wavelengths

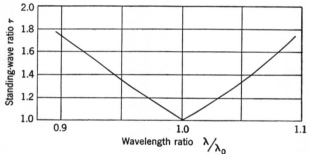

FIG. 11·4.—Variation of standing-wave ratio with wavelength for a dielectric plate one-half wavelength thick.

thick is reflectionless. The standing-wave ratio introduced by the plate is

$$r = \frac{|(Z^2 - 1) \tan \beta t| + \sqrt{4Z^2 + (Z^2 + 1)^2 \tan^2 \beta t}}{|(Z^2 - 1) \tan \beta t| - \sqrt{4Z^2 + (Z^2 + 1)^2 \tan^2 \beta t}}. \quad (25)$$

Figure 11·4 shows the variation in r with wavelength in the neighborhood of the wavelength λ_0 when the plate is a half wavelength thick. The dielectric constant is $\epsilon/\epsilon_0 = 2.45$, and $\lambda_0/\lambda_c = 1/\sqrt{2}$. If the dielectric is lossy, there is no plate thickness for which the reflection is zero. For small values of βt,

$$Z_{\text{in}} = 1 - j\beta t \frac{1 - Z^2}{Z},$$

$$r = 1 + \beta t \frac{1 - Z^2}{Z}.$$

[1] S. Roberts and A. von Hippel, "A New Method for Measuring Dielectric Constant and Loss in the Range of Centimeter Waves," Contribution from Department of Massachusetts Institute of Technology Electrical Engineering, March, 1941.

SEC. 11·3]   *DIELECTRIC PLATES IN WAVEGUIDES*   375

The thin dielectric plate is thus equivalent to a small shunt capacitance across the waveguide. By means of an inductive iris placed at the face of the plate, the circuit may be made resonant and reflectionless.

It is possible to construct a tuning device of two movable dielectric slabs each one-quarter wavelength thick. When the two slabs are in contact, the combination is reflectionless. When they are separated by

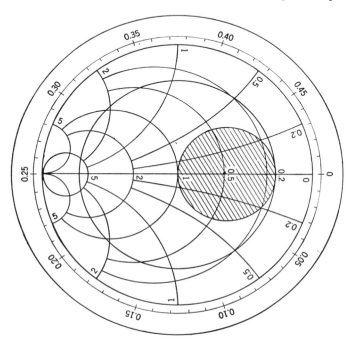

FIG. 11·5.—Impedance chart illustrating the use of two movable quarter-wavelength dielectric slabs in waveguide as a tuning device.

a quarter wavelength, the reflection is a maximum and corresponds to a resistance R, given by

$$R = \left[\frac{1 - \left(\frac{\lambda_0}{\lambda_c}\right)^2}{\frac{\epsilon}{\epsilon_0} - \left(\frac{\lambda_0}{\lambda_c}\right)^2} \right]^2,$$

for H-modes, at the face of the first dielectric slab. Thus, referring to Fig. 11·5, the impedance at the face of the first slab can have values along the boundary of the shaded circle. If the combination is moved, as a whole, along the guide, all impedances within the larger circle can be attained at a given point.

It is also possible to insert a quarter-wavelength transformer to match from an empty guide to one filled with a dielectric, the transformer section being composed of the guide filled with a dielectric having a dielectric constant of an intermediate value. This intermediate value, for H-modes, is such as to make the guide wavelength the geometric mean of the guide wavelengths in the full and empty guides. It should be noted that no end corrections are necessary, since the junction effect is absent.

A plate composed of a lossy dielectric can be treated in a manner similar to that for the lossless plate. The expressions become much more complicated in form, but their derivation is straightforward. The general nature of the behavior may be seen from Fig. 11·6 which shows some experimental values of the transmitted and reflected power as a function of the thickness of a piece of plywood in waveguide. The observations were taken at a wavelength of 10 cm in 1.5- by-3-in. waveguide.

11·4. The Nature of Dielectric Phenomena.—In a homogeneous isotropic dielectric medium, the electric displacement differs from its value in free space by the polarization **P** which is the electric moment per unit volume in the medium;

$$\mathbf{P} = \mathbf{D} - \epsilon_0 \mathbf{E} = (\epsilon - \epsilon_0)\mathbf{E}. \quad (26)$$

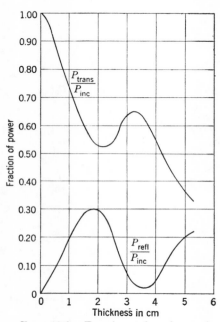

Fig. 11·6.—Experimental values of transmitted and reflected power as a function of the thickness of a plywood plug in waveguide.

The electric susceptibility χ_e is related to P by

$$\chi_e = \frac{P}{\epsilon_0 E} = \frac{\epsilon}{\epsilon_0} - 1 = k_e - 1. \quad (27)$$

If there are N molecules per unit volume and the electric moment of one molecule is m, then $P = Nm$. To explain the observed value of the dielectric constant and its variation with frequency, it is necessary to consider the nature of the mechanisms whereby molecules can acquire an electric moment. If a small conducting sphere of radius a is placed in an electric field F, the conduction electrons will distribute themselves in such a way that the sphere acquires an electric moment equal to $4\pi\epsilon_0 a^3 F$. In a similar manner the electrons in a molecule will redistribute themselves in such a manner that the molecule will acquire an electric moment whose magnitude is proportional to F. We write $m = \alpha_1 F$ and call α_1

the polarizability of the molecule. This polarizability will be independent of temperature. On the other hand, the molecule may have, by virtue of its structure, a permanent electric moment of magnitude m. When the field is applied, the molecule will tend to turn and align itself with the moment in the direction of the field. This alignment will be destroyed by the collisions and other random forces that the molecule experiences. Since the energy of the electric moment when it makes an angle θ with the field F is $-mF\cos\theta$, the mean value of the moment may be calculated by means of Boltzmann's distribution law,

$$\bar{m} = \frac{\int e^{m\cos\theta(F/kT)} m \cos\theta\, d\Omega}{\int e^{m\cos\theta(F/kT)}\, d\Omega},$$

where $d\Omega$ is an element of solid angle, k is Boltzmann's constant, T is the absolute temperature, and the integrals are taken over all directions. For values of T such that $kT \gg mF$, it is found that

$$\bar{m} = \frac{m^2 F}{3kT}. \qquad (28)$$

Thus a permanent electric moment results in a polarizability that is inversely proportional to the absolute temperature. Strictly speaking, $\bar{m}$ should be calculated not from the classic Boltzmann law but from the corresponding quantum-theory expression. However, it is easily shown that for the case of high temperatures, an identical result[1] is obtained.

The expression for the total average moment per molecule is, then,

$$\bar{m} = \left(\alpha_1 + \frac{m^2}{3kT}\right) F. \qquad (29)$$

It is now necessary to find the value of F, the total field acting upon the molecule. This field is made up of a contribution from the external applied field E plus the contributions from the fields of the other dipole moments in the medium. The contribution of the dipoles is very difficult to estimate. This is evident from the fact that the number of dipoles at a distance r from the point under consideration is proportional to $4\pi r^2\, dr$, and the field of a dipole is proportional to $(\cos\theta)/r^3$. Hence the total effect is proportional to $\int (4\pi \cos\theta/r)dr$. For large r this integral vanishes, since there are equal contributions from those regions where $\cos\theta$ has opposite signs. For small r, however, the integral diverges and the value of the field is extremely sensitive to the particular assumptions made about the nearest neighbors of the dipole under consideration.

[1] For a more complete discussion of the details of the quantum-theory calculation, the reader is referred to J. H. Van Vleck, *Electric and Magnetic Susceptibilities*, Oxford, New York, 1932.

In the case of a gas, no very large error is made by neglecting the field of the other dipoles entirely and putting $F = E$. For more concentrated substances, the next approximation may be considered to be the classic one of Clausius. Clausius assumed that the dipole could be thought of as being within a small spherical cavity within the medium. In this case

$$F = E + \frac{P}{3\epsilon_0}. \tag{30}$$

The use of this expression for F, together with the equations already obtained, to find ϵ results in the relation

$$\frac{\epsilon}{\epsilon_0} - 1 = \frac{\frac{N}{\epsilon_0}\left(\alpha_1 + \frac{m^2}{3kT}\right)}{1 - \frac{N}{3\epsilon_0}\left(\alpha_1 + \frac{m^2}{3kT}\right)},$$

or, as it is more usually written,

$$\frac{\frac{\epsilon}{\epsilon_0} - 1}{\frac{\epsilon}{\epsilon_0} + 2} = \frac{N}{3\epsilon_0}\left(\alpha_1 + \frac{m^2}{3kT}\right). \tag{31}$$

This approximation for F is not a very good one, and more exact expressions have been given by Onsager[1] and Kirkwood.[2] We have, however, established an important fact which is true regardless of the expression for F, namely, that the polarization consists of two parts, one part that is independent of temperature and depends on the shift of the charge within the molecule and one part whose contribution decreases with increasing temperature and is caused by the permanent electric moment of the molecule. The dependence of the dielectric constant upon temperature will be that of the temperature dependence of the polarizability, in general, since the effect of the local field would not be expected to be greatly dependent upon temperature.

The effect of frequency on the polarizability is again a twofold one. Since, in the microwave region, the natural frequencies of the molecule are large compared with the frequency of the radiation, the molecular polarizability α_1 is independent of frequency. When the frequency of the radiation approaches a natural frequency of the molecule, then α_1 changes and gives the familiar anomalous-dispersion curve for the frequency variation. The lowest natural frequencies of most molecules lie in the infrared and do not influence the values of the dielectric constant at microwave frequencies. The effect of the rotation of the electric moment, however, will be strongly dependent upon frequency. This

---

[1] L. Onsager, *J. Am. Chem. Soc.*, **58**, 1486 (1936).
[2] J. G. Kirkwood, *J. Chem. Phys.*, **7**, 911 (1939).

effect was first explained by Debye,[1] who showed that the transient part of the effect of collisions is similar in character to the effect of viscous forces that impede rotation. At low frequencies, these viscous forces would be small and the dielectric constant high. At high frequencies, the forces would be so large that, effectively, the molecules would be prevented from aligning themselves and the dielectric constant would be low. Thus, if the viscous forces are proportional to the rate of change of moment,

$$\alpha = \frac{\alpha_0}{1 + j\omega\tau}, \tag{32}$$

where τ is a "relaxation time" that is characteristic of the material and α_0 is the value of α at $\omega = 0$, namely,

$$\alpha_0 = \frac{m^2}{3kT}. \tag{33}$$

If the Clausius hypothesis is used for obtaining the value of F, then

$$\frac{\frac{\epsilon}{\epsilon_0} - 1}{\frac{\epsilon}{\epsilon_0} + 2} = \frac{N}{3\epsilon_0}\left(\alpha_1 + \frac{\frac{m^2}{3kT}}{1 + j\omega\tau}\right). \tag{34}$$

A rearrangement of this expression, and separation of the real and imaginary parts of ϵ, gives

$$\epsilon' = \epsilon_\infty + \frac{\epsilon_s - \epsilon_\infty}{1 + y^2},$$

$$\epsilon'' = \frac{\epsilon_s - \epsilon_\infty}{1 + y^2} y, \tag{35}$$

where

$$y = \frac{\epsilon_s + 2\epsilon_0}{\epsilon_\infty + 2\epsilon_0} \omega\tau \tag{36}$$

and ϵ_s and ϵ_∞ are the values of ϵ at zero frequency (static value) and infinite frequency (optical value), respectively. In terms of the molecular constants,

$$\epsilon_s = \epsilon_0 \frac{\frac{2N}{3}\left(\alpha_1 + \frac{m^2}{3kT}\right) + 1}{1 - \frac{N}{3}\left(\alpha_1 + \frac{m^2}{3kT}\right)}, \tag{37}$$

and

$$\epsilon_\infty = \epsilon_0 \frac{\frac{2N}{3}\alpha_1 + 1}{1 - \frac{N\alpha_1}{3}}.$$

[1] P. Debye, *Polar Molecules*, Chemical Catalog Co., 1929.

Thus we see that ϵ' decreases monotonically from ϵ_s to ϵ_∞ whereas ϵ'' increases from zero, passes through a maximum, and again decreases to zero. The maximum value of ϵ'' is $(\epsilon_s - \epsilon_\infty)/2$ and occurs when $y = 1$. At this frequency $\epsilon' = (\epsilon_s + \epsilon_\infty)/2$. The quantity ϵ is a relatively slowly varying function of frequency, and the change from ϵ_s to ϵ_∞ takes place over a range of at least a factor of 100 in frequency. Debye has compared this expression with experimental determinations and finds good agreement for substances in which Clausius' hypothesis may be expected to hold. He finds values of the relaxation time of the order

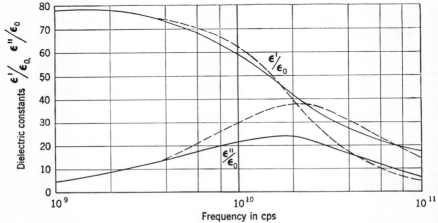

FIG. 11·7.—Variation of the dielectric constant of water with frequency.

of 10^{-10} sec for liquids and 10^{-5} to 10^{-6} sec for a solid, ice. These values are entirely reasonable judged from crude estimates made from the known values of the viscous forces. A viscous force of this kind would be expected to be strongly dependent on the temperature at all frequencies at which the polarity of the molecule contributes to ϵ, and indeed this is the case. For a strongly polar liquid, such as water, the agreement is not exact, but the general nature of the frequency variation is unaltered. Figures 11·7 and 11·8 show some experimental values[1] for the dielectric

[1] The observations from which the curve was drawn were taken from E. L. Younker, "Dielectric Properties of Water and Ice at K-band," RL Report No. 644, December 1944, A. von Hippel, "Progress Report on Ultrahigh-frequency Dielectrics," OSRD Report No. 1197, December 1942, and the references cited in these reports. Somewhat different conclusions have been reached by J. A. Saxton, "The Dielectric Properties of Water at Wavelengths from 2 Mm to 10 Cm, and over the Temperature Range 0° to 40°C," Paper No. RRB/C115, April 1945. Saxton concludes that the experimental evidence indicates that the Debye theory correctly represents the facts for water. If this is the case, water is very exceptional, since most dielectric liquids seem to possess a whole range of relaxation times.

constant of water as a function of frequency for a temperature of 25°C. The dotted lines show the values predicted by the Debye theory. For higher temperatures the maximum value of ϵ'' occurs at a higher frequency corresponding to a shorter relaxation time. It is seen that although the observations on water agree qualitatively with Debye's theory, the quantitative agreement is rather poor. The observed dielectric constant seems to change considerably more slowly with frequency than the theory predicts, and this is true for the imaginary as well as the real part. The maximum value of ϵ'' is observed to be lower than the theoretical value. Both of these deviations are what would be expected if not one but many values of the relaxation time τ exist[1] whose mean value corresponds to the frequency at which ϵ'' is a maximum. A valuable summary of the work in this field has been given by Kauzmann.[2]

If the dielectric medium has, in addition to the above-mentioned properties, a conductivity

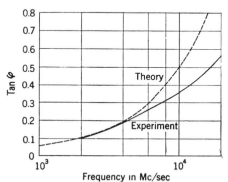

FIG. 11·8.—Variation with frequency of the loss tangent of water.

that is not negligible, this may be described, as has been shown, by an imaginary portion of ϵ which varies with frequency as

$$\epsilon'' = \frac{\sigma}{\omega}.$$

In all known cases, σ is independent of frequency, and this introduces a frequency variation for ϵ'' different from that caused by polar relaxation. For most dielectrics, σ is so small that the contribution to ϵ'' from this effect is negligible at microwave frequencies.

In Table 11·1 are given selected values of measured dielectric constants for three frequencies. It will be noted that ϵ'/ϵ_0 usually decreases slightly and that tan φ increases as ν increases. This is good evidence for the presence of permanent electric moments. However, it obviously is possible to have relatively loss-free materials even at very high frequencies. Care must be taken to choose a substance that contains polar molecules only as impurities in the dielectric. These impurities can be removed or at least minimized by carefully controlled manufacturing

[1] If there is a distribution of relaxation times, Eq. (32) is to be replaced by an integral expression similar to that given in Eq. (68) in Sec. 11·9. Several forms of the distribution law have been proposed, all agreeing equally well with experiment.

[2] W. Kauzmann, "Dielectric Relaxation as a Chemical Rate Process," *Rev. Mod. Phys.*, **14**, 12–44 (1942).

processes. It is more difficult to exclude the presence of moisture. Water molecules have such a large electric moment that even a small percentage of moisture is sufficient to cause an objectionable amount of loss. Many plastic materials contain sufficient moisture to affect the amount of dielectric loss, and this loss becomes larger the higher the frequency. The large values of tan φ at 60 cps are caused chiefly by the presence of a real conductivity.

TABLE 11·1.—DIELECTRIC CONSTANTS

Substance	$\nu = 60$ cps;		$\nu = 10^6$ cps;		$\nu = 10^{10}$ cps	
	$\dfrac{\epsilon'}{\epsilon_0}$	tan φ	$\dfrac{\epsilon'}{\epsilon_0}$	tan φ	$\dfrac{\epsilon'}{\epsilon_0}$	tan φ
Steatite ceramic, Alsimag 243	6.3	0.0015	6.2	0.0004	5.4	0.0002
Ruby mica	5.43	0.005	5.40	0.0004	5.4	0.0003
Quartz, fused	3.85	0.0009	3.82	0.0002	3.80	0.0001
Corning glass—702P	4.75	0.009	4.55	0.002	4.40	0.006
Corning glass—705AO	5.00	0.03	4.75	0.004	4.70	0.007
Corning glass—707DG	4.00	0.0006	4.00	0.0008	3.90	0.001
Black Bakelite	5.0	0.10	4.9	0.03	4.7	0.05
Lucite	3.3	0.07	2.6	0.015	2.5	0.005
Plexiglas	3.4	0.06	2.7	0.015	2.5	0.005
Polystyrene	2.51	0.0002	2.51	0.0003	2.45	0.0005
Polyethylene	2.25	0.0001	2.25	0.0001	2.25	0.0002
Apiezon W	2.80	0.022	2.65	0.0025	2.62	0.002
Paraffin	2.25	0.0002	2.25	0.0002	2.20	0.0002
Mahogany plywood, dry	2.4	0.01	2.4	0.02	2.0	0.02
Water, 25°C	79.	(3000)	79.	0.03	59.	0.46

11·5. Ferromagnetism at Microwave Frequencies.—Most substances are characterized by a value of the magnetic permeability that is inappreciably different from the permeability of free space, the ratio being 1 ± 10^{-6}. For nearly all purposes this small difference can be neglected. However, the ferromagnetic substances, iron, nickel, the Heusler alloys, and a few others, have permeabilities relative to free space that are large compared with unity. In addition, all these substances are characterized by hysteresis, and no simple relation such as $B = \mu H$ obtains. Nevertheless, for small amplitudes of a sinusoidally varying field it can be said that the amplitude of B is proportional to the amplitude of H. Because of the energy loss caused by hysteresis, it is necessary to have a permeability that is complex, exactly analogous to a complex dielectric constant. This may be written

$$\mu = \mu' - j\mu'', \qquad \frac{\mu''}{\mu'} = \tan \zeta. \tag{38}$$

Sec. 11·5] FERROMAGNETISM AT MICROWAVE FREQUENCIES

Likewise the expression for the propagation constant and wave impedance in a ferromagnetic medium is more complicated;

$$\left.\begin{array}{l} \gamma^2 = k_c^2 - \omega^2 \epsilon \mu' + j\omega^2 \epsilon \mu'', \\ Z_H = \dfrac{j\omega\mu' + \omega\mu''}{\gamma}. \end{array}\right\} \quad (39)$$

It is seen that γ has a real part and Z_H an imaginary part, both of which are representative of the energy loss from hysteresis. Since, however, most ferromagnetics are metals or at least semiconductors, an imaginary part of the dielectric constant must also be included. The result is

$$\gamma^2 = k_c^2 - \omega^2(\epsilon'\mu' - \epsilon''\mu'') + j\omega^2(\epsilon''\mu' + \epsilon'\mu''). \quad (40)$$

It is found that

$$\left.\begin{array}{l} \alpha = \omega \sqrt{\dfrac{\epsilon'\mu'}{2}} (\tan \varphi + \tan \zeta) R^{-\frac{1}{2}}, \\ \beta = \omega \sqrt{\dfrac{\epsilon'\mu'}{2}} R^{\frac{1}{2}}, \end{array}\right\} \quad (41)$$

where

$$R = 1 - \tan \varphi \tan \zeta \, \frac{k^2}{\omega^2 \epsilon' \mu'} + \left[\sec^2 \varphi \sec^2 \zeta - 2(1 - \tan \varphi \tan \zeta) \frac{k^2}{\omega^2 \epsilon' \mu'} + \frac{k^4}{\omega^4 \epsilon'^2 \mu'^2} \right]^{\frac{1}{2}}.$$

In terms of the other variables,

$$\left.\begin{array}{l} \alpha = \dfrac{2\pi}{\lambda_0} \dfrac{\epsilon''\mu' + \epsilon'\mu''}{\sqrt{2} \, \epsilon_0 \mu_0} S^{-\frac{1}{2}}, \\ \beta = \dfrac{2\pi}{\lambda_0} \dfrac{1}{\sqrt{2}} S^{\frac{1}{2}}, \end{array}\right\} \quad (42)$$

where

$$S = \frac{\epsilon'\mu' - \epsilon''\mu''}{\epsilon_0 \mu_0} - \left(\frac{\lambda_0}{\lambda_c}\right)^2 + \left[\frac{|\epsilon|^2 |\mu|^2}{(\epsilon_0 \mu_0)^2} - 2 \frac{\epsilon'\mu' - \epsilon''\mu''}{\epsilon_0 \mu_0} \left(\frac{\lambda_0}{\lambda_c}\right)^2 + \left(\frac{\lambda_0}{\lambda_c}\right)^4 \right]^{\frac{1}{2}}.$$

The wave impedance can be calculated from

$$Z_H = \frac{\alpha\omega\mu'' + \beta\omega\mu' + j(\alpha\omega\mu' - \beta\omega\mu'')}{\alpha^2 + \beta^2}. \quad (43)$$

The denominator in Eq. (43) can be written

$$\alpha^2 + \beta^2 = \left(\frac{2\pi}{\lambda_0}\right)^2 \left[\frac{|\epsilon|^2 |\mu|^2}{\epsilon_0^2 \mu_0^2} - 2 \frac{\epsilon'\mu' - \epsilon''\mu''}{\epsilon_0 \mu_0} \left(\frac{\lambda_0}{\lambda_c}\right) + \left(\frac{\lambda_0}{\lambda_c}\right)^4 \right]^{\frac{1}{2}}.$$

For a ferromagnetic metal, ϵ' may be neglected in comparison with ϵ'', and the wave impedance reduces to

$$Z_m = \sqrt{\frac{\omega}{2\sigma}} \, (\sqrt{|\mu| + \mu''} - j \sqrt{|\mu| - \mu''}). \quad (44)$$

The losses in a ferromagnetic metal can thus be expressed in terms of a skin depth δ as in the case of ordinary metals,

$$\delta = \sqrt{\frac{2}{\omega\sigma\mu_{\text{eff}}}}, \quad (45)$$

where the effective permeability μ_{eff} is

$$\mu_{\text{eff}} = |\mu| + \mu''. \quad (46)$$

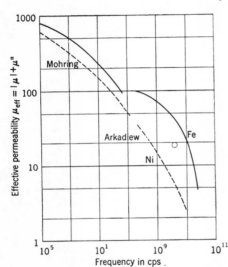

Fig. 11·9.—Effective permeability of iron and nickel vs. frequency.

Few experimental investigations of the properties of ferromagnetic metals have been made, and very little is known about them.[1] In Fig. 11·9 are reproduced some observations of Arkadiew[2] and Möhrning[3] of the values of μ_{eff} for iron and nickel.

An example of a nonconducting ferromagnetic substance is a finely powdered iron dust bound together by an insulating plastic such as is frequently used for cores in high-frequency transformers. The sample[4] of material that has been measured was approximately 50 per cent iron by volume. For a frequency of 3000 Mc/sec, it was found that

$$\frac{\epsilon'}{\epsilon_0} = 20, \qquad \frac{\mu'}{\mu_0} = 3.2,$$

$$\frac{\epsilon''}{\epsilon_0} = 1.4, \qquad \frac{\mu''}{\mu_0} = 4.2,$$

$$\tan \varphi = 0.07, \qquad \tan \zeta = 1.3.$$

---

[1] A recent summary has been prepared by J. T. Allanson, "The Permeability of Ferromagnetic Materials at Frequencies Greater than 10^5 cps," Central Radio Bureau 2545, WR-1157, JEIA 4281, Apr. 21, 1944.

[2] W. Arkadiew, *Physik Z.*, **14**, 561 (1913).

[3] N. Möhrning, *Hochfrequenztechnik u. Elektakus*, **53**, 196 (1939).

[4] The material in question was an experimental sample of polyiron furnished by H. L. Crowley and Co., Inc., West Orange, N. J.

These values were calculated from measurements of the impedance of small pieces of the material placed in a coaxial transmission line, as described earlier in this chapter for dielectric plates. The value of ϵ'/ϵ_0 is high because of the presence of the conducting iron particles which are polarized under the influence of the field. The point corresponding to these observations is plotted on Fig. 11·9 for comparison with the results for the solid material.

It may be seen by reference to the expression for Z_H in Eq. (43) that mixtures of this sort can be compounded to make Z_H have any arbitrary real value. If the imaginary part of Z_H is set equal to zero,

$$\frac{\alpha}{\beta} = \tan \zeta.$$

From Eqs. (41), for $k_c = 0$, is found the condition that φ must equal ζ. If this is true, then

$$Z_w = \sqrt{\frac{\mu'}{\epsilon'}}, \qquad (47)$$

$$\alpha + j\beta = \frac{2\pi}{\lambda_0}\sqrt{\frac{\epsilon'\mu'}{\epsilon_0\mu_0}}(\tan \varphi + j). \qquad (48)$$

11·6. Guides Partially Filled with Dielectric.—Let us consider the case of a waveguide containing two dielectric mediums, the boundary between the mediums being along the axis of the guide. The propagation constant and the impedance relative to that of empty waveguide specify completely the properties of such a configuration, and the method of calculating these quantities will be shown. Let us consider only rectangular waveguide in the lowest H-mode. It would be possible to follow the usual procedure, described in Chap. 2, of choosing the solution of the wave equation for H_z that satisfies the boundary conditions. These conditions include not only the usual one that the normal derivative of H_z vanish on the wall but also an additional condition which must be satisfied at the boundary of the two dielectric mediums. This procedure is straightforward, but there is an easier method of obtaining the propagation constant. As has already been shown, any one of the three directions may be regarded as the direction of propagation of the waves in rectangular guide. Then an equation for the propagation constant in the z-direction can be found from the condition that a standing wave must exist, for example, in the x-direction. Since expressions for the impedance of a guide completely full of dielectric have already been obtained, it is possible to write down this condition immediately if the direction perpendicular to the dielectric interface is chosen. This method of approach was first devised by Frank,[1] and some numerical results obtained by him will now be discussed.

[1] N. H. Frank, RL Report No. T-9, 1942.

Let us consider the configuration shown in Fig. 11·10. Here a longitudinal slab of dielectric of thickness d is placed in the center section of a rectangular guide of width a. The electric field E is a maximum at the center of the guide. The admittance looking from the center of the guide in the x-direction must therefore be zero. If the losses in the dielectric are negligible, the admittance at the center will be

$$Y_{\text{in}} = Y_0^{(2)} \frac{Y' + jY_0^{(2)} \tan\left(\kappa_x^{(2)} \frac{d}{2}\right)}{Y_0^{(2)} + jY' \tan\left(\kappa_x^{(2)} \frac{d}{2}\right)},$$

where Y' is the admittance looking to the left, at the left boundary of the

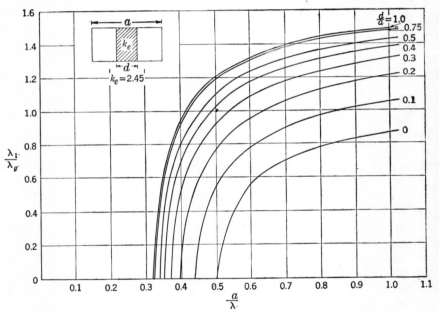

Fig. 11·10.—Variation of λ_1/λ_g with a/λ_1 for various values of d/a, when $\epsilon_2/\epsilon = 2.45$. Case I: Dielectric in center of guide.

dielectric. Just as in the case of a guide completely filled with dielectric, a small amount of loss produces only a second-order change in λ_g. If $Y_{\text{in}} = 0$, then

$$Y' = jY_0^{(1)} \tan\left(\kappa_x^{(1)} \frac{d-a}{2}\right),$$

where $Y_0^{(1)}$ and $Y_0^{(2)}$ are the characteristic admittances of portions (1) and (2), respectively. Setting $Y_{\text{in}} = 0$, we have

GUIDES PARTIALLY FILLED

$$\frac{Y_0^{(1)}}{Y_0^{(2)}} \tan \kappa_x^{(1)} \frac{d-a}{2} = -\tan \kappa_x^{(2)} \frac{d}{2}. \quad (49)$$

We know, however, that

$$\begin{rcases}(\kappa_x^{(1)})^2 = \left(\frac{2\pi}{\lambda_1}\right)^2 - \left(\frac{2\pi}{\lambda_g}\right)^2, \\ (\kappa_x^{(2)})^2 = \frac{\epsilon_2}{\epsilon_1}\left(\frac{2\pi}{\lambda_1}\right)^2 - \left(\frac{2\pi}{\lambda_g}\right)^2, \\ \frac{Y_0^{(1)}}{Y_0^{(2)}} = \frac{\kappa_x^{(1)}}{\kappa_x^{(2)}},\end{rcases} \quad (50)$$

where λ_1 is the wavelength of a plane wave in medium (1). Equation (49) is thus a transcendental equation for λ_g and can be solved numerically.

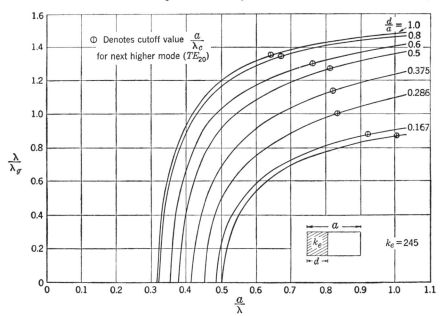

Fig. 11·11.—Variation of λ_1/λ_g with a/λ_1 for various values of d/a, when $\epsilon_2/\epsilon = 2.45$. Case II: Dielectric at edge of guide.

The results are given in Fig. 11·10, which shows λ_1/λ_g as a function of a/λ_1 for a series of values of d/a for the case where $\epsilon_2/\epsilon_1 = 2.45$. It is to be noted that for small values of d/a there is a large change in λ_g whereas the change in λ_g between $d/a = 0.75$ and $d/a = 1.0$ is very small. This is obviously because, for small d, dielectric has been added where the field is high and the effect is much larger than when the dielectric is added where the field is weak.

A second simple case is shown in Fig. 11·11. The equation for λ_g

now represents the condition that the impedance looking to the left vanish on the right-hand boundary of the waveguide. We have

$$\frac{Z_0^{(2)}}{Z_0^{(1)}} \tan \kappa_x^{(2)} d = - \tan \kappa_x^{(1)}(d - a), \tag{51}$$

where $\kappa_x^{(1)}$ and $\kappa_x^{(2)}$ are defined by Eqs. (50) and

$$\frac{Z_0^{(2)}}{Z_0^{(1)}} = \frac{Y_0^{(1)}}{Y_0^{(2)}}.$$

For $\epsilon_2/\epsilon_1 = 2.45$ the results of the calculation are as shown. It is seen from the curves that for small values of d the effect is small. As the dielectric interface approaches the center of the waveguide, the effect becomes much larger and then decreases again as the region of weak

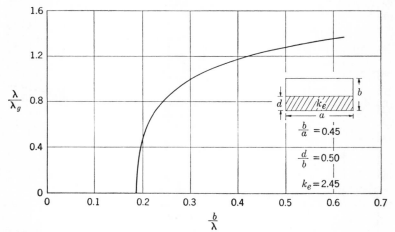

Fig. 11·12.—Variation of λ/λ_g with b/λ. Case III: Dielectric at bottom of guide.

fields near the right-hand wall of the guide is approached. The circles on the curve indicate the values of a/λ for which the next H-mode can propagate. The losses have again been neglected.

Figure 11·12 shows a somewhat more interesting example. The preceding cases involved a mode of transmission that was transverse-electric both in the direction normal to the dielectric interface and in the direction of propagation. In the present case, if the interface normal is chosen as the reference direction, the field configuration may be considered to be that of an E-mode. The field has components E_x, E_y, E_z, H_y, and H_z. Hence, with respect to the z-axis, the mode is neither a pure E-mode nor a pure H-mode but must be a combination of the two. The impedance method of calculation is still valid, but now the E-mode impedance [Eq. (11)] is used as the characteristic impedance of the lines.

The results for a particular case are shown in Fig. 11·12, for the values $\epsilon_2/\epsilon_1 = 2.45$ and $b/a = 0.45$, for a guide half full of dielectric. Figure 11·13 shows the variation with d/b for two values of b/λ. This case has been treated in a more general fashion by Pincherle,[1] who discusses other modes in rectangular guide. Pincherle also examines a waveguide of circular cross section with a dielectric rod down the center. This case can be considered from an impedance point of view by the methods of Chap. 8.

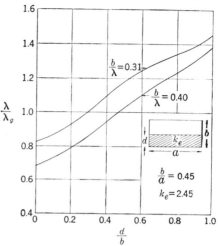

Fig. 11·13.—Variation of λ/λ_g with d/b for two values of b/λ.

11·7. Dielectric Post in Waveguide.—If there is a cylindrical dielectric post of circular cross section in rectangular guide operating in the H_{10}-mode, which extends in the direction of E at the center of the guide, the relative admittance Y can be expressed by the simple formula of Frank,[2]

$$\frac{1}{Y} = j \frac{a}{2\lambda_g} \left[\ln \frac{\lambda}{R} + 1.775 \frac{a}{\lambda} - \frac{\epsilon_0}{2\epsilon} \left(\frac{\lambda}{\pi R} \right)^2 - 1.91 \right],$$

where R is the radius of the post and a the width of the waveguide. This expression was derived for the case for which $|\epsilon/\epsilon_0| (2\pi R/\lambda)^2 \ll 1$ and the series arms of the equivalent T-network have a negligibly small impedance. The expression is valid to within 3 per cent for the range of wavelengths given by $\frac{1}{2} < a/\lambda < 1$, provided that the radius of the post is small enough.

The expression holds for a complex ϵ as well as for real values. It is possible to solve for ϵ in terms of a measured Y and in this way measure dielectric constants. For example, it was found for $a = 0.420$ in. and $\lambda = 1.25$ cm, a column of water for which R was 0.009 in. had a measured admittance of $1 - j$. The value of ϵ/ϵ_0 deduced from this was

$$\frac{\epsilon}{\epsilon_0} = 39 - j17.$$

For real values of ϵ, $Y = jB$, where B is positive and hence a capacitance. It is evident from an inspection of the formula that the frequency

[1] L. Pincherle, *Phys. Rev.*, **66**, 118 (1944).
[2] N. H. Frank, RL Report No. T-9, 1942, Sec. V, p. 32.

variation of the admittance is not at all that of a capacitance in the low-frequency approximation. For posts of larger radius the above expression is not accurate and a much better formula is given in Vol. 10 of this series.

11·8. Cavities Containing Dielectrics.—When a resonant cavity is filled with a dielectric material, both the resonant frequency and the Q of the cavity are changed. Conversely, if it is desired to maintain the same value of the resonant frequency, the dimensions of the cavity must be altered. If the losses are assumed to be small, then it is easy to see what this change in size must be. The form of the fields should remain the same under such a change, with the result that if ϵ is altered, the operator $\nabla^2 + \gamma^2 + \omega^2 \epsilon \mu$ must change only by a constant factor. If ϵ is changed by a factor f^2, then all the linear dimensions must evidently be changed by a factor $1/f$ if ω is to be kept constant. The losses in the metal walls will be proportional to $1/Q$ or to δ/l, where δ is the skin depth and l some dimension of the cavity. Thus it is seen that Q also changes by a factor $1/f$.

If the loss in the dielectric is to be included, then to the value of $1/Q$ for the metal losses must be added a quantity $1/Q_1$ corresponding to the dielectric loss. However, the dielectric loss per cycle is proportional to the square of the electric field and so is the stored energy. The conclusion is, therefore, that $1/Q_1$ is independent of the mode and of the size of the cavity and is equal to $\epsilon''/\epsilon' = \tan \varphi$. Thus, for the cavity and dielectric

$$\frac{1}{Q} = \frac{1}{Q_{\text{metal}}} + \tan \varphi.$$

This equation is correct as long as the losses in the dielectric are not large enough to alter the resonant frequency of the cavity. From the table of dielectric constants given above, it is seen that the Q of the dielectric can be as large as several thousand at microwave frequencies. In general, the metal losses may be neglected in comparison with the dielectric loss.

Consider now a cavity only partially filled with dielectric. The cases that can be treated simply are those in which the surface of the dielectric is perpendicular to an axis of the cylindrical cavity and parallel to the end plates. The wavelength in waveguides partially filled with dielectric have already been dealt with, and the results are immediately applicable here. Thus Figs. 11·10 to 11·13 inclusive make it possible to find the resonant frequency in these cases by direct utilization of the condition that the length l of the cavity must be

$$l = n \frac{\lambda_g}{2},$$

where n is an integer. The information given in these figures does not make it possible to find the Q of the cavity. We notice that Q_1, the dielectric Q, will depend on the mean value of E^2 in the dielectric, relative to the mean value of E^2 in the whole cavity, and hence Q_1, as well as the resonant frequency, depends on the position of the dielectric within the cavity.

To find an expression for Q_1 it is convenient to regard as the direction of propagation the direction of the normal to the dielectric interface as before and let this be the z-axis. The metal losses will be neglected. Let the cavity walls be located at $z = 0$ and $z = l$ and the dielectric interface at $z = a$, as shown in Fig. 11·14. The tangential electric field in region (1) is $A \sin \beta_1 z$; and in region (2), $B \sin \beta_2(l - z)$. At the interface

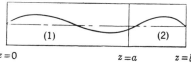

Fig. 11·14.—Field distribution in a cavity containing a dielectric material.

$$A \sin \beta_1 a = B \sin \beta_2(l - a). \tag{52}$$

The condition of resonance is determined from the continuity condition of the derivatives, or

$$\beta_1 A \cos \beta_1 a = -\beta_2 B \cos \beta_2(l - a). \tag{53}$$

If the value of B from Eq. (52) is inserted, then

$$\cot \beta_2(l - a) = -\frac{\beta_1}{\beta_2} \cot \beta_1 a. \tag{54}$$

The Q of the cavity is then given by

$$Q = \frac{\epsilon_1' \int_0^a A^2 \sin^2 \beta_1 z \, dz + \epsilon_2' \int_a^l B^2 \sin^2 \beta_2(l - z) \, dz}{\epsilon_1'' \int_0^a A^2 \sin^2 \beta_1 z \, dz + \epsilon_2'' \int_a^l B^2 \sin^2 \beta_2(l - z) \, dz}. \tag{55}$$

Substitution for B from Eq. (52) and the use of Eq. (54) reduces this to

$$Q = \frac{\epsilon_1' M + \epsilon_2' N}{\epsilon_1'' M + \epsilon_2'' M}, \tag{56}$$

where

$$M = a - \frac{1}{\beta_1} \sin \beta_1 a \cos \beta_1 a,$$

$$N = (l - a) \sin^2 \beta_1 a + (l - a) \frac{\beta_1^2}{\beta_2^2} \omega^2 \beta_1 a + \frac{\beta_1}{\beta_2^2} \sin \beta_1 a \cos \beta_1 a.$$

A calculation similar to this may be made for radial cavities. For this problem, it is convenient to employ the radial transmission-line theory, a discussion of which is found in Chap. 8. Feenberg[1] has made accurate calculations for the case of a cylindrical rod of dielectric in the center of a pillbox-shaped cavity operated in the mode in which the electric field is perpendicular to the end plates. The results are expressed in tables and curves for convenient use.

One other important example of the use of dielectrics in cavities remains to be discussed, namely, a dielectric material filling a hole that is used for coupling to the cavity. For example, a glass window sealed to the metal cavity might be employed as a pressure seal in the coupling aperture. Unfortunately, little is known in detail about such configurations, and the discussion must be confined to a few general remarks. The somewhat simpler case of a diaphragm, the opening of which is filled with dielectric, placed across a rectangular waveguide operated in the dominant mode may be considered. The loss in the dielectric is proportional to the total electric field in the aperture. A portion of this field, namely, the dominant-mode portion, is given by the equivalent-circuit arguments. The circuit, shown in Fig. 11·15, is driven by the constant-current generator I, the iris being represented by the admittance Y. The current through Y will be

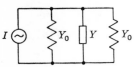

Fig. 11·15.—Equivalent circuit for the dominant-mode current in a dielectric-filled iris.

$$\frac{IY}{(2Y_0 + Y)},$$

and the voltage across Y will be

$$\frac{I}{(2Y_0 + Y)}.$$

This quantity is proportional to the dominant-mode field; therefore its square is proportional to the dielectric loss. It is evident that the loss decreases monotonically as Y is increased. To this must be added the loss produced by the higher-mode fields. If the aperture is completely open, no higher-mode fields are excited. This is also true when the aperture has no opening. Thus the dielectric loss caused by the higher modes will be zero when Y is zero; and as Y increases, the loss will increase, pass through a maximum, and then decrease again, approaching zero as Y approaches infinity. The total loss may then be represented as in Fig. 11·16.

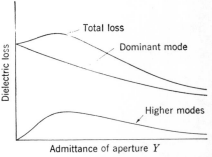

Fig. 11·16.—Loss vs. admittance for a dielectric-filled coupling aperture.

[1] E. Feenberg, "Use of Cylindrical Resonator to Measure Dielectric Properties at Ultrahigh Frequencies," Sperry Gyroscope Co., July 1942.

To obtain quantitative values, it is necessary to find the exact value of the field at all points in the aperture. This will, of course, depend on the shape as well as on the admittance of the diaphragm.

11·9. Propagation in Ionized Gases.—In the preceding sections, the effects of a complex dielectric constant or a complex permeability were considered. The problem can equally well be formulated in terms of a conductivity of the medium, this conductivity being a complex quantity. For certain applications the formulation in terms of conductivity has a more direct physical interpretation and is often useful. A case in point is the problem of the effect of an ionized gas on the passage of electromagnetic waves through it. The general expressions for γ and Z_H will first be derived, then specific application to ionized gases will be made.

For simplicity let us consider the dominant mode in rectangular waveguide, where $k_c = \pi/a$. The propagation constant γ takes the form

$$\gamma^2 = \left(\frac{\pi}{a}\right)^2 + j\omega\mu\sigma - \omega^2\epsilon\mu, \tag{57}$$

when the medium filling the waveguide has a conductivity σ. If σ is a complex quantity, it may be written

$$\sigma = \sigma' - j\sigma''. \tag{58}$$

The propagation constant is then

$$\gamma^2 = \left(\frac{\pi}{a}\right)^2 + j\omega\mu\sigma' - \omega^2\mu\left(\epsilon - \frac{\sigma''}{\omega}\right). \tag{59}$$

Thus σ''/ω is the contribution to the dielectric constant of the conductivity of the medium. It will be assumed that ϵ and μ are real. If it is remembered that $\gamma = \alpha + j\beta$, then

$$\begin{aligned}2\alpha^2 &= \left\{\left[\left(\frac{\pi}{a}\right)^2 + \omega\mu\sigma'' - \omega^2\epsilon\mu\right]^2 + \omega^2\mu^2\sigma'^2\right\}^{1/2} + \left[\left(\frac{\pi}{a}\right)^2 + \omega\mu\sigma'' - \omega^2\epsilon\mu\right], \\ 2\beta^2 &= \left\{\left[\left(\frac{\pi}{a}\right)^2 + \omega\mu\sigma'' - \omega^2\epsilon\mu\right]^2 + \omega^2\mu^2\sigma'^2\right\}^{1/2} - \left[\left(\frac{\pi}{a}\right)^2 + \omega\mu\sigma'' - \omega^2\epsilon\mu\right].\end{aligned} \tag{60}$$

If the substitution of

$$\omega^2\epsilon\mu = \left(\frac{2\pi}{\lambda}\right)^2$$

and

$$\left(\frac{2\pi}{\lambda}\right)^2 - \left(\frac{\pi}{a}\right)^2 = \left(\frac{2\pi}{\lambda_g}\right)^2$$

is made, then

$$\begin{aligned}2\alpha^2 &= \left\{\left[\omega\mu\sigma'' - \left(\frac{2\pi}{\lambda_g}\right)^2\right]^2 + \omega^2\mu^2\sigma'^2\right\}^{1/2} + \left[\omega\mu\sigma'' - \left(\frac{2\pi}{\lambda_g}\right)^2\right], \\ 2\beta^2 &= \left\{\left[\omega\mu\sigma'' - \left(\frac{2\pi}{\lambda_g}\right)^2\right]^2 + \omega^2\mu^2\sigma'^2\right\}^{1/2} - \left[\omega\mu\sigma'' - \left(\frac{2\pi}{\lambda_g}\right)^2\right].\end{aligned} \tag{61}$$

It should be pointed out that λ and λ_g are the wavelengths in the medium and in the waveguide, respectively, when the conductivity σ is zero.

The expression for β may be written as

$$2\beta^2 = \left|\omega\mu\sigma'' - \left(\frac{2\pi}{\lambda_g}\right)^2 + j\omega\mu\sigma'\right| - \left[\omega\mu\sigma'' - \left(\frac{2\pi}{\lambda_g}\right)^2\right]. \tag{62}$$

In the special case that $\sigma' = 0$,

$$2\beta^2 = \left|\omega\mu\sigma'' - \left(\frac{2\pi}{\lambda_g}\right)^2\right| - \left[\omega\mu\sigma'' - \left(\frac{2\pi}{\lambda_g}\right)^2\right].$$

Thus if

$$\left(\frac{2\pi}{\lambda_g}\right)^2 > \omega\mu\sigma'',$$

$$\beta^2 = \left(\frac{2\pi}{\lambda_g}\right)^2 - \omega\mu\sigma'' > 0;$$

and if

$$\left(\frac{2\pi}{\lambda_g}\right)^2 < \omega\mu\sigma'',$$

$$\beta^2 = 0.$$

In the latter case, the waveguide is beyond cutoff. The cutoff wavelength in the waveguide is

$$(\lambda_g)_{\text{cutoff}} = \frac{2\pi}{\sqrt{\omega\mu\sigma''}}. \tag{63}$$

The following approximate expressions are useful. If

$$\omega\mu\sigma'' - \left(\frac{2\pi}{\lambda_g}\right)^2 \gg \omega\mu\sigma',$$

$$\alpha = \sqrt{\omega\mu\sigma'' - \left(\frac{2\pi}{\lambda_g}\right)^2}\left\{1 + \frac{1}{8}\frac{\omega^2\mu^2\sigma'^2}{\left[\omega\mu\sigma'' - \left(\frac{2\pi}{\lambda_g}\right)^2\right]^2} + \cdots\right\}, \tag{64}$$

and

$$\beta = \frac{\omega\mu\sigma'}{2\sqrt{\omega\mu\sigma'' - \left(\frac{2\pi}{\lambda_g}\right)^2}}.$$

If the opposite situation is true and

$$\left.\begin{array}{c}\omega\mu\sigma' \\ \omega\mu\sigma''\end{array}\right\} \ll \left(\frac{2\pi}{\lambda_g}\right)^2,$$

then

$$\alpha = \frac{\omega\mu\sigma'}{2\sqrt{\left(\frac{2\pi}{\lambda_g}\right)^2 - \omega\mu\sigma''}}, \tag{65}$$

and

$$\beta = \sqrt{\left(\frac{2\pi}{\lambda_g}\right)^2 - \omega\mu\sigma''}\left[1 + \frac{1}{8}\frac{\omega^2\mu^2\sigma'^2}{\left[\left(\frac{2\pi}{\lambda_g}\right)^2 - \omega\mu\sigma''\right]^2} + \cdots\right].$$

In all the above expressions, the corresponding formulas for uniform plane waves may be obtained by the substitution $\lambda_g = \lambda$.

The wave impedance Z_H can be found by making substitutions in Eq. (15),

$$Z_H = \frac{4\omega\mu\beta^3}{4\beta^4 + \sigma'^2\mu^2}\left(1 + j\frac{\sigma'\mu}{2\beta^2}\right). \tag{66}$$

When $\beta = 0$,

$$Z_H = \frac{j\omega\mu}{\alpha}. \tag{67}$$

To illustrate the application of these formulas, let us consider a gas containing positive ions and electrons and having a net charge of zero. The complex conductivity has been calculated by Margenau[1] by kinetic-theory methods. Margenau finds

$$\sigma = \frac{4}{3}\frac{e^2 \ln}{(2\pi mkT')^{1/2}} \int_0^\infty \frac{t^{3/2}e^{-t}\,dt}{t^{1/2} + jp\omega}, \tag{68}$$

where l is the mean free path, n the density of charged particles of charge e and mass m, k is Boltzmann's constant, and T' is an effective temperature defined by

$$T' = T + \frac{Me^2E^2}{6m^2\omega^2 k}, \tag{69}$$

where M is the mass of the heavy particles and E is the amplitude of the electric fields. The quantity p is a mean free time, and

$$p = l\left(\frac{m}{2kT}\right)^{1/2}. \tag{70}$$

If only small fields are considered, the difference between T and T' may be neglected. It is seen immediately that the effects of the positive ions may be neglected compared with those of the electrons because of the occurrence of m in the denominator of the expression for σ. This expression reduces to simple form when the frequency is very low or very high. For low frequencies, that is, for

$$\omega^2 \ll \frac{ml^2}{2kT},$$

$$\sigma = \frac{4}{3}\frac{e^2 \ln}{(2\pi mkT)^{1/2}} - j\frac{\omega e^2 l^2 n}{3kT}. \tag{71}$$

[1] H. Margenau, *Phys. Rev.*, **69**, 508 (1946); RL Report No. 836, Oct. 26, 1945.

The real part of σ is the Langevin formula usually written in terms of the mobility. The imaginary part corresponds to a change in dielectric constant that is independent of ω. For high frequencies, $\omega^2 \gg ml^2/2kT$,

$$\sigma = \frac{16}{3} \frac{e^2 n}{m\omega^2 l} \left(\frac{kT}{2\pi m}\right)^{\frac{1}{2}} - j \frac{e^2 n}{m\omega}. \qquad (72)$$

The imaginary portion of σ is the familiar expression for entirely free electrons. Since σ'' is inversely proportional to ω and σ' is inversely proportional to ω^2, at sufficiently high frequencies $\sigma \approx -j\sigma''$. The effective dielectric constant $\epsilon - \sigma''/\omega$ [see Eq. (59)] may be positive or negative depending on whether n is small or large. The behavior of the ionosphere may be explained in terms of these expressions. As the altitude is increased, n increases, and there is a cutoff condition if σ' is sufficiently small. From Eqs. (63) and (72),

$$\lambda_c = \frac{2\pi}{\sqrt{\omega\mu\sigma''}} = 2\pi \sqrt{\frac{m}{e^2 n \mu}}. \qquad (73)$$

For this wavelength

$$\epsilon - \frac{\sigma''}{\omega} = 0.$$

If the losses, as represented by σ', cannot be neglected, the exact theory must be utilized. No true cutoff phenomenon occurs, because energy is lost in the medium.

A similar situation exists in the action of a gas-discharge switch (TR switch) in a waveguide circuit. During the discharge proper the above formula for σ does not hold, because ionizing collisions were neglected in the derivation. However, during the period of recovery after a discharge has taken place, the formula is valid. At microwave frequencies and for the gas pressures used, $\sigma'' \gg \sigma'$. The attenuation is therefore of the nature of attenuation in a waveguide beyond cutoff.

11·10. Absorbing Materials for Microwave Radiation.—An interesting example of the application of the principles discussed in this chapter is afforded by certain materials that absorb electromagnetic radiation without producing much reflection. Such materials were developed by the Germans during the war for radar camouflage.[1] They were used principally to cover the breathing tubes that extended above the surface of the ocean from submerged submarines. Two principal varieties were developed, one employing poorly conducting materials, the other using lossy dielectric materials that have also a complex permeability.

An absorbing material of the first type consists of alternate layers of dielectric and thin sheets of poorly conducting material. The structure

[1] G. G. Macfarlane, "Radar Camouflage, Research and Development by the Germans," T. 1905, M/99, TRE, July 23, 1945.

is shown schematically in Fig. 11·17. The dielectric material is a foam of polymicrylchloride with the low value of 1.3 for ϵ/ϵ_0 and a negligible loss tangent. Each layer of dielectric is about 7 mm thick. The conducting sheets, each about 0.1 mm thick, are made of paper impregnated with lampblack. The whole structure is glued together, and the outside is coated with a thin layer of water-repellent wax.

The design of a reflectionless structure is equivalent to the problem of matching from free space to a short circuit by means of a lossy transmission line. It is obviously impossible to obtain a perfectly reflectionless matching transformer, and an approximation must be chosen. The approximation employed here is to use a number of lumped elements to introduce the loss instead of a continuous lossy line. The equivalent circuit of the arrangement is shown in Fig. 11·18. The lossy line is made to behave in the manner of an exponentially tapered line[1] by using sheets

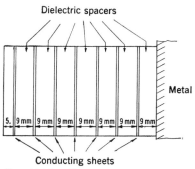

FIG. 11·17.—Construction of absorbing sheet containing conducting layers.

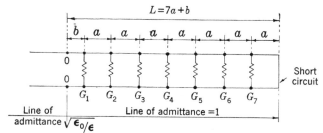

FIG. 11·18.—Equivalent circuit of absorbing sheet in Fig. 11·17.

of conducting material whose surface resistivity varies by a constant factor from one sheet to the next.

Table 11·2 gives the values of the resistivities of the conducting layers.

TABLE 11·2.—SURFACE RESISTIVITIES OF CONDUCTING SHEETS OF MULTILAYER ABSORBER

Sheet number (from front surface)...........	1	2	3	4	5	6	7
Surface resistivity, ohms per square...........	30,000	14,000	6500	3000	1400	650	300
Equivalent conductance relative to dielectric line.	0.011	0.024	0.051	0.110	0.24	0.51	1.10

[1] J. C. Slater, *Microwave Transmission*, McGraw-Hill, New York, 1942, Chap. 1, p. 75.

The behavior of the absorber may be easily calculated from the equivalent circuit. Even without calculation the approximate action is readily seen. At long wavelengths the absorber acts as a large inductive susceptance with a very small conductance. At shorter wavelengths the conductance is larger and the susceptance smaller, and the admittance of the combination traces out on an admittance diagram a spiral approaching the center. The admittance makes one revolution on the diagram each time the phase change through the absorber is 180°. When the

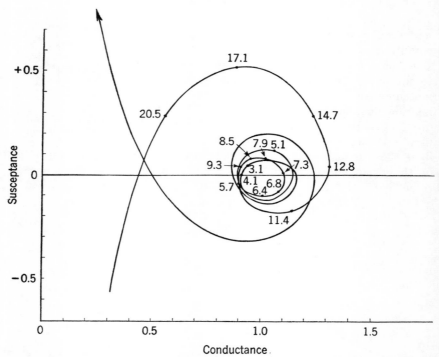

Fig. 11·19.—Impedance chart for the absorbing sheet of Fig. 11·17. The wavelengths are indicated on the curve.

spacing between the layers becomes an appreciable fraction of a wavelength, the spiral expands until the conducting layers are a half wavelength apart. At this frequency the conductances are short-circuited by the metal surface, and the admittance is pure imaginary. This behavior is illustrated in Fig. 11·19 where the points on the spiral are labeled with the wavelength in centimeters. Figure 11·20 shows the calculated reflection coefficient as a function of wavelength. The actual absorbers do not give as good results as calculated because of variations in the various parameters in manufacture. The reflection coefficient was observed to have a variation of 5 to 10 per cent in the wavelength range of 7 to 15 cm.

A second variety of absorber, composed of a synthetic rubber impregnated with iron powder, makes use of both dielectric and magnetic losses. For uniform plane waves, the equations of Sec. 11·5 are considerably simplified. The wave impedance may be written

$$Z_H = \sqrt{\frac{\mu' - j\mu''}{\epsilon' - j\epsilon''}} = \sqrt{\frac{\mu'(1 - j\tan\zeta)}{\epsilon'(1 - j\tan\varphi)}} = \sqrt{\frac{\mu'\cos\varphi}{\epsilon'\cos\zeta}}\, e^{-j\frac{\zeta-\varphi}{2}}$$

and depends, therefore, for loss angles that are not too large, essentially

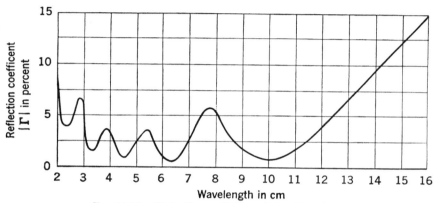

FIG. 11·20.—Reflection coefficient of absorbing sheet.

on the difference $\zeta - \varphi$. The propagation constant from Eqs. (41) becomes

$$\gamma = j\sqrt{\frac{\epsilon'\mu'}{\cos\varphi\cos\zeta}}\, e^{-j\frac{\varphi+\zeta}{2}}$$

which depends upon the sum of the loss angles. An efficient ferromagnetic absorber should have large and equal values of φ and ζ and large values of ϵ' and μ' subject to the condition that μ'/ϵ' be nearly equal to μ_0/ϵ_0.

The construction of an absorber utilizing these principles is shown in Fig. 11·21. The synthetic rubber is impregnated with iron powder, prepared from iron carbonyl, of particle size less than 10μ. The material has a specific gravity of about 4. At a wavelength of 10 cm the dielectric constant ϵ'/ϵ_0 is approximately 25, μ'/μ_0 varies from 3 to 4, $\tan\varphi$ is approximately equal to $\tan\zeta$, and $\tan\zeta$ and $\tan\varphi$ lie between 0.3 to 0.4. The intrinsic impedance is therefore real but rather small, and a resonant construction has been adopted to match into the absorbing material. The waffle construction, at wavelengths long compared with the grid spacing, acts as a shunt inductance. A thin layer of low-loss dielectric acts as a spacer between the rubber and the metal to be camouflaged. The equivalent circuit of the absorber is shown in Fig. 11·22. The resonant nature of the device indicates that it is effective only over a

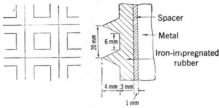

Fig. 11·21.—Absorbing sheet with resonant construction.

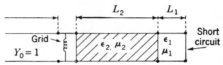

Fig. 11·22.—Equivalent circuit of absorbing sheet of Fig. 11·21.

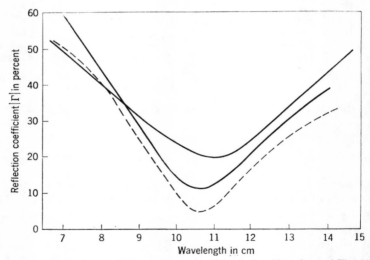

Fig. 11·23.—Reflection coefficient of three samples of absorbing sheet of Fig. 11·21.

narrow frequency band. Some observations showing the effectiveness of this material are shown in Fig. 11·23. Other materials with values of $\sqrt{\mu'/\epsilon'}$ more nearly equal to the impedance of free space can probably be obtained. Experiments with iron oxide (Fe_2O_3, γ-phase) and magnesium ferrite ($MgOFe_2O_3$), both of which are ferromagnetic, show promise in this direction.

CHAPTER 12

THE SYMMETRY OF WAVEGUIDE JUNCTIONS

By R. H. Dicke

In this chapter the properties of symmetrical junctions of two or more transmission lines will be investigated. As these junctions will be assumed to be lossless, all the results found for lossless junctions (Chap. 5) will be assumed to apply here. In particular a junction will be represented by an impedance, admittance, or scattering matrix of a lossless generalized waveguide junction. Terminal voltages and currents will be normalized in such a way as to make the characteristic impedance of all transmission lines unity.

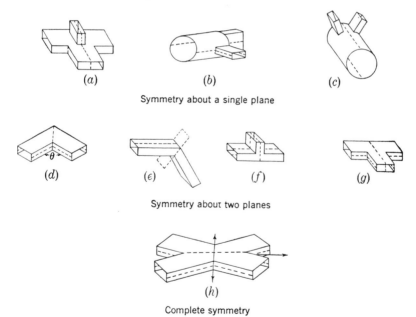

FIG. 12·1.—Junctions having reflection symmetries only. Symmetry of one or more planes.

12·1. Classes of Symmetry.—A number of symmetrical junctions are illustrated in Figs. 12·1 to 12·3, inclusive. It is evident that this collection is by no means complete and that there is an unlimited number of possible symmetrical junctions. Nearly all junctions of transmission lines encountered in practice have some sort of symmetry. Because of symmetry, a junction may have rather unusual properties. This will be brought out later when examples are discussed.

A symmetrical junction is characterized by the fact that it is left unchanged by a symmetry operation. For example, in Fig. 12·2a the symmetry operation is a rotation of the structure by 180° about the symmetry axis. This operation turns the figure back into itself. The junction is said to be "invariant" under this symmetry operation.

In Fig. 12·1 the junctions are characterized by their invariance under reflection in one or more planes. When there are two and only two

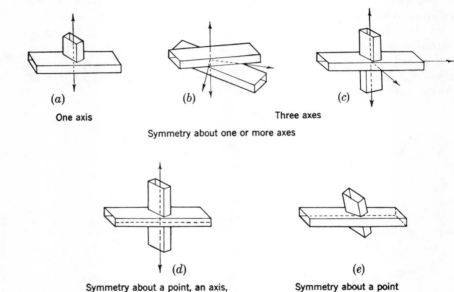

Fig. 12·2.—Junctions having reflection symmetries only, continuation.

symmetry planes, they must intersect normally. Their intersection is a symmetry axis, as may be seen by reflecting first in one plane and then in another. This is an example of the interrelation between symmetry operations. This matter will be discussed later in more detail.

Since space has no preferred directions, it is evident that there is nothing unique about the choice of the x-, y-, and z-axes in Maxwell's electromagnetic equations. In fact, a rotation of the coordinate frame and field quantity to a new position is an operation under which Maxwell's equations should be invariant. This rotation is again a symmetry operation.

Although Maxwell's equations are invariant under a symmetry

operation, any given solution need not be. For instance, a wave moving to the right can be transformed under a reflection into a wave moving to the left. However, a standing wave with the symmetry plane at a node or loop is left unchanged by the reflection. Such a solution is said to be invariant under the symmetry operation.

Symmetrical junctions will be investigated by looking for symmetrical solutions of Maxwell's equations that satisfy the boundary conditions of the junction. Any solution can then be expressed as a linear combination

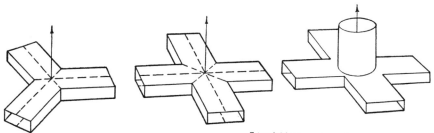

Three-fold symmetry axis   Four-fold symmetry axis

FIG. 12·3.—Higher-order symmetries.

of these symmetrical solutions. A detailed solution of the boundary-value problem is outside the scope of this book.[1] Instead, general conditions which result from symmetry will be investigated.

A useful method for obtaining the properties of symmetrical junctions is found in the theory of eigenvalue equations. This theory is developed sufficiently to make the subsequent treatment of special problems intelligible. However, before the general theory is developed, a simple special case will be considered as an illustration of the type of problem to be considered.

12·2. Symmetry of the Thin Iris.—An iris across a rectangular waveguide is a geometrical configuration with a single plane of symmetry and

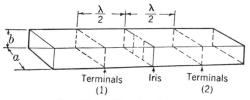

FIG. 12·4.—The thin iris.

represents one of the simplest examples of a symmetrical junction. Such a junction is represented in Fig. 12·4. Let terminal planes be chosen as indicated in the figure. The region between the two terminal planes

[1] J. Schwinger has treated these aspects of many of the problems in detail. This work is as yet unpublished.

may be regarded as a part of the lossless junction, and an impedance matrix Z (pure imaginary) may be defined such that

$$\left.\begin{array}{l} e_1 = Z_{11}i_1 + Z_{12}i_2, \\ e_2 = Z_{21}i_1 + Z_{22}i_2. \end{array}\right\} \quad (1)$$

where the e's and i's are the currents and voltages at the junction.

The first observation that can be made from symmetry is that $Z_{22} = Z_{11}$. This is evident because a reflection of the waveguide through the plane of symmetry leaves the guide unchanged. However, this reflection interchanges the field quantities at terminals (1) and (2). Thus an interchange of (1) and (2) in the elements of the impedance matrix should leave it unchanged. This is possible only if

$$\left.\begin{array}{l} Z_{12} = Z_{21}, \\ Z_{22} = Z_{11}. \end{array}\right\} \quad (2)$$

The first of conditions (2) will be recognized as the usual impedance-matrix symmetry condition and is valid independently of the existence of geometrical symmetry. The second condition is imposed by the geometrical symmetry.

Equation (1) is valid for any choice of i_1 and i_2. In particular, it holds for $i_2 = -i_1$. If this substitution is made in Eq. (1),

$$\left.\begin{array}{l} e_1 = i_1(Z_{11} - Z_{12}), \\ e_2 = i_2(Z_{11} - Z_{12}), \\ e_2 = -e_1. \end{array}\right\} \quad (3)$$

For this particular antisymmetrical solution the impedance seen looking into terminals (1) or (2) is

$$Z = \frac{i_1}{e_1} = \frac{i_2}{e_2} = Z_{11} - Z_{12}. \quad (4)$$

If $e_1 = -e_2$, the electric field at the terminals (1) is opposite in sign to that at the terminals (2). It is easily seen that this solution of Maxwell's equations has an electric field which is an odd function of position along the axis of the guide (see Fig. 12·5). Note that the field becomes zero at the symmetry plane; and since the iris is thin and effectively embedded in this plane, the field distribution is exactly the same as though the symmetry plane had become an electric wall.

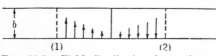

Fig. 12·5.—Field distribution for antisymmetrical solution.

As the field distribution is that which would be produced by a metal wall at the symmetry plane, the impedance Z is that of a short circuit transformed down the line a half wavelength; that is, $Z = 0$. Therefore, from Eq. (4), $Z_{11} = Z_{12}$. This is just the condition that the iris be a shunt susceptance at the symmetry plane. Perhaps the easiest way to see this is from the T-equivalent of the impedance Eq. (1) (see Chap. 4). If $Z_{11} = Z_{22} = Z_{12}$, the circuit becomes a pure shunt element of impedance Z_{12} across the junction.

The odd distribution in electric field of Fig. 12·5 is one symmetrical solution of Maxwell's equations. It is evident that the even solution is another.

$$e_1 = e_2,$$
$$\frac{i_1}{e_1} = 2Z_{12}. \quad (5)$$
$$i_1 = i_2,$$

These are the only two solutions which are symmetrical about the symmetry plane. It is evident that any other solution of Eq. (1) can be obtained as a linear combination of the solutions given in Eqs. (5) and (3).

MATRIX ALGEBRA

12·3. The Eigenvalue Problem.—The problem just considered was so simple that it could be seen by inspection that an odd or even distribution of fields about the symmetry plane was a symmetrical solution of Maxwell's equations. In more complicated cases involving many waveguides in complicated configurations the intuitive approach may not be sufficient to obtain a correct solution. It is the purpose of the next sections to develop formal methods that are applicable to these more complicated cases.

A formalism that is useful in the discussion of symmetrical junctions is that provided by the theory of the eigenvalue equations. This theory is developed here only to the point actually needed in the subsequent problems. A more complete treatment can be found in any of the standard works on matrix algebra. There are also introductory treatments for the reader unfamiliar with this field.[1]

The Eigenvalue Equations.—For a square matrix **P**, a column vector **a**, and a number p, the equation

$$\mathbf{Pa} = p\mathbf{a}, \quad (6)$$

is called an eigenvalue equation. The quantity p is called an eigenvalue, and **a** an eigenvector. As an example of an eigenvalue equation, with reference to Eq. (4), Eq. (3) may be written as

$$\mathbf{Zi} = z\mathbf{i}, \quad (7)$$

[1] Birkhoff and MacLane, *A Survey of Modern Algebra*, Macmillan, New York, 1941; H. Margenau and G. M. Murphy, *The Mathematics of Physics and Chemistry*, Van Nostrand, New York, 1943.

where

$$\mathbf{i} = i_1 \begin{pmatrix} 1 \\ -1 \end{pmatrix}. \tag{8}$$

Equation (6) can be satisfied only for certain discrete values of p. To see this, Eq. (6) may be written as

$$(\mathsf{P} - p\mathsf{I})\mathbf{a} = 0, \tag{9}$$

where I is the unit matrix. Equation (9) is a homogeneous set of n equations in n unknowns and has a nonvanishing solution for **a** only if the determinant of the coefficient vanishes,[1]

$$\det (\mathsf{P} - p\mathsf{I}) = 0. \tag{10}$$

By expansion of this determinant, a polynomial in p of degree n is obtained. This polynomial, called the characteristic polynomial, has n roots, some of which may be equal. When m of these roots are equal, the eigenvalue is said to have a degeneracy of order m. Equation (10) is called the characteristic equation of P.

Nondegenerate Eigenvalues.—The vectors $\mathbf{a}_1 \ldots \mathbf{a}_n$ are said to be linearly independent if for numbers c_j there is no solution of the equation

$$\sum_j c_j \mathbf{a}_j = 0 \tag{11}$$

other than the trivial one, $c_j = 0$ for all j.

Theorem 1.—The n eigenvectors of P corresponding to the n nondegenerate eigenvalues are linearly independent. Let the eigenvalue equation be

$$\mathsf{P}\mathbf{a}_j = p_j \mathbf{a}_j, \tag{12}$$

where the n eigenvalues p_j are all different. Let

$$\sum c_j \mathbf{a}_j = 0. \tag{13}$$

Multiply Eq. (13) by

$$(\mathsf{P} - p_2\mathsf{I})(\mathsf{P} - p_3\mathsf{I}) \cdots (\mathsf{P} - p_n\mathsf{I}). \tag{14}$$

From Eq. (12) the result is

$$c_1(p_1 - p_2)(p_1 - p_3) \cdots (p_1 - p_n)\mathbf{a}_1 = 0. \tag{15}$$

All other terms in the sum vanish. Equation (15) can be satisfied only by $c_1 = 0$. In a similar way each of the other c's can be shown to vanish, and the n eigenvectors are linearly independent.

[1] Margenau and Murphy, *op. cit.*, Chap. 10, p. 299.

There can be no more than n linearly independent n-dimensional vectors. To show this let a_0 be any n-dimensional vector. Then the equation

$$\sum_{j=0}^{n} c_j \mathbf{a}_j = 0 \tag{16}$$

must have a nonvanishing solution for c_j, since it consists of n equations in $n + 1$ unknowns. Also in such a nontrivial solution $c_0 \neq 0$; for if $c_0 = 0$, all the c's must vanish.

Theorem 2.—Any n-dimensional vector may be expressed as a linear combination of n linearly independent vectors,

$$\mathbf{a}_0 = \sum_{j=1}^{n} b_j \mathbf{a}_j. \tag{17}$$

The proof of this theorem is immediately evident if Eq. (16) is divided through by c_0.

From Theorems 1 and 2 it can be seen that there is only one linearly independent eigenvector for each eigenvalue of P. Any other eigenvector can be obtained from this eigenvector by multiplying by a constant.

Degenerate Eigenvalues.—If m of the n roots of the characteristic equation are equal, then there are m linearly independent eigenvectors associated with this eigenvalue. An easy way to see this is to form the matrix

$$\mathsf{P} + \epsilon \mathsf{Q} = \mathsf{T}, \tag{18}$$

where ϵ is a number and Q is a matrix so chosen that T has no degeneracies. Therefore Theorem 1 applies to T. In the limit, as $\epsilon \to 0$, $\mathsf{T} \to \mathsf{P}$ and m of the roots coalesce. The m eigenvectors associated with these roots become associated with this degenerate eigenvalue. It is evident that any linear combination of these m eigenvectors is also an eigenvector. Hence by taking linear combinations of the eigenvectors a new set of linearly independent eigenvectors can be obtained. Therefore the basic set of eigenvectors associated with a degenerate eigenvalue is not unique. There is an infinite number of possible sets.

12·4. Symmetrical Matrices.—The transpose of a matrix is obtained by changing rows into columns keeping the order the same. Note that the transpose of a column vector $\mathbf{a}$ is the row vector $\tilde{\mathbf{a}}$. The transpose of the product of two matrices is

$$\widetilde{\mathsf{P} \cdot \mathsf{Q}} = \tilde{\mathsf{Q}} \cdot \tilde{\mathsf{P}}. \tag{19}$$

This is easily seen from the definition of the product. A symmetrical matrix is by definition one that is equal to its transpose. Symbolically,

$$\mathsf{P} = \tilde{\mathsf{P}}. \tag{20}$$

Two column vectors a_i and a_j are said to be orthogonal when

$$\tilde{\mathsf{a}}_i \mathsf{a}_j = \tilde{\mathsf{a}}_j \mathsf{a}_i = 0. \tag{21}$$

Theorem 3.—If P is symmetrical, the eigenvectors associated with different eigenvalues are orthogonal to each other. Let

$$\mathsf{P}\mathsf{a}_j = p_j \mathsf{a}_j. \tag{22}$$

Taking the transpose of Eq. (22) and changing the index,

$$\tilde{\mathsf{a}}_k \tilde{\mathsf{P}} = p_k \tilde{\mathsf{a}}_k = \tilde{\mathsf{a}}_k \mathsf{P}. \tag{23}$$

Multiplying Eq. (23) on the right by a_j, Eq. (22) on the left by $\tilde{\mathsf{a}}_k$, and subtracting,

$$(p_k - p_j)\tilde{\mathsf{a}}_k \mathsf{a}_j = 0. \tag{24}$$

If $p_k \neq p_j$, the two eigenvectors are orthogonal. If $p_j = p_k$, then the linearly independent eigenvectors need not be orthogonal. However, it is possible to choose a set that is orthogonal. In fact this can be done in an infinite number of ways.

Real Symmetrical Matrices.—A matrix is pure real or imaginary if all its elements are pure real or imaginary.

Theorem 4.—If P is real and symmetrical, all its eigenvalues are real and the eigenvectors may be so chosen as to be real. For if

$$\mathsf{P}\mathsf{a}_j = p_j \mathsf{a}_j, \tag{25}$$

then

$$\mathsf{P}\mathsf{a}_j^* = p_j^* \mathsf{a}_j^*, \tag{26}$$

since

$$\mathsf{P} = \mathsf{P}^*.$$

Also

$$\tilde{\mathsf{a}}_j^* \mathsf{P} = p_j^* \tilde{\mathsf{a}}_j^*, \tag{27}$$

since

$$\mathsf{P} = \tilde{\mathsf{P}}.$$

Multiplying Eq. (27) on the right by a_j and Eq. (25) on the left by $\tilde{\mathsf{a}}_j^*$ and subtracting,

$$(p_j^* - p_j)\tilde{\mathsf{a}}_j^* \mathsf{a}_j = 0. \tag{28}$$

Since $\tilde{\mathsf{a}}_j^* \mathsf{a}_j$ is nonvanishing,

$$p_j^* = p_j. \tag{29}$$

Thus all eigenvalues are pure real. If Eq. (29) is substituted in Eq. (26), it is seen that both a_j and a_j^* are eigenvectors of the same eigenvalue.

Sec. 12·5] RATIONAL MATRIX FUNCTIONS, DEFINITIONS 409

The eigenvectors with some eigenvalue p_k may be divided into two classes, those for which a_k and a_k^* are linearly independent and those for which a_k and a_k^* are linearly dependent. If a_k and a_k^* are linearly dependent, then $a_k + a_k^*$ is a real eigenvector. If a_k and a_k^* are linearly independent, then $a_k + a_k^*$ and $j(a_k - a_k^*)$ are linearly independent and real eigenvectors. Thus it is always possible to pick a complete set of real linearly independent eigenvectors.

Corollary.—The eigenvalues of a pure imaginary symmetrical matrix are pure imaginary, and the eigenvectors may be so chosen as to be pure real. An example of the corollary is afforded by the problem of the thin iris. The odd and even solutions of Eqs. (3) and (5) may be written

$$Za_j = z_j a_j,$$

where

$$a_1 = \begin{pmatrix} 1 \\ -1 \end{pmatrix}, \quad a_2 = \begin{pmatrix} 1 \\ 1 \end{pmatrix}$$

and

$$z_1 = 0, \quad z_2 = 2Z_{12}.$$

It should be noted that z_1 and z_2 are pure imaginary and a_1 and a_2 are real and orthogonal.

12·5. Rational Matrix Functions, Definitions.—Any expression of the form

$$f(P) = c_0(P - c_1 I)(P - c_2 I) \cdots$$
$$(P - c_{-1} I)^{-1}(P - c_{-2} I)^{-1} \cdots , \quad (30)$$

where the c's are constants, is called a rational function of P.

Two matrices P and Q are said to commute if

$$PQ - QP = 0.$$

It can be shown easily that any two factors in Eq. (30) commute, and thus the factors may be taken in any order.

Theorem 5.—If

$$Pa_j = p_j a_j, \qquad (31)$$

then

$$f(P)a_j = f(p_j)a_j, \qquad (32)$$

where $f(P)$ is a rational function of P. This theorem is proved by adding the identity $c_k I a_j = c_k a_j$ to Eq. (31), where c_k is a constant,

$$(P + c_k I)a_j = (p_j + c_k)a_j. \qquad (33)$$

If Eq. (33) is multiplied by $(P + c_k I)^{-1}$, the result is

$$(P + c_k I)^{-1} a_j = (p_j + c_k)^{-1} a_j. \qquad (34)$$

The product

$$f(P)a_j = c_0(P - c_1 I) \cdots (P - c_{-m} I)^{-1} a_k$$

may be evaluated by taking the product of $\mathbf{a}_k$ by the last factor, then the product by the second last factor and so on. The result is, using Eqs. (33) and (34),
$$f(\mathsf{P})\mathbf{a}_j = f(p_j)\mathbf{a}_j.$$

An application of this theorem may be found in the scattering and impedance matrices that are connected by the equation
$$\mathsf{S} = (\mathsf{Z} - \mathsf{I})(\mathsf{Z} + \mathsf{I})^{-1}. \tag{35}$$

Therefore, if $\mathsf{Z}\mathbf{a}_j = z_j\mathbf{a}_j$, then from Theorem 5
$$\mathsf{S}\mathbf{a}_j = s_j\mathbf{a}_j,$$
where
$$s_j = \frac{z_j - 1}{z_j + 1}.$$

The general result has been proved that the impedance, admittance and scattering matrices have common eigenvectors. It should be noted that since $z_j^* = -z_j$,
$$|s_j|^2 = \frac{z_j - 1}{z_j + 1}\frac{z_j^* - 1}{z_j^* + 1} = 1.$$

12·6. Commuting Matrices.—An important theorem may be proved regarding the eigenvectors of two matrices that commute.

Theorem 6.—If P and Q commute and $\mathsf{P}\mathbf{a}_j = p_j\mathbf{a}_j$, where p_j is nondegenerate, then $\mathbf{a}_j$ is an eigenvector of Q. To prove this theorem, it should be noticed that
$$\mathsf{Q}\mathsf{P}\mathbf{a}_j = p_j\mathsf{Q}\mathbf{a}_j$$
or
$$\mathsf{P}(\mathsf{Q}\mathbf{a}_j) = p_j(\mathsf{Q}\mathbf{a}_j).$$

The vector $\mathsf{Q}\mathbf{a}_j$ is therefore an eigenvector of P corresponding to the nondegenerate eigenvalue p_j. Therefore $\mathsf{Q}\mathbf{a}_j$ can differ from $\mathbf{a}_j$ at most by a multiplicative constant, or
$$\mathsf{Q}\mathbf{a}_j = q_j\mathbf{a}_j.$$

Hence $\mathbf{a}_j$ is an eigenvector of Q. In a similar way it can be seen that if p_j is degenerate, $\mathsf{Q}\mathbf{a}_j$ is a linear combination of all the linearly independent eigenvectors of this eigenvalue.

12·7. Cayley-Hamilton's Theorem.—The characteristic equation of a matrix was defined by Eq. (10) of Sec. 12·3.

Theorem 7.—Every matrix satisfies its characteristic equation. To prove this, let the characteristic equation of P be
$$p^n + c_1 p^{n-1} + \cdots = 0,$$

and let the n roots of this equation be p_k. Form the matrix

$$\mathsf{M} = \mathsf{P}^n + c_1 \mathsf{P}^{n-1} + \cdots + c_n.$$

Any nth-order vector can be expanded in terms of eigenvectors a_k of P. Let

$$\mathsf{a} = \sum_k d_k \mathsf{a}_k.$$

Then

$$\mathsf{M}\mathsf{a} = \sum_k d_k (p_k^n + c_1 p_k^{n-1} \cdots + c_n) \mathsf{a}_k.$$

The expression in parentheses on the right of this equation vanishes. Therefore

$$\mathsf{M}\mathsf{a} = 0,$$

for any vector a. Therefore $\mathsf{M} = 0$, and P satisfies its characteristic equation, which proves the theorem.

The spur or trace of a matrix is defined as the sum of its diagonal elements. The last theorem needed for the discussion to follow will now be stated.

Theorem 8.—The spur of a matrix P is equal to the sum of its eigenvalues p_k. Let the characteristic equation of P be

$$p^n + c_1 p^{n-1} + \cdots = 0.$$

The sum of the roots of the polynomial is equal to c_1. The characteristic equation is the expansion of $\det(\mathsf{P} - p\mathsf{I})$. In the expansion of the determinant, the coefficient multiplying p^{n-1} is $\sum_n P_{nn}$. Therefore

$$\sum_n P_{nn} = \sum_k p_k.$$

SYMMETRIES OF MAXWELL'S EQUATIONS

As pointed out earlier, the fact that there are no preferred directions in space indicates that a particular choice of the x-, y-, and z-axes in Maxwell's equations is not unique. It should be possible to introduce a rotation in the geometrical axes and a transformation in the field quantities such that the new coordinates and field quantities satisfy Maxwell's equations. In such a case Maxwell's equations are said to be "invariant" under the transformation.

Rotations are not the only transformations that leave Maxwell's equations invariant. Reflections are also permissible transformations.

Then again several transformations may be applied in succession. The resulting transformation is also one under which Maxwell's equations are invariant. The general transformations will not be needed for the examples to be discussed. The transformations that will be needed are the reflections of various kinds and rotations about a single coordinate axis by fractional parts of 360°. The reflections are the transformations that will be required most often, and their theory will be developed in detail.

12·8. The Symmetry of a Reflection in a Plane.—Under a reflection in the yz-plane, x is transformed into $-x$, and the other space coordinates are left unchanged.

$$\left.\begin{array}{l} x \to x' = -x, \\ y \to y' = y, \\ z \to z' = z. \end{array}\right\} \tag{36}$$

A transformation on the components of electric and magnetic field that leaves Maxwell's equations invariant is desired. One such transformation, as can be seen from the inspection of Maxwell's equations, is

$$\left.\begin{array}{llll} x \to x' = -x, & E_x \to E'_x = -E_x, & H_x \to H'_x = H_x, \\ y \to y' = +y, & E_y \to E'_y = E_y, & H_y \to H'_y = -H_y, \\ z \to z' = z, & E_z \to E'_z = E_z, & H_z \to H'_z = -H_z, \\ \omega \to \omega' = \omega, & & \\ & J_x \to J'_x = -J_x, & \\ & J_y \to J'_y = J_y, & \rho \to \rho' = \rho. \\ & J_z \to J'_z = J_z, & \end{array}\right\} \tag{37}$$

It is well to ponder the meaning of this invariance of Maxwell's equations under the transformation (37). Stated in words the transformation replaces the electric field at the point x, y, z by that at the point $-x$, y, z, changing the sign of the x-component. The invariance of Maxwell's equations implies that this new field distribution is also a solution. Thus for any one solution, another can be obtained by applying the transformation (37). It should be noted that in general this new solution will not satisfy the original boundary conditions. Only when the geometrical structure is invariant under the reflection does this new transformed solution also satisfy the same boundary conditions as does the old. In this case, any permissible solution can be transformed into another permissible solution by the transformation (37).

Let us introduce a formal operator $\boldsymbol{F}_x$ to represent the transformation (37). $\boldsymbol{F}_x$ will be thought of as an operator that can operate on any coordinate to change it into its transformed value. For example,

$$\left.\begin{array}{l} \boldsymbol{F}_x \cdot x = -x, \\ \boldsymbol{F}_x \cdot E_x(x,y,z,\omega) = -E_x(-x,y,z,\omega). \end{array}\right\} \tag{38}$$

It is evident that
$$\boldsymbol{F}_x \cdot (CE_x) = C(\boldsymbol{F}_x \cdot E_x), \tag{39}$$
where C is a constant. Also
$$\boldsymbol{F}_x \cdot (x + y) = \boldsymbol{F}_x \cdot x + \boldsymbol{F}_x \cdot y. \tag{40}$$
An operator is said to be linear when conditions (39) and (40) are satisfied.

Information concerning general properties of the junction can often be obtained by searching for solutions of Maxwell's equations that are invariant under the symmetry transformation. It is desired to find solutions of Maxwell's equations that are left unchanged (except for a possible change in phase) by the symmetry operator. A change in phase can be compensated by a change in time zero, and such a change in solution is not significant.

If E_y is a symmetrical solution, then
$$\boldsymbol{F}_x \cdot E_y = fE_y, \tag{41}$$
where f is a number with unit modulus. Equation (41) will be recognized as a type of eigenvalue equation. Operating on Eq. (41) by $\boldsymbol{F}_x$,
$$\boldsymbol{F}_x^2 \cdot E_y = f\boldsymbol{F}_x \cdot E_y = f^2 E_y. \tag{42}$$
A reflection applied twice, however, leaves everything unchanged and therefore
$$\boldsymbol{F}_x^2 \cdot E_y = E_y. \tag{43}$$
From Eqs. (42) and (43),
$$\left.\begin{array}{c} f^2 = 1, \\ f = \pm 1. \end{array}\right\} \tag{44}$$

Solutions with $f = +1$ are even functions of x. If E_y and E_z are even functions of x, the solution of Maxwell's equation will be called "even." If $f = -1$ the solution will be called "odd." Note that if E_y and E_z are even functions of x, E_x is odd, and conversely. The symmetries of the even and odd solution are summarized in Table 12·1.

TABLE 12·1.—Even and Odd Solutions of Maxwell's Equations

Even	Odd
$E_x(x,y,z) = -E_x(-x,y,z)$	$E_x(x,y,z) = E_x(-x,y,z)$
$E_y(x,y,z) = E_y(-x,y,z)$	$E_y(x,y,z) = -E_y(-x,y,z)$
$E_z(x,y,z) = E_z(-x,y,z)$	$E_z(x,y,z) = -E_z(-x,y,z)$
$H_x(x,y,z) = H_x(-x,y,z)$	$H_x(x,y,z) = -H_x(-x,y,z)$
$H_y(x,y,z) = -H_y(-x,y,z)$	$H_y(x,y,z) = H_y(-x,y,z)$
$H_z(x,y,z) = -H_z(-x,y,z)$	$H_z(x,y,z) = H_z(-x,y,z)$

If the solution is continuous across the symmetry plane, that is, if the symmetry plane does not contain a metallic sheet at the point in question, then setting $x = 0$ in Table 12·1 yields

$$\begin{array}{ll} \text{Even} & \text{Odd} \\ E_x = 0, & \left.\begin{array}{l} E_y = 0, \\ E_z = 0, \\ H_x = 0. \end{array}\right\} \end{array} \qquad (45)$$

Conditions (45) for the odd case are just the ones that must be satisfied by the field quantities at the surface of an electric wall. In other words the field distribution is the same as though the symmetry plane were replaced by a perfectly conducting metallic film. In a similar way the even solutions correspond to a magnetic wall at the symmetry plane.

12·9. Symmetry Operators.—Operators F_y and F_z can be introduced in a similar manner to represent reflections in the xy- and the xz-planes. All the above results follow exactly as for the operator F_x. These reflection operators may be applied in combinations to the various coordinates. For instance a reflection in the yz-plane followed by a reflection in the xz-plane is equivalent to a reflection in the z-axis (or a rotation of 180° about the z-axis). A new operator R_z may be introduced to represent this reflection in the z-axis. Formally,

$$R_z = F_x F_y = F_y F_x. \qquad (46)$$

In a similar way

$$\left.\begin{array}{l} R_x = F_y F_z = F_z F_y, \\ R_y = F_z F_x = F_x F_z. \end{array}\right\} \qquad (47)$$

Another symmetry operator is

$$P = F_x F_y F_z. \qquad (48)$$

This is a reflection in each of the coordinate planes and is equivalent to a reflection in the origin. There is one other symmetry operator of importance in this set, namely, the identity operator. Let I represent the operator that leaves the coordinates unchanged. A multiplication table for these operators can now be constructed. This is shown in Table 12·2.

TABLE 12·2.—MULTIPLICATION TABLE FOR THE REFLECTION GROUP

	I	R_x	R_y	R_z	F_x	F_y	F_z	P
I	I	R_x	R_y	R_z	F_x	F_y	F_z	P
R_x	R_x	I	R_z	R_y	P	F_z	F_y	F_x
R_y	R_y	R_z	I	R_x	F_z	P	F_x	F_y
R_z	R_z	R_y	R_x	I	F_y	F_x	P	F_z
F_x	F_x	P	F_z	F_y	I	R_z	R_y	R_x
F_y	F_y	F_z	P	F_x	R_z	I	R_x	R_y
F_z	F_z	F_y	F_x	P	R_y	R_x	I	R_z
P	P	F_x	F_y	F_z	R_x	R_y	R_z	I

It can be verified from the table that a 180° rotation about the z-axis followed by a reflection in the yz-plane is equivalent to a reflection in the xz-plane. Symbolically,

$$F_x R_z = F_y. \tag{49}$$

These eight reflection operators together with their rules of multiplication given in Table 12·2 are an example of a group. A group is a set of elements with a law of multiplication such that for any three elements of the group A, B, and C,
1. $A(BC) = (AB)C$.
2. There is an element I such that for any element A, $IA = A$.
3. For every element A there is an element A^{-1} such that $AA^{-1} = I$.

It should be noted that for the reflection group every element commutes with every other element. For example

$$F_x R_z = R_z F_x.$$

A group all elements of which commute with one another is said to be Abelian. The elements I, R_x, R_y, and R_z form a group called a subgroup of the reflection group. The generators of a group are elements from which any element of the group can be obtained as a product. The elements F_x, F_y, F_z are generators of the reflection group. Another set is R_x, R_y, and P.

Table 12·2 may be used to construct a table of symmetry types. Each symmetry type is a subgroup of the reflection group. In Table 12·3, seven types of symmetry are listed. They are illustrated by the waveguide structures of Figs. 12·1 to 12·3. To illustrate how this table was constructed, notice that there is no symmetry with two symmetry axes. From Table 12·3, two symmetry axes automatically require a third.

TABLE 12·3.—SYMMETRY TYPES UNDER REFLECTION GROUP

Type	Number of symmetry elements			Type of symmetry
	Planes	Axes	Points	
1	1	0	0	Planar
2	2	1	0	
3	3	3	1	Complete
4	0	1	0	Axial
5	0	3	0	
6	1	1	1	Mixed
7	0	0	1	Point

12.10. Field Distributions Invariant under Axial and Point Reflections.—In Table 12·2 it should be noted that $R_z^2 = I$ and $P^2 = I$, so the arguments that led to even and odd field distributions as the only solutions of Maxwell's equations invariant under reflection in a plane are valid for reflection in an axis and a point.

The solutions with R_z as the symmetry operator are given in Table 12·4.

TABLE 12·4.—SOLUTIONS SYMMETRICAL UNDER R_z

Even	Odd
$E_x(x,y,z) = -E_x(-x,-y,z)$	$E_x(x,y,z) = E_x(-x,-y,z)$
$E_y(x,y,z) = -E_y(-x,-y,z)$	$E_y(x,y,z) = E_y(-x,-y,z)$
$E_z(x,y,z) = E_z(-x,-y,z)$	$E_z(x,y,z) = -E_z(-x,-y,z)$
$H_x(x,y,z) = -H_x(-x,-y,z)$	$H_x(x,y,z) = H_x(-x,-y,z)$
$H_y(x,y,z) = -H_y(-x,-y,z)$	$H_y(x,y,z) = H_y(-x,-y,z)$
$H_z(x,y,z) = H_z(-x,-y,z)$	$H_z(x,y,z) = -H_z(-x,-y,z)$

Along the symmetry axis ($x = y = 0$), therefore, the solutions are

$$\begin{array}{ll} \text{Even} & \text{Odd} \\ E_x(0,0,z) = 0, & E_z(0,0,z) = 0, \\ E_y(0,0,z) = 0, & H_z(0,0,z) = 0. \\ H_x(0,0,z) = 0, & \\ H_y(0,0,z) = 0, & \end{array} \qquad (50)$$

The solutions of Maxwell's equations symmetrical under P are given in Table 12·5.

TABLE 12·5.—SOLUTIONS SYMMETRICAL UNDER P

Even	Odd
$E_x(x,y,z) = E_x(-x,-y,-z)$	$E_x(x,y,z) = -E_x(-x,-y,-z)$
$E_y(x,y,z) = E_y(-x,-y,-z)$	$E_y(x,y,z) = -E_y(-x,-y,-z)$
$E_z(x,y,z) = E_z(-x,-y,-z)$	$E_z(x,y,z) = -E_z(-x,-y,-z)$
$H_x(x,y,z) = -H_x(-x,-y,-z)$	$H_x(x,y,z) = H_x(x,y,z)$
$H_y(x,y,z) = -H_y(-x,-y,-z)$	$H_y(x,y,z) = H_y(x,y,z)$
$H_z(x,y,z) = -H_z(-x,-y,-z)$	$H_z(x,y,z) = H_z(x,y,z)$

At the origin, $x = y = z = 0$, these solutions are characterized by the vanishing of the field components, thus

$$\begin{array}{ll} \text{Even} & \text{Odd} \\ H_x(0,0,0) = 0, & E_x(0,0,0) = 0, \\ H_y(0,0,0) = 0, & E_y(0,0,0) = 0, \\ H_z(0,0,0) = 0, & E_z(0,0,0) = 0. \end{array} \qquad (51)$$

The invariance of Maxwell's equations under the reflection group has been examined in some detail. These are not the only symmetry operators which will be encountered, but other types of symmetry will be discussed in connection with the particular problems to which they apply. As an example of another type of symmetry, the problem of the symmetrical H-plane Y-junction will be considered later.

WAVEGUIDE JUNCTIONS WITH TWO OR THREE ARMS

12·11. The Thick Iris.—The symmetry of the thick iris is the same as that of the thin iris. The main reason for presenting this type of problem again is to introduce with a simple illustration the formal methods of solution. The junction is shown in Fig. 12·6. The iris may have an aperture of any shape in a metal plate of uniform thickness d.

The junction has a symmetry plane through the iris, and all the results of Sec. 12·2 apply to this problem. The junction is invariant under the reflection operator F_x; and to any solution of Maxwell's equations satisfying the boundary conditions imposed by the waveguide and iris, there is another obtainable by operating with F_x on the solution.

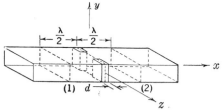

Fig. 12·6.—The thick iris.

If e and i are the voltage and current vectors

$$\mathsf{e} = \begin{pmatrix} e_1 \\ e_2 \end{pmatrix} \qquad \mathsf{i} = \begin{pmatrix} i_1 \\ i_2 \end{pmatrix}$$

of the junction, then e_1 is a measure of E_z at junction (1) and i_1 is a measure of H_y at junction (1). Under F_x, which symbolizes the transformation (37),

$$\left. \begin{array}{l} e_1 \rightarrow e_1' = e_2, \\ e_2 \rightarrow e_2' = e_1; \end{array} \right\} \tag{52}$$

$$\left. \begin{array}{l} i_1 \rightarrow i_1' = i_2, \\ i_2 \rightarrow i_2' = i_1. \end{array} \right\} \tag{53}$$

The sign convention on the junction currents (into the network) results in no sign reversal in Eq. (53). The reflection operator F_x takes the form

$$\mathsf{F} = \begin{pmatrix} 0 & 1 \\ 1 & 0 \end{pmatrix} \tag{54}$$

when operating on i and e. Thus,

$$\mathsf{Fi} = \mathsf{i}', \qquad \mathsf{Fe} = \mathsf{e}'. \tag{55}$$

The transformation given in Eqs. (52) and (53) may be made by performing the matrix operation of Eq. (55). The matrix F is said to represent the operator F_x.

If the impedance matrix of the junction is Z, then

$$e = Zi. \tag{56}$$

But the transformed voltages and currents also satisfy Eq. (56)

$$\left.\begin{aligned} e' &= Zi', \\ Fe &= ZFi, \\ FZi &= ZFi, \\ (FZ - ZF)i &= 0. \end{aligned}\right\} \tag{57}$$

Equation (57) is valid for any current vector i. Hence

$$FZ - ZF = 0. \tag{58}$$

Equation (58) is important. It is a direct link between the symmetry operator F and the impedance matrix.

The only solutions invariant under F_x are the "even" and "odd" solutions (Table (12·1). For these solutions

$$\begin{array}{ll} \text{Even} & \text{Odd} \\ e_1 = e_2, & \left.\begin{aligned} e_1 &= -e_2, \\ i_1 &= -i_2. \end{aligned}\right\} \end{array} \tag{59}$$

Conditions (59) can also be obtained from the eigenvalue equation

$$Fa = fa. \tag{60}$$

The characteristic equation is [see Eq. (10), Sec. 12·3]

$$\det \begin{pmatrix} f & 1 \\ 1 & f \end{pmatrix} = f^2 - 1 = 0.$$

The eigenvalues are

$$f_1 = +1, \quad f_2 = -1. \tag{61}$$

[See also Eq. (44)].

Eigenvectors of Eq. (61) can be found by inspection and can be written

$$a_1 = \frac{1}{\sqrt{2}} \begin{pmatrix} 1 \\ 1 \end{pmatrix}, \quad a_2 = \frac{1}{\sqrt{2}} \begin{pmatrix} 1 \\ -1 \end{pmatrix}.$$

Note that a_1 and a_2 are linearly independent (Theorem 1), orthogonal (Theorem 3), and pure real (Theorem 4) and are normalized to unity.

From Eq. (58) and Theorem 6, a_1 and a_2 are also eigenvectors of Z. Thus the eigenvalue equations of Z can be written as

$$Za_k = z_k a_k; \tag{62}$$

z_k is pure imaginary (corollary, Theorem 4).

From Theorem 5, a_1 and a_2 are also eigenvectors of Y and S.

THE THICK IRIS

As was discussed previously (Sec. 12·8) the even and odd solutions are those for which the symmetry plane becomes a magnetic and an electric wall respectively. Figure 12·7 is a cross section of Fig. 12·6 for these two cases, showing one side only.

Nothing very much can be said about the eigenvalue z_1 for the even case without a solution of the boundary-value problem. However, it is clear that the obstacle for the odd case will reflect in such a way that an effective short circuit lies somewhere between the symmetry plane and the left side of the iris. The eigenvalue z_2 will be capacitive, because the short circuit lies between one-quarter and one-half guide wavelength from the terminals. If the thickness of the iris, d, is small compared with the guide wavelength λ_g, then

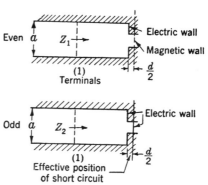

FIG. 12·7.—Boundary conditions for symmetrical and antisymmetrical solutions for the thick iris.

$$0 < jz_2 < \pi \frac{d}{\lambda_g}.$$

The impedance matrix can be written down directly in terms of z_1 and z_2. However, it may be obtained formally by the following procedure. Equation (62) may be combined to form the single equation

$$\mathsf{ZA} = \mathsf{AZ}_d, \tag{63}$$

where

$$\mathsf{A} = \frac{1}{\sqrt{2}} \begin{pmatrix} 1 & 1 \\ 1 & -1 \end{pmatrix},$$
$$\mathsf{Z}_d = \begin{pmatrix} z_1 & 0 \\ 0 & z_2 \end{pmatrix}.$$

Note that A has a_1 and a_2 as columns. The matrix A has columns that are orthogonal and normalized to unity. Such a matrix is said to be orthogonal. It has the property that

$$\tilde{\mathsf{A}} = \mathsf{A}^{-1}.$$

The fact that A is symmetrical is not very significant. If z_1 and z_2 were interchanged, the new A would not be symmetrical. From Eq. (63),

$$\mathsf{Z} = \mathsf{AZ}_d\mathsf{A}^{-1} = \mathsf{AZ}_d\tilde{\mathsf{A}}.$$

If this equation is multiplied out,

$$Z = \tfrac{1}{2}\begin{pmatrix} (z_1 + z_2) & (z_1 - z_2) \\ (z_1 - z_2) & (z_1 + z_2) \end{pmatrix}. \tag{64}$$

If a T-section equivalent is made for Eq. (64), it takes the form of Fig. 12·8.

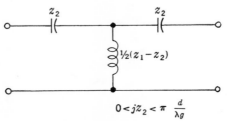

Fig. 12·8.—Equivalent circuit of a thick iris.

12·12. The Symmetrical Y-junction.—Figure 12·9 shows an H-plane symmetrical Y, and Fig. 12·10 is a diagram to illustrate its symmetries. There are three symmetry planes F_1, F_2, and F_3 intersecting in a threefold symmetry axis. The structure is invariant under rotations of 120° and 240° about the symmetry axis.

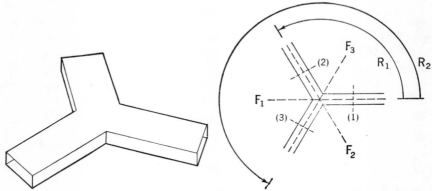

Fig. 12·9.—H-plane Y-junction.   Fig. 12·10.—Symmetries of the H-plane Y-junction.

The symmetries illustrated in Fig. 12·10 are not the only ones. The plane containing the axes of the three guides is a symmetry plane, and the intersection of this plane with the other planes (that is, the z-axes of the guides) are symmetry axes. However, these symmetries do not play an important role in the properties of the Y-junction. In fact, these extraneous symmetries will later be removed by placing a post, along the three-fold axis, that does not extend completely across the guide.

The unit operator together with R_1, R_2, F_1, F_2, and F_3 form a group whose multiplication table is Table 12·6. It should be emphasized that

TABLE 12·6.—SYMMETRY GROUP

	I	R_1	R_2	F_1	F_2	F_3
I	I	R_1	R_2	F_1	F_2	F_3
R_1	R_1	R_2	I	F_2	F_3	F_1
R_2	R_2	I	R_1	F_3	F_1	F_2
F_1	F_1	F_3	F_2	I	R_2	R_1
F_2	F_2	F_1	F_3	R_1	I	R_2
F_3	F_3	F_2	F_1	R_2	R_1	I

this table states that, for example,

$$F_3 F_1 = R_1.$$

Stated in words, a reflection in the plane F_1 (Fig. 12·10) followed by a reflection in F_3 is equivalent to the rotation R_1.

Note that

$$F_3 F_1 \neq F_1 F_3$$

and the group is noncommutative. The operators I, R_1, and R_2 form an Abelian subgroup.

Note that any element of the group may be generated as a product of one or more terms in R_1 and F_1. For instance,

$$F_3 = F_1 R_1^2. \tag{65}$$

Therefore R_1 and F_1 are generators of the group.

The rotation R_1 rotates the junction by 120°. The currents and voltages at a terminal plane become replaced by those at another terminal. We shall always use the convention that the terminal number is fixed in space but that the structure itself is transformed. Thus the currents and voltages at terminal 1 are replaced by those at terminal 3. Let

$$\mathsf{R}_1 = \begin{pmatrix} 0 & 0 & 1 \\ 1 & 0 & 0 \\ 0 & 1 & 0 \end{pmatrix}. \tag{66}$$

Then the terminal currents and voltages i and e are transformed into

$$\left. \begin{array}{l} \mathsf{R}_1 \mathsf{i} = \mathsf{i}', \\ \mathsf{R}_1 \mathsf{e} = \mathsf{e}'. \end{array} \right\} \tag{67}$$

This may be compared with the result given in Eq. (55).

Since the transformed currents and voltages are permissible solutions, one obtains in the usual way [see Eq. (58), for example]

$$\mathsf{R}_1 \mathsf{Z} - \mathsf{Z} \mathsf{R}_1 = 0. \tag{68}$$

with similar relations for the admittance and scattering matrices. In a similar way a matrix may be introduced to represent each of the other symmetry operators. However, for reasons that will be indicated below, it is necessary to introduce a matrix for only one more such operator, namely, F_1, the remaining generator of the group. Let

$$F_1 = \begin{pmatrix} 1 & 0 & 0 \\ 0 & 0 & 1 \\ 0 & 1 & 0 \end{pmatrix}. \tag{69}$$

Note that Eq. (69) interchanges fields at terminals 2 and 3 without affecting terminals 1. The impedance matrix must also commute with that given in Eq. (69) because of the symmetry of the junction. Thus

$$F_1 Z - Z F_1 = 0. \tag{70}$$

The commutation of the generators R_1 and F_1 with the impedance, admittance, or scattering matrix automatically guarantees the commutation of every symmetry matrix. To see this, note that from Eq. (68),

$$R_1^2 Z = R_1 Z R_1 = Z R_1^2 \tag{71}$$

and

$$F_1 R_1^2 Z = F_1 Z R_1^2 = Z F_1 R_1^2. \tag{72}$$

From Eqs. (72) and (65),

$$F_3 Z - Z F_3 = 0. \tag{73}$$

In a similar way each of the symmetry operators can be shown to commute with Z.

Note that the commutation relation [Eq. (73)] is the only connection between the impedance and symmetry operators. Thus if the generators of the symmetry group commute with Z, then every symmetry operation commutes and all the conditions that symmetry imposes on Z are fulfilled.

The Eigenvalue Problem.—Introduce the eigenvalue equation for R_1,

$$R_1 a_j = r_j a_j. \tag{74}$$

From Table 12·6 it can be seen that

$$R_1^3 = I; \tag{75}$$

physically, it is clear that three rotations of 120° return the figure to its initial position. Combining Eqs. (74) and (75),

$$R_1^3 a_j = I a_j = r_j^3 a_j.$$

Thus

$$r_j^3 = 1. \tag{76}$$

Let

$$r_1 = 1,$$
$$r_2 = -\frac{1}{2} + \frac{\sqrt{3}}{2j} = \alpha_1,$$
$$r_3 = -\frac{1}{2} - \frac{\sqrt{3}}{2j} = \alpha_2.$$
(77)

The three eigenvalues of R_1 are all different and therefore nondegenerate.

The three eigenvectors of R_1 can be found by inspection (remembering that $\alpha_1^2 = \alpha_2$, $\alpha_2^2 = \alpha_1$, $\alpha_1\alpha_2 = 1$).
Let

$$\mathsf{a}_1 = \begin{pmatrix} 1 \\ 1 \\ 1 \end{pmatrix},$$
$$\mathsf{a}_2 = \begin{pmatrix} 1 \\ \alpha_2 \\ \alpha_1 \end{pmatrix},$$
$$\mathsf{a}_3 = \begin{pmatrix} 1 \\ \alpha_1 \\ \alpha_2 \end{pmatrix}.$$
(78)

Since R_1 commutes with Z, Y, and S, the eigenvectors given in Eqs. (78) are, by Theorem 6, eigenvectors of $\mathsf{Z}, \mathsf{Y},$ and S. Thus, for instance,

$$\mathsf{Z}\mathsf{a}_k = z_k \mathsf{a}_k. \tag{79}$$

By Theorem 4 it must be possible to choose eigenvectors of Z that are pure real. This is possible only if $z_2 = z_3$. This can be seen in another way. Operating on Eq. (79) with F_1, remembering Eq. (70),

$$\mathsf{Z}(\mathsf{F}_1 \mathsf{a}_j) = z_j(\mathsf{F}_1 \mathsf{a}_j).$$

Thus $\mathsf{F}_1 \mathsf{a}_j$ is an eigenvector of z_j.
But

$$\mathsf{F}_1 \mathsf{a}_2 = \mathsf{a}_3,$$
$$\mathsf{F}_1 \mathsf{a}_3 = \mathsf{a}_2.$$

Therefore

$$z_2 = z_3.$$

By taking a_1 and linear combinations of a_2 and a_3, three linearly independent, orthogonal, real eigenvectors can be obtained. Let

$$\mathsf{b}_1 = \mathsf{a}_1,$$
$$\mathsf{b}_2 = \mathsf{a}_2 + \mathsf{a}_3,$$
$$\mathsf{b}_3 = \frac{1}{\sqrt{3}} j(\mathsf{a}_2 - \mathsf{a}_3).$$
(80)

Note that b_2 and b_3 are eigenvectors of Z. Also,

$$\left. \begin{array}{l} F_1 b_1 = b_1, \\ F_1 b_2 = b_2, \\ F_1 b_3 = -b_3. \end{array} \right\} \qquad (81)$$

Thus the b_j's are simultaneously eigenvectors of Z and F_1. By Theorem 3, since Z and F_1 are symmetric, b_1, b_2, and b_3 are mutually orthogonal. By Theorem 4, b_1, b_2, and b_3 must be, except for a possible multiplicative constant, pure real,

$$\left. \begin{array}{l} b_1 = \begin{pmatrix} 1 \\ 1 \\ 1 \end{pmatrix}, \\ b_2 = \begin{pmatrix} 2 \\ -1 \\ -1 \end{pmatrix}, \\ b_3 = \begin{pmatrix} 0 \\ 1 \\ -1 \end{pmatrix}. \end{array} \right\} \qquad (82)$$

It can be seen by inspection that the b's are real, orthogonal, and linearly independent.

By Theorem 5 (see example at end) the a's and b's are also eigenvectors of S and Y. Let

$$S a_j = s_j a_j, \qquad Y a_j = y_j a_j,$$

where

$$s_j = \frac{z_j - 1}{z_j + 1}, \qquad |s_j| = 1, \qquad y_j = z_j^{-1}.$$

Field Distribution.—Note that the eigensolutions always have standing waves in all of the joining waveguides, since $|s_j| = 1$. However, in connecting guides that are part of the junction or in the interior of the junction, there may be running waves. As an example of this a_2 and a_3, as can be seen by inspection, are solutions with a different time phase at each of the junctions. In the case of a_2 the voltage is a maximum first at terminals (1), then (2), then (3). In the vicinity of the symmetry axis the electromagnetic field rotates about the axis once per cycle. The solution a_3 is the same except for the opposite direction of rotation. These solutions might be called three-phase solutions because they are analagous to the types of field distributions obtained in three-phase a-c machinery.

Since these rotating fields have a time phase that varies with the angle of rotation, the components of E and H parallel to the symmetry axis must vanish along that axis [see Eq. (50) for the analogous twofold

SEC. 12·12]   THE SYMMETRICAL Y-JUNCTION

symmetry axis]. Since the other components of **E** vanish anyway, **E** must completely vanish along the symmetry axis.

The field distributions for b_1, b_2, and b_3 can be easily obtained. It is to be noted that from Eq. (80), b_1 is an eigenvector of F_1 for which the eigenvalue is $+1$. It is also an eigenvector of F_2 and of F_3. For all three reflection operators the solution b_1 is an even solution, and therefore the field distribution is characterized by a magnetic wall along each of

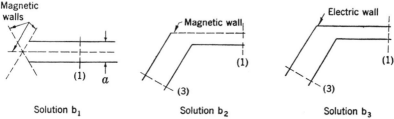

FIG. 12·11.—Boundary conditions for the various eigensolutions.

the symmetry planes. Thus arm (1) is terminated in a V-shaped magnetic wall. The solutions b_2 and b_3 are eigensolutions of F_1, even and odd respectively, and are therefore characterized by a magnetic and an electric wall in the symmetry plane (see Fig. 12·11).

The Scattering Matrix.—The eigenvalue equations

$$\mathsf{S}\mathsf{b}'_k = s_k \mathsf{b}'_k,$$

where the eigenvectors b'_k are the b's normalized to unity,

$$\mathsf{b}'_1 = \frac{1}{\sqrt{3}} \begin{pmatrix} 1 \\ 1 \\ 1 \end{pmatrix}, \qquad \mathsf{b}'_2 = \frac{1}{\sqrt{6}} \begin{pmatrix} 2 \\ -1 \\ -1 \end{pmatrix}, \qquad \mathsf{b}'_3 = \frac{1}{\sqrt{2}} \begin{pmatrix} 0 \\ 1 \\ -1 \end{pmatrix},$$

can be written

$$\mathsf{SB} = \mathsf{BS}_d,$$

[see Eq. (63)] where

$$\mathsf{B} = \begin{pmatrix} \dfrac{1}{\sqrt{3}} & \dfrac{2}{\sqrt{6}} & 0 \\ \dfrac{1}{\sqrt{3}} & -\dfrac{1}{\sqrt{6}} & \dfrac{1}{\sqrt{2}} \\ \dfrac{1}{\sqrt{3}} & -\dfrac{1}{\sqrt{6}} & -\dfrac{1}{\sqrt{2}} \end{pmatrix}$$

and

$$\mathsf{S}_d = \begin{pmatrix} s_1 & 0 & 0 \\ 0 & s_2 & 0 \\ 0 & 0 & s_2 \end{pmatrix}.$$

The matrix **B** is orthogonal. Hence,

$$\mathsf{S} = \mathsf{B}\mathsf{S}_d\tilde{\mathsf{B}}. \tag{83}$$

If **S** is multiplied out, there is obtained

$$\mathsf{S} = \begin{pmatrix} \alpha & \beta & \beta \\ \beta & \alpha & \beta \\ \beta & \beta & \alpha \end{pmatrix}, \tag{84}$$

where

$$\left. \begin{array}{l} \alpha = \tfrac{1}{3}(s_1 + 2s_2), \\ \beta = \tfrac{1}{3}(s_1 - s_2). \end{array} \right\} \tag{85}$$

Equation (84) also gives the impedance and admittance matrices **Z** and **Y**, provided that in Eqs. (85) s_j is replaced by z_j and y_j respectively, where

$$z_j = \frac{1 + s_j}{1 - s_j}, \qquad y_j = \frac{1 - s_j}{1 + s_j}. \tag{86}$$

Note that the sum of the diagonal elements, or spur, of the matrix given by Eq. (84) is equal to the sum of the eigenvalues (Theorem 8).

Power Division.—A junction is said to be matched when all the diagonal elements in the scattering matrix are zero. Clearly a necessary condition for a matched junction is that the sum of all eigenvalues of **S** be zero. In Eqs. (85) the phases of s_1 and s_2 can never be of such values that $\alpha = 0$. Hence the symmetrical Y-junction can never be matched.

This property is much more general than appears above. In fact no junction of three transmission lines can be matched. To show this, assume that such a junction has been matched. Its scattering matrix is

$$\mathsf{S} = \begin{pmatrix} 0 & S_{12} & S_{13} \\ S_{21} & 0 & S_{23} \\ S_{31} & S_{32} & 0 \end{pmatrix}. \tag{87}$$

But **S** is unitary and symmetrical; the product of any column by the complex and conjugate of any other column is therefore zero. If the first column is multiplied by the complex conjugate of the second, the result is

$$S_{31}S_{32}^* = 0.$$

In a similar way the other two products give

$$S_{12}S_{13}^* = 0, \qquad S_{21}S_{23}^* = 0.$$

These three equations cannot be satisfied unless two of the three quantities S_{12}, S_{23}, S_{13} are zero. But in this case there is a column of the matrix that is zero. This is impossible, since the product of every column by the complex conjugate of itself is unity. Thus it is impossible to match a T-junction or any other junction of three guides.

It is to be noted that the best match which can be obtained with the symmetrical Y-junction is when

$$s_1 = -s_2. \tag{88}$$

In this case the scattering matrix of Eq. (84) becomes

$$S = \tfrac{1}{3}s_1 \begin{pmatrix} -1 & 2 & 2 \\ 2 & -1 & 2 \\ 2 & 2 & -1 \end{pmatrix}. \tag{89}$$

With two of the waveguides terminated in their characteristic impedances, eight-ninths of the power entering the third arm goes into these terminating loads. The remaining ninth is reflected back to the generator. In order to satisfy the condition (88), it is necessary to adjust the phases of s_1 and s_2 relative to each other. One way in which this can be done is to insert a pin in the guide along the symmetry axis. It will be remembered that the electric field is zero along the axis for the modes whose eigenvalue is s_2. The pin does not affect these modes at all. The electric field is a maximum at this point for the eigenvector a_1 with the eigenvalue s_1. Hence as the pin is extended across the guide, the phase of s_1 would be expected to change without altering s_2. It does not necessarily follow that by such an adjustment, the phase of s_1 can be made to have any desired value, but it is to be expected that a sizable variation can be obtained in this way.

12·13. Experimental Determination of s_1 and s_2.—If plungers are inserted in two of the arms in symmetrical positions, then power entering the third arm will set up standing waves in the system. The plungers are adjusted until the nodal points come in the same place in each of the three arms; then the position of the nodal point is measured. The phase determined in this way is the phase of one of the eigenvalues.

The procedure just outlined is correct in principle but would be very difficult in practice. A procedure that involves a single plunger is much better in many ways. It is to be noticed that the eigenvector b_3 has no fields in one of the arms. If there are no fields, the plunger can be omitted in this arm. In fact, the condition of no power in this arm is a convenient test for determining when the remaining plunger is in the correct position. To see this algebraically, let a plunger be placed in arm (3) of the Y-junction and let a matched load terminate arm (1). Then the scattering equation is

$$Sa = b,$$

where

$$\left. \begin{array}{l} a_1 = 0, \\ a_3 = b_3 e^{-j\phi} \end{array} \right\}, \tag{90}$$

and ϕ depends upon the position of the plunger. If it is assumed that the reflection coefficient in arm (2) is s_2, then

$$a_2 s_2 = b_2. \tag{91}$$

Combining Eqs. (90) and (91), using the notation of Eq. (84),

$$b_1 = \beta a_2 + \beta a_3, \tag{92}$$

$$\left. \begin{array}{l} b_2 = \alpha a_2 + \beta a_3 = a_2 s_2, \\ b_3 = \beta a_2 + \alpha a_3 = a_3 e^{j\phi}. \end{array} \right\} \tag{93}$$

Equations (93) have a solution only if

$$\begin{vmatrix} (\alpha - s_2) & \beta \\ \beta & (\alpha - e^{j\phi}) \end{vmatrix} = 0. \tag{94}$$

Solving Eq. (94) for $e^{j\phi}$ using Eq. (84),

$$e^{j\phi} = s_2. \tag{95}$$

Thus if the reflection coefficient in arm (2) is s_2, the plunger is in such a position that Eq. (95) holds. The ratio of a_2 to a_3 is the ratio of the minors of a column of the determinant in Eq. (94).

$$\frac{a_2}{a_3} = \frac{\alpha - e^{j\phi}}{-\beta} = -1,$$

$$a_2 + a_3 = 0.$$

Substituting in Eq. (92),

$$b_1 = 0.$$

Under these conditions, therefore, no power enters the load on arm (1). Conversely, if the plunger is adjusted until no power enters arm (1), then s_2 is given by relation (95).

The determination of s_1 is the next step. One way in which it can be determined is to measure α by measuring the reflection coefficient at one of the arms with the other arms matched. Then, making use of Eqs. (85), s_1 can be determined. Another method, which is capable of greater accuracy, will now be outlined.

It is to be noted that the eigenvector b$_3$ is odd with respect to F but that b$_1$ and b$_2$ are even. The eigenvectors b$_1$ and b$_2$ have different eigenvalues with respect to S, and the positions of nodes are therefore different for b$_1$ and b$_2$. A linear combination of b$_1$ and b$_2$ with the same time phase is a new standing-wave solution. However, the nodes occur at different places. In particular, by taking the right combination, the nodes in arms (2) and (3) can be made to occur $\frac{1}{4}\lambda_g$ away from the nodes of b$_3$. A linear combination of this standing-wave solution with b$_3$, since the time phase of b$_3$ is in quadrature and its amplitude is equal to the other solu-

SEC. 12·13]   EXPERIMENTAL DETERMINATION   429

tion, can be made such that there are pure running waves in arms (2) and (3).

To recapitulate, there is a solution that corresponds to running waves in arms (2) and (3) and a standing wave in (1). This solution can be set up by a plunger in arm (1). The position of this plunger can be adjusted until there is no reflection by the Y-junction. The position of the plunger is then an accurate measure of a combination of s_1 and s_2. Since s_2 is known accurately, s_1 can be determined.

It will now be shown that the linear combination

$$g = \frac{1}{s_1 + s_2} b_1 + \frac{1}{2s_2} b_2 + \frac{s_2 - s_1}{2s_2(s_1 + s_2)} b_3 \tag{96}$$

s a set of incident waves which results in pure running waves in arms (2) and (3). The components of Eq. (96) are, from Eqs. (82),

$$g_1 = \frac{2s_2 + s_1}{s_2(s_1 + s_2)},$$

$$g_2 = \frac{s_2 - s_1}{s_2(s_1 + s_2)},$$

$$g_3 = 0.$$

Thus this solution corresponds to waves incident upon the junction from arms (1) and (2). After scattering by the junction, the waves are given by

$$h = Sg,$$

where

$$h = \frac{s_1}{s_1 + s_2} b_1 + \frac{s_2}{2s_2} b_2 + \frac{s_2(s_2 - s_1)}{2s_2(s_1 + s_2)} b_3,$$

since b_1, b_2, and b_3 are eigenvectors with eigenvalues s_1, s_2, and s_2. The components of h are

$$h_1 = \frac{2s_1 + s_2}{s_1 + s_2} = \frac{2s_1 + s_2}{2s_2 + s_1} \cdot s_2 g_1,$$

$$h_2 = 0,$$

$$h_3 = \frac{s_1 - s_2}{s_1 + s_2} = -s_2 g_2.$$

It should be noted that $h_2 = 0$ and there is no reflected wave in arm (2). In other words, power enters by arms (1) and (2) and leaves by arms (1) and (3). It should be noticed also that

$$|h_1| = |g_1|.$$

Thus there is a pure standing wave in arm (1). If this standing wave is set up by a plunger in the correct position, the position of this plunger is a measure of the eigenvalues of S. If a plunger in arm (1) is in such a position that its reflection coefficient is

$$e^{j\phi} = \frac{g_1}{h_1} = s_2^{-1} \frac{2s_2 + s_1}{2s_1 + s_2}, \tag{97}$$

then power is matched through the Y-junction. In practice the Y-junction might be measured on an impedance bridge to obtain an adjustment of this plunger for which Eq. (97) holds. Since s_2 is known from other measurements, Eq. (97) can be solved for s_1 in terms of $e^{j\phi}$ and s_2.

To recapitulate, it is found that for a plunger in one of the arms of the Y-junction, there is one position for which there is no coupling between the remaining two arms and there is another position for which the power is matched through the Y-junction without reflection. Both of these conditions are more general than appears here. Consider the second part: The symmetry of the Y-junction can be partially removed by including, as part of the Y-junction, a transformer in arm (1). The resulting junction has only a single plane of symmetry, namely, the plane including arm (1). However, this arm backed up by a plunger is still effectively a pure reactance. There is now a new position of the plunger for which power is matched through the Y-junction. A three-armed junction with a single plane of symmetry is usually called a T-junction. Thus power can be matched through a T-junction by means of a plunger placed at the proper position in the symmetry arm.

A property of all three-armed junctions, symmetrical or not, concerns the stopping of transmitted power. Let us consider a plunger to be placed in one of the arms in such a position that there is no coupling between the two remaining arms. If transformers are inserted in these arms, there will still be no coupling but the symmetry has been completely destroyed. Thus we have the result that for any junction of three transmission lines, there is a position of a plunger in any arm for which the remaining two arms are decoupled.

12·14. Symmetrical T-junctions.—Figure 12·12 is an illustration of three different types of symmetrical T-junctions. Two of the T-junctions have a single symmetry plane; the third has an axis of symmetry. It should be noted that the series T-junction is a special case of the axial T-junction.

The terminal planes in each of the arms are illustrated and numbered in the figure. The reflection operator applied to the shunt T-junction has the form

$$\mathsf{F} = \begin{pmatrix} 0 & 1 & 0 \\ 1 & 0 & 0 \\ 0 & 0 & 1 \end{pmatrix}. \tag{98}$$

SYMMETRICAL T-JUNCTIONS

Since **F** commutes with the scattering matrix,

$$\mathsf{S} = \mathsf{FSF}^{-1} = \mathsf{FSF}. \tag{99}$$

If Eq. (98) is substituted in Eq. (99) and the matrix product performed, the elements of **S** must satisfy

$$\begin{aligned} S_{11} &= S_{22}, \\ S_{13} &= S_{23}. \end{aligned} \tag{100}$$

In a similar way the reflection operator applied to the series T-junction takes the form

$$\mathsf{F} = \begin{pmatrix} 0 & 1 & 0 \\ 1 & 0 & 0 \\ 0 & 0 & -1 \end{pmatrix}.$$

The rotation operator applied to the axial T-junction takes the form

$$\mathsf{R} = \begin{pmatrix} 0 & 1 & 0 \\ 1 & 0 & 0 \\ 0 & 0 & -1 \end{pmatrix}.$$

Thus the rotation operator is quite equivalent to the reflection operator applied to the series T-junction, and the properties of the axial

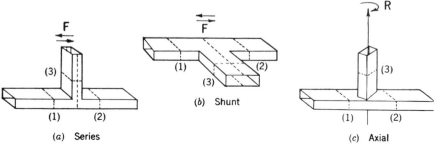

FIG. 12·12.—Symmetrical T-junction.

T-junction will be essentially the same as those of the series T-junction. Note that

$$\mathsf{RSR} = \mathsf{S}.$$

This condition leads to

$$\begin{aligned} S_{11} &= S_{22}, \\ S_{13} &= -S_{23}. \end{aligned}$$

These relations, except for the change of sign, are the same as Eqs. (100). For this reason the properties of all the T-junctions will be very similar. In order to avoid duplication, only the shunt T-junction will be discussed in detail.

12·15. The Shunt T-junction.

The symmetry operator of the shunt T-junction is the reflection operator given in Eq. (98). The eigenvalue equation of **F** is

$$\mathbf{F}\mathbf{a}_k = f_k \mathbf{a}_k.$$

Since

$$\mathbf{F}^2 = \mathbf{I},$$

the only eigenvalues are ± 1. If the characteristic equation of **F**,

$$\det(\mathbf{F} - f\mathbf{I}) = 0,$$

is solved, the roots are

$$f_1 = +1, \qquad f_2 = +1, \qquad f_3 = -1.$$

Thus there are two positive eigenvalues and one negative. The negative eigenvalue is nondegenerate; its eigenvector is therefore unique, except for the usual multiplicative constant. The eigenvectors of the degenerate positive eigenvalue are not unique, but the following set of eigenvectors are orthogonal and real:

$$\mathbf{a}_1 = \begin{pmatrix} 1 \\ 1 \\ \sqrt{2} \end{pmatrix}, \qquad \mathbf{a}_2 = \begin{pmatrix} 1 \\ 1 \\ -\sqrt{2} \end{pmatrix}, \qquad \mathbf{a}_3 = \begin{pmatrix} 1 \\ -1 \\ 0 \end{pmatrix}.$$

Let the eigenvalue equation for the scattering matrix of a shunt T-junction be

$$\mathbf{S}\mathbf{b}_j = s_j \mathbf{b}_j.$$

It must be assumed that the three eigenvalues of **S** are unequal unless there is some symmetry operator that requires equality. Since **S** and **F** commute, the **b**'s are also eigenvectors of **F** (Theorem 6). Thus $\mathbf{b}_1$ must be a linear combination of $\mathbf{a}_1$ and $\mathbf{a}_2$,

$$\mathbf{b}_1 = \begin{pmatrix} 1 \\ 1 \\ \sqrt{2}\alpha \end{pmatrix}.$$

Note that α is real by Theorem 4. By Theorem 3, $\mathbf{b}_2$ must have the form

$$\mathbf{b}_2 = \begin{pmatrix} 1 \\ 1 \\ -\dfrac{\sqrt{2}}{\alpha} \end{pmatrix}.$$

By Theorem 6, $\mathbf{b}_3 = \mathbf{a}_3$. The form of the **b**'s is not particularly simple, because the positions of the reference planes in the arms 1, 2, and 3 have been chosen in an arbitrary way. We shall now indicate how these planes may be chosen so that $\alpha = 1$.

THE SHUNT T-JUNCTION

As was pointed out previously, eigensolutions are always standing-wave solutions. A linear combination of two standing-wave solutions with equal time phase is again a standing-wave solution. It is desired to take a linear combination of b_1 and b_2 to produce a standing-wave solution in which the amplitude of the wave in guide (3) is $\sqrt{2}$ times that in guides (1) and (2). This is always possible, and we shall assume, without going into details, that it has been done. New reference planes are now chosen to occur at the voltage loops in guides (1), (2), and (3). The reference planes in (1) and (2) are, of course, symmetrically placed. The standing-wave solution obtained is, with respect to these new reference planes, an eigensolution with voltage loops at the reference planes. Its eigenvalue is clearly

$$s_1 = 1.$$

Its eigenvector may be either

$$\mathbf{a}_1 = \begin{pmatrix} 1 \\ 1 \\ \sqrt{2} \end{pmatrix}$$

or the same thing with the opposite sign for the third element. It is always possible to choose the original linear combination in such a way that $\mathbf{a}_1$ is the correct eigenvector. A comparison with $\mathbf{b}_1$ shows that $\alpha = 1$. Thus for these new reference planes,

$$\mathsf{S}\mathbf{a}_j = s_j \mathbf{a}_j,$$

where $s_1 = 1$. The eigenvectors $\mathbf{a}_j$ may be combined to form an orthogonal matrix

$$\mathsf{A} = \tfrac{1}{2}\begin{pmatrix} 1 & 1 & \sqrt{2} \\ 1 & 1 & -\sqrt{2} \\ \sqrt{2} & -\sqrt{2} & 0 \end{pmatrix}. \tag{101}$$

Then

$$\mathsf{S}\mathsf{A} = \mathsf{A}\mathsf{S}_d,$$

where

$$\mathsf{S}_d = \begin{pmatrix} s_1 & 0 & 0 \\ 0 & s_2 & 0 \\ 0 & 0 & s_3 \end{pmatrix}.$$

From Eq. (101)

$$\mathsf{S} = \mathsf{A}\mathsf{S}_d\tilde{\mathsf{A}}. \tag{102}$$

If the product on the right of Eq. (102) is formed, the result is

$$\mathsf{S} = \begin{pmatrix} \alpha & \delta & \gamma \\ \delta & \alpha & \gamma \\ \gamma & \gamma & \beta \end{pmatrix}, \tag{103}$$

where
$$\begin{aligned}\alpha &= \tfrac{1}{4}(1 + s_2 + 2s_3), \\ \beta &= \tfrac{1}{2}(1 + s_2), \\ \gamma &= \frac{\sqrt{2}}{4}(1 - s_2), \\ \delta &= \tfrac{1}{4}(1 + s_2 - 2s_3).\end{aligned} \quad (104)$$

Note that
$$2\alpha + \beta = 1 + s_2 + s_3$$
as it should (Theorem 8).

It will be remembered that it is impossible to match a T-junction. As a check on this, for a matched T-junction,
$$\alpha = \beta = 0.$$
This is clearly impossible as an inspection of Eqs. (104) will show.

It will now be shown that if $\alpha = 0$, the trivial case for which guide (3) is completely decoupled is obtained. If
$$\alpha = 0,$$
then
$$s_2 = -s_3 = 1. \quad (105)$$
Substituting Eq. (105) in Eq. (104),
$$\begin{aligned}\alpha &= 0, \\ \beta &= 1, \\ \gamma &= 0, \\ \delta &= 1.\end{aligned}$$

This result is more general, applying to unsymmetrical junctions as well. To see this let the scattering matrix be
$$\mathsf{S} = \begin{pmatrix} 0 & S_{21} & S_{31} \\ S_{21} & 0 & S_{32} \\ S_{31} & S_{32} & S_{33} \end{pmatrix}.$$

Since S is unitary, the product of the first column by the complex conjugate of the second vanishes;
$$S_{31} S_{32}^* = 0.$$
Either S_{31} or S_{32} must vanish. In either case,
$$|S_{21}| = 1,$$
since the square of each column must be unity. Then
$$\begin{aligned}S_{31} &= S_{32} = 0, \\ |S_{21}| &= |S_{33}| = 1.\end{aligned}$$

The guide (3) is completely decoupled. Note that it is consequently impossible to have all three diagonal elements vanish.

12·16. The Use of the T-junction as an Element of a Tuner.—It will now be shown that if a plunger is inserted in guide (3) to short-circuit it, the remainder of the guide acts as a line with a shunt susceptance at the symmetry plane. The electrical distance to this susceptance will not, in general, be the same as the geometrical distance.

A shunt susceptance in a transmission line sets up equal waves traveling in opposite directions. This can be seen from the scattering-matrix analogue of Eq. (64).

$$\mathsf{S} = \begin{pmatrix} S_{11} & S_{21} \\ S_{21} & S_{11} \end{pmatrix}, \quad (106)$$

where

$$S_{11} = \tfrac{1}{2}(s_1 + s_2),$$
$$S_{21} = \tfrac{1}{2}(s_1 - s_2).$$

(Reference planes are assumed to be chosen as in Fig. 12·6.) For a thin iris, $s_2 = -1$. When the iris is completely absent, $s_1 = +1$ and $\alpha = 0$; S becomes

$$\mathsf{S} = \begin{pmatrix} 0 & 1 \\ 1 & 0 \end{pmatrix}. \quad (107)$$

With the iris present, S is given by Eq. (106) which, it is to be noted, can be obtained from Eq. (107) by adding $\tfrac{1}{2}(s_1 + s_2)$ to every element of Eq. (107). Thus a pure susceptance generates waves of equal amplitude in either direction. Another way of expressing this relation is

$$|S_{11} - S_{21}| = 1 \quad (108)$$

provided the junction is a pure susceptance.

In order to check this relation for the T-junction with a plunger in guide (3), it is necessary to obtain a solution with a pure standing wave

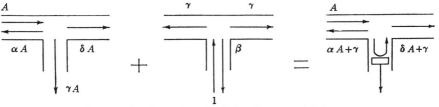

FIG. 12·13.—Operation of a T-junction as a stub tuner.

in (3) but with traveling waves in the other guides. To be more explicit, let a solution with power entering guide (1) be combined with a solution with power flowing into guide (3). This is illustrated in Fig. 12·13, in which the small Greek letters have the same meaning as in Eq. (103).

The amplitude A has to be chosen in such a way that there is a standing wave in arm (3). The condition for this is

$$\gamma A + \beta = e^{j\phi},$$

where ϕ is a phase determined by the position of the plunger in guide (3). If the junction acts as a shunt susceptance at the symmetry plane, Eq. (108) must be satisfied; therefore

$$\left|\frac{(\alpha A + \gamma)}{A} - \frac{(\delta A + \gamma)}{A}\right| = 1,$$

from which

$$|\alpha - \delta| = 1.$$

It is apparent from Eq. (104) that this relation is correct. Hence the junction acts as a shunt susceptance. From Eq. (103),

$$\alpha - \delta = s_3.$$

Thus the electrical length of line between the two terminals is s_3. This electrical line length may be determined experimentally by inserting a plunger in arm (2) and adjusting until there is no coupling between (1) and (3). Under these conditions the plunger is electrically an integral number of half wavelengths from the effective susceptance. Another method that can be used to determine this length will be taken up in the next section.

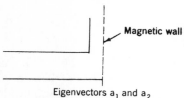

Fig. 12·14.—Boundary conditions for eigenvectors a_1 and a_2.

It should be noted that the reflection coefficient of the junction is

$$S_\phi = \alpha + \frac{\gamma}{A} = \alpha + \frac{\gamma^2}{e^{j\phi} - \beta}.$$

This reflection coefficient is zero when

$$e^{j\phi} = \beta - \frac{\gamma^2}{\alpha}.$$

It may be verified, by substituting from Eq. (103), that this equation has a solution with real ϕ.

The eigenvectors a_1 and a_2 are also eigenvectors of F with eigenvalue $+1$. This is the usual even solution which is equivalent to the field distribution when the symmetry plane is replaced by a magnetic wall (see Fig. 12·14). The magnetic wall reduces the junction to a two-terminal junction for which there are two eigenvectors a_1 and a_2. The eigenvector a_3 is odd about the symmetry plane. The fields are those with the symmetry plane replaced by a metal wall (see Fig. 12·15). To measure the properties of a right-angle bend with a magnetic wall, it would be necessary to apply equal incident waves to the symmetrical arms of the T-junction. The right-angle bend of Fig. 12·15, however, can be constructed, and its properties measured directly. Note that the eigenvalue of S for a_3 (Fig. 12·15) is s_3 which is also the distance along the line to the effective position of the susceptance when the junction is used as a tuner. This, then, is another way in which this line length can be determined. The series T-junction is analyzed in a similar way. If a plunger is inserted in guide (3), it acts as a series reactor in the line.

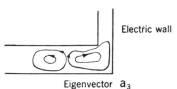

FIG. 12·15.—Boundary conditions for eigenvector a_3 showing magnetic field lines.

12·17. Directional Couplers.—Most of the directional couplers in use have a symmetry such that their scattering matrices have the form

$$S = \begin{pmatrix} \alpha & \beta & \gamma & \delta \\ \beta & \alpha & \delta & \gamma \\ \hdashline \gamma & \delta & \alpha & \beta \\ \delta & \gamma & \beta & \alpha \end{pmatrix}. \tag{109}$$

It will now be shown that if this junction is matched ($\alpha = 0$), it is also a directional coupler; and conversely if it is a directional coupler (say $\beta = 0$), then it is matched. First let $\alpha = 0$. Then, since S is unitary,

$$\text{Re}(\beta\delta^*) = \text{Re}(\gamma\delta^*) = \text{Re}(\gamma\beta^*) = 0. \tag{110}$$

Equation (110) states that plotted in the complex plane, the three vectors γ, δ, and β are mutually at right angles. This is possible only if one of them vanishes. If one vanishes, however, the device is a directional coupler.

The converse of this theorem is also easily proved. If the device is a directional coupler, then either β, γ, or δ must vanish. Assume that $\beta = 0$. The three remaining complex quantities must again be at right angles to one another, and one must vanish. Assume first that $\alpha \neq 0$. In this case either γ or δ must vanish, but this corresponds to the limiting case of a directional coupler with zero coupling. Except for this case, $\alpha = 0$ and the coupler is matched. Thus a directional coupler with a nonvanishing amount of coupling and with the symmetry given by Eq. (109) is automatically matched.

12·18. The Single-hole Directional Coupler.—The single-hole directional coupler is an example of a junction having three symmetry axes but no symmetry planes.

In Fig. 12·16 the symmetry axes are designated by R_1, R_2, and R_3. The terminal planes are numbered (1), (2), (3), and (4), and the direction of positive electric field in each plane is indicated by an arrow. Figure 12·17 is a representation of the common metal wall between the two

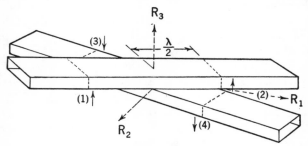

FIG. 12·16.—Directional coupler.

waveguides. The coupling hole is shown as well as the symmetry axes. The coupling hole may have any shape consistent with the symmetry conditions.

Symmetry Operators.—Designate the operation of 180° rotation about the various symmetry axes by the three symmetry operators R_1, R_2, and R_3. These three operators have the multiplicative properties

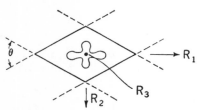

FIG. 12·17.—Wall common to the two waveguides of Fig. 12·16 showing coupling hole.

$$\left.\begin{array}{l} R_1^2 = R_2^2 = R_3^2 = I, \\ R_1 R_2 = R_3, \\ R_2 R_3 = R_1, \\ R_3 R_1 = R_2, \end{array}\right\} \quad (111)$$

where I is, as usual, the identity operator. It should be noted that all four operations can be generated by R_1 and R_2. Let these symmetries be the generators of the group.

The matrix representation of R_1 can be seen by inspection to be

$$R_1 = \begin{pmatrix} 0 & 0 & 1 & 0 \\ 0 & 0 & 0 & 1 \\ \hdashline 1 & 0 & 0 & 0 \\ 0 & 1 & 0 & 0 \end{pmatrix}. \quad (112)$$

The operator R_1, operating on the terminal quantities

$$a = \begin{pmatrix} a_1 \\ a_2 \\ a_3 \\ a_4 \end{pmatrix} \quad (113)$$

interchanges the first and third as well as the second and fourth components. In a similar way the operator R_2 takes the form

$$R_2 = \begin{pmatrix} 0 & 0 & 0 & 1 \\ 0 & 0 & 1 & 0 \\ \hline 0 & 1 & 0 & 0 \\ 1 & 0 & 0 & 0 \end{pmatrix}. \tag{114}$$

Both R_1 and R_2 have doubly degenerate eigenvalues. However, it is possible to take a linear combination of R_1 and R_2 whose eigenvalues are nondegenerate. Let

$$M = \epsilon_1 R_1 + \epsilon_2 R_2, \tag{115}$$

where

$$\begin{aligned} \epsilon_1 &= \tfrac{1}{2}(1 + j), \\ \epsilon_2 &= \tfrac{1}{2}(1 - j). \end{aligned} \tag{116}$$

Note that

$$\begin{aligned} M^2 &= R_1 R_2, \\ M^4 &= I. \end{aligned} \tag{117}$$

The Eigenvalue Equations.—The eigenvalues of M satisfy the characteristic equation

$$m^4 - 1 = 0 \tag{118}$$

and are

$$\begin{aligned} m_1 &= +1, \\ m_2 &= j, \\ m_3 &= -1, \\ m_4 &= -j. \end{aligned}$$

The eigenvalue equation of M takes the form

$$M a_j = m_j a_j. \tag{119}$$

The easiest way to obtain the eigenvectors of Eq. (119) is to note that M commutes with R_1 and R_2. Since M is nondegenerate, a_k must be an eigenvector of both R_1 and R_2 (Theorem 6). The eigenvalues of R_1 and R_2 are ± 1.

To illustrate the procedure for finding the eigenvectors, let a_1 be an eigenvector of M with eigenvalue $+1$; that is, $M a_1 = a_1$. Thus,

$$\begin{aligned} a_1 &= (\epsilon_1 R_1 + \epsilon_2 R_2) a_1 \\ &= \epsilon_1 R_1 a_1 + \epsilon_2 R_2 a_1 \\ &= \epsilon_1 r_1 a_1 + \epsilon_2 r_2 a_1 \\ &= (\epsilon_1 r_1 + \epsilon_2 r_2) a_1, \end{aligned}$$

where r_1 and r_2 are eigenvalues of R_1 and R_2 and must be either ± 1. From Eqs. (116), $\epsilon_1 + \epsilon_2 = 1$; thus for the eigenvalue $m_1 = 1$, the eigenvalues of R_1 and R_2 must be $+1$. Thus a_1 is an eigenvector of R_1 and R_2 with $+1$ as eigenvalue. From Eq. (112), the following conditions are imposed on a_1:

$$a_1^{(1)} = a_3^{(1)}, \\ a_2^{(1)} = a_4^{(1)}. \tag{120}$$

The matrix R_2 imposes the conditions

$$a_1^{(1)} = a_4^{(1)}, \\ a_2^{(1)} = a_3^{(1)}. \tag{121}$$

Conditions (120) and (121) require that

$$a_1^{(1)} = a_2^{(1)} = a_3^{(1)} = a_4^{(1)}. \tag{122}$$

Clearly a possible eigenvector is

$$a_1 = \tfrac{1}{2} \begin{pmatrix} 1 \\ 1 \\ 1 \\ 1 \end{pmatrix}. \tag{123}$$

In a completely analogous way the remainder of the eigenvectors can be shown to be

$$a_2 = \tfrac{1}{2} \begin{pmatrix} 1 \\ -1 \\ 1 \\ -1 \end{pmatrix}, \quad a_3 = \tfrac{1}{2} \begin{pmatrix} 1 \\ 1 \\ -1 \\ -1 \end{pmatrix}, \quad a_4 = \tfrac{1}{2} \begin{pmatrix} 1 \\ -1 \\ -1 \\ 1 \end{pmatrix}. \tag{124}$$

Note that M is symmetric, and by Theorem 3 the a's are all mutually orthogonal. They have been normalized to unity. The eigenvalues of the three symmetry operators R_1, R_2, and R_3 for the four eigenvectors $a_1 \ldots a_4$ are given in Table 12·7.

TABLE 12·7.—EIGENVALUES FOR THE SINGLE-HOLE DIRECTIONAL COUPLER

Symmetry operators	Eigenvectors				
		a_1	a_2	a_3	a_4
	R_1	1	1	-1	-1
	R_2	1	-1	-1	1	Eigenvalues
	R_3	1	-1	1	-1

Since the scattering matrix must commute with R_1 and R_2, it also commutes with M. By Theorem 6, the a's are also eigenvectors of S,

$$Sa_k = s_k a_k. \tag{125}$$

In Eq. (124), the eigenvectors of S are pure real, a condition required by

THE SINGLE-HOLE DIRECTIONAL COUPLER

Theorems 4 and 5 if the s_k are nondegenerate. Equation (125) can be written in the unified form

$$ST = TS_d, \qquad (126)$$

where

$$T = \tfrac{1}{2}\begin{pmatrix} 1 & 1 & 1 & 1 \\ 1 & -1 & 1 & -1 \\ 1 & 1 & -1 & -1 \\ 1 & -1 & -1 & 1 \end{pmatrix},$$

$$S_d = \begin{pmatrix} s_1 & 0 & 0 & 0 \\ 0 & s_2 & 0 & 0 \\ 0 & 0 & s_3 & 0 \\ 0 & 0 & 0 & s_4 \end{pmatrix}.$$

Since T is symmetrical and orthogonal, Eq. (126) may be written as

$$S = TS_dT. \qquad (127)$$

The product on the right-hand side of Eq. (127) may be expanded to give

$$S = \begin{pmatrix} \alpha & \beta & \gamma & \delta \\ \beta & \alpha & \delta & \gamma \\ \gamma & \delta & \alpha & \beta \\ \delta & \gamma & \beta & \alpha \end{pmatrix}, \qquad (128)$$

where

$$\left.\begin{aligned} \alpha &= \tfrac{1}{4}(s_1 + s_2 + s_3 + s_4), \\ \beta &= \tfrac{1}{4}(s_1 - s_2 + s_3 - s_4), \\ \gamma &= \tfrac{1}{4}(s_1 + s_2 - s_3 - s_4), \\ \delta &= \tfrac{1}{4}(s_1 - s_2 - s_3 + s_4). \end{aligned}\right\} \qquad (129)$$

As a check on the correctness of Eqs. (129), note that by Theorem 8,

$$\text{spur } S = s_1 + s_2 + s_3 + s_4.$$

As another check assume that $\alpha = 0$. Since the four eigenvalues s_k have unit moduli, this condition can be satisfied only if the four s's are paired, with each pair consisting of two eigenvalues of opposite sign. There are three possibilities

$$\left.\begin{aligned} s_1 &= -s_2 \\ s_3 &= -s_4 \end{aligned}\right\} \quad \gamma = 0, \qquad (130)$$

$$\left.\begin{aligned} s_1 &= -s_3 \\ s_2 &= -s_4 \end{aligned}\right\} \quad \beta = 0, \qquad (131)$$

$$\left.\begin{aligned} s_1 &= -s_4 \\ s_2 &= -s_3 \end{aligned}\right\} \quad \delta = 0. \qquad (132)$$

Thus the general condition is satisfied: A matched junction is automatically a directional coupler. The three conditions given by Eqs. (130), (131), and (132) show that in so far as symmetry is concerned, it should be possible to match the junction in such a way as to have guide (1) decoupled from any one of the other three guides.

Field Distribution of the Eigenvectors.—For this discussion, it will be assumed that the wall between the two guides is thin (Fig. 12·17). However, there will be no assumption about the shape or size of the coupling hole other than that it satisfy the proper symmetry requirements. Equations (50) show the conditions that the fields must satisfy along a symmetry axis for the "odd" and "even" solutions. These conditions may be combined with the results of Table 12·7 to obtain information concerning the electromagnetic field in the coupling hole.

For example, consider the eigenvector a_2. This eigenvector is even under R_1 but odd under R_2 and R_3. From Eq. (50), the field normal to R_1 must vanish along the axis R_1. Also the field parallel to R_2 and R_3 must vanish along R_2 and R_3. These conditions are compatible with a field in the center of the hole in the R_1 direction only. The components of the electromagnetic field at the center of the hole, for each of the eigenvectors, are

$$a_1 \rightarrow \text{no field,}$$
$$a_2 \rightarrow E_1, H_1,$$
$$a_3 \rightarrow E_3, H_3,$$
$$a_4 \rightarrow E_2, H_2.$$

The subscripts of **E** and **H** correspond to components along the three symmetry axes.

Symmetry with $\theta = 0$.—Symmetry cannot be used to give quantitative information concerning the relative strengths of the electric and magnetic fields in the coupling hole except for one special case. When $\theta = 0°$ or $180°$, new symmetries appear. In fact, the symmetry of the junction is then complete with three symmetry planes, three symmetry axes, and a symmetry point. The plane of the coupling hole becomes a symmetry plane. Let **F** represent the operation of reflection in the plane of the hole. The eigenvectors a_k are also eigenvectors under **F**. The eigenvectors a_1 and a_2 correspond to the eigenvalue $+1$, and a_3 and a_4 to the eigenvalue -1. It is to be remembered that for the eigenvalue -1, the field quantities satisfy the boundary conditions of an electric wall in the plane of the hole. Thus, since the wall is thin, the hole effectively disappears. The eigenvectors a_3 and a_4 have field distributions characteristic of two independent waveguides. Thus the hole has no effect on the fields for these two eigenvectors for $\theta = 0$. It should be emphasized that this is true independently of the size of

Sec. 12·18]   THE SINGLE-HOLE DIRECTIONAL COUPLER   443

the hole. Thus for $\theta = 0$,

$$s_3 = +1$$
$$s_4 = -1.$$

Operation as a Directional Coupler.—It will be assumed that the coupling produced by the coupling hole is small. However, nothing will be assumed about the shape of the coupling hole other than it have the correct symmetry. The four terminal planes in Fig. 12·16 were chosen to be one-half wavelength from the R_3-axis. For the problem of most interest, namely, that of the small coupling hole, the effect of the hole on the field quantities may be considered as a perturbation. The zero-order approximation is obtained from the field distribution without the coupling hole. In this case the problem reduces to two independent guides and the eigenvalues are easily seen to be

$$\left.\begin{array}{l} s_1^{(0)} = +1, \\ s_2^{(0)} = -1, \\ s_3^{(0)} = +1, \\ s_4^{(0)} = -1. \end{array}\right\} \quad (133)$$

Equations (133) will be called the unperturbed eigenvalues.

It is to be noted that the eigenvalues given in Eqs. (133) are pure real; and for small coupling, the perturbations on these eigenvalues will be essentially imaginary. The conditions of Eqs. (133) are independent of the angle between the two waveguides, but the perturbations will, in general, be a function of this angle. Let the perturbations be denoted by $s_k^{(1)}(\theta)$. From the previous discussion of the fields associated with $\mathbf{a}_3$ and $\mathbf{a}_4$ for $\theta = 0$ it is clear that

$$s_3^{(1)}(0) = s_4^{(1)}(0) = 0. \quad (134)$$

Also for similar reasons

$$s_3^{(1)}(\pi) = s_2^{(1)}(\pi) = 0. \quad (135)$$

The conditions of Eq. (135) can be shown formally to follow from Eq. (134). Let us introduce the operator $\mathbf{R}_\theta$ which rotates θ through 180°. This has the effect of interchanging the terminals (1) and (2). This transformation, in acting on the terminals, takes the form of the matrix

$$\mathbf{R}_\theta = \left(\begin{array}{cc|cc} 0 & 1 & 0 & 0 \\ 1 & 0 & 0 & 0 \\ \hline 0 & 0 & 1 & 0 \\ 0 & 0 & 0 & 1 \end{array}\right).$$

When R_θ operates on the a's, the result is

$$R_\theta a_1 = a_1,$$
$$R_\theta a_2 = -a_4,$$
$$R_\theta a_3 = a_3,$$
$$R_\theta a_4 = -a_2.$$

The eigenvalue equation $Sa_k = s_k a_k$ is transformed by R_θ into

$$R_\theta S R_\theta^{-1} R_\theta a_k = s_k R_\theta a_k,$$
$$S'(R_\theta a_k) = s_k(R_\theta a_k),$$

where $S' = R_\theta S R_\theta^{-1}$ is the scattering matrix of the transformed junction. But

$$S' a_k = s'_k a_k, \tag{136}$$

where

$$s'_k(\theta) = s_k(\pi + \theta). \tag{137}$$

Thus

$$\left.\begin{array}{l} s_1(\theta + \pi) = s_1(\theta), \\ s_2(\theta + \pi) = s_4(\theta), \\ s_3(\theta + \pi) = s_3(\theta), \\ s_4(\theta + \pi) = s_2(\theta). \end{array}\right\} \tag{138}$$

Also it may be seen that

$$s_k(\theta) = s_k(-\theta),$$

from Eqs. (138) and (134).

Remembering Eqs. (134), (135), and (137), typical perturbations can

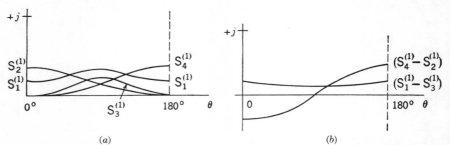

Fig. 12·18.—Perturbations of eigenvalue of S as a function of a.

be plotted as a function of θ, as in Fig. 12·18a. By adding these curves together correctly, the values in Fig. 12·18b are obtained. It should be noted that the two curves will always cross provided that for $\theta = 0$,

$$|s_2^{(1)}| > |s_1^{(1)}|. \tag{139}$$

Referring to Eq. (129) it can be seen that $\delta = 0$ at the crossover point. Thus the device will act as a directional coupler at some angle provided that Eq. (139) is satisfied. It should be pointed out that the device is not a perfect directional coupler in that, in general, $\alpha \neq 0$ at the crossover point. This seems to contradict the results previously obtained, but there is no real paradox. The relation $\delta = 0$ is only an approximation based upon the small-coupling assumption that $s_k^{(1)}$ is pure imaginary, which is, of course, not exactly correct.

12·19. The Biplanar Directional Coupler.—The structure of the biplanar directional coupler shown in Fig. 12·19 is characterized by two symmetry planes F_1 and F_2 and a symmetry axis that is the line of intersection of these two planes. The plane that contains the axes of all four guides will not be regarded as a symmetry plane. This particular symmetry group can be generated by the symmetry operators F_1 and F_2.

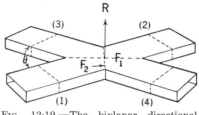

Fig. 12·19.—The biplanar directional coupler.

In operating on terminal quantities (current, voltage, incident-wave amplitude, etc.), F_1 and F_2 take the form

$$F_1 = \begin{pmatrix} 0 & 0 & 1 & 0 \\ 0 & 0 & 0 & 1 \\ \hline 1 & 0 & 0 & 0 \\ 0 & 1 & 0 & 0 \end{pmatrix}, \quad F_2 = \begin{pmatrix} 0 & 0 & 0 & 1 \\ 0 & 0 & 1 & 0 \\ \hline 0 & 1 & 0 & 0 \\ 1 & 0 & 0 & 0 \end{pmatrix}.$$

Note that although the symmetries of Figs. 12·19 and 12·16 are quite different, the generators of each of the symmetry groups take exactly the same form when written as matrices. A comparison may be made with Eqs. (111) and (113). F_1 and R_1 are quite equivalent to each other, and F_2 is equivalent to R_2. As a result of this fact, except for those results connected with the symmetries of the field, namely, the field distributions for the various eigenvectors, all the results obtained for the single-hole directional coupler also apply to Fig. 12·19. In particular, the eigenvectors of Eqs. (123) and (124) are eigenvectors of the scattering matrix that takes the form of Eq. (128). Table 12·7 is still valid if R_1 and R_2 are replaced by F_1 and F_2 and R_3 by the rotation R.

Fields Associated with the Four Eigenvectors.—From Table 12·7 it can be seen that the eigenvalues of F_1 and F_2 for the eigenvector a_1 are $+1$. From Eq. (45) it is seen that the fields at the two symmetry planes satisfy the conditions imposed by magnetic walls along the symmetry planes. In Fig. 12·20, these boundary conditions are shown. The figure also shows the boundary conditions that the remaining eigenvectors must satisfy. Note that if a thin metal plate is inserted in the plane F_1 in such a way as not to destroy the symmetry, the phases of s_2 and s_1

will be changed but not those of s_3 or s_4. Also s_1 or s_4 can be changed by a symmetrical metal wall in F_2. Moreover, it can be seen that a metal pin inserted in the guide along the symmetry axis will change s_1 but leave the other eigenvalues of S unchanged.

The insertion of irises in the two symmetry planes and a pin along the symmetry axis are three adjustments that can be used to obtain any scattering matrix consistent with this symmetry, for example, that for a directional coupler. It must be assumed, however, that these adjustments have sufficient range to obtain the desired results.

In order to be a directional coupler, the junction must be matched. The condition for a match is the vanishing of the sum of the eigenvalues

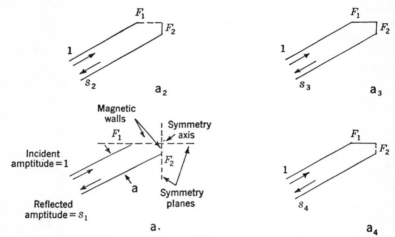

FIG. 12·20.—Boundary conditions for eigenvector solutions a_1, a_2, a_3, a_4.

of S in Eq. (128). Let us try by means of obstacles along the plane F_1 and the symmetry axis to cause α, in that matrix, to vanish. The phase of s_2 can be adjusted by means of the obstacle in F_1 in such a way that $s_2 = -s_3$. Then the axial pin can be used to set $s_1 = -s_4$. These two conditions, substituted in Eqs. (129), result in

$$\left.\begin{array}{l}\alpha = 0, \\ \delta = 0.\end{array}\right\} \quad (140)$$

Another possible procedure is to adjust s_1 and s_2 to satisfy

$$\left.\begin{array}{l}s_1 = -s_3, \\ s_2 = -s_4.\end{array}\right\} \quad (141)$$

These conditions, substituted in Eq. (128) yield

$$\begin{array}{l}\alpha = 0, \\ \beta = 0.\end{array}$$

Figure 12·21 shows the form that the obstacles might take. The power distribution is that given by Eqs. (140). Notice that the distribution is just opposite that which one might expect from simple optics.

If in Fig. 12·19 the angle θ is very small, the two eigenvalues s_3 and s_4 are very nearly equal. This can be seen by reference to Fig. 12·20. In this case the wall F_1 together with the remainder of the guide forms a tapered waveguide that at some position reaches the cutoff width. The amplitude of the wave dies down quickly after this. As a result there is very little electromagnetic field at the wall F_2. Hence, it matters little whether F_2 is electric or magnetic. Thus it is seen that for small θ the wave incident in guide (1) goes only into guide (3) when the junction is matched as in Fig. 12·21. This result is even more general. In fact it can be seen by the inspection of Eq. (129) that if $s_3 = s_4$, $\beta = \delta$. If in addition $\alpha = 0$, then $\beta = \delta = 0$ and $|\gamma| = 1$.

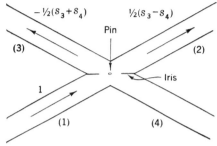

Fig. 12·21.—One method of matching the junction of Fig. 12·19.

Thus independently of the mechanism used to match the junction, γ is the only nonvanishing matrix element.

12·20. The Magic T.—The "magic T" may be defined as a directional coupler with equal power division. Clearly the scattering matrix of a magic T may always be written in the form

$$S = \begin{pmatrix} 0 & 0 & e & f \\ 0 & 0 & g & h \\ \hline e & g & 0 & 0 \\ f & h & 0 & 0 \end{pmatrix}. \quad (142)$$

The elements e, f, g, and h are not completely independent but must satisfy the unitary conditions

$$\left. \begin{array}{l} ef^* + gh^* = 0, \\ eg^* + fh^* = 0, \\ |e|^2 + |g|^2 = |f|^2 + |h|^2 = 1, \\ |e|^2 + |f|^2 = |g|^2 + |h|^2 = 1. \end{array} \right\} \quad (143)$$

In addition, the equal-coupling condition requires that

$$|e| = |f| = |g| = |h|.$$

By the correct choice of the positions of the terminals in three of the four guides, three of the four parameters e, f, g, and h can be made to have any phase angle desired. The fourth argument is then determined through Eqs. (143). For example, any magic T may, by proper selection of reference terminals, be made to have a scattering matrix with one of the following forms,

$$S = \frac{1}{\sqrt{2}} \begin{pmatrix} 0 & 0 & j & j \\ 0 & 0 & j & -j \\ \hline j & j & 0 & 0 \\ j & -j & 0 & 0 \end{pmatrix}, \quad (144)$$

$$S = \frac{1}{\sqrt{2}} \begin{pmatrix} 0 & 0 & 1 & j \\ 0 & 0 & j & 1 \\ \hline 1 & j & 0 & 0 \\ j & 1 & 0 & 0 \end{pmatrix}. \quad (145)$$

Biplanar Symmetry.—It should be noticed that Eqs. (145) and (128) are of the same form, and it might be expected that the biplanar directional coupler of Fig. 12·19 could be tuned in such a way as to make a magic T out of it. This tuning would require three adjustments which could be the irises in the two symmetry planes and the pin along the symmetry axis. The phases of s_1, s_2, and s_4 could be adjusted until

$$s_1 = -s_3,$$
$$s_2 = -s_4,$$
$$s_1 = js_2.$$

If these values are substituted in Eq. (129), the elements of the scattering matrix are found to be

$$\left.\begin{aligned} \alpha &= 0, \\ \beta &= 0, \\ \gamma &= \tfrac{1}{2}s_1(1 - j), \\ \delta &= \tfrac{1}{2}s_1(1 + j). \end{aligned}\right\} \quad (146)$$

Except for a phase factor, γ and δ are the same as in Eq. (145). In fact by choosing the terminals in a new symmetrical set of positions Eqs. (146) become

$$\gamma = \frac{1}{\sqrt{2}},$$
$$\delta = \frac{1}{\sqrt{2}} j.$$

12·21. The Synthesis Problem.—A few words regarding the synthesis problem at microwave frequencies are necessary. At microwave fre-

SEC. 12·21] THE SYNTHESIS PROBLEM

quencies there are a large number of problems in which the frequency dependence of a device is only of incidental interest. This is because the frequency is so high that the bandwidths are small compared with the frequencies.

Wideband systems or junctions are of importance, but only because it is desirable to have components that are fixed-tuned. In any case, there is a synthesis problem regarding the attainment of certain properties at one frequency. Additional properties may be desirable, such as power-handling capacity or large bandwidth, but these may be regarded as secondary in importance.

At low frequencies, lumped elements of inductive or capacitive characteristics are used as the building blocks from which a network is synthesized. At high frequencies this type of technique is not particularly useful because of the size of the resulting components. The transmission line is a convenient element for microwave circuits.

To understand how the transmission line may be used to synthesize a network, consider the admittance matrix

$$\mathbf{Y} = \begin{pmatrix} y_{11} & y_{12} & \cdots \\ y_{21} & y_{22} & \cdots \\ \cdot & & \cdots \\ \cdot & & \cdots \\ \cdot & & \cdots \end{pmatrix}. \tag{147}$$

This may be written as

$$\mathbf{Y} = \begin{pmatrix} y_{11} & 0 & 0 & 0 & \cdots \\ 0 & 0 & 0 & 0 & \cdots \\ 0 & 0 & 0 & 0 & \cdots \\ \cdot & \cdot & \cdot & \cdots \\ \cdot & \cdot & \cdot & \cdots \\ \cdot & \cdot & \cdot & \cdots \end{pmatrix} + \begin{pmatrix} 0 & y_{12} & 0 & \cdots \\ y_{21} & 0 & 0 & \cdots \\ 0 & 0 & 0 & \cdots \\ \cdot & \cdot & \cdots \\ \cdot & \cdot & \cdots \\ \cdot & \cdot & \cdots \end{pmatrix}$$

$$+ \begin{pmatrix} 0 & 0 & 0 & \cdots \\ 0 & y_{22} & 0 & \cdots \\ 0 & 0 & 0 & \cdots \\ \cdot & \cdot & \cdots \\ \cdot & \cdot & \cdots \\ \cdot & \cdot & \cdots \end{pmatrix} + \cdots. \tag{148}$$

Each of these submatrices may be examined individually. The first matrix has zeros in all rows and columns except the first. This implies that if voltages are applied to all the terminals, the only terminal influenced by the first submatrix is terminal (1). At this terminal a current proportional to y_{11} flows. In other words, the first matrix in Eq. (148)

may be represented by a shunt susceptance y_{11} connected to the terminals (1). In a similar way the second matrix can be represented by a transmission line one-quarter or three-quarters wavelength long depending on the sign of y_{12}, of characteristic admittance $|y_{12}|$, connected between the terminals (1) and (2). The remainder of the submatrices follow in a similar way. The end result is illustrated for a three-terminal-pair device in Fig. 12·22.

An alternative arrangement is a synthesis of the impedance matrix in terms of lines connected in series. This is a convenient way of synthesizing a junction by means of lengths of waveguide connected in series at the various junctions.

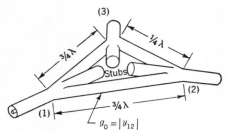

Fig. 12·22.—Synthesis of a three-terminal-pair junction in coaxial line.

In order to synthesize a particular circuit whose scattering matrix is given, it is necessary to solve for the equivalent impedance or admittance matrix and then use the above synthesis procedure. It should be pointed out that by a discreet choice of the location of the reference planes the impedance matrix can often be made to take a simplified form with a resulting simplicity in brass.

Synthesis of the Biplanar Magic T.—It will be noticed that Eq. (145) is very closely related to the symmetry matrix **M** given by Eq. (115). In fact,

$$\mathbf{S} = \frac{1-j}{\sqrt{2}} \mathbf{M}. \tag{149}$$

Substitution of Eq. (149) in the second of Eqs. (117) gives

$$\mathbf{S}^4 + \mathbf{I} = 0. \tag{150}$$

The admittance matrix is given by

$$\mathbf{Y} = (\mathbf{I} - \mathbf{S})(\mathbf{I} + \mathbf{S})^{-1}.$$

This equation may be written

$$\mathbf{Y} = (\mathbf{I} - \mathbf{S})^2(\mathbf{I} + \mathbf{S}^2)(\mathbf{I} + \mathbf{S}^2)^{-1}(\mathbf{I} - \mathbf{S})^{-1}(\mathbf{I} + \mathbf{S})^{-1}$$
$$= (\mathbf{I} - 2\mathbf{S} + 2\mathbf{S}^2 - 2\mathbf{S}^3 + \mathbf{S}^4)(\mathbf{I} - \mathbf{S}^4)^{-1}.$$

Substituting from Eq. (150),

$$\mathbf{Y} = -\mathbf{S}(\mathbf{I} - \mathbf{S} + \mathbf{S}^2). \tag{151}$$

Substituting from Eq. (145) in Eq. (151),

$$Y = j \begin{pmatrix} 0 & 1 & 0 & -\sqrt{2} \\ 1 & 0 & -\sqrt{2} & 0 \\ \hline 0 & -\sqrt{2} & 0 & 1 \\ -\sqrt{2} & 0 & 1 & 0 \end{pmatrix}. \quad (152)$$

The synthesis, by coaxial lines, of a junction whose admittance matrix is given by Eq. (152) follows in the same way as before. Note that the lines with a characteristic conductance of unity are one-quarter wavelength long. Those whose characteristic conductance is $\sqrt{2}$ are three-quarters wavelength long.

From a practical point of view a three-quarter-wavelength line is more frequency-sensitive than a quarter-wavelength line, and it is desirable, if possible, to choose reference planes in such a way that all elements in the matrix of Eq. (152) are positive. If the reference planes are moved back one-quarter wavelength, S in Eq. (151) changes sign and, as may be verified, the admittance matrix of this new junction is

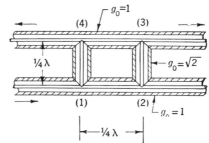

FIG. 12·23.—Synthesis of a magic T in coaxial line.

$$Y = j \begin{pmatrix} 0 & 1 & 0 & +\sqrt{2} \\ 1 & 0 & +\sqrt{2} & 0 \\ \hline 0 & +\sqrt{2} & 0 & 1 \\ +\sqrt{2} & 0 & 1 & 0 \end{pmatrix}.$$

This junction may be synthesized by the circuit of Fig. 12·23. It should be noticed that power entering arm (1) is split equally between arms (3) and (4) and no power leaves arm (2).

12·22. Coupling-hole Magic T's.—As another example of a magic T with a scattering matrix of the form of Eq. (145), consider a directional coupler of the type shown in Figs. 12·16 and 12·17 with $\theta = 0$. The coupling hole or holes will be assumed to be large enough to produce equal power division. One possible arrangement is a set of two holes about a quarter-wavelength apart. Another possibility is a large slot or oblong hole in the direction of the guide axis. As was pointed out previously, the symmetry of the junction becomes complete for $\theta = 0$. The symmetry plane containing the coupling holes becomes effectively an electric wall for the eigenvectors a_3 and a_4. Consequently the coupling hole has no effect upon these standing waves, and the field distributions are identical with those of two independent waveguides. As a result, because of the location of reference planes, the eigenvalues of the scattering matrix for these eigenvectors are (see paragraph entitled *Symmetry with $\theta = 0$*, Sec. 12·18).

$$s_3 = 1, \quad s_4 = -1.$$

The eigenvalues s_1 and s_2 are dependent upon the size and shape of the coupling hole or holes. However, it is to be noted that these two eigenvalues are the only parameters of the junction left open. Thus it requires only two adjustments to convert the junction into a magic T. For example, if the coupling is produced by two circular coupling holes, there may exist a particular diameter for the holes and a distance between the holes for which the device is a magic T.

With reference to Eqs. (129), it is evident that for the matched junction for which $\alpha = 0$, the conditions of Eq. (130) require that $\gamma = 0$. Thus, if the junction is a magic T (for which $\alpha = 0$), it must be one for which there is no coupling between guides (1) and (3) or between (2) and (4). It is to be noted that this is just the opposite of the behavior of the small-single-hole coupler for which the coupling between (1) and (4) is the least (see last paragraph of Sec. 12·18). Referring again to Eqs. (129), it is evident that the conditions to be satisfied in order that the junction be a magic T are

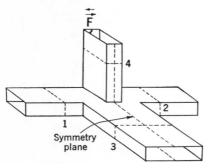

Fig. 12·24.—Magic T.

$$s_1 = -s_2 = \pm j.$$

It should be noted that only one condition must be satisfied in order that the junction be a directional coupler, namely,

$$s_1 = -s_2.$$

A suggested means of designing such a magic T is to vary one of the parameters of the system, for instance, the length of a wide slot, until the device is a directional coupler (matched). The power-division ratio is measured, and the procedure repeated with another slot width. These two sets of data are used to predict the correct width. This procedure is refined by successive approximations.

12·23. Magic T with a Single Symmetry Plane.—The junction of Fig. 12·24 has a single symmetry plane. For the terminal planes num-

SEC. 12·23] MAGIC T WITH A SINGLE SYMMETRY PLANE 453

bered as shown, a reflection in this plane can be represented by the matrix

$$F = \begin{pmatrix} 0 & 1 & 0 & 0 \\ 1 & 0 & 0 & 0 \\ \hline 0 & 0 & 1 & 0 \\ 0 & 0 & 0 & -1 \end{pmatrix}.$$

As F commutes with the scattering matrix and is its own inverse,

$$S = FSF.$$

Multiplying out the right-hand side and setting the product equal to the left-hand side,

$$\left.\begin{aligned} S_{11} &= S_{22}, \\ S_{31} &= S_{32}, \\ S_{41} &= -S_{42}, \\ S_{34} &= 0. \end{aligned}\right\} \quad (153)$$

It can be seen also from the field distributions in the junction that the above conditions are correct. The junction is not yet a magic T, as there is still coupling between (1) and (2); moreover it is not matched.

Assume that matching transformers are inserted in such a way as to make

$$S_{33} = S_{44} = 0.$$

The scattering matrix can then be written as

$$S = \begin{pmatrix} \alpha & \beta & \gamma & \delta \\ \beta & \alpha & \gamma & -\delta \\ \hline \gamma & \gamma & 0 & 0 \\ \delta & -\delta & 0 & 0 \end{pmatrix}. \quad (154)$$

The scattering matrix is unitary, and the absolute square of each column must be unity; that is,

$$\left.\begin{aligned} |\alpha|^2 + |\beta|^2 + |\gamma|^2 + |\delta|^2 &= 1, \\ 2|\gamma|^2 &= 1, \\ 2|\delta|^2 &= 1. \end{aligned}\right\} \quad (155)$$

From Eqs. (155),

$$|\alpha|^2 + |\beta|^2 = 0.$$

This is possible only if both α and β vanish. Thus with $\alpha = \beta = 0$ the scattering matrix of a magic T is given by Eq. (154). If the positions of the reference planes in guides (3) and (4) are chosen correctly, the scattering matrix can be made to take the form

$$S = \frac{-j}{\sqrt{2}} \begin{pmatrix} 0 & 0 & 1 & 1 \\ 0 & 0 & 1 & -1 \\ 1 & 1 & 0 & 0 \\ 1 & -1 & 0 & 0 \end{pmatrix} \quad (156)$$

which is, except for sign, identical with Eq. (144).

It should be emphasized that the lack of coupling between guides (1) and (2) results from the matching of the device into arms (3) and (4). However, the zero coupling between arms (3) and (4) results from symmetry. It is this symmetry property which makes the junction shown in Fig. 12·22 a particularly desirable type of magic T.

Matching Conditions.—It should be noted that it takes four conditions to match the junction of Fig. 12·24. Two conditions are required for each of the terminals (3) and (4). The biplanar magic T required three adjustments, and the large-hole type only two. Thus the greater the number of symmetries of a junction, in general, the fewer conditions required for a match.

If power enters guide (3) only, the field distribution is even about the symmetry plane. This may be seen from Eqs. (153). If power enters guide (4), the field is odd about the symmetry plane. These distributions are eigenvectors of F. Thus at the symmetry plane the fields satisfy the boundary conditions of a magnetic or an electric wall.

It is highly desirable to have methods of adjusting for match that are independent for guides (3) and (4). One way in which this could be done is to introduce magnetic and electric conductors in the symmetry plane. Each of these materials would affect only one of the field distributions. For instance, a metal sheet in the symmetry plane has no effect on the odd field distributions or on the power entering arm (4).

Of course, the lack of a good magnetic conductor is a difficulty, and it is necessary to use some other method. In practice, the junction has been matched by a combination of a metal object in the symmetry plane and an inductive iris in guide (4). This iris is far enough inside the junction so that fringing fields from the even distribution are negligibly small.

12·24. Synthesis of Magic T with a Single Symmetry Plane in Coaxial Lines.—Before proceeding, it should be noticed that there are several possible locations of a symmetry plane that lead to a scattering matrix of the form given by Eq. (156). For instance, instead of the plane's cutting terminals (3) and (4), it could have cut terminals (1) and (2). Another possibility is a symmetry plane that reflects (1) and (4) into each other and (2) and (3) into each other. It will be this type of symmetry which will appear in the coaxial magic T.

Notice that the matrix of Eq. (154) is pure imaginary. Therefore, since S is unitary,

$$S^2 = -I. \tag{157}$$

The relation between the admittance and scattering matrix becomes

$$\begin{aligned} Y &= (1 - S)(1 + S)^{-1} \\ &= (1 - S)^2(1 - S)^{-1}(1 + S)^{-1} \\ &= (1 - 2S + S^2)(1 - S^2)^{-1}. \end{aligned} \tag{158}$$

Substituting from Eq. (157), Eq. (158) becomes

$$Y = -S. \tag{159}$$

Using the methods outlined previously, the junction whose matrix has the form of Eq. (159) would be synthesized by three quarter-wavelength lines and one three-quarter-wavelength line. This is illustrated in Fig. 12·25.

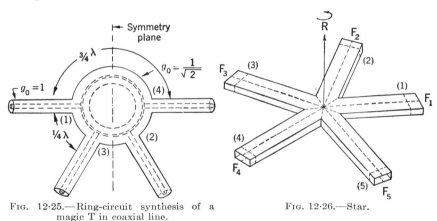

Fig. 12·25.—Ring-circuit synthesis of a magic T in coaxial line.

Fig. 12·26.—Star.

12·25. The Star.—The junction illustrated in Fig. 12·26 has a fivefold axis of symmetry and five symmetry planes. The plane including the axes of all five guides will not be regarded as a symmetry plane.

Let a rotation of 72° counterclockwise be designated by R_1, one of 144° by R_2, and so forth. Then the five rotations are I, R_1, R_2, R_3, and R_4. The five symmetry planes (see Fig. 12·26) are designated by

$$F_1 \ldots F_5.$$

These ten symmetry operators form a group with the following multiplication table:

TABLE 12·8.—MULTIPLICATION TABLE FOR THE GROUP OF SYMMETRY OPERATORS FOR A STAR JUNCTION

	I	R_1	R_2	R_3	R_4	F_1	F_2	F_3	F_4	F_5
I	I	R_1	R_2	R_3	R_4	F_1	F_2	F_3	F_4	F_5
R_1	R_1	R_2	R_3	R_4	I	F_3	F_4	F_5	F_1	F_2
R_2	R_2	R_3	R_4	I	R_1	F_5	F_1	F_2	F_3	F_4
R_3	R_3	R_4	I	R_1	R_2	F_2	F_3	F_4	F_5	F_1
R_4	R_4	I	R_1	R_2	R_3	F_4	F_5	F_1	F_2	F_3
F_1	F_1	F_4	F_2	F_5	F_3	I	R_2	R_4	R_1	R_3
F_2	F_2	F_5	F_3	F_1	F_4	R_3	I	R_2	R_4	R_1
F_3	F_3	F_1	F_4	F_2	F_5	R_1	R_3	I	R_2	R_4
F_4	F_4	F_2	F_5	F_3	F_1	R_4	R_1	R_3	I	R_2
F_5	F_5	F_3	F_1	F_4	F_2	R_2	R_4	R_1	R_3	I

It may be seen by inspection of Table 12·8 that R_1 and F_1 are generators of the group. In matrix form the operators R_1 and F_1 are

$$\mathsf{R}_1 = \begin{pmatrix} 0 & 0 & 0 & 0 & 1 \\ 1 & 0 & 0 & 0 & 0 \\ 0 & 1 & 0 & 0 & 0 \\ 0 & 0 & 1 & 0 & 0 \\ 0 & 0 & 0 & 1 & 0 \end{pmatrix}, \quad (160)$$

$$\mathsf{F}_1 = \begin{pmatrix} 1 & 0 & 0 & 0 & 0 \\ 0 & 0 & 0 & 0 & 1 \\ 0 & 0 & 0 & 1 & 0 \\ 0 & 0 & 1 & 0 & 0 \\ 0 & 1 & 0 & 0 & 0 \end{pmatrix}. \quad (161)$$

By referring to Table 12·8 it may be seen that R_1 satisfies the equation

$$\mathsf{R}_1^5 = \mathsf{I}.$$

It also can be seen by inspection of the characteristic determinant that the characteristic equation of R_1 is

$$r^5 - 1 = 0$$

and that the eigenvalues of R_1 are the five fifth roots of 1 (see Theorem 7). Let

$$r_1 = 1,$$
$$r_2 = e^{j\phi_1}, \quad \phi_1 = \tfrac{2}{5}\pi,$$
$$r_3 = e^{j\phi_2}, \quad \phi_2 = \tfrac{4}{5}\pi,$$
$$\cdot \qquad \cdot$$
$$\cdot \qquad \cdot$$
$$\cdot \qquad \cdot$$

Let the eigenvalue equation for R_1 be

$$\mathsf{R}_1 \mathsf{a}_k = r_k \mathsf{a}_k.$$

THE STAR

The five eigenvectors can be written by inspection

$$\mathbf{a}_1 = \begin{pmatrix} 1 \\ 1 \\ 1 \\ 1 \\ 1 \end{pmatrix}, \quad \mathbf{a}_2 = \begin{pmatrix} 1 \\ r_5 \\ r_4 \\ r_3 \\ r_2 \end{pmatrix}, \quad \mathbf{a}_3 = \begin{pmatrix} 1 \\ r_4 \\ r_2 \\ r_5 \\ r_3 \end{pmatrix}, \quad \mathbf{a}_4 = \begin{pmatrix} 1 \\ r_3 \\ r_5 \\ r_2 \\ r_4 \end{pmatrix}, \quad \mathbf{a}_5 = \begin{pmatrix} 1 \\ r_2 \\ r_3 \\ r_4 \\ r_5 \end{pmatrix}.$$

(162)

Since the eigenvalues of R_1 are nondegenerate, the eigenvectors of Eqs. (162) are, by Theorem 6, eigenvectors of the scattering matrix S.

From Theorem 4 the eigenvalues of S associated with $\mathbf{a}_2 \ldots \mathbf{a}_5$ are degenerate, since the r's are complex numbers. The nature of these degeneracies can be found from Eq. (161). If

$$\mathsf{S}\mathbf{a}_k = s_k \mathbf{a}_k,$$
$$\mathsf{S}(\mathsf{F}_1 \mathbf{a}_k) = s_k (\mathsf{F}_1 \mathbf{a}_k). \tag{163}$$

Thus $\mathsf{F}_1 \mathbf{a}_k$ is an eigenvector of S with an eigenvalue s_k. But

$$\begin{rcases} \mathsf{F}_1 \mathbf{a}_1 = \mathbf{a}_1, \\ \mathsf{F}_1 \mathbf{a}_2 = \mathbf{a}_5, \\ \mathsf{F}_1 \mathbf{a}_3 = \mathbf{a}_4, \\ \mathsf{F}_1 \mathbf{a}_4 = \mathbf{a}_3, \\ \mathsf{F}_1 \mathbf{a}_5 = \mathbf{a}_2. \end{rcases} \tag{164}$$

Substituting Eqs. (164) in Eq. (163), we have

$$s_2 = s_5$$
$$s_3 = s_4$$

There are only three independent eigenvalues of S. Since one of these can always be adjusted by location of the terminal planes, there are only two important independent parameters of the star which may be adjusted by matching transformers.

Field Distributions of Eigenvectors.—The field distributions are very similar to those of the symmetrical Y-junction. The components of the electric and magnetic fields perpendicular to the symmetry axis vanish along the axis for a_1. All the remaining eigenvectors are characterized by fields rotating about the axis. The electric and magnetic fields parallel to the axis must vanish along the axis.

It is apparent that a pin along the symmetry axis will affect s_1 but not s_2 or s_3.

The Scattering Matrix.—The scattering matrix may be obtained from the three eigenvalues in the usual way. Since the eigenvectors $\mathbf{a}_k$ are not all orthogonal to each other, it is convenient to pick an orthogonal set. It may be verified that the set

$$\left.\begin{aligned} b_1 &= \frac{1}{\sqrt{5}} a_1, \\ b_2 &= \frac{1}{\sqrt{10}} (a_2 + a_5), \\ b_3 &= \frac{1}{\sqrt{10}} (a_3 + a_4), \\ b_4 &= \frac{j}{\sqrt{10}} (a_3 - a_4), \\ b_5 &= \frac{j}{\sqrt{10}} (a_2 - a_5) \end{aligned}\right\} \quad (165)$$

is orthogonal, real, and normalized to unity.

The eigenvalue equation for **S** may as usual be written

$$\mathbf{SB} = \mathbf{S}_d \mathbf{B},$$

where **B** is a matrix with the eigenvectors of Eq. (165) as columns and where

$$\mathbf{S}_d = \begin{pmatrix} s_1 & 0 & 0 & 0 & 0 \\ 0 & s_2 & 0 & 0 & 0 \\ 0 & 0 & s_3 & 0 & 0 \\ 0 & 0 & 0 & s_3 & 0 \\ 0 & 0 & 0 & 0 & s_2 \end{pmatrix}.$$

Since **B** is orthogonal,

$$\mathbf{S} = \mathbf{B} \mathbf{S}_d \mathbf{B}.$$

If this equation is multiplied out, it is found that

$$\mathbf{S} = \begin{pmatrix} \alpha & \beta & \gamma & \gamma & \beta \\ \beta & \alpha & \beta & \gamma & \gamma \\ \gamma & \beta & \alpha & \beta & \gamma \\ \gamma & \gamma & \beta & \alpha & \beta \\ \beta & \gamma & \gamma & \beta & \alpha \end{pmatrix}, \quad (166)$$

where

$$\left.\begin{aligned} \alpha &= \tfrac{1}{5}(s_1 + 2s_2 + 2s_3), \\ \beta &= \tfrac{1}{5}[s_1 + (r_2 + r_5)s_2 + (r_3 + r_4)s_3], \\ \gamma &= \tfrac{1}{5}[s_1 + (r_3 + r_4)s_2 + (r_2 + r_5)s_3]. \end{aligned}\right\} \quad (167)$$

It should be noticed that α is one-fifth of the sum of the eigenvalues, in accordance with Theorem 8. As another check on the correctness of Eqs. (167) note that if

$$s_1 = s_2 = s_3 = 1,$$

then

$$\begin{aligned} \alpha &= 1, \\ \beta &= 0, \\ \gamma &= 0. \end{aligned}$$

The star may be adjusted in such a way as to be matched. This may be done by inserting any obstacle that does not ruin the symmetry, for instance, a cylindrical rod along the symmetry axis. In general, this will change the phases of all three eigenvalues with respect to one another. The second adjustment could be a pin along the symmetry axis. This changes the phase of s_1 only. Thus these two adjustments are independent; and if there is sufficient range in the adjustments, the junction can be matched. If $\alpha = 0$,

$$|\beta| = |\gamma|,$$

and β and γ are at an angle of 120° with respect to each other. Thus the matched junction distributes the power equally among the remaining four guides. Thus if the five star is matched, power entering one of the arms is split equally among the other four arms.

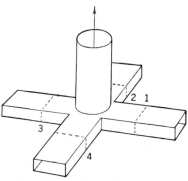

FIG. 12·27.—Turnstile junction.

12·26. The Turnstile Junction.—The turnstile junction shown in Fig. 12·27 is a six-terminal-pair device. The two polarizations in the round guide furnish two of the terminal pairs. Figure 12·28 shows the numbering scheme of the terminal planes.

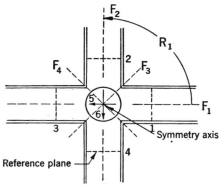

FIG. 12·28.—Symmetry properties of the turnstile junction. The symmetry planes are F_1, F_2, F_3, F_4; the terminal planes in the rectangular waveguide are 1, 2, 3, 4; the terminals in the round wavoguide for the two polarizations are 5 and 6. A rotation of 90° about the symmetry axis is R_1.

The junction has a fourfold symmetry axis and four symmetry planes. The symmetry planes in Fig. 12·28 are designated by F_1, F_2, F_3, and F_4. Let a counterclockwise rotation of the fields by 90° be designated by R_1. Let R_2, R_3, and I represent rotations of 180°, 270°, and 0° respectively. Terminal numbers are kept in a fixed position under any symmetry operation.

The rotations and reflections form a group with the following multiplication table.

TABLE 12·9.—MULTIPLICATION TABLE OF THE SYMMETRY OPERATORS FOR THE TURNSTILE JUNCTION

	I	R_1	R_2	R_3	F_1	F_2	F_3	F_4
I	I	R_1	R_2	R_3	F_1	F_2	F_3	F_4
R_1	R_1	R_2	R_3	I	F_4	F_3	F_1	F_2
R_2	R_2	R_3	I	R_1	F_2	F_1	F_4	F_3
R_3	R_3	I	R_1	R_2	F_3	F_4	F_2	F_1
F_1	F_1	F_3	F_2	F_4	I	R_2	R_1	R_3
F_2	F_2	F_4	F_1	F_3	R_2	I	R_3	R_1
F_3	F_3	F_2	F_4	F_1	R_3	R_1	I	R_2
F_4	F_4	F_1	F_3	F_2	R_1	R_3	R_2	I

It may be seen by inspection of Table 12·9 that any rotation (other than I) and any reflection are generators of the group. In particular, R_1 and F_1 are generators of the group. In operating on the terminal quantities, R_1 and F_1 take the form

$$\mathsf{R}_1 = \begin{pmatrix} 0 & 0 & 0 & 1 & 0 & 0 \\ 1 & 0 & 0 & 0 & 0 & 0 \\ 0 & 1 & 0 & 0 & 0 & 0 \\ 0 & 0 & 1 & 0 & 0 & 0 \\ \hline 0 & 0 & 0 & 0 & 0 & -1 \\ 0 & 0 & 0 & 0 & 1 & 0 \end{pmatrix},$$

$$\mathsf{F}_1 = \begin{pmatrix} 1 & 0 & 0 & 0 & 0 & 0 \\ 0 & 0 & 0 & 1 & 0 & 0 \\ 0 & 0 & 1 & 0 & 0 & 0 \\ 0 & 1 & 0 & 0 & 0 & 0 \\ \hline 0 & 0 & 0 & 0 & 1 & 0 \\ 0 & 0 & 0 & 0 & 0 & -1 \end{pmatrix}.$$

It should be noticed that both R_1 and F_1 can be divided, as shown by the dotted lines, into two square submatrices. The rectangular sections contain only zeros. The determinant of such a matrix is the product of the determinants of the submatrices.

Let the eigenvalue equation of R_1 be

$$\mathsf{R}_1 \mathsf{a}_k = r_k \mathsf{a}_k,$$

where r_k is a root of the characteristic equation

$$\det(\mathsf{R}_1 - \mathsf{I}r) = 0$$

or

$$(r^4 - 1)(r^2 + 1) = 0.$$

The roots of this equation are

$$\begin{aligned} r_1 &= 1, \\ r_2 &= j \text{ (doubly degenerate)}, \\ r_3 &= -1, \\ r_4 &= -j \text{ (doubly degenerate)}. \end{aligned}$$

A set of eigenvectors is

$$\mathsf{a}_1 = \begin{pmatrix} 1 \\ 1 \\ 1 \\ 1 \\ 0 \\ 0 \end{pmatrix}, \quad \mathsf{a}_2 = \begin{pmatrix} 1 \\ -j \\ -1 \\ j \\ 0 \\ 0 \end{pmatrix}, \quad \mathsf{a}_3 = \begin{pmatrix} 1 \\ -1 \\ +1 \\ -1 \\ 0 \\ 0 \end{pmatrix},$$

$$\mathsf{a}_4 = \begin{pmatrix} 1 \\ j \\ -1 \\ -j \\ 0 \\ 0 \end{pmatrix}, \quad \mathsf{a}_5 = \begin{pmatrix} 0 \\ 0 \\ 0 \\ 0 \\ 1 \\ -j \end{pmatrix}, \quad \mathsf{a}_6 = \begin{pmatrix} 0 \\ 0 \\ 0 \\ 0 \\ 1 \\ j \end{pmatrix}.$$

The eigenvectors a_5 and a_6 are eigenvectors of the eigenvalues r_2 and r_4 respectively. It may be seen by inspection that

$$\left. \begin{aligned} \mathsf{F}_1 \mathsf{a}_1 &= \mathsf{a}_1, \\ \mathsf{F}_1 \mathsf{a}_2 &= \mathsf{a}_4, \\ \mathsf{F}_1 \mathsf{a}_3 &= \mathsf{a}_3, \\ \mathsf{F}_1 \mathsf{a}_4 &= \mathsf{a}_2, \\ \mathsf{F}_1 \mathsf{a}_5 &= \mathsf{a}_6, \\ \mathsf{F}_1 \mathsf{a}_6 &= \mathsf{a}_5. \end{aligned} \right\} \quad (168)$$

Since the scattering matrix S commutes with R_1, the eigenvectors a_1 and a_3 are also eigenvectors of S (Theorem 6). Linearly independent eigenvectors of S can be formed by taking linear combinations of a_2 and a_5 and of a_4 and a_6. Let

$$\left. \begin{aligned} \mathsf{S}(\mathsf{a}_2 + \alpha \mathsf{a}_5) &= s_2(\mathsf{a}_2 + \alpha \mathsf{a}_5), \\ \mathsf{S}(\mathsf{a}_2 + \beta \mathsf{a}_5) &= s_4(\mathsf{a}_2 + \beta \mathsf{a}_5), \end{aligned} \right\} \quad (169)$$

where α and β are numbers. If Eqs. (169) are multiplied by F_1, then using the conditions of Eqs. (168),

$$S(a_4 + \alpha a_6) = s_2(a_4 + \alpha a_6),$$
$$S(a_4 + \beta a_6) = s_4(a_4 + \beta a_6).$$

It is seen that the eigenvalues s_2 and s_4 are both doubly degenerate. It should be possible to take linear combinations of the eigenvectors of both s_2 and s_4 in such a way that the resulting eigenvectors are pure real. It may be readily verified that the set

$$b_1 = a_1,$$
$$b_2 = \tfrac{1}{2}(a_2 + a_4) + \tfrac{1}{2}\alpha(a_5 + a_6),$$
$$b_3 = a_3,$$
$$b_4 = \tfrac{1}{2}(a_2 + a_4) + \tfrac{1}{2}\beta(a_5 + a_6),$$
$$b_5 = \frac{j}{2}(a_2 - a_4) + \frac{j}{2}\alpha(a_5 - a_6),$$
$$b_6 = \frac{j}{2}(a_2 - a_4) + \frac{j}{2}\beta(a_5 - a_6)$$

is pure real if α and β are real. Written as column vectors, this set becomes

$$b_1 = \begin{pmatrix} 1 \\ 1 \\ 1 \\ 1 \\ 0 \\ 0 \end{pmatrix}, \quad b_2 = \begin{pmatrix} 1 \\ 0 \\ -1 \\ 0 \\ \alpha \\ 0 \end{pmatrix}, \quad b_3 = \begin{pmatrix} 1 \\ -1 \\ 1 \\ -1 \\ 0 \\ 0 \end{pmatrix},$$

$$b_4 = \begin{pmatrix} 1 \\ 0 \\ -1 \\ 0 \\ \beta \\ 0 \end{pmatrix}, \quad b_5 = \begin{pmatrix} 0 \\ 1 \\ 0 \\ -1 \\ 0 \\ \alpha \end{pmatrix}, \quad b_6 = \begin{pmatrix} 0 \\ 1 \\ 0 \\ -1 \\ 0 \\ \beta \end{pmatrix}.$$

The eigenvectors b_j satisfy the eigenvalue equations

$$Sb_j = s_j b_j,$$

where

$$s_5 = s_2,$$
$$s_6 = s_4.$$

It should be noted that b_2 and b_4 are eigenvectors of different eigenvalues (namely, s_2 and s_4) and consequently must be orthogonal to each other, by Theorem 3. Therefore

$$\alpha\beta = -2.$$

In b_2 and b_4 only three of the elements do not vanish. If the vanishing terms are disregarded, b_2 and b_4 are similar to the b_1 and b_2 of Sec.

12·15 and are those of a series T-junction. It is possible to choose the position of the reference planes in such a way that $\alpha = \sqrt{2}$. Actually, for any symmetrical position of the planes in guides (1) to (4) there is some reference position in the round guide such that α has this value. Assume that this has been done.

The above eigenvalue equation for S can be written

$$\mathsf{SB} = \mathsf{BS}_d, \tag{170}$$

where

$$\mathsf{S}_d = \begin{pmatrix} s_1 & 0 & 0 & 0 & 0 & 0 \\ 0 & s_2 & 0 & 0 & 0 & 0 \\ 0 & 0 & s_3 & 0 & 0 & 0 \\ 0 & 0 & 0 & s_4 & 0 & 0 \\ \hline 0 & 0 & 0 & 0 & s_2 & 0 \\ 0 & 0 & 0 & 0 & 0 & s_4 \end{pmatrix},$$

$$\mathsf{B} = \tfrac{1}{2} \begin{pmatrix} 1 & 1 & 1 & 1 & 0 & 0 \\ 1 & 0 & -1 & 0 & 1 & 1 \\ 1 & -1 & 1 & -1 & 0 & 0 \\ 1 & 0 & -1 & 0 & -1 & -1 \\ \hline 0 & \sqrt{2} & 0 & -\sqrt{2} & 0 & 0 \\ 0 & 0 & 0 & 0 & \sqrt{2} & -\sqrt{2} \end{pmatrix}.$$

The columns of B are the eigenvectors $\mathbf{b}_k$; therefore B is orthogonal. Solving Eq. (170) for S,

$$\mathsf{S} = \mathsf{B} \mathsf{S}_d \tilde{\mathsf{B ```}},$$

or

$$\mathsf{S} = \begin{pmatrix} \alpha & \gamma & \delta & \gamma & \epsilon & 0 \\ \gamma & \alpha & \gamma & \delta & 0 & \epsilon \\ \delta & \gamma & \alpha & \gamma & -\epsilon & 0 \\ \gamma & \delta & \gamma & \alpha & 0 & -\epsilon \\ \hline \epsilon & 0 & -\epsilon & 0 & \beta & 0 \\ 0 & \epsilon & 0 & -\epsilon & 0 & \beta \end{pmatrix},$$

where

$$\alpha = \tfrac{1}{4}(s_1 + s_2 + s_3 + s_4),$$
$$\beta = \tfrac{1}{2}(s_2 + s_4),$$
$$\gamma = \tfrac{1}{4}(s_1 - s_3),$$
$$\delta = \tfrac{1}{4}(s_1 - s_2 + s_3 - s_4),$$
$$\epsilon = \frac{\sqrt{2}}{4}(s_2 - s_4).$$

The spur of S is

$$4\alpha + 2\beta = s_1 + 2s_2 + s_3 + 2s_4,$$

as is necessary to satisfy Theorem 8. If the turnstile is matched,

$$\alpha = \beta = 0.$$

The conditions for this are

$$\left.\begin{matrix} s_1 = -s_3, \\ s_2 = -s_4, \end{matrix}\right\} \quad (171)$$

which lead to

$$\delta = 0,$$
$$|\gamma| = \tfrac{1}{2},$$
$$|\epsilon| = \frac{1}{\sqrt{2}}.$$

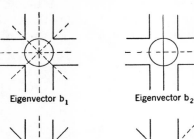

Fig. 12·29.—Eigenvector solutions b_0 and b_0', b_1, b_2, and b_3.

Therefore, if the turnstile is matched, power entering guide (1) leaves by guides (2), (4), and (5). One-half of the power leaves by the round guide; the remainder divides equally between guides (2) and (4).

Field Distributions.—The eigenvalues of the four reflection operators are indicated in Table 12·10 for the six eigenvectors b_k.

TABLE 12·10.—EIGENVALUES FOR THE REFLECTION OPERATORS OF THE TURNSTILE JUNCTION

	b_1	b_2	b_3	b_4	b_5	b_6	
F_1	$+1$	$+1$	$+1$	$+1$	-1	-1	
F_2	$+1$	-1	$+1$	-1	$+1$	$+1$	eigenvalues
F_3	$+1$	—	-1	—	—	—	
F_4	$+1$	—	-1	—	—	—	

Note that only b_1 and b_3 are eigenvectors of F_3 and F_4. As the boundary conditions on the symmetry planes are determined by the eigenvalues of the symmetry operators, Table 12·10 can be used to verify the correctness of the diagrams in Fig. 12·29. The remainder of the eigenvectors satisfy boundary conditions similar to those of b_2.

As b_2 and b_5 are eigenvectors of S with the same eigenvalue s_2, any linear combination is also an eigenvector. Let

$$b_0 = b_2 + b_5 = \begin{Bmatrix} 1 \\ 1 \\ -1 \\ -1 \\ \sqrt{2} \\ \sqrt{2} \end{Bmatrix}. \quad (172)$$

Also the vector
$$b_0' = b_4 + b_6$$
is an eigenvector of S with the eigenvalue s_4.

$$b_0' = b_4 + b_6 = \begin{pmatrix} 1 \\ 1 \\ -1 \\ -1 \\ -\sqrt{2} \\ -\sqrt{2} \end{pmatrix}. \tag{173}$$

The eigenvector given in Eq. (172) is introduced because it satisfies different boundary conditions from those of either b_2 or b_5 (b_4 or b_6).

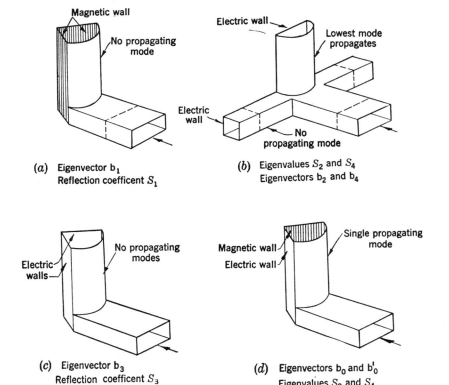

FIG. 12·30.—Junction partitioned by electric and magnetic walls for the various eigenvector solutions.

Note that b_0 and b_0' are eigenvectors of F_3 and F_4 but not of F_1 or F_2. The boundary conditions satisfied by b_0 or b_0' are illustrated also in Fig. 12·29.

Since the magnetic and electric walls shown in Fig. 12·29 divide the junction into separable parts, it is possible to illustrate the results in a more graphic form. This is done in Fig. 12·30. It should be noted that although the structures shown in parts b and d of this figure look very different, they have the same eigenvalues; this means that they are electrically equivalent. This equivalence, depending as it does only on symmetry, is independent of frequency. Figure 12·30a is the only structure that does not have at least one electric wall passing through the symmetry axis. If a pin is inserted along the symmetry axis, the only eigenvalue affected thereby is s_1.

Matching the Turnstile.—It should be noted that there are four parameters associated with the turnstile, namely, the four eigenvalues. Positions of reference planes are not important in the consideration of power division; and since reference planes can always be chosen in such a way as to make one of the eigenvalues $+1$, only three parameters of importance remain.

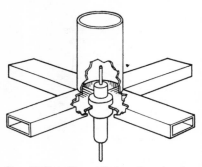

FIG. 12·31.—Turnstile junction showing triple plunger in position.

The matching of the turnstile requires the adjustment of three of the parameters until the conditions of Eqs. (171) are satisfied. In practice this might be done by inserting a triple plunger in the bottom of the junction along the symmetry axis. A turnstile junction containing such a plunger is shown in Fig. 12·31. The plunger consists of a thin pin and two concentric sleeves. The matching procedure consists of inserting the two sleeves into the guide and adjusting them until a match is obtained looking into arm (5). The conditions for this are

$$\beta = 0,$$
$$s_2 = -s_4.$$

After this the pin is inserted. It will be remembered from the previous discussion that this affects only s_1. Its position is adjusted until $s_1 = -s_3$. Then $\alpha = 0$, and the junction is completely matched.

12·27. Purcell's Junction.—The device shown in Fig. 12·32 is a junction of six rectangular guides. It is completely symmetrical in the sense that all the waveguides are equivalent. The junction may be regarded as one generated by a cube, each guide being mounted on a face of a cube.

The terminal planes are shown as dotted lines in Fig. 12·32 and are numbered according to a rule that allows a regular progression from one number to the next. Each terminal plane has an arrow assigned to it that represents the direction of positive electric field. The way in which the arrows are assigned is evident from the figure.

PURCELL'S JUNCTION

It may be seen by inspection of the figure that a rotation of 180° about the axis marked R_1 turns the figure back into itself. Hence the rotation R_1 is a symmetry operator. The effect of the rotation R_1 on the terminal voltages may be seen from the figure. First it is seen that the voltage

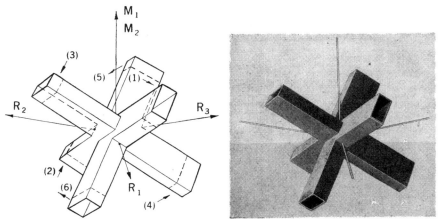

FIG. 12·32.—Purcell's junction.

of terminal (1) goes into that of terminal (6) without a change in sign. This, together with the other permutations of the terminals, are

$$\left.\begin{array}{l}(1) \rightarrow (6) \\ (2) \rightarrow (5) \\ (3) \rightarrow (4) \\ (4) \rightarrow (3) \\ (5) \rightarrow (2) \\ (6) \rightarrow (1)\end{array}\right\} \quad \begin{array}{c}\text{Symmetry operator} \\ R_1.\end{array}$$

This permutation may be induced by the matrix

$$R_1 = \left(\begin{array}{ccc|ccc} & & & 0 & 0 & 1 \\ & 0 & & 0 & 1 & 0 \\ & & & 1 & 0 & 0 \\ \hline 0 & 0 & 1 & & & \\ 0 & 1 & 0 & & 0 & \\ 1 & 0 & 0 & & & \end{array}\right) \tag{174}$$

operating on a current or voltage column vector.

In a similar way the 180° rotations about the axes R_2 and R_3 are symmetry operators and have, as matrices representing the corresponding permutations of the terminal pairs,

$$R_2 = \left(\begin{array}{ccc|ccc} 0 & 0 & 0 & 1 & 0 & 0 \\ 0 & 0 & 1 & 0 & 0 & 0 \\ 0 & 1 & 0 & 0 & 0 & 0 \\ \hline 1 & 0 & 0 & 0 & 0 & 0 \\ 0 & 0 & 0 & 0 & 0 & 1 \\ 0 & 0 & 0 & 0 & 1 & 0 \end{array}\right),$$

$$R_3 = \left(\begin{array}{ccc|ccc} 0 & 1 & 0 & 0 & 0 & 0 \\ 1 & 0 & 0 & 0 & 0 & 0 \\ 0 & 0 & 0 & 0 & 0 & 1 \\ \hline 0 & 0 & 0 & 0 & 1 & 0 \\ 0 & 0 & 0 & 1 & 0 & 0 \\ 0 & 0 & 1 & 0 & 0 & 0 \end{array}\right).$$

At first glance it might be thought that these three rotations are the only symmetry operators, except for the identity operator I. If this were true, however, the product of any two ought to yield the third matrix. It is found that, on the contrary, the product of two of them, R_1 and R_2 for example, yields new permutations

$$M_1 = R_1 R_2$$

and

$$M_2 = R_2 R_1,$$

where

$$M_1 = \left(\begin{array}{ccc|ccc} 0 & 0 & 0 & 0 & 1 & 0 \\ 0 & 0 & 0 & 0 & 0 & 1 \\ 1 & 0 & 0 & 0 & 0 & 0 \\ \hline 0 & 1 & 0 & 0 & 0 & 0 \\ 0 & 0 & 1 & 0 & 0 & 0 \\ 0 & 0 & 0 & 1 & 0 & 0 \end{array}\right)$$

and

$$M_2 = \left(\begin{array}{ccc|ccc} 0 & 0 & 1 & 0 & 0 & 0 \\ 0 & 0 & 0 & 1 & 0 & 0 \\ 0 & 0 & 0 & 0 & 1 & 0 \\ \hline 0 & 0 & 0 & 0 & 0 & 1 \\ 1 & 0 & 0 & 0 & 0 & 0 \\ 0 & 1 & 0 & 0 & 0 & 0 \end{array}\right).$$

These two operators are rotations about a symmetry axis designated in the figure by M_1 or M_2. The operator M_1 represents a rotation clockwise (looking out along the arrow) of 120°. The operator M_2 is a rotation of 240°.

The four symmetry axes of Fig. 12·32 are also symmetry axes of the cube upon which the junction is built. R_1, R_2, and R_3 are symmetry axes of the cube passing through midpoints of two opposite edges. The fourth axis is a threefold axis of the cube which passes through diagonally opposite corners of the cube.

It is evident from an inspection of the figure that there are no symmetry planes in the junction. Hence the symmetry operations are the six rotations I, R_1, R_2, R_3, M_1, and M_2. The multiplication properties of these operators are summarized in Table 12·11. Note that the group

TABLE 12·11.—MULTIPLICATIVE PROPERTIES OF THE OPERATORS FOR PURCELL'S JUNCTION

	R_1	R_2	R_3	M_1	M_2
I	R_1	R_2	R_3	M_1	M_2
R_1	I	M_1	M_2	R_2	R_3
R_2	M_2	I	M_1	R_3	R_1
R_3	M_1	M_2	I	R_1	R_2
M_1	R_3	R_1	R_2	M_2	I
M_2	R_2	R_3	R_1	I	M_1

whose multiplication table is Table 12·11 is of order 6 and has three subgroups of order 2 and one of order 3. It is a subgroup of the symmetry group of the cube.

As may be seen from Table 12·11, the group may be generated by the elements R_1 and M_1. Since $M_1^3 = I$, the eigenvalues of M_1 are the three cube roots of 1. Each of these roots is a doubly degenerate eigenvalue. Thus

$$M_1 a_k = m_k a_k, \qquad (175)$$

where

$$m_{1,2} = 1, \qquad \alpha_1 = \alpha_2^2 = \frac{1 + \sqrt{3}j}{2},$$

$$m_{3,4} = \alpha_1, \qquad \alpha_2 = \alpha_1^2 = \frac{1 - \sqrt{3}j}{2},$$

$$m_{5,6} = \alpha_2, \qquad \alpha_1 \alpha_2 = 1.$$

The eigenvectors a_k are not uniquely determined because of the degeneracy of the eigenvalues. However, a set may be easily written down.

Let

$$\mathbf{a}_1 = \begin{pmatrix} 1 \\ 1 \\ 1 \\ \hline 1 \\ 1 \\ 1 \end{pmatrix} \quad \mathbf{a}_2 = \begin{pmatrix} 1 \\ -1 \\ 1 \\ \hline -1 \\ 1 \\ -1 \end{pmatrix} \quad \mathbf{a}_3 = \begin{pmatrix} 1 \\ 0 \\ \alpha_2 \\ \hline 0 \\ \alpha_1 \\ 0 \end{pmatrix} \quad \mathbf{a}_4 = \begin{pmatrix} 0 \\ 1 \\ 0 \\ \hline \alpha_2 \\ 0 \\ \alpha_1 \end{pmatrix}$$

$$\underbrace{}_{m_1 = m_2 = 1} \qquad \underbrace{}_{m_3 = m_4 = \alpha_1}$$

$$\mathbf{a}_5 = \begin{pmatrix} 1 \\ 0 \\ \alpha_1 \\ \hline 0 \\ \alpha_2 \\ 0 \end{pmatrix} \quad \mathbf{a}_6 = \begin{pmatrix} 0 \\ 1 \\ 0 \\ \hline \alpha_1 \\ 0 \\ \alpha_2 \end{pmatrix}. \tag{176}$$

$$\underbrace{}_{m_5 = m_6 = \alpha_2}$$

If **S** is the scattering matrix of the junction,

$$\mathbf{M}_1 \mathbf{S} = \mathbf{S}\mathbf{M}_1.$$

Multiplication of Eq. (175) by **S** gives

$$\mathbf{M}_1(\mathbf{Sa}_k) = m_k(\mathbf{Sa}_k).$$

Thus $\mathbf{Sa}_k$ is an eigenvector of $\mathbf{M}_1$ with the eigenvalue m_k. But this implies that $\mathbf{Sa}_k$ is a linear combination of the two eigenvectors in Eqs. (176) that have m_k as eigenvalues. In particular

$$\mathbf{Sa}_1 = g_{11}\mathbf{a}_1 + g_{12}\mathbf{a}_2$$

and

$$\mathbf{Sa}_2 = g_{21}\mathbf{a}_1 + g_{22}\mathbf{a}_2, \tag{177}$$

where the g's are numbers.

Equations (177) may be combined to produce eigenvalue equations for **S**. Multiply these equations by the numbers k_1 and k_2 respectively, and add

$$\mathbf{S}(k_1\mathbf{a}_1 + k_2\mathbf{a}_2) = (g_{11}k_1 + g_{21}k_2)\mathbf{a}_1 + (g_{12}k_1 + g_{22}k_2)\mathbf{a}_2. \tag{178}$$

Assume that this is an eigenvalue equation for **S**. Then

$$\mathbf{S}(k_1\mathbf{a}_1 + k_2\mathbf{a}_2) = s(k_1\mathbf{a}_1 + k_2\mathbf{a}_2), \tag{179}$$

where s is the eigenvalue. Equating the right-hand sides of Eqs. (178)

and (179) results, because of the linear independence of a_1 and a_2, in the two equations
$$g_{11}k_1 + g_{21}k_2 = sk_1,$$
$$g_{21}k_1 + g_{22}k_2 = sk_2.$$
The permissible values of s are the two roots of the equation
$$\begin{vmatrix} g_{11} - s & g_{12} \\ g_{21} & g_{22} - s \end{vmatrix} = 0.$$
Let these roots be s_1 and s_2. There are two linearly independent solutions for k_1 and k_2 corresponding to each root. Let these solutions be k_{11} and k_{12} for $s = s_1$ and k_{21} and k_{22} for $s = s_2$. Then
$$\mathsf{S}\mathbf{b}_1 = s_1\mathbf{b}_1, \tag{180}$$
where
$$\mathbf{b}_1 = k_{11}\mathbf{a}_1 + k_{12}\mathbf{a}_2. \tag{181}$$

Since S and R_1 commute, Eq. (180) may be multiplied by R_1 to give
$$\mathsf{S}_1(\mathsf{R}_1\mathbf{b}_1) = s_1(\mathsf{R}_1\mathbf{b}_1).$$
However, as may be seen from Eqs. (176) and (174),
$$\mathsf{R}_1\mathbf{b}_1 = k_{11}\mathbf{a}_1 - k_{12}\mathbf{a}_2.$$
Thus both the sum and the difference of the terms in $\mathbf{a}_1$ and $\mathbf{a}_2$ are eigenvectors of S with eigenvalue s_1, and each of the terms $k_{11}\mathbf{a}_1$ and $k_{12}\mathbf{a}_2$ must be independently an eigenvector of S. Therefore, because k_{11} and k_{12} may not both vanish, either $\mathbf{a}_1$ or $\mathbf{a}_2$ is an eigenvector. If neither k_{11} nor k_{12} vanishes, both $\mathbf{a}_1$ and $\mathbf{a}_2$ are eigenvectors, which implies that s_1 is degenerate or that $s_1 = s_2$. It will be assumed, without information to the contrary, that s_1 and s_2 are nondegenerate, in which case k_{11} or k_{12} must vanish. Without loss in generality it may be assumed that k_{12} in Eq. (181) vanishes; thus
$$\mathsf{S}_1\mathbf{a}_1 = s_1\mathbf{a}_1.$$
In a similar way,
$$\mathsf{S}_2\mathbf{a}_2 = s_2\mathbf{a}_2.$$
In a completely analogous way, there are linear combinations of $\mathbf{a}_3$ and $\mathbf{a}_4$ and of $\mathbf{a}_5$ and $\mathbf{a}_6$ that are eigenvectors of S. Thus let
$$\left. \begin{array}{l} \mathbf{b}_3 = g_{33}\mathbf{a}_3 + g_{34}\mathbf{a}_4, \\ \mathbf{b}_4 = g_{43}\mathbf{a}_3 + g_{44}\mathbf{a}_4, \\ \mathbf{b}_5 = g_{55}\mathbf{a}_5 + g_{56}\mathbf{a}_6, \\ \mathbf{b}_6 = g_{65}\mathbf{a}_5 + g_{66}\mathbf{a}_6, \end{array} \right\} \tag{182}$$
where
$$\mathsf{S}\mathbf{b}_k = s_k\mathbf{b}_k. \tag{183}$$

Then
$$S(R_1 b_k) = s_k(R_1 b_k). \tag{184}$$

Thus $R_1 b_k$ is an eigenvector of S of eigenvalue s_k. However
$$R_1 a_3 = \alpha_1 a_6,$$
$$R_1 a_4 = \alpha_1 a_5,$$
$$R_1 a_5 = \alpha_2 a_4,$$
$$R_1 a_6 = \alpha_2 a_3.$$

Therefore
$$\left. \begin{array}{l} R_1 b_3 = \alpha_1(g_{33} a_6 + g_{34} a_5), \\ R_1 b_4 = \alpha_1(g_{43} a_6 + g_{44} a_5), \\ R_1 b_5 = \alpha_2(g_{55} a_4 + g_{56} a_3), \\ R_1 b_6 = \alpha_2(g_{65} a_4 + g_{66} a_3). \end{array} \right\} \tag{185}$$

Equations (185) are compatible with Eqs. (183) and (184) only provided that
$$s_3 = s_5 \text{ and/or } s_6,$$
$$s_4 = s_5 \text{ and/or } s_6.$$

As has always been done, it will be assumed that the degeneracy in S is the minimum consistent with symmetry conditions. Without loss in generality, therefore, it may be assumed that $s_3 = s_5$ and that $s_4 = s_6$. It requires only a renumbering of the s's to put them into this form.

By Theorem 4, it is possible to choose linear combinations of b_3 and b_5 and of b_4 and b_6 that are pure real. It may be seen by inspection of Eqs. (182) and (176) that this is possible only if
$$\left. \begin{array}{l} \dfrac{g_{33}}{g_{34}} = \dfrac{g_{55}}{g_{56}}, \\ \dfrac{g_{43}}{g_{44}} = \dfrac{g_{65}}{g_{66}}. \end{array} \right\} \tag{186}$$

However,
$$R_1 b_3 = \alpha_1 g_{33} \left(a_6 + \frac{g_{34}}{g_{33}} a_5 \right)$$
and
$$b_5 = g_{55} \left(a_5 + \frac{g_{56}}{g_{55}} a_6 \right)$$

are eigenvectors of S with the same eigenvalue. Either a_5 and a_6 are independently eigenvectors of S, which implies that
$$s_3 = s_4 = s_5 = s_6,$$
or else
$$\frac{g_{56}}{g_{55}} = \frac{g_{33}}{g_{34}}. \tag{187}$$

As usual the minimum condition on the eigenvalues is the one that will be assumed. Combining Eqs. (186) and (187),

In a similar way,
$$g_{33} = \pm g_{34}, \\ g_{55} = \pm g_{56}.$$ (188)

$$g_{43} = \mp g_{44}, \\ g_{65} = \mp g_{66}.$$ (189)

By the utilization of Eqs. (188) and (189), the b's may be redefined, without loss in generality, as

$$\begin{aligned} b_1 &= a_1, \\ b_2 &= a_2, \\ b_3 &= a_3 + a_4, \\ b_4 &= a_3 - a_4, \\ b_5 &= a_5 + a_6, \\ b_6 &= a_5 - a_6, \end{aligned}$$ (190)

where
$$Sb_k = s_k b_k, \\ s_3 = s_5, \\ s_4 = s_6.$$

The eigenvectors given in Eqs. (190) are not pure real. However a pure real set may be chosen. Let

$$\begin{aligned} e_1 &= b_1 = a_1, \\ e_2 &= a_2, \\ e_3 &= b_3 + b_5 = a_3 + a_4 + a_5 + a_6, \\ e_4 &= b_4 + b_6 = a_3 - a_4 + a_5 - a_6, \\ e_5 &= \frac{j}{\sqrt{3}} (b_3 - b_5) = \frac{j}{\sqrt{3}} (a_3 + a_4 - a_5 - a_6), \\ e_6 &= \frac{j}{\sqrt{3}} (b_4 - b_6) = \frac{j}{\sqrt{3}} (a_3 - a_4 - a_5 + a_6). \end{aligned}$$

From Eq. (176) the e's may be seen to be

$$e_1 = \begin{pmatrix} 1 \\ 1 \\ 1 \\ 1 \\ 1 \\ 1 \end{pmatrix}, \quad e_2 = \begin{pmatrix} 1 \\ -1 \\ 1 \\ -1 \\ 1 \\ -1 \end{pmatrix}, \quad e_3 = \begin{pmatrix} 2 \\ 2 \\ -1 \\ -1 \\ -1 \\ -1 \end{pmatrix}, \quad e_4 = \begin{pmatrix} 2 \\ -2 \\ -1 \\ 1 \\ -1 \\ 1 \end{pmatrix},$$

$$e_5 = \begin{pmatrix} 0 \\ 0 \\ 1 \\ 1 \\ -1 \\ -1 \end{pmatrix}, \quad e_6 = \begin{pmatrix} 0 \\ 0 \\ 1 \\ -1 \\ -1 \\ 1 \end{pmatrix}.$$

It may be seen by inspection that the e's are mutually orthogonal. The eigenvalue equation

$$Se_k = s_k e_k$$

may be written as

$$ST = TS_d,$$

where

$$S_d = \begin{pmatrix} s_1 & 0 & 0 & & & \\ 0 & s_2 & 0 & & 0 & \\ 0 & 0 & s_3 & & & \\ \hline & & & s_4 & 0 & 0 \\ & 0 & & 0 & s_3 & 0 \\ & & & 0 & 0 & s_4 \end{pmatrix}.$$

The matrix T has the vectors $e_1 \ldots e_6$ (normalized to unity) as columns. Since its columns are all mutually orthogonal, T is an orthogonal matrix and

$$T^{-1} = \tilde{T}.$$

Therefore,

$$S = T S_d \tilde{T} = \begin{pmatrix} \alpha & \beta & \gamma & \delta & \gamma & \delta \\ \beta & \alpha & \delta & \gamma & \delta & \gamma \\ \gamma & \delta & \alpha & \beta & \gamma & \delta \\ \hline \delta & \gamma & \beta & \alpha & \delta & \gamma \\ \gamma & \delta & \gamma & \delta & \alpha & \beta \\ \delta & \gamma & \delta & \gamma & \beta & \alpha \end{pmatrix}$$

where

$$\alpha = \tfrac{1}{6}(s_1 + s_2 + 2s_3 + 2s_4),$$
$$\beta = \tfrac{1}{6}(s_1 - s_2 + 2s_3 - 2s_4),$$
$$\gamma = \tfrac{1}{6}(s_1 + s_2 - s_3 - s_4),$$
$$\delta = \tfrac{1}{6}(s_1 - s_2 - s_3 + s_4).$$

It is seen that the spur of S is the sum of the eigenvalues, in compliance with Theorem 8. It should be noted that

$$S_{14} = S_{16},$$

a condition that can hardly be said to be obvious from an inspection of Fig. 12·32. That condition

$$S_{13} = S_{15}$$

is necessary can, however, be seen by inspection of the figure.

There are several special cases of interest. First assume that the junction is matched (i.e., $\alpha = 0$). This requires that

$$s_1 + s_2 + 2(s_3 + s_4) = 0.$$

It is apparent that this condition is not sufficient to determine completely the power division. Consider then the following cases.

Case 1: $\begin{cases} \alpha = 0, \\ \beta = 0. \end{cases}$

This requires that $s_1 = 2s_3$, which is impossible, since $|s_k| = 1$.

Case 2: $\begin{cases} \alpha = 0, \\ \gamma = 0. \end{cases}$

This requires that
$$\beta = \tfrac{1}{3}(s_1 + 2s_3),$$
$$\delta = \tfrac{1}{3}(s_1 - s_3),$$

or clearly
$$1 \geqq |\beta| \geqq \tfrac{1}{3},$$
$$0 \leqq |\delta| \leqq \tfrac{2}{3},$$

since $|s_1| = |s_3| = 1$.

Case 3: $\begin{cases} \alpha = 0, \\ \delta = 0. \end{cases}$

This requires that
$$2s_2 + 3s_3 + s_4 = 0,$$
$$2s_1 + s_3 + 3s_4 = 0,$$

which is possible only if
$$s_1 = s_3 = -s_4 = -s_2.$$

Then
$$\gamma = 0,$$
$$|\beta| = 1.$$
$$\beta = s_1,$$

Stated in words, if the junction is matched in such a way that there is no coupling between terminals (1) and (6), then all the incident power is transmitted out the arm opposite the one into which it is injected.

Case 4: For this case, relax the condition that the junction be matched, but let
$$\gamma = 0.$$

This requires that
$$s_1 + s_2 - s_3 - s_4 = 0.$$

This equation can be satisfied in three distinct ways:

1. $\left.\begin{array}{l} s_1 = -s_2 \\ s_3 = -s_4 \end{array}\right\}$ $\begin{array}{l} \alpha = 0 \\ \gamma = 0. \end{array}$

2. $\left.\begin{array}{l} s_1 = s_3 \\ s_2 = s_4 \end{array}\right\}$ $\begin{array}{l} \delta = 0 \\ \gamma = 0. \end{array}$

3. $\left.\begin{array}{l} s_1 = s_4 \\ s_2 = s_3 \end{array}\right\}$ $\begin{array}{l} \delta = -2\beta \\ \gamma = 0. \end{array}$

Case 5: $\delta = 0$.
This requires that
$$s_1 - s_2 - s_3 + s_4 = 0.$$
This equation, too, can be satisfied in three distinct ways

1. $\left.\begin{array}{l} s_1 = s_2 \\ s_3 = s_4 \end{array}\right\}$ $\begin{array}{l} \beta = 0 \\ \delta = 0. \end{array}$

2. $\left.\begin{array}{l} s_1 = s_3 \\ s_2 = s_4 \end{array}\right\}$ $\begin{array}{l} \delta = 0 \\ \gamma = 0. \end{array}$

3. $\left.\begin{array}{l} s_1 = -s_4 \\ s_2 = -s_3 \end{array}\right\}$ $\begin{array}{l} \delta = 0 \\ \gamma = -2\alpha. \end{array}$

FREQUENCY DEPENDENCE OF SYMMETRICAL JUNCTIONS

12·28. The Eigenvalue Formulation.—From Eq. (5·126), the rate of change of impedance or admittance matrices with respect to angular frequency is given by

$$\tilde{\imath}Z'i = 4jW \tag{191}$$

or

$$\tilde{e}Y'e = 4jW,$$

where

$$Z' = \frac{dZ}{d\omega},$$

$$Y' = \frac{dY}{d\omega},$$

and W is the total average electromagnetic energy in the junction when excited by i or e which are assumed to be pure real. In a similar way the rate of change of the scattering matrix S is given by

$$\tilde{a}S^*S'a = -2jW, \tag{192}$$

subject to the condition that the incident wave vector a satisfy

$$a = e^{j\beta}S^*a^* \tag{193}$$

for some real β.

The application of these equations directly to the impedance, admittance, or scattering matrices leads to rather complicated results but the eigenvalue formalism greatly simplifies everything. Assume that

$$\left.\begin{array}{l} Za_k = z_k a_k, \\ Ya_k = y_k a_k, \\ Sa_k = s_k a_k, \end{array}\right\} \tag{194}$$

where

$$z_k = \frac{1}{y_k} = \frac{1 + s_k}{1 - s_k} \tag{195}$$

and a_k are pure real, orthogonal, and normalized to unity. It should be noted that $\mathbf{a}_k$ is a permissible wave column vector for Eq. (192), since Eq. (193) is satisfied by $\mathbf{a}_k$ with

$$e^{j\beta} = s_k.$$

Substituting in Eq. (192) and making use of Eqs. (194),

$$\tilde{\mathbf{a}}_k S^* S' \mathbf{a}_k = -2j W_k,$$
$$\tilde{\mathbf{a}}_k S' \mathbf{a}_k = -2j W_k s_k. \tag{196}$$

If Eq. (194) is differentiated with respect to frequency, the result is

$$S'\mathbf{a}_k + S\mathbf{a}_k' = s_k' \mathbf{a}_k + s_k \mathbf{a}_k'. \tag{197}$$

Many of the junctions that have been considered have enough symmetry so that $\mathbf{a}_k$ is completely determined by symmetry. In this case $\mathbf{a}_k$ is independent of frequency and Eq. (197) becomes

$$S'\mathbf{a}_k = s_k' \mathbf{a}_k. \tag{198}$$

Equation (198) is satisfied in the case of the symmetrical Y-junction but is not satisfied in the case of the shunt T-junction. Only those examples for which Eq. (198) is satisfied will be considered in this section. Since S' and S have the same eigenvectors, they commute. This may be seen as follows

$$\left.\begin{array}{l} S\mathbf{a}_k = s_k \mathbf{a}_k, \\ S'S\mathbf{a}_k = s_k S' \mathbf{a}_k = s_k s_k' \mathbf{a}_k \\ \qquad = s_k' s_k \mathbf{a}_k = s_k' S \mathbf{a}_k = S s_k' \mathbf{a}_k = SS' \mathbf{a}_k, \\ (S'S - SS')\mathbf{a}_k = 0. \end{array}\right\} \tag{199}$$

Since the $\mathbf{a}_k$'s form a complete set, Eq. (199) is satisfied for any vector and hence

$$S'S - SS' = 0.$$

If Eq. (198) is substituted in Eq. (196), it is seen that

$$s_k' = -2j W_k s_k. \tag{200}$$

This equation is important. It can be written

$$\frac{d}{d\omega} \ln s_k = -2j W_k$$

or

$$\frac{d\phi_k}{d\omega} = -2W_k, \tag{201}$$

where

$$s_k = e^{j\phi_k}.$$

Since W_k is the stored energy associated with the eigenvector $\mathbf{a}_k$ and is positive, the phase angle ϕ_k always decreases with frequency. Since $\mathbf{a}_k$ is normalized to unity, the total incident power on the junction is

$$P = \tfrac{1}{2}\tilde{\mathbf{a}}_k^* \mathbf{a}_k = \tfrac{1}{2} \qquad \text{watt}.$$

Equation (201) may be written

$$\frac{d\phi_k}{d\omega} = -\left(\frac{W_k}{P}\right). \tag{202}$$

Equations (194) and (196) may be written as

$$\left.\begin{array}{l} \mathbf{SA} = \mathbf{AS}_d, \\ \mathbf{S'A} = \mathbf{AS}'_d, \end{array}\right\} \tag{203}$$

where $\mathbf{A}$ as usual has the vectors $\mathbf{a}_k$ as columns and

$$\mathbf{S}_d = \begin{pmatrix} s_1 & 0 & 0 & \cdots & \cdots & \cdot \\ 0 & & & & & \\ 0 & \cdot & & & & \\ \cdot & & & & & \\ \cdot & & \cdot & & & \\ \cdot & & & & & \\ \cdot & & & & & \\ \cdot & & & & s_n \end{pmatrix},$$

$$\mathbf{S}'_d = \begin{pmatrix} s'_1 & 0 & 0 & \cdots & \cdots & \cdot \\ 0 & & & & & \\ 0 & \cdot & & & & \\ \cdot & & & & & \\ \cdot & & \cdot & & & \\ \cdot & & & & & \\ \cdot & & & & & \\ \cdot & & & & s'_n \end{pmatrix}.$$

Equation (200) becomes

$$\mathbf{S}'_d = -2j\mathbf{W}_d\mathbf{S}_d, \tag{204}$$

where

$$\mathbf{W}_d = \begin{pmatrix} W_1 & 0 & 0 & \cdots & \cdots & \cdot \\ 0 & & & & & \\ 0 & \cdot & & & & \\ \cdot & & & & & \\ \cdot & & \cdot & & & \\ \cdot & & & & & \\ \cdot & & & & & \\ \cdot & & & & W_n \end{pmatrix}.$$

Substituting Eq. (204) in Eq. (203), there results

$$S'A = -2j A W_d S_d,$$
$$S' = -2j A W_d \tilde{A} A S_d \tilde{A} \quad (205)$$
$$= -2j W S,$$

where
$$W = A W_d \tilde{A}.$$

It should be noted that S', S, and W all have the same eigenvectors and hence commute with each other.

In a completely similar way Eq. (191) becomes

$$\tilde{a}_k Z' a_k = 4j W_k. \quad (206)$$

Again assuming that symmetry is sufficient to make a_k independent of frequency, a_k is an eigenvector of Z' and Eq. (206) becomes

$$z'_k = 4j W_k, \quad (207)$$

where z'_k is the eigenvalue of Z'. Note that $-j z'_k$ is always positive. Z' is obtained in a way completely analogous to that of Eq. (205) as

$$Z' = 4j A W_d A = 4j W. \quad (208)$$

It should be pointed out that the W_k of Eq. (207) is different from the W_k of Eq. (200), since an eigencurrent normalized to unity is physically different from a unity eigenwave. The rate of change of Y may be determined in a completely analogous way.

12·29. Wideband Symmetrical Junctions.—A junction is said to be wideband when the power distribution by the junction is insensitive to frequency. The wideband junction is an ideal that is seldom achieved. It is possible, however, to state the conditions under which a junction is wideband. If, in Eq. (205), W has the form

$$W = W_0(\omega) I, \quad (209)$$

then the modulus of each element of S is independent of frequency, or the junction is wideband. The necessary and sufficient condition for W to have the form of Eq. (209) is for W_k of Eq. (196) to satisfy

$$W_k = W_0(\omega).$$

Stated in words, this requires that the electromagnetic energy stored in the junction be the same for all the various eigensolutions.

Index

A

Admittance, 84
  intrinsic, 19
Admittance matrix, 89, 140
Admittances, combination of, 70
Allanson, J. T., 384
Antennas, minimum-scattering, 329–333
  radiation by, 317–333
  scattering by, 317–333
Arkadiew, W., 384
Attenuation constant, 18

B

Babinet's principle, 28–30, 256
Bandwidth of cavity, 239
Baños, A., 219
Bartlett, A. C., 110
Bessel functions, roots of, 41
  table of, 247
Bethe, H. A., 176
Bethe-hole coupler, 313
Birkhoff, G. D., 405
Bisection theorem, 110
Boundary conditions, 12
Brainerd, J. G., 335
Branch points, 85
Branches, 85
  in coaxial lines, 193–195
Brillouin, L., 53
Broadband stub support, 194
Broadbanding, 203–206

C

Capacitance, 84
Cavity, bandwidth of, 239
  capacitively loaded, 274
    with change in height, 275–277
  coaxial, 273
  cylindrical, 48
  with dielectrics, 390–393
  formed by shunt reactances, 182–186

Cavity, loop-coupled, equivalent circuit of, 218–225
  resonant, 207–239
  with two loop-coupled lines, representation of, 237
Cavity-coupling systems, equivalent circuit of, 208
  with two emergent transmission lines, 234–239
  two-line, transmission through, 237–239
Cayley-Hamilton's theorem, 410
Choke joints, in coaxial line, 196
  in waveguide, 197
Chu, L. J., 42
Column vectors, 87
Compensation theorem, 92
Condon, E. U., 219
Conductance, 84
Coupling from TE_{10}-mode to TE_{20}-mode, 337
Coupling coefficients, 228–230
Cross section, resonant change in, 189
Crout, P. D., 219
cst (x,y), 268
ct (x,y), 259
Ct (x,y), 259
Currents, circulating, 86
Cutoff frequency, 34, 116
Cutoff wavelength, 35
  for TE_{10}-mode between coaxial cylinders, 42

D

Darlington, S., 99, 215
Debye, P., 379
Diaphragms, capacitive, 166
  interaction between, 173
  as shunt reactances, 163–179
Dielectric constant, 382
  phase angle of, 365
  reflection from change in, 369–374
Dielectric plates in waveguide, 374–376

Dielectrics, 365–400
  cavity with, 390–393
  waveguide partially filled with, 385–389
Directional couplers, 299–301, 437
  biplanar, 445–447
  scattering matrix of, 301–303
  Schwinger reverse-coupling, 312
  single-hole, 437–445
  two-hole, 312
Duality principle, 85

E

E-modes, 253
  equivalent transmission lines for, 119
E-plane bifurcation of waveguide, 293
E-plane T-junction, equivalent circuit of, 289
  at high frequencies, 291–294
    equivalent circuits for, 292
  at long wavelengths, 288–291
  with three-winding transformer, equivalent circuit of, 294
E-waves, 17
E_{01}-mode in round waveguide, 57
E_{11}-mode in rectangular waveguide, 56
  in round waveguide, 59
Edson, W. A., 335
Eigenvalue equations, 405
Electrical quantities, dimensions of, 12
  units of, 12
Electromagnetic waves, longitudinal, 30
Energy, impedance and, 135–137
Energy density in waveguide, 50–54
Equivalent circuit, choice of, 286–288
  in radial lines, 267–271
  of two apertures, 174

F

Farr, H. K., 362
Feenberg, E., 392
Ferromagnetism at microwave frequencies, 382–385
Field representation by characteristic modes, 252–256
Filters, 115–119
  mode, 347–349
Foster's theorem, 97, 156–158
Four-junctions, degenerate, 313–315
  equivalent circuit of, 298
  with small holes, 311–313

Fox, A. G., 355
Frank, N. H., 173, 385, 389
Frequency dependence, of lossless junctions, 151
  of symmetrical junctions, 476–479

G

Green's theorem, 44, 51
Guillemin, E. A., 120, 209, 214, 216, 234

H

H-plane cross, 315
H-plane T-junction, 294
  equivalent circuit for, 295
H-waves, 17
H_{01}-mode in round waveguide, 58
H_{10}-mode in rectangular waveguide, 54
H_{11}-mode in round waveguide, 56
H_{20}-mode in rectangular waveguide, 55
H_{21}-mode in round waveguide, 57
H-type modes, 253
Hankel function, amplitude and phase of, 246
Harrison, J. R., 296
Hertz vector, 21
Hybrid coil, 307

I

Impedance, 66, 84
  characteristic, 66, 115, 116
    changes in, 187–193
  and energy, 135–137
  image, 114
  intrinsic, 19
  iterative, 113
  mutual, 86
  transformation of, 67–69
  wave, 18
Impedance chart with rectangular coordinates, 74
  Smith, 73, 226
Impedance description, of radial lines, 256–265
  of uniform lines, 248–252
Impedance matching, 179–187
Impedance matrix, 87, 140
  symmetry of, 141
Index of refraction, 366
Inductance, 84

INDEX 483

Inductive slit, 164–166
Inductive wire, thin, 167
Ionized gases, 393–396
Iris, 162
   dielectric-filled, 392
   resonant, 169–171
   thick, 417–420

J

Jamieson, H. W., 172
Johnson, K. S., 112
Junction effect, 187
Junctions connected in cascade, 150
   four-, degenerate, 313–315
   lossless, frequency dependence of, 151
   symmetrical, frequency dependence of, 476–479
   turnstile, 459–466
   wideband symmetrical, 479

K

Kauzmann, W., 381
Kirchhoff's laws, 85
Kirkwood, J. G., 378
Koehler, G., 335
Kuhn, S., 315
Kyhl, R. L., 206, 303

L

Lattice network, 110
Lippmann, B. A., 310, 315
Loaded Q, 228
Loss, relative, of various metals, 47

M

Macfarlane, G. G., 396
MacLane, S., 405
Magic T, 306–308, 447
   coupling-hole, 451
   equivalent circuit of, 308
   with single symmetry plane, 452–454
Magnetic wall, 13
Marcuvitz, N., 109
Margenau, H., 320, 395, 405
Matrix, admittance, 89, 140
   commuting, 410
   impedance (*see* Impedance matrix)
   scattering (*see* Scattering matrix)

Matrix, symmetrical, 407–409
   unit, 88
   zero, 88
Matrix algebra, 405–411
Matrix functions, rational, 409
Maxwell's equations, 10–14
   symmetries of, 411–417
Meshes, 85
Microwaves, 3
Microwave-circuit analysis, 8
Microwave-circuit theorems, 130–161
Microwave frequencies, ferromagnetism at, 382–385
Microwave region, 1
Mks system of units, 11
Mode absorbers, 347–349
Mode filters, 347–349
Mode results, summary of, 54–59
Mode transducers, 335–340
   properties of, measurement of, 343–347
   with taper, 339
Mode transformations, 334–364
Modes, characteristic, field representation by, 252–256
   in coaxial cylinders, 41
   dominant, 35
   normal, in rectangular pipes, 33–38
   in round pipes, 38–40
Möhrning, N., 384
Murphy, G. M., 320, 405

N

N terminal pairs, circuits with, 124
Network parameters of lines, 78
Network theorems, 90–95
Network theory, 83–129
Networks, in cascade, 112
   matching, 95
   with one terminal pair, 95–99
   parallel connection of, 119, 120
   series connection of, 119, 120
   three-terminal-pair, 121–124
   two-terminal-pair, 99–104
   equivalent circuits of, 104–110
   symmetrical, 110–112
Nodes, 85

O

Onsager, L., 378
Oscillator cavity coupled to rectangular waveguide, 277–282

P

Π-network, 100
Phase angle of dielectric constant, 365
Phase constant, 18
Phase shifter, rotary, 355–358
Physical realizability, 142
Pickering, W., 202
Pincherle, L., 389
Pipes, half-wave, 355
  quarter-wave, 354
  rectangular, normal modes in, 33–38
  round, normal modes in, 38–40
Plane waves, nonuniform, 19–21
  uniform, 17–19
Plates, parallel, TEM-waves between, 22
Plungers for rectangular waveguide, 198
Polarizations, basic, 350
  transformation of, by rotation, 351
Post, thick, 199
Power, available, 95
Power flow in waveguide, 50–54
Power transfer, maximum, 94
Poynting energy theorems, 14–16, 132–134
Poynting vector, complex, 16
Poynting vector theorems, 14–16
Propagation constant, 18
  iterative, 113
Purcell's junction, 466–476

Q

Q, 48, 127, 230
  loaded, 228
  unloaded, 228
Q_0, 48

R

Radar camouflage, 396
Radial cotangents, large, 259
  small, 259
Radial lines, equivalent circuit in, 267–271
  impedance description of, 256–265
Radial transmission lines, 240–282
Radiation, by antennas, 317–333
  from thick holes, 201
Ramo, S., 252
Reactances, 84
  shunt (see Shunt reactances)

Reciprocity theorem, 90
Reflection, from change in dielectric constant, 369–374
  in plane, symmetry of, 412–414
Reflection coefficient, 63, 137
  current, 67, 266
  in radial lines, 265–267
  and stored energy, 138
  voltage, 67, 266
Reflection group, multiplication table for, 414
  symmetry types under, 415
Refraction, index of, 366
Reich, H. J., 335
Relaxation time, 379
Resistance, 84
Resonance in closed circular waveguide, 361–364
Resonant cavities, 207–239
Resonant circuits, 127–129
Ring circuits, 308–311
Robbins, T. E., 172
Roberts, S., 374
Row vector, 89

S

Sarbacher, R. I., 335
Saxon, David S., 163, 174
Saxton, J. A., 380
Scattering, by antennas, 317–333
Scattering matrix, 146–149
  of directional coupler, 301–303
  of free space, 324
  of simple electric dipole, 325
  transformation of, 149
Schelkunoff, S. A., 19, 43, 202, 209, 211, 216, 252
Schwinger, J., 45, 82, 163, 174, 318, 403
Schwinger reverse-coupling directional coupler, 312
Screw tuners, 181
Series elements, discontinuities with, 198–206
Shunt elements, discontinuities with, 198–206
Shunt reactances, 163
  cavity formed by, 182–186
  diaphragms as, 163–179
Skin depth, 46
  of various metals, 47
Slater, J. C., 170, 190, 219, 315, 397

INDEX

Slit, inductive, 164–166
   thick capacitive, 200
Smith, P. H., 73
Smith impedance chart, 73, 226
Spence, R. D., 43
Standing-wave ratio, 62
Standing waves, 61
Star, 455–459
Starr, A. T., 97
Starr, Frank M., 127
Starr, M. A., 122
Stub support, broadband, 194
Stub tuners, 195
Superposition theorem, 87
Susceptance, 84
Symmetry, classes of, 401–403
   of waveguide junctions, 401–478
Symmetry operators, 414

T

T-junction, 283–297
   coaxial-line, 295
   E-plane (*see* E-plane T-junction)
   general theorems of, 283–286
   H-plane (*see* H-plane T-junction)
   120°, 292
   with small hole, 296
   symmetrical, 430
   transformer representation of, 122
T-network, 99
   equivalent, of length of waveguide, 77
Tapered lines, 191
TE waves, 17
TE_{10}-mode, between coaxial cylinders, cutoff wavelength for, 42
   coupling from, to TE_{20}-mode, 337
   in rectangular waveguide, 54
   transition from, to TE_{20}-mode, 339
TE_{11}-mode, in rectangular waveguide, 56
   in round waveguide, 56, 349–351
TE_{20}-mode, excited from coaxial line, 336
   in rectangular waveguide, 55
TE_{21}-mode, in round waveguide, 57
TEM-mode, coaxial, 54
TEM waves, 17–30
   between coaxial cylinders, 23–25
   between parallel plates, 22
   spherical, 25
Terman, F. E., 335
Terminal currents, definition of, 144–146
   uniqueness of, 134, 135

Terminal voltages, definition of, 144–146
   uniqueness of, 134, 135
Thévenin's theorem, 90
TM waves, 17
TM_{01}-mode, in round waveguide, 57
TM_{11}-mode, in round waveguide, 59
tn (x,y), 259
Tn (x,y), 259
Transducer, for converting from TE_{20}-mode in rectangular guide to TE_{01}-mode in round guide, 340
   mode (*see* Mode transducers)
   rectangular-to-round, 358
   from TE_{10}-mode to TM_{01}-mode, **339**
Transfer constant, 113
   image, 115
Transformer, ideal, 103
   quarter-wavelength, 189–191
Transformer representation of T-junction, 122
Transition, rectangular-to-round, 339
   from TE_{10}-mode to TE_{20}-mode, 339
Transmission line, branched, 193–198
   equivalent, for E-modes, 119
   nonuniform, 242
   radial, 240–282
   single, termination of, 132–138
Transmission-line charts, 71–75
Transmission-line equations, 65, 79–82
Transmission-line termination, 154–156
Transmission losses, 45–48
Transverse-electric (TE) waves, 17
Transverse-electromagnetic waves (*see* TEM waves)
Transverse-magnetic (TM) waves, 17
Tuning screw, capacitive, 168
Turner, L. B., 315
Turnstile junction, 459–466

U

Uniqueness theorem, extension of, 139
Unloaded Q, 228

V

Van Vleck, J. H., 377
Variational energy integral, 151–153
Vector, column, 87
   Hertz, 21
   row, 89
Vector potential, 21
Von Hippel, A., 374, 380

W

Walker, R. M., 202
Watson, G. N., 246
Wave impedance, 18
Wave number, 18
Waveguide, capacitively loaded, 192
  choke joints in, 197
  closed circular, resonance in, 361–364
  dielectric plates in, 374–376
  E-plane bifurcation of, 293
  energy density in, 50–54
  height of, change in, 188
  iris-coupled, short-circuited, 231–234
  length of, equivalent T-network of, 77
  partially filled with dielectrics, 385–389
  power flow in, 50–54
  rectangular, bends and corners in, 201–203
    E_{11}-mode in, 56
    equivalent circuit for, 119
    H_{10}-mode in, 54
    H_{20}-mode in, 55
    oscillator cavity coupled to, 277–282
    plungers for, 198
    TE_{10}-mode in, 54
    TE_{11}-mode in, 56
    TE_{20}-mode in, 55
  round, E_{01}-mode in, 57
Waveguide, round, E_{11}-mode in, 59
  H_{01}-mode in, 58
  H_{11}-mode in, 56
  H_{21}-mode in, 57
  principal axes in, 360
  TE_{01}-mode in, 58
  TE_{11}-mode in, 56, 349–351
  TE_{21}-mode in, 57
  TM_{01}-mode in, 57
  TM_{11}-mode in, 59
  width of, change in, 188
Waveguide circuit elements, 162–206
Waveguide junctions, 130–132
  with four arms, 298–317
  with several arms, 283–333
  symmetry of, 401–478
Waves, circularly polarized, 350
  uniform cylindrical, 26–28
Weissfloch, Albert, 206
Wells, C. P., 43
Whinnery, J. R., 172, 252
Wideband symmetrical junctions, 479
Woodruff, L. F., 335

Y

Y-junction, symmetrical, 420–427
Younker, E. L., 380
$\zeta(x,y)$, 259